T0211595

Lecture Notes in Computer Science   **10236**

Commenced Publication in 1973
Founding and Former Series Editors:
Gerhard Goos, Juris Hartmanis, and Jan van Leeuwen

More information about this series at http://www.springer.com/series/7407

Dimitris Fotakis · Aris Pagourtzis
Vangelis Th. Paschos (Eds.)

Algorithms and Complexity

10th International Conference, CIAC 2017
Athens, Greece, May 24–26, 2017
Proceedings

 Springer

Editors
Dimitris Fotakis ⓘ
National Technical University of Athens
Zografou, Athens
Greece

Aris Pagourtzis ⓘ
National Technical University of Athens
Zografou, Athens
Greece

Vangelis Th. Paschos
Université Paris-Dauphine
Paris
France

ISSN 0302-9743   ISSN 1611-3349 (electronic)
Lecture Notes in Computer Science
ISBN 978-3-319-57585-8   ISBN 978-3-319-57586-5 (eBook)
DOI 10.1007/978-3-319-57586-5

Library of Congress Control Number: 2017937695

LNCS Sublibrary: SL1 – Theoretical Computer Science and General Issues

Printed on acid-free paper

This Springer imprint is published by Springer Nature
The registered company is Springer International Publishing AG
The registered company address is: Gewerbestrasse 11, 6330 Cham, Switzerland

Preface

This volume contains the papers selected for presentation at CIAC 2017, the 10th International Conference on Algorithms and Complexity, held at the National Technical University of Athens, Greece, during May 24–26, 2017. This series of conferences presents original research contributions in the theory and applications of algorithms and computational complexity.

The volume begins with abstracts of the invited lectures, continues with contributed papers, arranged alphabetically by the last names of their authors, and concludes with an article devoted to the celebration of the 70th birthday of Stathis Zachos. The volume contains 36 accepted papers, selected by the Program Committee from 90 submissions received. Each submission was reviewed by at least three Program Committee members. Paper selection was based on originality, technical quality, and relevance.

We thank all the authors who submitted papers, the members of the Program Committee, and the external reviewers who assisted the Program Committee in the evaluation process. We are grateful to the three invited speakers, Giuseppe Italiano (Università di Roma "Tor Vergata", Italy), Klaus Jansen (University of Kiel, Germany), and Christos Papadimitriou (University of California at Berkeley, USA), who kindly accepted our invitation to give plenary lectures at CIAC 2017.

Springer sponsored a CIAC 2017 best paper award, which was shared by two papers: one by Hans L. Bodlaender and Tom C. van der Zanden on "Improved Lower Bounds for Graph Embedding Problems" and the other by Robert Bredereck, Christian Komusiewicz, Stefan Kratsch, Hendrik Molter, Rolf Niedermeier, and Manuel Sorge on "Assessing the Computational Complexity of Multi-Layer Subgraph Detection." Our warmest congratulations to all of them for these achievements! We also thank the members of the Best Paper Award Committee for selecting these papers.

We gratefully acknowledge the support from the National Technical University of Athens and its School of Electrical and Computer Engineering, the Institute of Communications and Computer Systems, Springer, and the European Association for Theoretical Computer Science (EATCS).

We would also like to thank Euripides Markou, Ioannis Milis, Dimitris Sakavalas, and Vassilis Zissimopoulos, who served in the Organizing Committee, as well as the local Arrangements Committee, and in particular, Alexandros Angelopoulos, Antonis Antonopoulos, Aggeliki Chalki, Eleni Iskou, Stratis Skoulakis, and Lydia Zakynthinou for their active participation in several organization tasks.

March 2017

Dimitris Fotakis
Aris Pagourtzis
Vangelis Th. Paschos

Organization

Program Committee

Vincenzo Bonifaci	IASI-CNR, Rome, Italy
Jarek Byrka	University of Wroclaw, Poland
Tiziana Calamoneri	Università di Roma I, "La Sapienza", Italy
Éric Colin de Verdière	CNRS and Université Paris-Est Marne-la-Vallée, France
Thomas Erlebach	University of Leicester, UK
Irene Finocchi	Università di Roma I, "La Sapienza", Italy
Dimitris Fotakis	National Technical University of Athens, Greece
Evangelos Kranakis	Carleton University, Canada
Dieter Kratsch	University of Lorraine, France
Michael Lampis	University Paris-Dauphine, France
Vangelis Markakis	Athens University of Economics and Business, Greece
Daniel Marx	Hungarian Academy of Science, Hungary
Monaldo Mastrolilli	IDSIA, Switzerland
Aris Pagourtzis	National Technical University of Athens, Greece
Vangelis Th. Paschos	University Paris-Dauphine, France (Chair)
Francesco Pasquale	Università di Roma "Tor Vergata", Italy
Giuseppe Persiano	Università di Salerno, Italy
Tomasz Radzik	King's College London, UK
Adi Rosén	CNRS and Université Paris Diderot, France
Guido Schäfer	CWI, The Netherlands
Maria Serna	Universitat Politècnica de Catalunya, Spain
Paul Spirakis	University of Liverpool, UK and CTI, Greece
Angelika Steger	ETH Zurich, Switzerland
Ioan Todinca	University of Orleans, France
Andreas Wiese	Universidad de Chile, Chile

Best Paper Award Committee

Ljiljana Brankovic	University of Newcastle, Australia
Ralf Klasing	CNRS and University Bordeaux 1, France (Chair)
Cécile Murat	University Paris-Dauphine, France
Vangelis Th. Paschos	University Paris-Dauphine, France
Peter Widmayer	ETH Zurich, Switzerland

Steering Committee

Giorgio Ausiello	Università di Roma I, "La Sapienza", Italy
Vangelis Paschos	University Paris-Dauphine, France
Rossella Petreschi	Università di Roma I, "La Sapienza", Italy

Paul Spirakis   University of Liverpool, UK and CTI, Greece
Peter Widmayer   ETH Zurich, Switzerland

Organizing Committee

Dimitris Fotakis   National Technical University of Athens, Greece
                         (Co-chair)
Euripides Markou   University of Thessaly, Greece
Ioannis Milis   Athens University of Economics and Business, Greece
Aris Pagourtzis   National Technical University of Athens, Greece
                         (Co-chair)
Dimitris Sakavalas   National Technical University of Athens, Greece
Vassilis Zissimopoulos   National and Kapodistrian University of Athens, Greece

Additional Reviewers

Abed, Fidaa
Achlioptas, Dimitris
Akrida, Eleni C.
Amanatidis, Georgios
Auletta, Vincenzo
Axiotis, Kyriakos
Azar, Yossi
Belmonte, Rémy
Bienkowski, Marcin
Bodlaender, Hans L.
Boeckenhauer, Hans-Joachim
Bonnet, Edouard
Casel, Katrin
Chang, Huilan
Chen, Lin
Cicalese, Ferdinando
Cohen-Addad, Vincent
Courtois, Nicolas
Croitoru, Cosmina
Da Lozzo, Giordano
De Keijzer, Bart
Deligkas, Argyrios
Dell'Orefice, Matteo
Della Croce, Federico
Dorbec, Paul
Dudycz, Szymon
Eiben, Eduard
Einarsson, Hafsteinn
Escoffier, Bruno

Ferraioli, Diodato
Fiala, Jiri
Galanis, Andreas
Gasieniec, Leszek
Ghosal, Pratik
Giannakos, Aristotelis
Golovach, Petr
Gourves, Laurent
Groenland, Carla
Gualà, Luciano
Göbel, Andreas
Heydrich, Sandy
Hoeksma, Ruben
Jeż, Łukasz
Kanj, Iyad
Kanté, Mamadou Moustapha
Karakostas, George
Keusch, Ralph
Khan, Arindam
Kim, Eun Jung
Kleer, Pieter
Kontogiannis, Spyros
Kotsialou, Grammateia
Koumoutsos, Grigorios
Kovacs, Annamaria
Kraska, Artur
Krizanc, Danny
Kurpisz, Adam
Lauria, Massimo

Lecroq, Thierry
Lengler, Johannes
Leppänen, Samuli
Letsios, Dimitrios
Lewandowski, Mateusz
Lianeas, Thanasis
Liedloff, Mathieu
Limouzy, Vincent
Matheus Gauy, Marcelo
Megow, Nicole
Meissner, Julie
Melissourgos, Themistoklis
Mertzios, George
Michail, Othon
Miltzow, Tillmann
Misra, Neeldhara
Mitsou, Valia
Mnich, Matthias
Montealegre, Pedro
Monti, Angelo
Mousset, Frank
Muller, Haiko
Mömke, Tobias
Nicholson, Patrick K.
Nomikos, Christos
Ordyniak, Sebastian
Otachi, Yota
Penna, Paolo
Pennarun, Claire
Pilipczuk, Michał

Piperno, Adolfo
Potika, Katerina
Pournin, Lionel
Psomas, Christos-Alexandros
Raptopoulos, Christoforos
Raymond, Jean-Florent
Renault, Marc
Rzażewski, Paweł
Sakavalas, Dimitris
Schewior, Kevin
Schmid, Andreas
Sirén, Jouni
Skoulakis, Stratis
Sornat, Krzysztof
Spoerhase, Joachim
Stamoulis, Georgios
Symvonis, Antonios
Thilikos, Dimitrios
Uehara, Ryuhei
Valiente, Gabriel
van Leeuwen, Erik Jan
Ventre, Carmine
Vigneron, Antoine
Watrigant, Rémi
Weissenberger, Felix
Weller, Mathias
Zhou, Hang
Zois, Georgios
Zwick, Uri

Invited Talks

TFNP: An Update

Paul W. Goldberg[1] and Christos H. Papadimitriou[2]

[1] University of Oxford, Oxford, UK
paul.goldberg@cs.ox.ac.uk
[2] University of California at Berkeley, Berkeley, USA
christos@cs.berkeley.edu

Abstract. The class TFNP was introduced a quarter of a century ago to capture problems in NP that have a witness for all inputs. A decade ago, this line of research culminated in the proof that the NASH equilibrium problem is complete for the subclass PPAD. Here we review some interesting developments since.

2-Edge and 2-Vertex Connectivity
in Directed Graphs

Giuseppe F. Italiano

University of Rome Tor Vergata, Italy
giuseppe.italiano@uniroma2.it

Abstract. We survey some recent results on 2-edge and 2-vertex connectivity problems in directed graphs. Despite being complete analogs of the corresponding notions on undirected graphs, in digraphs 2-vertex and 2-edge connectivity have a much richer and more complicated structure. It is thus not surprising that 2-connectivity problems on directed graphs appear to be more difficult than on undirected graphs. For undirected graphs it has been known for over 40 years how to compute all bridges, articulation points, 2-edge- and 2-vertex-connected components in linear time, by simply using depth-first search. In the case of digraphs, however, the very same problems have been much more challenging and required the development of new tools and techniques.

New Algorithmic Results for Bin Packing and Scheduling

Klaus Jansen

Institut für Informatik, Christian-Albrechts-Universität zu Kiel,
24098 Kiel, Germany
kj@informatik.uni-kiel.de

Abstract. In this paper we present an overview about new results for bin packing and related scheduling problems. During the last years we have worked on the design of efficient exact and approximation algorithms for packing and scheduling problems. In order to obtain faster algorithms we studied integer linear programming (ILP) formulations for these problems and proved structural results for optimum solutions of the corresponding ILPs.

New Algorithmic Results for Bin Packing and Scheduling

Abstract.

Contents

Paper Dedicated to Stathis Zachos on the Occasion of his 70th Birthday

Extended Abstracts

TFNP: An Update

Paul W. Goldberg[1] and Christos H. Papadimitriou[2](✉)

[1] University of Oxford, Oxford, UK
paul.goldberg@cs.ox.ac.uk
[2] University of California at Berkeley, Berkeley, USA
christos@cs.berkeley.edu

Abstract. The class TFNP was introduced a quarter of a century ago to capture problems in NP that have a witness for all inputs. A decade ago, this line of research culminated in the proof that the NASH equilibrium problem is complete for the subclass PPAD. Here we review some interesting developments since.

1 Introduction

Many apparently intractable problems in NP are *total*, that is, they are guaranteed to have a solution for all inputs. FACTORING (given a non-prime integer, find a prime factor) is perhaps the most accessible example of such a problem, and was the first to be identified, but by now many natural problems of this sort are known. (The hardness of FACTORING stands in contrast with primality testing, which is well known to be polynomial-time solvable [1].) This computational phenomenon is captured by the class TFNP (the initials stand for *total functions in NP*) [19,20].

How does one provide evidence of intractability for such a problem? Since problems in TFNP are unlikely to be NP-complete, and TFNP appears to have no complete problems ([22], Sect. 6 constructs an oracle relative to which there is no single TFNP problem to which all others reduce), focus quickly shifted to subclasses of TFNP with complete problems. To show that a problem is in TFNP, one must establish a theorem of the form $\forall x \exists y \Phi(x, y)$ stating that in all situations x of some kind (for example, in every bimatrix game) corresponding to the problem's input, a certain pattern y can be found (a solution, in this example a Nash equilibrium). If the proof of this theorem is constructive in a computationally meaningful sense, then the problem is in P. Hence, all intractable problems in TFNP must harbor an exponentially non-constructive step in their proof, presumably a combinatorial lemma guaranteeing the existence of a particular kind of element in an exponential structure. A productive classification of the problems in TFNP is in terms of the particular combinatorial lemma of the form $\forall x \exists y \Phi(x, y)$ employed in their proof. The following subclasses of TFNP had been known since the early 1990s.

1. PPP, for polynomial pigeonhole principle, the combinatorial lemma stating that *"for every function $f : \{0,1\}^n \mapsto \{0,1\}^n$ there must be either an $x \in \{0,1\}^n$ such that $f(x) = 0^n$, or $x, y \in \{0,1\}^n, x \neq y$ such that $f(x) = f(y)$."*

© Springer International Publishing AG 2017
D. Fotakis et al. (Eds.): CIAC 2017, LNCS 10236, pp. 3–9, 2017.
DOI: 10.1007/978-3-319-57586-5_1

2. PPA, for polynomial parity argument: *"In every finite graph containing an odd-degree node, a second such node must exist."*
3. PPAD, for polynomial parity argument for directed graphs: *"In every finite directed graph containing an unbalanced (in-degree ≠ out-degree) node, a second such node must exist."* It is not hard to see that PPAD is a subset of PPA, and also of PPP.
4. PPADS. A variant of PPAD, based on a slightly stronger lemma stating that *"a second oppositely unbalanced node must exist."* PPADS includes PPAD.
5. PLS, for polynomial local search: *"Every dag has a sink."*

It is often quite nontrivial to capture these classes in terms of a concrete "basic complete problem," but in all these cases it can be done [16,21]; typically, the class is then defined as all search problems reducible to the basic complete problem. In the case of PPAD, the basic complete problem is called END OF THE LINE, seeking a second degree-one node in a directed graph where no node has in-degree or out-degree greater than one, and 0^n has in-degree zero and out-degree one.

But are these classes different from P and one another? In [3] and elsewhere evidence was provided, through oracle constructions, that essentially all these classes are "compellingly different" from P and from each other, and that no easy inclusions seem to hold beyond the ones noted above.

Arguably, the whole TFNP research direction was motivated by one main quest, establishing that the NASH problem mentioned above is intractable; this was resolved in 2006 [7,8], by showing that NASH is PPAD-complete.

2   Recent Developments

Rogue Problems. The FACTORING problem does not immediately belong to any of these classes, as its totality seems to draw from the Fundamental Theorem of Arithmetic: *"any non-prime has a prime divisor that is smaller"*. What is the relation between this important total problem and the subclasses of TFNP already defined? Recently, Emil Jeřábek [15] employed elementary algebraic number theory to show that FACTORING belongs to *both* PPAD and PPP through randomized reductions. *Combinatorial Nullstellensatz*, another "rogue" problem, was recently shown to be in PPA [24]. There are further rogue problems still defying classification within TFNP. Two examples are RAMSEY: *"Given a Boolean circuit encoding the edges of a graph with 4n nodes, find n nodes that are either a clique or an independent set"* and BERTRAND-CHEBYSHEV: *"Given n, produce a prime between n and 2n"*, both embodying important and classical eponymous existence theorems in combinatorics and number theory, respectively.

The Class CLS. There had been a number of interesting total problems which have, frustratingly, defied polynomial-time algorithms for too long, and which were known to be inside *both* PPAD and PLS. Examples: finding an approximate fixpoint of a contraction map; finding the max-min of a simple stochastic game; finding a solution of a linear complementarity problem with a P-matrix; finding a Nash equilibrium in a network coordination, or congestion, game; or finding a stationary point of a multivariate polynomial. A new class called CLS (for *"continuous local search"*) lying within (and probably well within) the intersection of the two classes PLS and PPAD was defined in [9]; there are no known non-generic complete problems for this class (roughly, generic problems are ones whose instances contain generic boolean circuits). Very recently, a new problem was added to this collection, a version of the END OF THE LINE problem in which there is only one line starting at zero (no floating cycles or paths) and the nodes of the line are numbered $0, 1, 2, 3, \ldots$, while the remaining nodes are numbered ∞. This new problem, END OF THE METERED LINE, renders existing black-box complexity lower bounds for PPAD and PLS [25], as well as cryptographic lower bounds (see next), applicable to CLS as well.

Cryptographic Assumptions. There are obvious connections between TFNP classes and Cryptography, as cryptosystems can be based on intractable TFNP problems such as FACTORING. In the other direction, if there are hashing functions that are secure with respect to collisions, then PPP is clearly not P. Recently, sophisticated arguments were articulated establishing hardness for other subclasses of TFNP, based on other standard (if not universally accepted) cryptographic assumptions. In [5, 11] it is shown that, if indistinguishability obfuscation of software is possible, then PPAD (and therefore NASH) is intractable. Such results can be extended to CLS, as noted above. Furthermore, it was recently established in [14] that, under the assumption that NP has a problem that is hard on the average for some distribution (an assumption that has many consequences of interest to cryptography but is not per se cryptographic), then PPAD also has such a problem. A new paper [17] shows that hardness of RAMSEY follows from the existence of collision-resistant hash functions.

Approximate NASH. The major concrete problem in complexity left open by the establishment of the PPAD-completeness of NASH had been whether there is a PTAS for approximate Nash equilibria, or whether there is a finite ε such that finding an ε-approximate Nash equilibrium is PPAD-complete. This was recently resolved by Rubinstein [23] in favor of the latter eventuality through a brilliant PCP construction for PPAD.

Finitary Lemmata. One of the curiosities of the class TFNP and its subclasses is the surprising poverty (or, depending on your point of view, parsimony) of proof techniques needed to establish totality of functions in NP: in a quarter of a century, no new classes, no new combinatorial lemmata, have been added to the original five (we are not counting CLS, which lies within the intersection of all five classes). A look at the list of the five "combinatorial lemmata" reveals that they share an intriguing property: *They are all finitary*, that is, they fail to hold if the underlying structure is infinite. It has been recently pointed out in [12], by using a classical theorem from Logic [13], that this is necessary: Any combinatorial lemma that is not finitary must yield a subclass of TFNP that is necessarily a subset of P — an observation that may help focus the pursuit of a sixth combinatorial lemma.

Provable TFNP and WRONG PROOF

The fragmented nature of TFNP has been quite productive over the years, but it is also intriguing, and a little disturbing. Is there a way to unify all these genres of totality? Are there problems, perhaps complete for a master class, that generalize all total problems known to be complete for the subclasses?[1] Can we think of the phenomenon of total problems as a whole?

We have recently proposed such a unified theory [12]. The idea is to define a new class that we call *provable TFNP*, or PTFNP, which includes all total problems whose totality is proven within some minimalistic logical framework. The logical framework should be strong enough to supply proofs of the five lemmata, and versatile enough for the other important mission, namely identifying a complete problem for the whole class.

In [12] this was achieved through the framework of a first-order propositional logic. The language has a polynomial (in the underlying complexity parameter n) collection of Boolean variables of the form x_i, as well as the Boolean constants 0–1, connectives such as $\vee, \rightarrow$ and quantifiers $\forall, \exists$, all with the standard meaning. Importantly, the framework also allows an exponential collection of n-ary Boolean function symbols, represented as $f_i(x_1, \ldots, x_n)$, where "f" is a symbol of the language and the index i is expressed in binary (it is at present an open question whether these function symbols are necessary). The framework also includes a rather standard axiom system, encompassing a complete semantic understanding of the roles of connectives, functions, and quantifiers (see [12] for details).

Once we have all that, and having fixed the complexity parameter n, we can now have *succinctly represented proofs* (a concept studied earlier in [18]) in our system; these proofs play an important role in providing us with a complete problem (and from that, as it is common with total functions, a definition of the class PTFNP). In particular, consider a Boolean circuit C, with n input gates and of size polynomial in n, which maps an input j (a binary integer with n bits) to $C(j)$ where $C(j)$ can be of one of two forms:

Either $C(j)$ is a sentence that holds due to an axiom (and this can be checked easily), or it is of the form $(F(j), k, \ell)$, where $F(j)$ is (the Boolean encoding of) a logical formula in the language, and k, ℓ are integers smaller than j, such that $F(j)$ follows from $F(k)$ and $F(\ell)$ due to an inference rule.

Call the above condition "correctness of C at j." Note that a circuit C satisfying the correctness condition at all $j = 0, \ldots, 2^n - 1$ encodes a *proof* of length 2^n in our system.

But suppose now that we are given such a circuit C standing for a purported proof in our language (it will be an actual proof if it is correct at all j), and we

---

[1] Notice immediately that there is a trivial way of combining any finite number of classes with complete problems via some kind of "direct product" construction to obtain one all-encompassing class and complete problem. The challenge is to do this in a way that does not explicitly refer to the parts.

notice that the last line $C(2^n - 1)$ is of the form $(F(2^n - 1), k, \ell)$ *with* $F(2^n - 1) =$ FALSE. Since our logical system is consistent, it must be the case that there is a $j < 2^n$ such that C is not correct at j. The point is that *finding this j may be nontrivial!*

This is the total problem which we call WRONG PROOF:. Given n and a Boolean circuit C polynomial in n such that $F(2^n - 1) =$ FALSE, find a j such that C is not correct at j. Finally, we define PTFNP (for *provable* TFNP) as the class of all search problems in NP that reduce to WRONG PROOF.

The main theorem in [12] is the following:

Theorem 1. *PTFNP contains the five classes.*

We note that, in related work, Arnold Beckmann and Sam Buss prove in a recent paper [4] certain results that appear to be closely related to Theorem 1. They define two problems similar to WRONG PROOF, one corresponding to Frege systems, and another to extended Frege, and show these complete for two classes of total function problems in NP whose totality is provable within the bounded arithmetic [6] systems U_2^1 and V_2^1, respectively. Theorem 1 differs from [4] in that we reduce TFNP problems to a propositional proof system capable of encoding concisely instances of these problems, without resorting to bounded arithmetic.

3 Open Problems

There are many proof systems, some of them more powerful than others. It is possible that the one we propose in [12] can be simplified, while continuing to generalise PPP and the related complexity classes. So, one direction of future research is to look for minimal proof systems that continue to have this capability. One can also look in the opposite direction, and study more powerful proof systems, which may allow us to construct a hierarchy of increasingly general TFNP problems.

Some more concrete open problems include the following:

- *Is* FACTORING *in PPAD?* is a consequential and tempting conjecture, as the problem is in both PPP and PPA, the two important classes containing PPAD. Note that inclusion in PPAD would probably be through randomized reductions, as such are the reductions of [15] to PPP and PPA.
- *How about the remaining "rogue problems"* BERTRAND-CHEBYSHEV *and* RAMSEY, discussed in the introduction? We conjecture that they are both in PPP.
- Are there natural complete problems for PPA? Even though several discrete fixpoint-type problems are by now known to be PPA-complete, they all contain in their input the description of a computational device such as a circuit or a Turing machine; see [10] for an intriguing possibility.
- *How about PPP?* It is shown in [2] that several natural problems in PPP such as EQUAL SUMS (given n integers, find two subsets with the same sum modulo $2^n - 1$) and DIRICHLET (Given n rational numbers and integer n find $\frac{1}{Nq}$

approximations for some denominator q) can be reduced to MINKOWSKI (given an matrix A with determinant less than one, find a nontrivial combination of its rows within the unit square) and thus the latter becomes an interesting target in which to encode all of PPP.

- *How robust is PPP?* The class PPP can be parametrised as $PPP_K(n)$, for any function K: given $f : \{0,1\}^n \mapsto \{0,1\}^n$ identify either a collision $x \neq y$ with $f(x) = f(y)$, or a $x \in \{0,1\}^n$ such that $f(x) \leq K(n)$, where $\leq$ is meant in the standard binary notation. Obviously, PPP_0 is the standard PPP, and it is easy to see that $PPP_{p(n)} = PPP$, and $PPP_{2^n - p(n)} = P$ for all polynomials $p(n)$. But what happens for faster growing $K(n)$, for example if $K(n) = 2^{n-1}$ (the case known as "weak pigeonhole principle")? Are there oracle separation results?

- *Prove that one of the problems known to be in CLS is CLS-complete.* Alternatively, show that CLS is in P.

Acknowledgments. Many thanks to the "PPAD-like classes reading group" at the Simons Institute during the Fall 2015 program on Economics and Computation for many fascinating interactions. We also thank Arnold Beckmann, Pavel Pudlák, and Sam Buss for helpful discussions. Work supported by NSF grant CCF-1408635.

References

1. Agrawal, M., Kayal, N., Saxena, N.: PRIMES is in P. Ann. Math. **160**(2), 781–793 (2004)
2. Ban, F., Jain, K., Papadimitriou, C.H., Psomas, C.A., Rubinstein, A.: Reductions in PPP. Manuscript (2016)
3. Beame, P., Cook, S., Edmonds, J., Impagliazzo, R., Pitassi, T.: The relative complexity of NP search problems. In: 27th ACM Symposium on Theory of Computing, pp. 303–314 (1995)
4. Beckmann, A., Buss, S.: The NP Search Problems of Frege and Extended Frege Proofs, preliminary draft, December 2015
5. Bitansky, N., Paneth, O., Rosen, A.: On the cryptographic hardness of finding a Nash equilibrium. In: FOCS 2015, pp. 1480–1498 (2015)
6. Buss, S.: Bounded Arithmetic, Bibliopolis, Naples, Italy (1986). www.math.ucsd.edu/~sbuss/ResearchWeb/BAthesis/
7. Chen, X., Deng, X., Teng, S.-H.: Settling the complexity of computing two-player Nash equilibria. J. ACM **56**(3), 1–57 (2009)
8. Daskalakis, C., Goldberg, P.W., Papadimitriou, C.H.: The complexity of computing a Nash equilibrium. SIAM J. Comput. **39**(1), 195–259 (2009)
9. Daskalakis, C., Papadimitriou, C.H.: Continuous local search. In: SODA 2011, pp. 790–804 (2011)
10. Filos-Ratsikas, A., Frederiksen, S.K.S., Goldberg, P.W., Zhang, J.: Hardness Results for Consensus-Halving. Corr, arXiv:1609.05136 [cs.GT] (2016)
11. Garg, S., Pandey, O., Srinivasan, A.: Revisiting the cryptographic hardness of finding a Nash equilibrium. In: Robshaw, M., Katz, J. (eds.) CRYPTO 2016. LNCS, vol. 9815, pp. 579–604. Springer, Heidelberg (2016). doi:10.1007/978-3-662-53008-5_20

12. Goldberg, P.W., Papadimitriou, C.H.: Towards a Unified Complexity Theory of Total Functions (2017). Submitted

13. Herbrand, J.: Récherches sur la théorie de la démonstration. Ph.D. thesis, Université de Paris (1930)

14. Hubáček, P., Naor, M., Yogev, E.: The journey from NP to TFNP hardness. In: 8th ITCS (2017)

15. Jeřábek, E.: Integer factoring and modular square roots. J. Comput. Syst. Sci. **82**(2), 380–394 (2016)

16. Johnson, D.S., Papadimitriou, C.H., Yannakakis, M.: How easy is local search? J. Comput. Syst. Sci. **37**(1), 79–100 (1988)

17. Komargodski, I., Naor, M., Yogev, E.: White-Box vs. Black-Box Complexity of Search Problems: Ramsey and Graph Property Testing. ECCC Report 15 (2017)

18. Krajíček, J.: Implicit proofs. J. Symbolic Logic **69**(2), 387–397 (2004)

19. Megiddo, N.: A note on the complexity of P-matrix LCP and computing an equilibrium. Res. Rep. RJ6439, IBM Almaden Research Center, San Jose, pp. 1–5 (1988)

20. Megiddo, N., Papadimitriou, C.H.: On total functions, existence theorems and computational complexity. Theoret. Comput. Sci. **81**(2), 317–324 (1991)

21. Papadimitriou, C.H.: On the complexity of the parity argument and other inefficient proofs of existence. J. Comput. Syst. Sci. **48**, 498–532 (1994)

22. Pudlák, P.: On the complexity of finding falsifying assignments for Herbrand disjunctions. Arch. Math. Logic **54**, 769–783 (2015)

23. Rubinstein, A.: Settling the complexity of computing approximate two-player Nash equilibria. In: FOCS (2016)

24. Varga, L.: Combinatorial Nullstellensatz modulo prime powers and the parity argument. arXiv:1402.4422 [math.CO] (2014)

25. Zhang, S.: Tight bounds for randomized and quantum local search. SIAM J. Comput. **39**(3), 948–977 (2009)

New Algorithmic Results for Bin Packing
and Scheduling

Klaus Jansen[✉]

Institut für Informatik, Christian-Albrechts-Universität zu Kiel, 24098 Kiel, Germany
kj@informatik.uni-kiel.de

Abstract. In this paper we present an overview about new results for
bin packing and related scheduling problems. During the last years we
have worked on the design of efficient exact and approximation algorithms for packing and scheduling problems. In order to obtain faster
algorithms we studied integer linear programming (ILP) formulations
for these problems and proved structural results for optimum solutions
of the corresponding ILPs.

1 Introduction

In the first part of the paper we focus on the running times of approximation
schemes for scheduling problems. A problem admits a polynomial-time approximation scheme (PTAS) for a minimization problem, if there is a family of algo-
rithms $\{A_\varepsilon \mid \varepsilon > 0\}$ such that for any $\varepsilon > 0$ and any instance I, A_ε produces
a $(1 + \varepsilon)$-approximate solution in time polynomial in the size of the input. Two
important restricted classes of approximation schemes were defined to distinguish
the running times. An efficient polynomial-time approximation scheme (EPTAS)
is a PTAS with running time of the form $f(1/\epsilon) \cdot poly(|I|)$, while a fully time
polynomial time approximation scheme (FPTAS) runs in time $poly(|I|, 1/\epsilon)$.

In the second part we consider the classical bin packing problem and focus
on the design of fixed parameter tractable (FPT) algorithms where the running
time of the algorithm on an instance I is at most $f(k(I)) \cdot poly(|I|)$. Here f is
a computable function and k the parameterization. A natural parameter k for
bin packing is e.g. the optimum number $OPT(I)$ of bins used or the number d
of different bin sizes.

2 Scheduling Problems

Minimum makespan scheduling is one of the fundamental problems in the literature on approximation algorithms [7,8]. In the *identical machine* setting the
problem asks for an assignment of a set of n jobs $\mathcal{J}$ to a set of m identical

---

Research supported by the Deutsche Forschungsgemeinschaft (DFG), Project Ja
612/14-2, Entwicklung und Analyse von effizienten polynomiellen Approximationss-
chemata für Scheduling- und verwandte Optimierungsprobleme.

© Springer International Publishing AG 2017
D. Fotakis et al. (Eds.): CIAC 2017, LNCS 10236, pp. 10–15, 2017.
DOI: 10.1007/978-3-319-57586-5_2

machines $\mathcal{M}$. Each job $j \in \mathcal{J}$ is characterized by a non-negative processing time $p_j \in \mathbb{Z}_{>0}$. The load of a machine is the total processing time of jobs assigned to it, and our objective is to minimize the *makespan*, that is, the maximum machine load. This problem is usually denoted $P||C_{\max}$.

For the setting with uniform machines the problem $Q||C_{max}$ is defined as follows. Suppose that we are given a set $\mathcal{J}$ of n independent jobs J_j with processing time p_j and a set $\mathcal{M}$ of m non-identical machines M_i that run at different speeds s_i. If job J_j is executed on machine M_i, the machine needs p_j/s_i time units to complete the job. The problem is to find an assignment $a : \mathcal{J} \to \mathcal{M}$ for the jobs to the machines that minimizes the total execution time $\max_{i=1,\ldots,m} \sum_{J_j:a(J_j)=P_i} p_j/s_i$. This is the minimum time needed to complete the execution of all jobs on the processors.

2.1 Known Results

It is well known that $P||C_{\max}$ admits a *polynomial time approximation scheme* (PTAS) [10], and there has been many subsequent works improving the running time or deriving PTAS's for more general settings. The first PTAS was found by Hochbaum and Shmoys [10] and had a running time of $(n/\varepsilon)^{O((1/\varepsilon)^2)} = n^{O((1/\varepsilon)^2 \log(1/\varepsilon))}$. This was improved to $n^{O((1/\varepsilon) \log^2(1/\varepsilon))}$ by Leung [17]. Subsequent articles improved further the running time. In particular Hochbaum and Shmoys (see [12]) and Alon et al. [1,2] obtained an *efficient PTAS* (EPTAS) with running time $2^{(1/\varepsilon)^{\mathrm{poly}(1/\varepsilon)}} + O(n \log n)$; doubly exponentional in $1/\epsilon$. The fastest previous known PTAS for $P||C_{\max}$ achieved a running time of $2^{O(1/\varepsilon^2) \log^3(1/\varepsilon)} + O(n \log n)$ for $(1+\varepsilon)$-approximate solutions [13].

For uniform processors, the decision problem for the scheduling problem with makespan at most T can be interpreted as a bin packing problem with different bin sizes. Using an ϵ-relaxed version of this bin packing problem, Hochbaum and Shmoys [11] were able to obtain a PTAS for scheduling jobs on uniform processors $Q||C_{max}$ with running time $(n/\epsilon)^{O(1/\epsilon^2)}$. The existence of an EPTAS for uniform processors was mentioned as an open problem by Epstein and Sgall [5]. Some years ago we found an EPTAS [13] with an improved running time for $Q||C_{max}$ based on an MILP formulation with a constant number of integral variables. For any $\epsilon > 0$ our algorithm A_ϵ produces a schedule for the jobs of length $A_\epsilon(I) \leq (1 + \epsilon)OPT(I)$. The running time of A_ϵ is $2^{O(1/\epsilon^2 \log(1/\epsilon)^3)} + poly(n)$.

Very recently, Chen et al. [3] showed that, assuming the *exponential time hypothesis* (ETH), there is no PTAS that yields $(1 + \varepsilon)$-approximate solutions for $\varepsilon > 0$ with running time $2^{(1/\varepsilon)^{1-\delta}} + poly(n)$ for any $\delta > 0$ [3].

2.2 New Results

We describe in this section the main new ideas for the identical machines setting; for the uniform setting we refer to [14]. Given a guess $T \in \mathbb{N}$ on the optimal makespan, which can be found with binary search, the problem reduces to deciding the existence of a packing of the jobs to m machines (or bins) of capacity T. If we aim for a $(1 + \varepsilon)$-approximate solution, for some $\varepsilon > 0$, we can assume that all processing times are integral and T is a constant number, namely $T \in O(1/\varepsilon^2)$. This can be achieved with well known rounding and scaling techniques [1,2,12]. Let $\pi_1 < \pi_2 < \ldots < \pi_d$ be the job sizes appearing in the instance after rounding, and let b_k denote the number of jobs of size π_k. The mentioned rounding procedure implies that the number of different job sizes is $d = O((1/\varepsilon) \log(1/\varepsilon))$. Hence, for large n we obtain a highly symmetric problem where several jobs will have the same processing time. Consider the *knapsack polytope* $\mathcal{P} = \{c \in \mathbb{R}_{\geq 0}^d : \pi \cdot c \leq T\}$. A packing on one machine can be expressed as a vector $c \in Q = \mathbb{Z}^d \cap \mathcal{P}$, where c_k denotes the number of jobs of size π_k assigned to the machine. Elements in $Q = \mathbb{Z}^d \cap \mathcal{P}$ are called *configurations*. Considering a variable $x_c \in \mathbb{Z}_{\geq 0}$ that decides the multiplicity of configuration c in the solution, our problem reduces to solving the following linear integer program (ILP):

$$[\text{conf} - \text{ILP}] \quad \sum_{c \in Q} c \cdot x_c = b, \tag{1}$$

$$\sum_{c \in Q} x_c = m, \tag{2}$$

$$x_c \in \mathbb{Z}_{\geq 0} \qquad \qquad \text{for all } c \in Q. \tag{3}$$

In this paper we derive new insights on this ILP that help us to design faster algorithms for $P||C_{\max}$. We prove the following result.

Theorem 1 [14]. *The scheduling problem on identical machines admits an efficient polynomial time approximation scheme (EPTAS) with running time*

$$2^{O((1/\varepsilon) \log^4(1/\varepsilon))} + O(n).$$

Hence, our algorithm is best possible up to polylogarithmic factors in the exponent assuming the ETH [3].

The support $supp(x)$ of a solution vector $x = (x_c)$ is defined as the set of non-negative components $x_c > 0$. Eisenbrand and Shmonin [4] proved that there always exists an optimum solution x of the configuration ILP with $|supp(x)| \leq 2(d + 1) \log(4(d + 1)T)$.

Our main technical contribution is a new structural result on the configuration ILP. More precisely, we show the existence of a highly symmetric and sparse optimal solution, in which all but a constant number of machines are assigned a configuration with small support. We say that a configuration c is simple if its support (or number of non-negative components $c_k > 0$) is of size at most $\log(T + 1)$, otherwise it is complex. Then we can prove the following structural result:

Theorem 2 [14] . *Suppose that the* [conf-ILP] *is feasible. Then there exists a feasible solution x to* [conf-ILP] *such that*

(1) *if $x_c > 1$ then the configuration c is simple,*
(2) *the support of x satisfies $|supp(x)| \leq 4(d+1)\log(4(d+1)T)$, and*
(3) *$\sum_{c \in Q_c} x_c \leq 2(d+1)\log(4(d+1)T)$, where Q_c denotes the set of complex configurations.*

This structure can then be exploited by integer programming techniques [16,18] and dynamic programming. Interestingly, the result can be generalized also to the uniform machine setting with the same running time $2^{O((1/\varepsilon)\log^4(1/\varepsilon))} + O(n)$ using an MILP formulation and the same structural result. We believe that our structural result is of independent interest and should find applications to other settings.

3 Bin Packing

We consider the classical bin packing problem with d different item sizes $s_1, \ldots, s_d$ and build upon the results by Goemans and Rothvoß [6] to obtain a new polynomial time algorithm for the bin packing problem when d is constant [15]. Therefore, we present new techniques on how solutions of an instance can be modified and we give a new structural theorem that relies on the set of vertices of the underlying integer polytope.

3.1 Known Results

Given a polytope $\mathcal{P} = \{x \in \mathbb{R}^d \mid Ax \leq c\}$ for some matrix $A \in \mathbb{Z}^{m \times d}$ and a vector $c \in \mathbb{Z}^d$. We consider the integer cone

$$int.cone(\mathcal{P} \cap \mathbb{Z}^d) = \{ \sum_{p \in \mathcal{P} \cap \mathbb{Z}^d} \lambda_p p \mid \lambda \in \mathbb{Z}_{\geq 0}^{\mathcal{P} \cap \mathbb{Z}^d} \}$$

of integral points inside the polytope $\mathcal{P}$. When we choose $\mathcal{P}$ to be the knapsack polytope, i.e. $\mathcal{P} = \{x \in \mathbb{Z}_{\geq 0}^d \mid s^T x \leq 1\}$, then each integral point of the polytope represents one possibility of packing a single bin with items from $s_1, \ldots, s_d$. Hence a vector $\lambda \in \mathbb{Z}_{\geq 0}^{\mathcal{P} \cap \mathbb{Z}^d}$ of $int.cone(\mathcal{P} \cap \mathbb{Z}^d)$ represents a packing of the bin packing problem.

A long standing open question was, whether the bin packing problem can be solved in polynomial time when the number of different item sizes d is constant. This problem was recently solved by Goemans and Rothvoß [6] using structural properties of the integer cone. Essentially, they proved the existence of a distinguished set $X \subseteq \mathcal{P} \cap \mathbb{Z}^d$ of bounded size $|X| \leq m^d d^{O(d)} (\log \Delta)^d$ such that for every vector $b \in int.cone(\mathcal{P} \cap \mathbb{Z}^d)$ there exists an integral vector $\lambda \in \mathbb{Z}_{\geq 0}^{\mathcal{P} \cap \mathbb{Z}^d}$ where most of the weight lies in X.

3.2 New Results

In this section we show that a similar structural theorem holds for a rather natural choice of the distinguished set X. Therefore, we consider the so called integer polytope $\mathcal{P}_I$. It is defined by the convex hull of all integer points inside $\mathcal{P}$, i.e. $\mathcal{P}_I = Conv(\mathcal{P} \cap \mathbb{Z}^d)$. Let V_I be the vertices of the integer polytope $\mathcal{P}_I$ i.e. $\mathcal{P}_I = Conv(V_I)$. Based on the set V_I, we can show the following structural result for solutions λ of $int.cone(\mathcal{P} \cap \mathbb{Z}^d)$:

Theorem 3 [15]. *Let $\mathcal{P} = \{x \in \mathbb{R}^d \mid Ax \leq c\}$ be a polytope with $A \in \mathbb{Z}^{m \times d}, c \in \mathbb{Z}^d_{\geq 0}$ and let $supp(\lambda)$ be the set of non-zero components of λ. Then for any vector $b \in int.cone(\mathcal{P} \cap \mathbb{Z}^d)$, there exists an integral vector $\lambda \in \mathbb{Z}^{\mathcal{P} \cap \mathbb{Z}^d}_{\geq 0}$ such that $b = \sum_{p \in \mathcal{P} \cap \mathbb{Z}^d} \lambda_p p$ and*

(1) $\lambda_p \leq 2^{2^{O(d)}} \quad \forall p \in (\mathcal{P} \cap \mathbb{Z}^d) \setminus V_I$,
(2) $|supp(\lambda) \cap V_I| \leq d \cdot 2^d$,
(3) $|supp(\lambda) \setminus V_I| \leq 2^{2d}$.

As a consequence of our structural result, we obtain an algorithm for the bin packing problem with a running time of $|V_I|^{2^{O(d)}} \cdot \log(\Delta)^{O(1)}$, where Δ is the maximum over all multiplicities b and denominators in s. Since $|V_I| \geq d + 1$ this is an FPT-algorithm parameterized by the number of vertices of the integer knapsack polytope V_I.

Theorem 4 [15]. *The bin packing problem can be solved in time $|V_I|^{2^{O(d)}} \cdot (\log \Delta)^{O(1)}$ and hence in FPT-time, parameterized by the number of vertices V_I.*

This algorithmic result shows that the bin packing problem can be solved efficiently when the underlying knapsack polytope has an easy structure, i.e. has not too many vertices. However, since the total number of vertices is bounded by $O(\log \Delta)^d$ (see also [9]) the algorithm has a worst case running time of $(\log \Delta)^{2^{O(d)}}$, which is identical to the running time of the algorithm by Goemans and Rothvoß [6].

Furthermore, we were able to complement this result by giving a matching lower bound. We prove that the double exponential bound of the structure theorem is actually tight, even in the mentioned special case of bin packing, when all items sizes $s_1, \ldots, s_d$ are of the form $s_i = \frac{1}{a_i}$ for some $a_i \in \mathbb{Z}_{\geq 1}$.

References

1. Alon, N., Azar, Y., Woeginger, G., Yadid, T.: Approximation schemes for scheduling. In: 8th ACM-SIAM Symposium on Discrete Algorithms (SODA), pp. 493–500 (1997)
2. Alon, N., Azar, Y., Woeginger, G., Yadid, T.: Approximation schemes for scheduling on parallel machines. J. Sched. **1**, 55–66 (1998)
3. Chen, L., Jansen, K., Zhang, G.: On the optimality of approximation schemes for the classical scheduling problem. In: 25th ACM-SIAM Symposium on Discrete Algorithms (SODA), pp. 657–668 (2014)
4. Eisenbrand, F., Shmonin, G.: Carathéodory bounds for integer cones. Oper. Res. Lett. **34**, 564–568 (2006)

5. Epstein, L., Sgall, J.: Approximation schemes for scheduling on uniformly related and identical parallel machines. Algorithmica **39**, 43–57 (2004)
6. Goemans, M.X., Rothvoß, T.: Polynomiality for bin packing with a constant number of item types. In: 25th ACM-SIAM Symposium on Discrete Algorithms (SODA), pp. 830–839 (2014)
7. Graham, R.L.: Bounds for certain multiprocessing anomalies. Bell Syst. Tech. J. **45**, 1563–1581 (1966)
8. Graham, R.L.: Bounds on multiprocessing timing anomalies. SIAM J. Appl. Math. **17**, 416–429 (1969)
9. Hayes, A.C., Larman, D.G.: The vertices of the knapsack polytope. Discret. Appl. Math. **6**(2), 135–138 (1983)
10. Hochbaum, D.S., Shmoys, D.B.: Using dual approximation algorithms for scheduling problems: theoretical and practical results. J. ACM **34**, 144–162 (1987)
11. Hochbaum, D.S., Shmoys, D.B.: A polynomial approximation scheme for scheduling on uniform processors: using the dual approximation approach. SIAM J. Comput. **17**, 539–551 (1988)
12. Hochbaum, D.S.: Approximation Algorithms for NP-Hard Problems. PWS Publishing Company, Boston (1997)
13. Jansen, K.: An EPTAS for scheduling jobs on uniform processors: using an MILP relaxation with a constant number of integral variables. SIAM J. Discret. Math. **24**, 457–485 (2010)
14. Jansen, K., Klein, K.M., Verschae, J.: Closing the gap for makespan scheduling via sparsification techniques. In: 43rd International Colloquium on Automata, Languages, Programming, (ICALP), pp. 72:1–72:13 (2016). arXiv:1604.07153
15. Jansen, K., Klein, K.M.: About the structure of the integer cone, its application to bin packing. In: 28th ACM-SIAM Symposium on Discrete Algorithms (SODA), pp. 1571–1581 (2017). arXiv:1604.07286
16. Lenstra, H.W.: Integer programming with a fixed number of variables. Math. Oper. Res. **8**, 538–548 (1983)
17. Leung, J.Y.-T.: Bin packing with restricted piece sizes. Inf. Process. Lett. **31**, 145–149 (1989)
18. Kannan, R.: Minkowski's convex body theorem and integer programming. Math. Oper. Res. **12**, 415–440 (1987)

Regular Papers

Scheduling Maintenance Jobs in Networks

Fidaa Abed[1], Lin Chen[2], Yann Disser[3], Martin Groß[4(✉)], Nicole Megow[5],
Julie Meißner[6], Alexander T. Richter[7], and Roman Rischke[8]

[1] University of Jeddah, Jeddah, Saudi Arabia
fabed@uj.edu.sa
[2] University of Houston, Houston, TX, USA
chenlin198662@gmail.com
[3] TU Darmstadt, Darmstadt, Germany
disser@mathematik.tu-darmstadt.de
[4] University of Waterloo, Waterloo, ON, Canada
mgrob@uwaterloo.ca
[5] University of Bremen, Bremen, Germany
nicole.megow@uni-bremen.de
[6] TU Berlin, Berlin, Germany
jmeiss@math.tu-berlin.de
[7] TU Braunschweig, Braunschweig, Germany
a.richter@tu-bs.de
[8] TU München, Munich, Germany
rischke@ma.tum.de

Abstract. We investigate the problem of scheduling the maintenance of edges in a network, motivated by the goal of minimizing outages in transportation or telecommunication networks. We focus on maintaining connectivity between two nodes over time; for the special case of path networks, this is related to the problem of minimizing the busy time of machines.

We show that the problem can be solved in polynomial time in arbitrary networks if preemption is allowed. If preemption is restricted to integral time points, the problem is NP-hard and in the non-preemptive case we give strong non-approximability results. Furthermore, we give tight bounds on the power of preemption, that is, the maximum ratio of the values of non-preemptive and preemptive optimal solutions.

Interestingly, the preemptive and the non-preemptive problem can be solved efficiently on paths, whereas we show that mixing both leads to a weakly NP-hard problem that allows for a simple 2-approximation.

Keywords: Scheduling · Maintenance · Connectivity · Complexity theory · Approximation algorithm

This work is supported by the German Research Foundation (DFG) under project ME 3825/1 and within project A07 of CRC TRR 154. It is partially funded in the framework of MATHEON supported by the Einstein Foundation Berlin and by the Alexander von Humboldt Foundation. A version with proofs and illustrations is available on arXiv.

D. Fotakis et al. (Eds.): CIAC 2017, LNCS 10236, pp. 19–30, 2017.
DOI: 10.1007/978-3-319-57586-5_3

1   Introduction

Transportation and telecommunication networks are important backbones of modern infrastructure and have been a major focus of research in combinatorial optimization and other areas. Research on such networks usually concentrates on optimizing their usage, for example by maximizing throughput or minimizing costs. In the majority of the studied optimization models it is assumed that the network is permanently available, and our choices only consist in deciding which parts of the network to use at each point in time.

Practical transportation and telecommunication networks, however, can generally not be used non-stop. Be it due to wear-and-tear, repairs, or modernizations of the network, there are times when parts of the network are unavailable. We study how to schedule and coordinate such maintenance in different parts of the network to ensure connectivity.

While network problems and scheduling problems individually are fairly well understood, the combination of both areas that results from scheduling network maintenance has only recently received some attention [1,2,4,11,16] and is theoretically hardly understood.

Problem Definition. In this paper, we study connectivity problems which are fundamental in this context. In these problems, we aim to schedule the maintenance of edges in a network in such a way as to preserve connectivity between two designated vertices. Given a network and maintenance jobs with processing times and feasible time windows, we need to decide on the temporal allocation of the maintenance jobs. While a maintenance on an edge is performed, the edge is not available. We distinguish between MINCONNECTIVITY, the problem in which we minimize the total time in which the network is disconnected, and MAXCONNECTIVITY, the problem in which we maximize the total time in which it is connected.

In both of these problems, we are given an undirected graph $G = (V, E)$ with two distinguished vertices $s^+, s^- \in V$. We assume w. l. o. g. that the graph is simple; we can replace a parallel edge $\{u, w\}$ by a new node v and two edges $\{u, v\}, \{v, w\}$. Every edge $e \in E$ needs to undergo $p_e \in \mathbb{Z}_{\geq 0}$ time units of maintenance within the time window $[r_e, d_e]$ with $r_e, d_e \in \mathbb{Z}_{\geq 0}$, where r_e is called the release date and d_e is called the deadline of the maintenance job for edge e. An edge $e = \{u, v\} \in E$ that is maintained at time t, is not available at t in the graph G. We consider preemptive and non-preemptive maintenance jobs. If a job must be scheduled non-preemptively then, once it is started, it must run until completion without any interruption. If a job is allowed to be preempted, then its processing can be interrupted at any time and may resume at any later time without incurring extra cost.

A *schedule* S for G assigns the maintenance job of every edge $e \in E$ to a single time interval (if non-preemptive) or a set of disjoint time intervals (if preemptive) $S(e) := \{[a_1, b_1], \ldots, [a_k, b_k]\}$ with $r_e \leq a_i \leq b_i \leq d_e$ for $i \in [k]$ and $\sum_{[a,b] \in S(e)} (b - a) = p_e$. If not specified differently, we define $T := \max_{e \in E} d_e$ as our *time horizon*. We do not limit the number of simultaneously maintained edges.

For a given maintenance schedule, we say that the network G is *disconnected at time t* if there is no path from s^+ to s^- in G at time t, otherwise we call the network G *connected at time t*. The goal is to find a maintenance schedule for the network G so that the total time where G is disconnected is minimized (MINCONNECTIVITY). We also study the maximization variant of the problem, in which we want to find a schedule that maximizes the total time where G is connected (MAXCONNECTIVITY).

Our Results. For *preemptive* maintenance jobs, we show that we can solve both problems, MAXCONNECTIVITY and MINCONNECTIVITY, efficiently in arbitrary networks (Theorem 1). This result crucially requires that we are free to preempt jobs at arbitrary points in time. Under the restriction that we can *preempt* jobs only at *integral points in time*, the problem becomes NP-hard. More specifically, MAXCONNECTIVITY does not admit a $(2 - \epsilon)$-approximation algorithm for any $\epsilon > 0$ in this case, and MINCONNECTIVITY is inapproximable (Theorem 2), unless P = NP. By inapproximable, we mean that it is NP-complete to decide whether the optimal objective value is zero or positive, leading to unbounded approximation factors.

This is true even for unit-size jobs. This complexity result is interesting and may be surprising, as it is in contrast to results for standard scheduling problems, without an underlying network. Here, the restriction to integral preemption typically does not increase the problem complexity when all other input parameters are integral. However, the same question remains open in a related problem concerning the busy-time in scheduling, studied in [7,8].

For *non-preemptive* instances, we establish that there is no $(c\sqrt[3]{|E|})$-approximation algorithm for MAXCONNECTIVITY for some constant $c > 0$ and that MINCONNECTIVITY is inapproximable even on disjoint paths between two nodes s and t, unless P = NP (Theorems 3 and 4). On the positive side, we provide an $(\ell + 1)$-approximation algorithm for MAXCONNECTIVITY in general graphs (Theorem 6), where ℓ is the number of distinct latest start times (deadline minus processing time) for jobs.

We use the notion *power of preemption* to capture the benefit of allowing arbitrary job preemption. The power of preemption is a commonly used measure for the impact of preemption in scheduling [6,10,18,19]. Other terms used in this context include *price of non-preemption* [9], *benefit of preemption* [17] and *gain of preemption* [12]. It is defined as the maximum ratio of the objective values of an optimal non-preemptive and an optimal preemptive solution. We show that the power of preemption is $\Theta(\log |E|)$ for MINCONNECTIVITY on a path (Theorem 7) and unbounded for MAXCONNECTIVITY on a path (Theorem 8). This is in contrast to other scheduling problems, where the power of preemption is constant, e. g. [10,18].

On paths, we show that *mixed* instances, which have both preemptive and non-preemptive jobs, are weakly NP-hard (Theorem 9). This hardness result is of particular interest, as both purely non-preemptive and purely preemptive instances can be solved efficiently on a path (see Theorem 1 and [14]).

Furthermore, we give a simple 2-approximation algorithm for mixed instances of MINCONNECTIVITY (Theorem 10).

Related Work. The concept of combining scheduling with network problems has been considered by different communities lately. However, the specific problem of only maintaining connectivity over time between two designated nodes has not been studied to our knowledge. Boland et al. [2–4] study the combination of non-preemptive arc maintenance in a transport network, motivated by annual maintenance planning for the Hunter Valley Coal Chain [5]. Their goal is to schedule maintenance such that the maximum s-t-flow over time in the network with zero transit times is maximized. They show strong NP-hardness for their problem and describe various heuristics and IP based methods to address it. Also, they show in [3] that in their non-preemptive setting, if the input is integer, there is always an optimal solution that starts all jobs at integer time points. In [2], they consider a variant of their problem, where the number of concurrently performable maintenances is bounded by a constant.

Their model generalizes ours in two ways – it has capacities and the objective is to maximize the total flow value. As a consequence of this, their IP-based methods carry over to our setting, but these methods are of course not efficient. Their hardness results do not carry over, since they rely on the capacities and the different objective. However, our hardness results – in particular our approximation hardness results – carry over to their setting, illustrating why their IP-based models are a good approach for some of these problems.

Bley et al. [1] study how to upgrade a telecommunication network to a new technology employing a bounded number of technicians. Their goal is to minimize the total service disruption caused by downtimes. A major difference to our problem is that there is a set of given paths that shall be upgraded and a path can only be used if it is either completely upgraded or not upgraded. They give ILP-based approaches for solving this problem and show strong NP-hardness for a non-constant number of paths by reduction from the linear arrangement problem.

Nurre et al. [16] consider the problem of restoring arcs in a network after a major disruption, with restoration per time step being bounded by the available work force. Such network design problems over time have also been considered by Kalinowski et al. [13].

In scheduling, minimizing the busy time refers to minimizing the amount of time for which a machine is used. Such problems have applications for instance in the context of energy management [15] or fiber management in optical networks [11]. They have been studied from the complexity and approximation point of view in [7,11,14,15]. The problem of minimizing the busy time is equivalent to our problem in the case of a path, because there we have connectivity at a time point when no edge in the path is maintained, i. e., no machine is busy.

Thus, the results of Khandekar et al. [14] and Chang et al. [7] have direct implications for us. They show that minimizing busy time can be done efficiently for purely non-preemptive and purely preemptive instances, respectively.

2 Preemptive Scheduling

In this section, we consider problem instances where all maintenance jobs can be preempted.

Theorem 1. *Both* MAXCONNECTIVITY *and* MINCONNECTIVITY *with preemptive jobs can be solved optimally in polynomial time on arbitrary graphs.*

Proof. We establish a linear program (LP) for MAXCONNECTIVITY. Let $TP = \{0\} \cup \{r_e, d_e : e \in E\} = \{t_0, t_1, \ldots, t_k\}$ be the set of all *relevant time points* with $t_0 < t_1 < \cdots < t_k$. We define $I_i := [t_{i-1}, t_i]$ and $w_i := |I_i|$ to be the length of interval I_i for $i = 1, \ldots, k$.

In our linear program we model connectivity during interval I_i by an (s^+, s^-)-flow $x^{(i)}$, $i \in \{1, \ldots, k\}$. To do so, we add for every undirected edge $e = \{u, v\}$ two directed arcs (u, v) and (v, u). Let A be the resulting arc set. With each edge/arc we associate a capacity variable $y_e^{(i)}$, which represents the fraction of availability of edge e in interval I_i. Hence, $1 - y_e^{(i)}$ gives the relative amount of time spent on the maintenance of edge e in I_i. Additionally, the variable f_i expresses the fraction of availability for interval I_i.

$$\max \qquad \sum_{i=1}^{k} w_i \cdot f_i \tag{1}$$

$$\text{s.t.} \quad \sum_{u:(v,u)\in A} x_{(v,u)}^{(i)} - \sum_{u:(u,v)\in A} x_{(u,v)}^{(i)} = \begin{cases} f_i & \forall i \in [k], \ v = s^+, \\ 0 & \forall i \in [k], \ v \in V \setminus \{s^+, s^-\}, \\ -f_i & \forall i \in [k], \ v = s^-, \end{cases} \tag{2}$$

$$\sum_{i:I_i \subseteq [r_e, d_e]} (1 - y_e^{(i)})w_i \geq p_e \qquad \forall e \in E, \tag{3}$$

$$x_{(u,v)}^{(i)}, x_{(v,u)}^{(i)} \leq y_{\{u,v\}}^{(i)} \qquad \forall i \in [k], \ \{u,v\} \in E, \tag{4}$$

$$f_i \leq 1 \qquad \forall i \in [k], \tag{5}$$

$$x_{(u,v)}^{(i)}, x_{(v,u)}^{(i)}, y_{\{u,v\}}^{(i)} \in [0,1] \qquad \forall i \in [k], \ \{u,v\} \in E. \tag{6}$$

Notice that the LP is polynomial in the input size, since $k \leq 2|E|$. We show in Lemma 1 that this LP is a relaxation of preemptive MAXCONNECTIVITY on general graphs and in Lemma 2 that any optimal solution to it can be turned into a feasible schedule with the same objective function value in polynomial time, which proves the claim for MAXCONNECTIVITY. For MINCONNECTIVITY, notice that any solution that maximizes the time in which s and t are connected also minimizes the time in which s and t are disconnected – thus, we can use the above LP there as well. □

Lemma 1. *The given LP is a relaxation of preemptive* MAXCONNECTIVITY *on general graphs.*

Lemma 2. *Any feasible LP solution can be turned into a feasible maintenance schedule at no loss in the objective function value in polynomial time.*

The statement of Theorem 1 crucially relies on the fact that we may preempt jobs arbitrarily. However, if preemption is only possible at integral time points, the problem becomes NP-hard even for unit-size jobs. This follows from the proof of Theorem 3 for $t_1 = 0$, $t_2 = 1$, and $T = 2$.

Theorem 2. MAXCONNECTIVITY *with preemption only at integral time points is NP-hard and does not admit a* $(2 - \epsilon)$-*approximation algorithm for any* $\epsilon > 0$, *unless* P = NP. *Furthermore,* MINCONNECTIVITY *with preemption only at integral time points is inapproximable.*

3   Non-preemptive Scheduling

We consider problem instances in which no job can be preempted. We show that there is no $(c\sqrt[3]{|E|})$-approximation algorithm for MAXCONNECTIVITY for some $c > 0$. We also show that MINCONNECTIVITY is inapproximable, unless P = NP. Furthermore, we give an $(\ell + 1)$-approximation algorithm, where $\ell :=$ $|\{d_e - p_e \mid e \in E\}|$ is the number of distinct latest start times for jobs.

To show the strong hardness of approximation for MAXCONNECTIVITY, we begin with a weaker result which provides us with a crucial gadget.

Theorem 3. *Non-preemptive* MAXCONNECTIVITY *does not admit a* $(2 - \epsilon)$-*approximation algorithm, for* $\epsilon > 0$, *and non-preemptive* MINCONNECTIVITY *is inapproximable, unless* P = NP. *This holds even for unit-size jobs.*

Proof (Sketch). This is shown by a reduction from 3SAT. We construct a network such that connectivity is possible only within two disjoint time slots $[t_1, t_1 + 1]$ and $[t_2, t_2 + 1]$.

We show that this network admits a schedule with total connectivity time greater than one if and only if the 3SAT-instance is a YES-instance. Furthermore, we show that if the total connectivity time is greater than one, then there is a schedule with maximum total connectivity time of two. For this, we distinguish between *variable paths* and *clause paths*. By construction, variable paths exist only in $[t_2, t_2 + 1]$ and clause paths only in $[t_1, t_1 + 1]$. These paths walk through variable gadgets which encapsulate the decision whether to set a variable to TRUE or FALSE. A variable path ensures that we have a valid variable assignment, and a clause path sets literals in a clause to TRUE. If and only if both types of paths exist, then the 3SAT-instance is a YES-instance.

For $t_1 = 0$, $t_2 = 1$, and $T = 2$, this construction uses only unit-size jobs, and in the MINCONNECTIVITY case YES-instances have an objective value of 0 and NO-instances a value of 1.   □

We reuse the construction in the proof of Theorem 3 repeatedly to obtain the following improved lower bound.

Theorem 4. *Unless* P = NP, *there is no* $(c\sqrt[3]{|E|})$-*approximation algorithm for non-preemptive* MAXCONNECTIVITY, *for some constant* $c > 0$.

Proof (Sketch). We show this by reduction from 3SAT. Let n be the number of variables in the given 3SAT instance. Using the construction from Theorem 3 repeatedly allows us to construct a network that has maximum connectivity time n if the given 3SAT instance is a YES-instance and maximum connectivity time 1 otherwise. This implies that there cannot be an $(n - \epsilon)$-approximation algorithm for non-preemptive MAXCONNECTIVITY, unless P = NP. Notice that the construction in the proof of Theorem 3 has $\Theta(n)$ maintenance jobs and we will introduce $\Theta(n^2)$ copies of the construction, yielding $|E| \leq c \cdot n^3$ for some $c > 0$. Hence, we have $n \geq c' \sqrt[3]{|E|}$ for some $c' > 0$.

For the construction, we use $n^2 - n$ copies of the 3SAT-network from the proof of Theorem 3, where each copy uses *different* (t_1, t_2)-combinations with $t_1, t_2 \in \{0, \ldots, n-1\}$ and $t_1 \neq t_2$. Considering special (s^+, s^-)-paths, a path labeled with k allows connectivity only during $[k, k + 1]$, $k = 0, \ldots, n - 1$, and passes through every 3SAT-network with $t_1 = k$ or $t_2 = k$. Notice that within a 3SAT-network we have connectivity during both time slots if and only if the corresponding 3SAT-instance is a YES-instance. Also, we know due to [3] that there is an optimal solution which starts all jobs at integral times. Now, if the 3SAT-instance is a YES-instance, there is a global schedule such that its restriction to every 3SAT-network allows connectivity during both intervals. Thus each path with label $k \in \{0, \ldots, n - 1\}$ allows connectivity during $[k, k + 1]$. This implies that the maximum connectivity time is n.

Conversely, suppose there exists a global schedule with connectivity during two time slots. Then there must exist two paths P_1, P_2 from s^+ to s^- with two distinct labels, each realizing connectivity during one of both intervals. By construction there is one 3SAT-network they both use. This implies by the proof of Theorem 3, that the global schedule restricted to this 3SAT-network corresponds to a satisfying truth assignment for the 3SAT-instance. □

The results above hold for general graph classes, but even for graphs as simple as disjoint paths between s and t, the problem remains strongly NP-hard.

Theorem 5. *Non-preemptive* MAXCONNECTIVITY *is strongly NP-hard, and non-preemptive* MINCONNECTIVITY *is inapproximable even if the given graph consists only of disjoint paths between* s *and* t.

We give an algorithm that computes an $(\ell + 1)$-approximation for non-preemptive MAXCONNECTIVITY, where $\ell \leq |E|$ is the number of different time points $d_e - p_e, e \in E$. The basic idea is that we consider a set of $\ell + 1$ feasible maintenance schedules, whose total time of connectivity upper bounds the maximum total connectivity time of a single schedule. Then the schedule with maximum connectivity time among our set of $\ell + 1$ schedules is an $(\ell + 1)$-approximation.

The schedules we consider start every job either immediately at its release date, or at the latest possible time. In the latter case it finishes exactly at the deadline. More precisely, for a fixed time point t, we start the maintenance of all edges $e \in E$ with $d_e - p_e \geq t$ at their latest possible start time $d_e - p_e$. All other edges start maintenance at their release date r_e. This yields at most $\ell + 1 \leq |E| + 1$ different schedules S_t, as for increasing t, each time point where $d_e - p_e$ is passed for some edge e defines a new schedule. Algorithm 1 formally describes this procedure, where $E(t) := \{e \in E : e \text{ is not maintained at } t\}$.

Algorithm 1. Approx. Algorithm for Non-preemptive MAXCONNECTIVITY

1: Let $t_1 < \cdots < t_\ell$ be all different time points $d_e - p_e, e \in E$, $t_0 = 0$ and $t_{\ell+1} = T$.
2: Let S_i be the schedule, where all edges e with $d_e - p_e < t_i$ start maintenance at r_e
     and all other edges at $d_e - p_e$, $i = 1, \ldots, \ell + 1$.
3: For each S_i, initialize total connectivity time $c(t_i) \leftarrow 0, i = 1, \ldots, \ell + 1$.
4: **for** $i = 1$ to $\ell + 1$ **do**
5:     Partition the interval $[t_{i-1}, t_i]$ into subintervals such that each time point $r_e, r_e +$
        $p_e, d_e, e \in E$, in this interval defines a subinterval bound.
6:     **for all** subintervals $[a, b] \subseteq [t_{i-1}, t_i]$ **do**
7:         **if** $(V, E(1/2 \cdot (a + b)))$ contains an (s^+, s^-)-path for S_i **then**
8:             Increase $c(t_i)$ by $b - a$.
9: **return** Schedule S_i for which $c(t_i), i = 1, \ldots, \ell + 1$, is maximized.

Algorithm 1 considers finitely many intervals, as all (sub-)interval bounds are defined by a time point $r_e, r_e + p_e, d_e - p_e$ or d_e of some $e \in E$. As we can check the network for (s^+, s^-)-connectivity in polynomial time, and the algorithm does this for each (sub-)interval, Algorithm 1 runs in polynomial time.

Theorem 6. *Algorithm 1 is an $(\ell + 1)$-approximation algorithm for non-preemptive* MAXCONNECTIVITY *on general graphs, with $\ell \leq |E|$ being the number of different time points $d_e - p_e, e \in E$.*

4   Power of Preemption

We first focus on MINCONNECTIVITY on a path and analyze how much we can gain by allowing preemption. First, we show that there is an algorithm that computes a non-preemptive schedule whose value is bounded by $O(\log |E|)$ times the value of an optimal preemptive schedule. Second, we argue that one cannot gain more than a factor of $\Omega(\log |E|)$ by allowing preemption.

Theorem 7. *The power of preemption is $\Theta(\log |E|)$ for* MINCONNECTIVITY *on a path.*

Proof. Observe that if at least one edge of a path is maintained at time t, then the whole path is disconnected at t. We give an algorithm for MINCONNECTIVITY on a path that constructs a non-preemptive schedule with cost at most $O(\log |E|)$ times the cost of an optimal preemptive schedule.

We first compute an optimal preemptive schedule. This can be done in polynomial time by Theorem 1. Let x_t be a variable that is 1 if there exists a job j

that is processed at time t and 0 otherwise. We shall refer to x also as the *maintenance profile*. Furthermore, let $a := \int_0^T x_t \, dt$ be the active time, i.e., the total time of maintenance. Then we apply the following *splitting procedure*. We compute the time point $\bar{t}$ where half of the maintenance is done, i.e., $\int_0^{\bar{t}} x_t \, dt = a/2$. Let $E(t) := \{e \in E \mid r_e \leq t \wedge d_e \geq t\}$ and $p_{\max} := \max_{e \in E(t)} p_e$. We reserve the interval $[\bar{t} - p_{\max}, \bar{t} + p_{\max}]$ for the maintenance of the jobs in $E(\bar{t})$, although we might not need the whole interval. We schedule each job in $E(\bar{t})$ around $\bar{t}$ so that the processing time before and after $\bar{t}$ is the same. If the release date (deadline) of a jobs does not allow this, then we start (complete) the job at its release date (deadline). Then we mark the jobs in $E(\bar{t})$ as scheduled and delete them from the preemptive schedule.

This splitting procedure splits the whole problem into two separate instances $E_1 := \{e \in E \mid d_e < \bar{t}\}$ and $E_2 := \{e \in E \mid r_e > \bar{t}\}$. Note that in each of these sub-instances the total active time in the preemptive schedule is at most $a/2$. We apply the splitting procedure to both sub-instances and follow the recursive structure of the splitting procedure until all jobs are scheduled.   □

Lemma 3. *For* MINCONNECTIVITY *on a path, the given algorithm constructs a non-preemptive schedule with cost $O(\log |E|)$ times the cost of an optimal preemptive schedule.*

Proof. The progression of the algorithm can be described by a binary tree in which a node corresponds to a partial schedule generated by the splitting procedure for a subset of the job and edge set E. The root node corresponds to the partial schedule for $E(\bar{t})$ and the (possibly) two children of the root correspond to the partial schedules generated by the splitting procedure for the two subproblems with initial job sets E_1 and E_2. We can cut a branch if the initial set of jobs is empty in the corresponding subproblem. We associate with every node v of this tree B two values (s_v, a_v) where s_v is the number of scheduled jobs in the subproblem corresponding to v and a_v is the amount of maintenance time spent for the scheduled jobs.

The binary tree B has the following properties. First, $s_v \geq 1$ holds for all $v \in B$, because the preemptive schedule processes some job at the midpoint $\bar{t}_v$ which means that there must be a job $e \in E$ with $r_e \leq \bar{t}_v \wedge d_e \geq \bar{t}_v$. This observation implies that the tree B can have at most $|E|$ nodes and since we want to bound the worst total cost we can assume w.l.o.g. that B has exactly $|E|$ nodes. Second, $\sum_{v \in B} a_v = \int_0^T y_t \, dt$ where y_t is the maintenance profile of the non-preemptive solution.

The cost a_v of the root node (level-0 node) is bounded by $2p_{\max} \leq 2a$. The cost of each level-1 node is bounded by $2 \cdot a/2 = a$, so the total cost on level 1 is also at most $2a$. It is easy to verify that this is invariant, i.e., the total cost at level i is at most $2a$ for all $i \geq 0$, since the worst node cost a_v halves from level i to level $i + 1$, but the number of nodes doubles in the worst case. We obtain the worst total cost when B is a complete balanced binary tree. This tree has at most $O(\log |E|)$ levels and therefore the worst total cost is $a \cdot O(\log |E|)$. The total cost of the preemptive schedule is a.   □

We now provide a matching lower bound for the power of preemption on a path.

Lemma 4. *The power of non-preemption is $\Omega(\log |E|)$ for* MINCONNECTIVITY *on a path.*

Proof. We construct a path with $|E|$ edges and divide the $|E|$ jobs into ℓ levels such that level i contains exactly i jobs for $1 \leq i \leq \ell$. Hence, we have $|E| = \ell(\ell+1)/2$ jobs. Let P be a sufficiently large integer such that all of the following numbers are integers. Let the jth job of level i have release date $(j-1)P/i$, deadline $(j/i)P$, and processing time P/i, where $1 \leq j \leq i$. Note that now no job has flexibility within its time window, and thus the value of the resulting schedule is P.

We now modify the instance as follows. At every time point t where at least one job has a release date and another job has a deadline, we stretch the time horizon by inserting a gap of size P. This stretching at time t can be done by adding a value of P to all time points after the time point t, and also adding a value of P to all release dates at time t. The deadlines up to time t remain the same. Observe that the value of the optimal preemptive schedule is still P, because when introducing the gaps we can move the initial schedule accordingly such that we do not maintain any job within the gaps of size P.

Let us consider the optimal non-preemptive schedule. The cost of scheduling the only job at level 1 is P. In parallel to this job we can schedule at most one job from each other level, without having additional cost. This is guaranteed by the introduced gaps. At level 2 we can fix the remaining job with additional cost $P/2$. As before, in parallel to this fixed job, we can schedule at most one job from each level i where $3 \leq i \leq \ell$. Applying the same argument to the next levels, we notice that for each level i we introduce an additional cost of value P/i. Thus the total cost is at least $\sum_{i=1}^{\ell} P/i \in \Omega(P \log \ell)$ with $\ell \in \Theta(\sqrt{|E|})$. $\square$

Theorem 8. *For non-preemptive* MAXCONNECTIVITY *on a path the power of preemption is unbounded.*

5 Mixed Scheduling

We know that both the non-preemptive and preemptive MAXCONNECTIVITY and MINCONNECTIVITY on a path are solvable in polynomial time by Theorem 1 and [14, Theorem 9], respectively. Notice that the parameter g in [14] is in our setting ∞. Interestingly, the complexity changes when mixing the two job types – even on a simple path.

Theorem 9. MAXCONNECTIVITY *and* MINCONNECTIVITY *with preemptive and non-preemptive maintenance jobs is weakly* NP-*hard, even on a path.*

Theorem 10. *There is a 2-approximation algorithm for* MINCONNECTIVITY *on a path with preemptive and non-preemptive maintenance jobs.*

6 Conclusion

Combining network flows with scheduling aspects is a very recent field of research. While there are solutions using IP based methods and heuristics, exact and approximation algorithms have not been considered extensively. We provide strong hardness results for connectivity problems, which is inherent to all forms of maintenance scheduling, and give algorithms for tractable cases.

In particular, the absence of $c\sqrt[3]{|E|}$-approximation algorithms for some $c > 0$ for general graphs indicates that heuristics and IP-based methods [2–4] are a good way of approaching this problem. An interesting open question is whether the inapproximability results carry over to series-parallel graphs, as the network motivating [2–4] is series-parallel. Our results on the power of preemption as well as the efficient algorithm for preemptive instances show that allowing preemption is very desirable. Thus, it could be interesting to study models where preemption is allowed, but comes at a cost to make it more realistic.

On a path, our results have implications for minimizing busy time, as we want to minimize the number of times where some edge on the path is maintained. Here, an interesting open question is whether the 2-approximation for the mixed case can be improved, e.g. by finding a pseudo-polynomial algorithm, a better approximation ratio, or conversely, to show an inapproximability result for it.

Acknowledgements. We thank the anonymous reviewers for their helpful comments.

References

1. Bley, A., Karch, D., D'Andreagiovanni, F.: WDM fiber replacement scheduling. Electron. Notes Discret. Math. **41**, 189–196 (2013). http://www.sciencedirect.com/science/article/pii/S1571065313000954
2. Boland, N., Kalinowski, T., Kaur, S.: Scheduling arc shut downs in a network to maximize flow over time with a bounded number of jobs per time period. J. Comb. Optim. **32**(3), 885–905 (2016)
3. Boland, N., Kalinowski, T., Kaur, S.: Scheduling network maintenance jobs with release dates and deadlines to maximize total flow over time: Bounds and solution strategies. Comput. Oper. Res. **64**, 113–129 (2015). http://www.sciencedirect.com/science/article/pii/S0305054815001288
4. Boland, N., Kalinowski, T., Waterer, H., Zheng, L.: Scheduling arc maintenance jobs in a network to maximize total flow over time. Discret. Appl. Math. **163**, 34–52 (2014). http://dx.doi.org/10.1016/j.dam.2012.05.027
5. Boland, N.L., Savelsbergh, M.W.P.: Optimizing the hunter valley coal chain. In: Gurnani, H., Mehrotra, A., Ray, S. (eds.) Supply Chain Disruptions: Theory and Practice of Managing Risk, pp. 275–302. Springer, London (2012). doi:10.1007/978-0-85729-778-5_10
6. Canetti, R., Irani, S.: Bounding the power of preemption in randomized scheduling. SIAM J. Comput. **27**(4), 993–1015 (1998). http://dx.doi.org/10.1137/S0097539795283292

7. Chang, J., Khuller, S., Mukherjee, K.: LP rounding and combinatorial algorithms for minimizing active and busy time. In: Blelloch, G.E., Sanders, P. (eds.) Proceedings of the 26th SPAA, pp. 118–127. ACM, New York (2014). http://doi.acm.org/10.1145/2612669.2612689

8. Chang, J., Khuller, S., Mukherjee, K.: Active and busy time minimization. In: Proceedings of the 12th MAPSP, pp. 247–249 (2015). http://feb.kuleuven.be/mapsp.2015/Proceedings%20MAPSP%202015.pdf

9. Cohen-Addad, V., Li, Z., Mathieu, C., Milis, I.: Energy-efficient algorithms for non-preemptive speed-scaling. In: Bampis, E., Svensson, O. (eds.) WAOA 2014. LNCS, vol. 8952, pp. 107–118. Springer, Cham (2015). doi:10.1007/978-3-319-18263-6_10

10. Correa, J.R., Skutella, M., Verschae, J.: The power of preemption on unrelated machines and applications to scheduling orders. Math. Oper. Res. **37**(2), 379–398 (2012). http://dx.doi.org/10.1287/moor.1110.0520

11. Flammini, M., Monaco, G., Moscardelli, L., Shachnai, H., Shalom, M., Tamir, T., Zaks, S.: Minimizing total busy time in parallel scheduling with application to optical networks. Theor. Comput. Sci. **411**(40–42), 3553–3562 (2010). http://www.sciencedirect.com/science/article/pii/S0304397510002926

12. Ha, S.: Compile-time scheduling of dataflow program graphs with dynamic constructs. Ph.D. thesis, University of California, Berkeley (1992). http://www.eecs.berkeley.edu/Pubs/TechRpts/1992/ERL-92-43.pdf

13. Kalinowski, T., Matsypura, D., Savelsbergh, M.W.: Incremental network design with maximum flows. Eur. J. Oper. Res. **242**(1), 51–62 (2015). http://www.sciencedirect.com/science/article/pii/S0377221714008078

14. Khandekar, R., Schieber, B., Shachnai, H., Tamir, T.: Real-time scheduling to minimize machine busy times. J. Sched. **18**(6), 561–573 (2015). http://dx.doi.org/10.1007/s10951-014-0411-z

15. Mertzios, G.B., Shalom, M., Voloshin, A., Wong, P.W.H., Zaks, S.: Optimizing busy time on parallel machines. In: Proceedings of the 26th IPDPS, pp. 238–248. IEEE (2012). http://ieeexplore.ieee.org/xpl/articleDetails.jsp?arnumber=6267839

16. Nurre, S.G., Cavdaroglu, B., Mitchell, J.E., Sharkey, T.C., Wallace, W.A.: Restoring infrastructure systems: an integrated network design and scheduling (INDS) problem. Eur. J. Oper. Res. **223**(3), 794–806 (2012). http://www.sciencedirect.com/science/article/pii/S0377221712005310

17. Parsons, E.W., Sevcik, K.C.: Multiprocessor scheduling for high-variability service time distributions. In: Feitelson, D.G., Rudolph, L. (eds.) JSSPP 1995. LNCS, vol. 949, pp. 127–145. Springer, Heidelberg (1995). doi:10.1007/3-540-60153-8_26

18. Schulz, A.S., Skutella, M.: Scheduling unrelated machines by randomized rounding. SIAM J. Discret. Math. **15**(4), 450–469 (2002). http://dx.doi.org/10.1137/S0895480199357078

19. Soper, A.J., Strusevich, V.A.: Power of preemption on uniform parallel machines. In: Proceedings of the 17th APPROX. LIPIcs, vol. 28, pp. 392–402. Schloss Dagstuhl-Leibniz-Zentrum fuer Informatik, Dagstuhl, Germany (2014). http://drops.dagstuhl.de/opus/volltexte/2014/4711

Paths to Trees and Cacti

Akanksha Agrawal[1]([✉]), Lawqueen Kanesh[2], Saket Saurabh[1,2],
and Prafullkumar Tale[2]

[1] Department of Informatics, University of Bergen, Bergen, Norway
akanksha.agrawal@uib.no
[2] The Institute of Mathematical Sciences, HBNI, Chennai, India
{lawqueen,saket,pptale}@imsc.res.in

Abstract. For a family of graphs $\mathcal{F}$, the $\mathcal{F}$-CONTRACTION problem takes as an input a graph G and an integer k, and the goal is to decide whether there exists $F \subseteq E(G)$ of size at most k such that G/F belongs to $\mathcal{F}$. When $\mathcal{F}$ is the family of paths, trees or cacti, then the corresponding problems are PATH CONTRACTION, TREE CONTRACTION and CACTUS CONTRACTION, respectively. It is known that TREE CONTRACTION and CACTUS CONTRACTION do not admit a polynomial kernel unless NP $\subseteq$ coNP/poly, while PATH CONTRACTION admits a kernel with $\mathcal{O}(k)$ vertices. The starting point of this article are the following natural questions: *What is the structure of the family of paths that allows* PATH CONTRACTION *to admit a polynomial kernel? Apart from the size of the solution, what other additional parameters should we consider so that we can design polynomial kernels for these basic contraction problems?* With the goal of designing polynomial kernels, we consider the family of trees with bounded number of leaves (note that the family of paths are trees with at most two leaves). In particular, we study BOUNDED TREE CONTRACTION (BOUNDED TC). Here, an input is a graph G, integers k and ℓ, and the goal is to decide whether or not, there exists a subset $F \subseteq E(G)$ of size at most k such that G/F is a tree with at most ℓ leaves. We design a kernel for BOUNDED TC with $\mathcal{O}(k\ell)$ vertices and $\mathcal{O}(k^2 + k\ell)$ edges. Finally, we study BOUNDED CACTUS CONTRACTION (BOUNDED CC) which takes as input a graph G and integers k and ℓ. The goal is to decide whether there exists a subset $F \subseteq E(G)$ of size at most k such that G/F is a cactus graph with at most ℓ leaf blocks in the corresponding block decomposition. For BOUNDED CC we design a kernel with $\mathcal{O}(k^2 + k\ell)$ vertices and $\mathcal{O}(k^2 + k\ell)$ edges. We complement our results by giving kernelization lower bounds for BOUNDED TC, BOUNDED OTC and BOUNDED CC by showing that unless NP $\subseteq$ coNP/poly the size of the kernel we obtain is optimal.

1 Introduction

Graph editing problems are one of the central problems in graph theory that have received lot of attention in the realm of parameterized complexity. Some

The research leading to these results has received funding from the European Research Council (ERC) via grant PARAPPROX, reference 306992.

© Springer International Publishing AG 2017
D. Fotakis et al. (Eds.): CIAC 2017, LNCS 10236, pp. 31–42, 2017.
DOI: 10.1007/978-3-319-57586-5_4

of the important graph editing operations are vertex deletion, edge deletion, edge addition and edge contraction. For a family of graphs $\mathcal{F}$, the $\mathcal{F}$-EDITING problem takes as an input a graph G and an integer k, and the objective is to decide if at most k edit operations can result in a graph that belongs to the graph family $\mathcal{F}$. In fact, the $\mathcal{F}$-EDITING problem, where the edit operations are restricted to vertex deletion or edge deletion or edge addition or edge contraction alone have also been studied extensively in parameterized complexity. When we just focus on deletion operation (vertex/edge deletion) then the corresponding problem is called $\mathcal{F}$-VERTEX (EDGE) DELETION problem. For instance, the $\mathcal{F}$-EDITING problems encompasses several NP-hard problems such as VERTEX COVER, FEEDBACK VERTEX SET, PLANAR $\mathcal{F}$-DELETION, INTERVAL VERTEX DELETION, CHORDAL VERTEX DELETION, ODD CYCLE TRANSVERSAL, EDGE BIPARTIZATION, TREE CONTRACTION, PATH CONTRACTION, SPLIT CONTRACTION, CLIQUE CONTRACTION etc. However, most of the study in paramterized complexity or classical complexity, have been restricted to combination of vertex deletion, edge deletion or edge addition [2,3,6–9,12–18,22,24,26,28,29]. Only recently, edge contraction as an edit operation has started to gain attention in the realm of parameterized complexity. In this paper we study three edge-contraction problems from the perspective of kernelization complexity – one of the established subarea in parameterized complexity.

In parameterized complexity each problem instance is accompanied by a parameter k. A central notion in this field is the one of *fixed parameter tractable* (FPT). This means, for a given instance (I, k), solvability in time $\mathcal{O}(f(k)|I|^{\mathcal{O}(1)})$ where f is some function of k. Other important notion in parameterized complexity is *kernelization*, which captures the efficiency of data reduction techniques. A parameterized problem Π admits a kernel of size $g(k)$ (or $g(k)$-kernel) if there is a polynomial time algorithm (called *kernelization algorithm*) which takes as an input (I, k), and in time $\mathcal{O}(|I|^{\mathcal{O}(1)})$ returns an equivalent instance (I', k') of Π such that $|I'| + k' \leq g(k)$. Here, $g(\cdot)$ is a computable function whose value depends only on k. Depending on whether the function $g(\cdot)$ is *linear, polynomial* or *exponential*, the problem is said to admit a *linear, polynomial* or *exponential kernel*, respectively. It turns out that linear and polynomial kernels are most interesting from the kernelization perspective, because any problem that is fixed-parameter tractable admits an exponential kernel [10]. In this paper whenever we say *kernel*, we will refer to polynomial or linear kernels.

For several families of graphs $\mathcal{F}$, early papers by Watanabe et al. [30,31] and Asano and Hirata [1] showed that $\mathcal{F}$-EDGE CONTRACTION is NP-complete. In the framework of parameterized complexity (or even the classical complexity), these problems exhibit properties that are quite different than those of problems where we only delete or add vertices and edges. For instance deleting k edges from a graph such that the resulting graph is a tree is polynomial time solvable. On the other hand, Asano and Hirata showed that TREE CONTRACTION is NP-hard [1]. Furthermore, a well-known result by Cai [4] states that in case $\mathcal{F}$ is a hereditary family of graphs with a finite set of forbidden induced subgraphs, then the graph modification problem defined by $\mathcal{F}$ and the edit operations restricted to vertex deletion, edge deletion and edge addition admits a simple FPT algorithm. Indeed, for these problems, the result by Cai [4] does not hold when the edit operation is edge contraction. In particular, Lokshtanov et al. [27] and Cai and Guo [5] independently showed that if $\mathcal{F}$ is either the family of P_ℓ-free graphs for some $\ell \geq 5$ or the family of C_ℓ-free graphs for some $\ell \geq 4$, then $\mathcal{F}$-EDGE CONTRACTION is W[2]-hard. To the best of our knowledge, Heggernes et al. [21] were the first to explicitly study $\mathcal{F}$-EDGE CONTRACTION from the viewpoint of Parameterized Complexity. They showed that in case $\mathcal{F}$ is the family of trees, $\mathcal{F}$-EDGE CONTRACTION is FPT but does not admit a polynomial kernel, while in case $\mathcal{F}$ is the family of paths, the corresponding problem admits a faster algorithm and an $\mathcal{O}(k)$-vertex kernel. Golovach et al. [19] proved that if $\mathcal{F}$ is the family of planar graphs, then $\mathcal{F}$-EDGE CONTRACTION is again FPT. Moreover, Cai and Guo [5] showed that in case $\mathcal{F}$ is the family of cliques, $\mathcal{F}$-EDGE CONTRACTION is solvable in time $2^{\mathcal{O}(k \log k)} \cdot n^{\mathcal{O}(1)}$, while in case $\mathcal{F}$ is the family of chordal graphs, the problem is W[2]-hard. Heggernes et al. [23] developed an FPT algorithm for the case where $\mathcal{F}$ is the family of bipartite graphs. Later, a faster algorithm was proposed by Guillemot and Marx [20].

It is evident from our discussion that the complexity of the graph editing problem when restricted to edge contraction seems to be more difficult than their vertex or edge deletion counterparts. The starting point of our research is the following result by Heggernes et al. [21] who showed that TREE CONTRACTION does not admit a polynomial kernel unless NP $\subseteq$ coNP/poly [21] and PATH CONTRACTION admits a linear vertex kernel.

We wanted to understand the structure of the family of paths that allows PATH CONTRACTION to admit a polynomial kernel. Apart from the size of the solution, what other additional parameters should we consider so that we can design polynomial kernels for these basic contraction problems? One of the natural candidate for such an extension is to consider family of trees with the bounded number of leaves. With the goal to apprehend the understanding on role the number of leaves plays in the kernelization complexity for contracting to "path-like" graph, we study the problem which we call as BOUNDED TREE CONTRACTION (BOUNDED TC). Formally, the problem is defined below.

BOUNDED TREE CONTRACTION (BOUNDED TC)   **Parameter:** $k + \ell$
Input: A graph G and integers k, ℓ
Question: Does there exist $F \subseteq E(G)$ of size at most k such that G/F is a tree with at most ℓ leaves?

We give a kernel for BOUNDED TC with $\mathcal{O}(k\ell)$ vertices and $\mathcal{O}(k^2 + k\ell)$ edges. The approach we follow is similar to the one Heggernes et al. [21] used to obtain a linear kernel for PATH CONTRACTION. We observe that our algorithms works even when the input is a directed graph. In particular, we consider BOUNDED OUT-TREE CONTRACTION (BOUNDED OTC), which is defined as follows.

BOUNDED OUT-TREE CONTRACTION (BOUNDED OTC) **Parameter:** $k + \ell$
Input: A digraph D and integers k, ℓ
Question: Does there exist $F \subseteq A(D)$ of size at most k such that D/A is an out-tree with at most ℓ leaves?

We give a kernel for BOUNDED OTC with $\mathcal{O}(k^2 + k\ell)$ vertices and arcs.

We also study the contraction problem for a class of graphs which generalizes trees – the family of cacti. Formally, the problem we study is defined as follows.

BOUNDED CACTUS CONTRACTION (BOUNDED CC)   **Parameter:** $k + \ell$
Input: A graph G and integers k, ℓ
Question: Does there exist $F \subseteq E(G)$ of size at most k such that G/F is a cactus with at most ℓ leaf blocks in its block decomposition?

For BOUNDED CC we give a kernel with $\mathcal{O}(k^2+k\ell)$ vertices and edges. Finally, we give kernelization lower bound result. We complement all our kernelization algorithms by giving a matching lower bound. In particular, we show that BOUNDED TC, BOUNDED OTC and BOUNDED CC do not admit better kernels unless NP $\subseteq$ coNP/poly.

2   Preliminaries

For an undirected graph G, by $V(G)$ and $E(G)$ we denote the set of vertices and edges in G respectively. For a directed graph (or digraph) D, by $V(D)$ and $A(D)$ we denote the sets of vertices and directed edges (arcs) in D, respectively. The neighbourhood of a vertex v, in G denoted by $N_G(v)$, is the set $\{u \in V(G) \mid uv \in E(G)\}$. For a vertex $v \in V(D)$, $N_D^-(v)$ denotes the set $\{u \in V(D) \mid uv \in A(D)\}$, of its in-neighbors and $N_D^+(v)$ denotes the set $\{u \in V(D) \mid vu \in A(D)\}$, of its out-neighbors. The neighbourhood of a vertex $v \in V(D)$ is the set $N_D(v) = N_D^+(v) \cup N_D^-(v)$. The closed neighbourhood of a vertex is $N_G[v] = N_G(v) \cup \{v\}$. Degree of a vertex $\deg_G(u)$, is the cardinality of the set $N_G(v)$. In case of digraphs, the in-degree and out-degree of a vertex v, denoted by $\deg_D^-(v)$, $\deg_D^+(v)$, is $|N_D^-(v)|$ and $|N_D^+(v)|$ respectively. The (total) degree of v, denoted by $\deg_G(v)$, is the sum of its in-degree and out-degree. The subscripts in the notation for neighbourhood and degree will be omitted if the context is clear. For $F \subseteq E(G)$, $V(F)$ denotes the set of endpoints of edges (or arcs) in F. For a subset $S \subseteq V(G)$, by $G - S$ and $G[S]$ we denote the graph obtained by deleting vertices in S from G and the graph obtained by removing vertices in $V(G) \setminus S$ from G, respectively. For $F \subseteq E(G)$, $G - F$ is graph obtained by deleting edges in F from G. For $X, Y \subseteq V(G)$, we say X, Y are adjacent if there exist an edge with one end point in X and other in Y.

A *leaf* is a vertex with $\deg_G(v) = 1$. An *out-tree* is a digraph where each vertex has in-degree at most one and underlying undirected graph is a tree. A vertex v of an out-tree T is called a *leaf* if $\deg^-(v) = 1$ and $\deg^+(v) = 0$. The *root* of an out-tree is the unique vertex that has no in-neighbours. A *cactus* is an undirected graph such that every edge part of at most one cycle.

A *component* of a graph is a maximal connected subgraph. A *cut-vertex* in G is a vertex v such that the number of components in $G \setminus \{v\}$ is strictly more than the number of components in G. A graph that has no cut-vertex is called a *2-connected* graph. An edge uv of a graph G is called a *cut-edge* if the number of connected components in $G - \{uv\}$ is more than the number of connected components in G. We note that the number of connected components after removal of an edge can increase by at most 1. A digraph D is connected (disconnected, 2-connected) if its underlying undirected graph is connected (disconnected, 2-connected). An arc uv of a digraph D is called a *cut-arc* if the number of connected components in $D - \{uv\}$ is more than the number of connected components in D.

A maximal 2-connected subgraph of a graph G is called a *block*. Two distinct blocks in G can intersect in at most one vertex. A vertex which is contained in at least two block must be a cut-vertex in G. Let K be the set of cut-vertices and $\mathcal{B}$ be the set of blocks in G. A *block-decomposition* of G is a bipartite graph $\mathcal{D}$ with the vertex set bipartitioned into K and $\mathcal{B}$. Furthermore, $aB \in E(\mathcal{D})$ for $a \in K$ and $B \in \mathcal{B}$ if and only if $a \in V(B)$. It is known that a block decomposition of a connected graph is unique and is a tree [11]. For the sake of clarity, we call vertices in $\mathcal{D}$ as nodes. A block in a cactus can be either a cycle or an edge or an isolated vertex. The *number of leaves* in a cactus is defined to be the number of leaves in its block decomposition.

For a digraph D, contracting an arc $e = uv$ in D results in a digraph with vertex set as $V' = (V(D) \setminus \{u, v\}) \cup \{w\}$ and arc set as $A(D/e) = \{xy \mid x, y \in V(D) \setminus \{u, v\}, xy \in A(D)\} \cup \{wx \mid x \in (N_D^+(u) \cup N_D^+(v)) \setminus \{u, v\}\} \cup \{xw \mid x \in (N_D^-(u) \cup N_D^-(v)) \setminus \{u, v\}\}$. For a set of edges $F \subseteq E(G)$, G/F denotes the graph obtained from G by sequentially contracting the edges in F. G/F is oblivious to the order in which edges in F are contracted. A graph G is *isomorphic* to a graph H if there exists a *one-to-one* and *onto* function $\varphi : V(G) \to V(H)$ such that for $u, v \in V(G)$, $(u, v) \in E(G)$ if and only if $\varphi(u)\varphi(v) \in E(H)$. A graph G is *contractible* to a graph H, if there exists $F \subseteq E(G)$ such that G/F is *isomorphic* to H. In other words, G is contractible to H if there exists a *onto* function $\psi : V(G) \to V(H)$ such that the following properties hold.

– For all $h \in V(H)$ with $W(h) = \{v \in V(G) \mid \psi(v) = h\}$, $G[W(h)]$ is connected
– For all $h, h' \in V(H)$, $hh' \in E(H)$ if and only if $W(h)$ and $W(h')$ in G are adjacent.

For digraphs, we define the notion of contraction in an analogous way. Let $\mathcal{W} = \{W(h) \mid h \in V(H)\}$. We call $\mathcal{W}$ an *H-witness structure* of G. The sets in $\mathcal{W}$ are called *witness sets*. If a *witness set* contains more than one vertex of G then it is a *big* witness-set, otherwise it is a *small* witness set. A graph G is said to be *k-contractible* to a graph H if there exists $F \subseteq E(G)$ such that G/F is isomorphic to H and $|F| \leq k$. We will use the following observation in designing our algorithms.

Observation 1. *Let G be a graph which is k-contractible to a graph H and $\mathcal{W}$ be an H-witness structure of G. Then,*

– $|V(G)| \leq |V(H)| + k$;
– *No witness set in $\mathcal{W}$ contains more than $k + 1$ vertices;*
– $\mathcal{W}$ *has at most k big witness sets;*
– *The union of big witness sets in $\mathcal{W}$ contains at most $2k$ vertices.*

Definition 1. *A polynomial compression of a parameterized language $Q \subseteq \Sigma^* \times \mathbb{N}$ into a language $\Pi \subseteq \Sigma^*$ is an algorithm that takes as input an instance $(x, k) \in \Sigma^* \times \mathbb{N}$, and in time polynomial in $|x| + k$ returns a string y such that:*

- *$|y| \leq p(k)$ for some polynomial $p(\cdot)$, and*
- *$y \in \Pi$ if and only if $(x, k) \in Q$.*

3 Kernel for Bounded Out-Tree Contraction

In this section we design a polynomial kernel for BOUNDED OUT-TREE CONTRACTION. Our algorithm is inspired by the kernelization algorithm for PATH CONTRACTION presented in [21]. We first give the following useful observation.

Observation 2. *Let T be an out-tree and T' be the digraph obtained from T by contracting an arc $v_1 v_2 \in A(T)$. If T is an out-tree with at most ℓ leaves then, T' is an out-tree with at most ℓ leaves.*

Let T be an out-tree, v be a vertex in T with w being its unique in-neighbor and L, R be a partition of $N^+(v)$ such that $R \neq \emptyset$. Let T' be the digraph obtained from T by replacing v by a cut-arc $v_1 v_2$ in T'. Formally, $V(T') = (V(T) \setminus \{v\}) \cup \{v_1, v_2\}$ and $A(T') = (A(T) \setminus (\{vu \mid u \in N^+(v)\} \cup \{wv\})) \cup \{v_1 u \mid u \in L\} \cup \{v_2 u \mid u \in R\} \cup \{wv_1, v_1 v_2\}$. The following lemma proves a property of T' that will be useful later in designing our algorithm.

Lemma 1. [*]^1 *Let T be an out-tree and T' be the out-tree obtained from T as described above. If T is an out-tree with at most ℓ leaves then, T' is an out-tree with at most ℓ leaves.*

We now move to the description of the kernelization algorithm. Let (D, k, ℓ) be an instance of BOUNDED OTC. Without loss of generality we assume that D is connected, otherwise, (D, k, ℓ) is a NO instance. Recall that by our definition, D is connected if its underlying graph G_D is connected. The algorithm has only one reduction rule. To state the reduction rule, we first define notions of a *nice path* and a *reducible-tuple*. An induced directed path P from s to z in D together with a distinguished arc xy of P, which is a cut-arc in D, is called as a *nice path* if the following conditions are satisfied.

1. For each $v \in V(P)$, $\deg_D^-(v) = \deg_D^+(v) = 1$;
2. The sub-path P_x of P from s to x has at least $k + 3$ vertices;
3. The sub-path P_y of P from y to z has at least $k + 3$ vertices.
4. In $D/(A(P) \setminus \{xy\})$, xy is a cut-arc.

We note that whenever we talk about the number of vertices in a path P from x to y, then x, y are *also considered as vertices of P. For a *nice path* P from s to z in D, with a distinguished arc xy, we refer to the tuple (P_x, x, y, P_y) as a *reducible-tuple*. Here, P_x and P_y are the sub-paths of P from s to x and y to z, respectively.

We are now ready to state the Reduction Rule.

---

^{1} Proofs of results marked with [*] are omitted due to space constraints.

Reduction Rule 1. *Let (P_x, x, y, P_y) be a reducible-tuple in D. Then contract the arc xy and let the resulting instance be (D', k, ℓ), where $D' = D/\{xy\}$.*

Lemma 2 proves that the Reduction Rule 1 is safe and can be applied in polynomial time.

Lemma 2. *Reduction rule 1 is safe and can be applied in polynomial time.*

Proof. Let (P_x, x, y, P_y) be a *reducible-tuple* in D, $D' = D/\{xy\}$ and x^* be the vertex obtained after contracting the arc xy. We need to show that (D, k, ℓ) is a YES instance of BOUNDED OTC if and only if (D', k, ℓ) is a YES instance of BOUNDED OTC. Clearly, in polynomial time, given D and (P_x, x, y, P_y) one can apply Reduction Rule 1 and also find a *reducible-tuple*, if it exists.

In the forward direction let (D, k, ℓ) be a YES instance of BOUNDED OTC and $F \subseteq A(D)$ such that $|F| \le k$ and $T = D/F$ is an out-tree with at most ℓ leaves. By Observation 2, we know that $D/(F \cup \{xy\})$ is also an out tree with at most ℓ leaves. However, $D/(F \cup \{xy\}) = (D/\{xy\})/(F \setminus \{xy\}) = D'/(F \setminus \{xy\})$ is an out-tree with at most ℓ leaves. This implies that $D'/(F \setminus \{xy\})$ is an our-tree with at most ℓ leaves and $|F \setminus \{xy\}| \le |F| \le k$. Hence, it follows that (D', k, ℓ) is a YES instance of BOUNDED OTC.

In the reverse direction let (D', k, ℓ) be a YES instance of BOUNDED OTC and let $F' \subseteq A(D')$ of size at most k such that $T' = D'/F'$ is an out-tree with at most ℓ leaves with $\mathcal{W}'$ being the underlying T'-witness structure of D'. Since xy is a cut-arc in D, x^* is a cut-vertex in D'. Let $t^* \in V(T')$ such that $x^* \in W(t^*)$. Consider the set $W(t) = (W(t^*) \setminus \{x^*\}) \cup \{x, y\}$. Observe that xy is a cut-arc in $D[W(t)]$. Let C_x and C_y be the connected components in $D[W(t)] - \{xy\}$ containing x and y, respectively. Further, we let $W_x = V(C_x)$, $W_y = V(C_y)$ and $\mathcal{W} = (\mathcal{W}' \setminus W(t^*)) \cup \{W_x, W_y\}$. Notice that $\mathcal{W}$ partitions $V(D)$ and for each $W \in \mathcal{W}$, $D[W]$ is connected. Let T be the digraph for which $\mathcal{W}$ is a T-witness structure of D. Let $t_x, t_y \in V(T)$ be the vertices such that $W(t_x) = W_x$ and $W(t_y) = W_y$, respectively. We now argue that T is out-tree with at most ℓ leaves. Towards this we prove the following claim.

Claim 1. $\deg_T^+(t_y) \ge 1$ and $\deg_T^-(t_x) \ge 1$.

Proof. In this proof whenever we use the term *nearest* or *farthest*, it is defined by traversing the path P from y to z. Suppose $\deg_T^+(t_y) = 0$, then we let $z' \in V(P_y)$ nearest to y such that $z' \notin W(t_y)$ and $W(t_{z'})$ to be the witness set containing z'. Existence of z' is guaranteed by following facts: (a) $|W(t_y)| \le k + 1$ which in turn is implied from Observation 1, (b) $|V(P_y)| \ge k + 3$, and (c) F' is a solution to (D', k, ℓ). Further, we let y' to be the farthest vertex of $V(P_y)$ in $W(t_y)$ (it can be same as y). Observe that $y'z' \in A(D)$ and $t_{z'} \ne t_y$. But then, t_y has an out-neighbor, namely $t_{z'}$ which contradicts out assumption that $\deg_T^+(t_y) = 0$. ◇

We are now ready to prove that T is an out-tree with at most ℓ leaves. We first show that T is an out-tree. It is sufficient to argue that underlying undirected graph of T is connected and every vertex of T except the root (the vertex of in-degree 0) has in-degree at most 1. Note that t^* has at most one in-neighbor in T'. From

Claim 1 it follows that t^* must have an in-neighbor in T'. Since T' is an out-tree, we let t_q to be the unique in-neighbor of t^*. From Claim 1 it holds that $t_q t_x \in A(T)$. Since $xy \in A(D)$ is a cut-arc, it holds that $t_q t_y \notin A(T)$. For any $t' \in N_{T'}^+(t^*)$, if $t_x t', t_y t' \in A(T)$ then $W(t_x), W(t')$ are adjacent in D, and $W(t_y), W(t')$ are adjacent in D, contradicting that xy is a cut-arc in D. This implies that $N_{T'}^+(t^*)$ can be partitioned into $L = N_T^+(t_x)$ and $R = N_T^+(t_y)$. From Claim 1 it follows that R is non-empty. Hence digraph T is obtained from out-tree T' by replacing t' by a cut-arc t_x, t_y such that the out-neighbors of t_x is L and out-neighbors of t_y is R. By Lemma 1, T is an out-tree with at most ℓ leaves. $\square$

Lemma 3. [*] *Let* (D, k, ℓ) *be a* YES *instance of* Bounded OTC *on which Reduction Rule 1 is not applicable. Then,* D *has at most* $\mathcal{O}(k^2 + k\ell)$ *vertices and* $\mathcal{O}(k^2 + k\ell)$ *arcs.*

Theorem 1. Bounded OTC *admits a kernel of size* $\mathcal{O}(k^2 + k\ell)$.

Proof. Given an instance (D, k, ℓ), the algorithm repeatedly applies Reduction Rule 1, if applicable. By Lemma 2, we know that Reduction Rule 1 is safe and can be applied in polynomial time. Each application of reduction rule decreases the number of arcs and thus it can be applied only $|A(D)|$ times. If Reduction Rule 1 is not applicable then either the size of the instance is bounded by $\mathcal{O}(k^2 + k\ell)$, in which case we return a kernel of desired size. Otherwise, the algorithm correctly concludes that the instance is a NO instance of Bounded OTC. The correctness of this step is given by Lemma 3. $\square$

Following a similar approach we show the following results.

Theorem 2. [*] Bounded TC *admits a kernel of size* $\mathcal{O}(k^2 + k\ell)$.

Theorem 3. [*] Bounded CC *admits a kernel of size* $\mathcal{O}(k^2 + k\ell)$.

4 Kernel Lower Bounds

In this section we show that the kernelization algorithm that we gave in Sect. 3 for Bounded OTC, Bounded TC and Bounded CC are optimal assuming NP $\not\subseteq$ coNP/poly.

The problem of Red-Blue Dominating Set (RBDS) takes as an input a bipartite graph G, with vertex bi-partitions as R, B and an integer k. The goal is to decide if there exists $R' \subseteq R$ of size at most k such that for each $b \in B$, $R' \cap N(B) \neq \emptyset$. The problem Dominating Set takes as an input a graph G and an integer k, and the goal is to decide whether there exists $X \subseteq V(G)$ of size at most k, such that for each $v \in V(G)$, $X \cap N[v] \neq \emptyset$. Jansen and Pieterse proved that Dominating Set does not admit a compression of bit size $\mathcal{O}(n^{2-\epsilon})$, for any $\epsilon > 0$ unless NP $\subseteq$ coNP/poly, where n is the number of vertices in the input graph [25]. As an immediate corollary to this we have the following Proposition;

Proposition 1. [∗] RED-BLUE DOMINATING SET *does not admit a polynomial compression of bit size* $\mathcal{O}(n^{2-\epsilon})$, *for any* $\epsilon > 0$ *unless* NP $\subseteq$ coNP/poly. *Here,* n *is the number of vertices in the input graph.*

In light Proposition 1, we show that the kernelization algorithms we designed for BOUNDED OTC, BOUNDED TC and BOUNDED CC are optimal. Given an instance (G, R, B, k) of RBDS, we create an instance (D, k', ℓ') of BOUNDED OTC. We will show that indeed (G'_D, k', ℓ'), where G'_D is the underlying undirected graph of D serves as an instance of BOUNDED TC and BOUNDED CC respectively, for proving the desired kernel lower bounds.

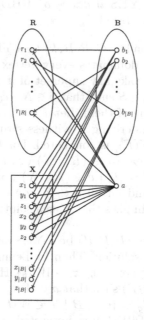

Fig. 1. Kernel lower bound for BOUNDED OTC.

Reduction. Let (G, R, B, k) be an instance of RBDS. We construct a digraph D as the following (refer Fig. 1). Initially, $V(D) = V(G)$ and $A(D) = \{br \mid b \in B, r \in R$ and $br \in E(G)\}$. We add a vertex a in $V(D)$ and for each $r \in R$, add the arc ar to $A(D)$. For each $b_i \in B$, we add three new vertices x_i, y_i, z_i to $V(D)$ and arcs $b_i x_i, b_i y_i, b_i z_i$ to $A(D)$. We let $X = \{x_i, y_i, z_i \mid b_i \in B\}$. For each $x \in X$, we add the arc ax to $A(D)$. Finally, we set $k' = |B| + k$ and $\ell' = |R| + 3|B| - k$. We let G'_D to be the underlying undirected graph of D.

4.1   Lower bound for Bounded OTC

We prove the following Lemmas that will be useful in establishing the equivalence of the instance.

Lemma 4. [*] *Let* (D, k', ℓ') *be a* YES *instance of* BOUNDED OTC. *For any a solution* $F \subseteq A(D)$ *of size at most* k', *for each* $b_i \in B$ *one of the following holds.*

- $b_i \in W(t_a)$;
- *All of* x_i, y_i, z_i *belong to* $W(t_a)$;
- *All of* x_i, y_i, z_i *belong to* $W(t_{b_i})$.

Here, $\mathcal{W}$ *is the underlying* D/F-*witness structure of* D *with* $W(t_a), W(t_b)$ *being the witness sets containing* a, b_i, *respectively.*

Lemma 5. (G, R, B, k) *is a* YES *instance of* RBDS *if and only if* (D, k', ℓ') *is a* YES *instance of* BOUNDED OTC.

Proof. In the forward direction let (G, R, B, k) be a YES instance of RBDS and $S \subseteq R$ of size k such that S dominates every vertex in B. Here, if S contains less than k vertices, then we take any superset of it which is of size exactly k. For each $b \in B$, there is an $r_b \in S$ such that $b \in N_G(r_b)$, if there are multiple r_bs then we arbitrarily choose one of them. Let $F = \{br_b \mid b \in B\} \cup \{ar \mid r \in S\}$, $T = D/F$ and $\mathcal{W}$ be the underlying T-witness structure of D. Observe that $|F| = |B| + k = k'$ and $D[V(F)]$ is connected. For $t_a \in V(T)$, such that $a \in W(t_a)$, $S \cup B \subseteq W(t_a)$. For each $v \in (R \cup X) \setminus S$, $t \in T$ such that $v \in W(t)$, $|W(t)| = 1$. Furthermore, $X' = \{t_v \mid v \in (R \cup X) \setminus S\}$ is an independent set T of size $|R| + 3|B| - k = \ell'$ and for all $v \in X'$, $av \in A(T)$. Therefore, T is an out-tree with ℓ' leaves. This implies that F is a solution to BOUNDED OTC for (D, k', ℓ').

In the reverse direction, let (D, k', ℓ') be a YES instance of BOUNDED OTC and $F \subseteq A(D)$ be one of its solution. Then by Lemma 4, for all $b_i \in B$, either $b_i \in W(t_a)$ or $x_i, y_i, z_i \in W(t_a)$ or $x_i, y_i, z_i \in W(t_b)$. Here, $\mathcal{W}$ is the D/F-witness structure of D and $t_a \in V(D/F)$ such that $a \in W(t_a)$. We partition B based on the cases in Lemma 4. Let $B_g = \{b_i \in B \mid b_i \in W(t_a)\}$. For each $b \in B_g$, since none of ba or ab is in $A(D)$, $D[W(t_a)]$ is connected and B is an independent set in D, therefore there must exist $r_b \in R \cup X$ such that $r_b \in W(t_a)$. We create a set S_g as the following, which initially is an empty set. For each $b \in B_g$, if there is $r_b \in R \cap W(t_a)$ such that $b \in N(r_b)$ then we add r_b to S_g. For each $b \in B$ such that there is no $r_b \in R \cap W(t_a)$ such that $b \in N(r_b)$, then we arbitrarily add a neighbor of b in R to S_g. We create another set S_w which initially is empty set. For each $b \in B \setminus B_g$, we add an arbitrary neighbor of b in R to S_w. Notice that for every $b \in B_g$, arcs br_b and ar_b are contained in the solution F. For every vertex $b_i \in B \setminus B_g$, either x_i, y_i, z_i are in $W(t_a)$ or x_i, y_i, z_i are in $W(t_b)$. By construction, x_i, y_i, z_i are adjacent to only a or b_i and hence $\{ax_i, ay_i, az_i\}$ or $\{b_ix_i, b_iy_i, b_iz_i\}$ are in solution F. Hence for every vertex $b_i \in B \setminus B_g$, there are three arcs in F. It follows that $|B_g| + |S_g| + 3|B \setminus B_g| \leq |F|$. We know that $|S_w| \leq |B \setminus B_g|$ and hence $|B_g| + |S_g| + 3|B \setminus B_g| = |S_g| + |B_g| + |S_w| + 2|B \setminus B_g| \leq |F| = k + |B|$. This implies that $|S_g| + |S_w| \leq k$. Furthermore, by choice of vertices in $S_g \cup S_w$, it dominates all the vertices in B. This concludes the proof. □

Theorem 4. [*] BOUNDED OUT-TREE CONTRACTION *does not admit a compression of size* $\mathcal{O}((k^2 + k\ell)^{1-\epsilon})$, *for any* $\epsilon > 0$.

Similarly, we can prove the following Theorems.

Theorem 5. [*] BOUNDED TREE CONTRACTION *does not admit a compression of size* $\mathcal{O}((k^2 + k\ell)^{1-\epsilon})$, *for any* $\epsilon > 0$.

Theorem 6. [*] BOUNDED CACTUS CONTRACTION *does not admit a compression of size* $\mathcal{O}((k^2 + k\ell)^{1-\epsilon})$, *for any* $\epsilon > 0$.

References

1. Asano, T., Hirata, T.: Edge-contraction problems. J. Comput. Syst. Sci. **26**(2), 197–208 (1983)
2. Bliznets, I., Fomin, F.V., Pilipczuk, M., Pilipczuk, M.: A subexponential parameterized algorithm for proper interval completion. In: Schulz, A.S., Wagner, D. (eds.) ESA 2014. LNCS, vol. 8737, pp. 173–184. Springer, Heidelberg (2014). doi:10.1007/978-3-662-44777-2_15
3. Bliznets, I., Fomin, F.V., Pilipczuk, M., Pilipczuk, M.: A subexponential parameterized algorithm for interval completion. In: SODA, pp. 1116–1131 (2016)
4. Cai, L.: Fixed-parameter tractability of graph modification problems for hereditary properties. Inf. Process. Lett. **58**(4), 171–176 (1996)
5. Cai, L., Guo, C.: Contracting few edges to remove forbidden induced subgraphs. In: Gutin, G., Szeider, S. (eds.) IPEC 2013. LNCS, vol. 8246, pp. 97–109. Springer, Cham (2013). doi:10.1007/978-3-319-03898-8_10
6. Cao, Y.: Unit interval editing is fixed-parameter tractable. In: Halldórsson, M.M., Iwama, K., Kobayashi, N., Speckmann, B. (eds.) ICALP 2015. LNCS, vol. 9134, pp. 306–317. Springer, Heidelberg (2015). doi:10.1007/978-3-662-47672-7_25
7. Cao, Y.: Linear recognition of almost interval graphs. In: SODA, pp. 1096–1115 (2016)
8. Cao, Y., Marx, D.: Chordal editing is fixed-parameter tractable. In: STACS, pp. 214–225 (2014)
9. Cao, Y., Marx, D.: Interval deletion is fixed-parameter tractable. ACM Trans. Algorithms (TALG) **11**(3), 21 (2015)
10. Cygan, M., Fomin, F.V., Kowalik, Ł., Marx, D., Pilipczuk, M., Pilipczuk, M., Saurabh, S.: Parameterized Algorithms. Springer, Cham (2015)
11. Diestel, R.: Graph theory. Grad. Texts Math. 101 (2005)
12. Drange, P.G., Pilipczuk, M.: A polynomial kernel for trivially perfect editing. In: Bansal, N., Finocchi, I. (eds.) ESA 2015. LNCS, vol. 9294, pp. 424–436. Springer, Heidelberg (2015). doi:10.1007/978-3-662-48350-3_36
13. Drange, P.G., Dregi, M.S., Lokshtanov, D., Sullivan, B.D.: On the threshold of intractability. In: Bansal, N., Finocchi, I. (eds.) ESA 2015. LNCS, vol. 9294, pp. 411–423. Springer, Heidelberg (2015). doi:10.1007/978-3-662-48350-3_35
14. Drange, P.G., Fomin, F.V., Pilipczuk, M., Villanger, Y.: Exploring subexponential parameterized complexity of completion problems. In: STACS, pp. 288–299 (2014)
15. Fomin, F.V.: Tight bounds for parameterized complexity of cluster editing with a small number of clusters. J. Comput. Syst. Sci. **80**(7), 1430–1447 (2014)
16. Fomin, F.V., Lokshtanov, D., Misra, N., Saurabh, S.: Planar F-deletion: approximation, kernelization and optimal FPT algorithms. In: FOCS (2012)

17. Fomin, F.V., Villanger, Y.: Subexponential parameterized algorithm for minimum fill-in. SIAM J. Comput. **42**(6), 2197–2216 (2013)

18. Ghosh, E., Kolay, S., Kumar, M., Misra, P., Panolan, F., Rai, A., Ramanujan, M.S.: Faster parameterized algorithms for deletion to split graphs. Algorithmica **71**(4), 989–1006 (2015)

19. Petr, A., van't Hof, P., Paulusma, D.: Obtaining planarity by contracting few edges. Theoret. Comput. Sci. **476**, 38–46 (2013)

20. Guillemot, S., Marx, D.: A faster FPT algorithm for bipartite contraction. Inf. Process. Lett. **113**(22–24), 906–912 (2013)

21. Heggernes, P., van't Hof, P., Lévêque, B., Lokshtanov, D., Paul, C.: Contracting graphs to paths and trees. Algorithmica **68**(1), 109–132 (2014)

22. Heggernes, P., van't Hof, P., Jansen, B.M.P., Kratsch, S., Villanger, Y.: Parameterized complexity of vertex deletion into perfect graph classes. In: FCT, pp. 240–251 (2011)

23. Heggernes, P., van't Hof, P., Lokshtanov, D., Paul, C.: Obtaining a bipartite graph by contracting few edges. SIAM J. Discrete Math. **27**(4), 2143–2156 (2013)

24. Jansen, B.M.P., Lokshtanov, D., Saurabh, S.: A near-optimal planarization algorithm. In: Proceedings of the Twenty-Fifth Annual ACM-SIAM Symposium on Discrete Algorithms, SODA 2014, Portland, Oregon, USA, 5–7 January 2014, pp. 1802–1811 (2014)

25. Jansen, B.M.P., Pieterse, A.: Sparsification upper and lower bounds for graphs problems and not-all-equal SAT. In: 10th International Symposium on Parameterized and Exact Computation, IPEC, pp. 163–174 (2015)

26. Kim, E.J., Langer, A., Paul, C., Reidl, F., Rossmanith, P., Sau, I., Sikdar, S.: Linear kernels and single-exponential algorithms via protrusion decompositions. In Proceedings of the 40th International Colloquium Automata, Languages, and Programming - ICALP , Riga, Latvia, 8–12 July, Part I, pp. 613–624 (2013)

27. Lokshtanov, D., Misra, N., Saurabh, S.: On the hardness of eliminating small induced subgraphs by contracting edges. In: Gutin, G., Szeider, S. (eds.) IPEC 2013. LNCS, vol. 8246, pp. 243–254. Springer, Cham (2013). doi:10.1007/978-3-319-03898-8_21

28. Marx, D.: Chordal deletion is fixed-parameter tractable. Algorithmica **57**(4), 747–768 (2010)

29. Reed, B.A., Smith, K., Adrian, V.: Finding odd cycle transversals. Oper. Res. Lett. **32**(4), 299–301 (2004)

30. Watanabe, T., Ae, T., Nakamura, A.: On the removal of forbidden graphs by edge-deletion or by edge-contraction. Discrete Appl. Math. **3**(2), 151–153 (1981)

31. Watanabe, T., Ae, T., Nakamura, A.: On the NP-hardness of edge-deletion and-contraction problems. Discrete Appl. Math. **6**(1), 63–78 (1983)

Temporal Flows in Temporal Networks

Eleni C. Akrida[1]([✉]), Jurek Czyzowicz[2], Leszek Gąsieniec[1], Łukasz Kuszner[3],
and Paul G. Spirakis[1,4]

[1] Department of Computer Science, University of Liverpool, Liverpool, UK
{E.Akrida,L.A.Gasieniec,P.Spirakis}@liverpool.ac.uk
[2] Dépt. d'informatique, Université du Québec en Outaouais, Gatineau, QC, Canada
jurek@uqo.ca
[3] Faculty of Electronics, Telecommunications and Informatics,
Gdańsk University of Technology, Gdańsk, Poland
kuszner@eti.pg.gda.pl
[4] Computer Technology Institute and Press "Diophantus" (CTI), Patras, Greece

Abstract. We introduce temporal flows on temporal networks [17,19], i.e., networks the links of which exist only at certain moments of time. Such networks are ephemeral in the sense that no link exists after some time. Our flow model is new and differs from the "flows over time" model, also called "dynamic flows" in the literature. We show that the problem of finding the maximum amount of flow that can pass from a source vertex s to a sink vertex t up to a given time is solvable in Polynomial time, even when node buffers are bounded. We then examine mainly the case of unbounded node buffers. We provide a simplified static *Time-Extended network* (STEG), which is of *polynomial size to the input* and whose static flow rates are equivalent to the respective temporal flow of the temporal network; using STEG, we prove that the maximum temporal flow is equal to the minimum *temporal s-t cut*. We further show that temporal flows can always be decomposed into flows, each of which moves only through a journey, i.e., a directed path whose successive edges have strictly increasing moments of existence. We partially characterise networks with random edge availabilities that tend to eliminate the $s \rightarrow t$ temporal flow. We then consider *mixed* temporal networks, which have some edges with specified availabilities and some edges with random availabilities; we show that it is **#P**-hard to compute the *tails and expectations of the maximum temporal flow* (which is now a random variable) in a mixed temporal network.

This work was partially supported by (i) the School of EEE and CS and the NeST initiative of the University of Liverpool, (ii) the NSERC Discovery grant, (iii) the Polish National Science Center grant DEC-2011/02/A/ST6/00201, and (iv) the FET EU IP Project MULTIPLEX under contract No. 317532.
Due to lack of space, an extended literature review and all missing proofs can be found in the full version of this paper at http://arxiv.org/abs/1606.01091 [2].

© Springer International Publishing AG 2017
D. Fotakis et al. (Eds.): CIAC 2017, LNCS 10236, pp. 43–54, 2017.
DOI: 10.1007/978-3-319-57586-5_5

1   Introduction and Motivation

1.1   Our Model, the Problem, and Our Results

It is generally accepted to describe a network topology using a graph, whose vertices represent the communicating entities and edges correspond to the communication opportunities between them. Consider a directed graph (network) $G(V, E)$ with a set V of n vertices (nodes) and a set E of m edges (links). Let $s, t \in V$ be two special vertices called the *source* and the *sink*, respectively; for simplicity, assume that no edge enters the source s and no edge leaves the sink t. We also assume that an infinite amount of a quantity, say, a liquid, is available in s at time zero. However, our network is *ephemeral*; each edge is available for use only at certain *days* in time, described by positive integers, and after some (finite) day in time, no edge becomes available again; the reader may think of these days as instances of availability of that edge. Our liquid, located initially at node s, can flow in this ephemeral network through edges only at days at which the edges are available.

Each edge $e \in E$ in the network is also equipped with a *capacity* $c_e > 0$ which is a positive integer, unless otherwise specified. We also consider each node $v \in V$ to have an internal buffer (storage) $B(v)$ of maximum size B_v; here, B_v is also a positive integer; initially, we shall consider both the case where $B_v = +\infty$, for all $v \in V$, and the case where all nodes have finite buffers. From Sect. 3 on, we only consider unbounded (infinite) buffers.

The *semantics* of the flow of our liquid within G are the following:

– Let an amount x_v of liquid be at node v, i.e., in $B(v)$, at the *beginning* of day l, for some $l \in \mathbb{N}$. Let $e = (v, w)$ be an edge that exists at day l. Then, v may *push* some of the amount x_v through e at day l, as long as that amount is at most c_e. This quantity will arrive to w at the *end* of *the same day*, l, and will be stored in $B(w)$.
– At the end of day l, for any node w, some flows may arrive from edges (v, w) that were available at day l. Since each such quantity of liquid has to be stored in w, the sum of all flows incoming to w plus the amount of liquid that is already in w at the end of day l, after w has sent any flow out of it at the beginning of day l, must not exceed B_w.
– Flow arriving at w at (the end of) day l can leave w only via edges existing at days $l' > l$.

Thus, our flows are not flow rates, but flow amounts (similar to considerations in *transshipment problems* [14,16]). Notice that we assume above that we have absolute knowledge of the days of existence of each edge. Admittedly, the encoding of the input in our temporal network problems is quite detailed but specific description of the edge availabilities (or lack thereof) may be required in a range of network infrastructure settings where there is a planned schedule of link existence, e.g., one may need to have detailed information on planned maintenance on pipe-sections in a water network to assure restoration of the network services.

On the positive side, some problems that are weakly NP-hard in similar dynamic flow models become polynomially solvable in our model.

Our Results. We provide polynomial-time solutions to the *Maximum temporal flow problem* (MTF): Given a directed graph G with edge availabilities, distinguished nodes s, t, edge capacities and node buffers as previously described, and also given a specific day $l' > 0$, find the maximum value of the quantity of liquid that can arrive to t by (the end of) day l'.

For the case of infinite buffers, we give a simplified static *Time-Extended network* (STEG) which, in contrast to all previous dynamic flows literature and due to the encoding of our input, is of *linear* size to the input, and *not exponential.* The static flow rates of STEG are equivalent to the respective temporal flow of the temporal network; using it, we prove that the maximum temporal flow is equal to the minimum *temporal s-t cut.* We also show that temporal flows can always be decomposed into flows, each of which moves only through a journey, i.e., a directed path whose successive edges have strictly increasing moments of existence.

In many practical scenarios it is reasonable to assume that not all edge availabilities are known in advance, e.g., in a water network where there may be unplanned disruptions at one or more pipe sections; in these cases, one may have statistical information on the pattern of link availabilities. We partially characterise networks with random edge availabilities that tend to eliminate the $s \to t$ temporal flow. We also introduce and study here flows in *mixed temporal* networks for the first time; these are networks in which the availabilities of some edges are random and the availabilities of some other edges are specified. In such networks, the value of the maximum temporal flow is a random variable. Consider, for example, the temporal flow network of Fig. 1 where there are n directed disjoint two-edge paths from s to t. Assume that *every* edge independently selects a *single* label uniformly at random from the set $\{1, \ldots, \alpha\}$, $\alpha \in \mathbb{N}^*$. The edge capacities are the numbers drawn in the boxes, with $w_i' \geq w_i$ for all i. Here, the value of the maximum $s \to t$ flow is a random variable that is the sum of Bernoulli random variables. This already indicates that the exact calculation of the maximum flow in mixed networks is a hard problem; we show for mixed networks that it is #**P**-hard to compute tails and expectations of the maximum temporal flow.

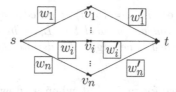

Fig. 1. A mixed temporal network

1.2 Previous Work

The traditional (static) network flows were extensively studied in the seminal book of Ford and Fulkerson [13] (see also Ahuja et al. [1]) and the relevant literature is vast. They have recently been re-examined for the purpose of approximating their maximum value or improving their time complexity [8,18,20,23,24]. *Dynamic network flows* (also called *flows over time*) [15] refer to *static* directed networks, the edges of which have capacities as well as transit times. Ford and Fulkerson [13] formulated and solved the dynamic maximum flow problem. For excellent surveys on dynamic network flows, the reader is also referred to the work of Aronson [6], the work of Powell [22], and the great survey by Skutella [25].

Temporal networks, defined by Kempe et al. [17], are graphs *the edges of which exist only at certain instants of time, called labels* (see also [19]). So, they are a type of *dynamic* networks. Various aspects of temporal (and other dynamic) networks were also considered in the work of Erlebach et al. [12] and in [4,5,7,9]; as far as we know, this is the first work to examine flows on temporal networks. There is also literature on models of temporal networks with random edge availabilities [3,10,11], but to the best of our knowledge, ours is the first work on flows in such temporal networks.

Perhaps the closest model in the flows literature to our model is the "*Dynamic*[1] dynamic network flows", studied by Hoppe in his PhD thesis [15, Chap. 8]. Hoppe introduces *mortal edges* that exist between a start and an end time; still, Hoppe assumes transmission rates on the edges and the ability to hold any amount of flow on a node (infinite node buffers). Thus, our model is an extreme case of the latter, since we assume that edges exist only at specific days (instants) and that our transit rates are virtually unbounded, since at one instant *any amount* of flow can be sent through an edge if the capacity allows.

1.3 Formal Definitions

Definition 1 ((Directed) Temporal Graph). *Let $G = (V, E)$ be a directed graph. A (directed) temporal graph on G is an ordered triple $G(L) = (V, E, L)$, where $L = \{L_e \subseteq \mathbb{N} : e \in E\}$ assigns a finite set L_e of discrete labels to every edge (arc) e of G. L is called the* labelling *of G. The labels, L_e, of an edge $e \in E$ are the* integer time instances *(e.g., days) at which e is available.*

Definition 2 (Time edge). *Let $e = (u, v)$ be an edge of the underlying digraph of a temporal graph and consider a label $l \in L_e$. The ordered triplet (u, v, l), also denoted as (e, l), is called* time edge. *We denote the set of time edges of a temporal graph $G(L)$ by E_L.*

A basic assumption that we follow is that when a (flow) entity passes through an available edge e at time t, then it can pass through a subsequent edge only at some time $t' \geq t + 1$ and only at a time at which that edge is available. In the tradition of assigning "transit times" in the dynamic flows literature,

---

[1] The first "dynamic" term refers to the dynamic nature of the underlying graph.

one may think that any edge e of the graph has some *transit time*, tt_e, with $0 < tt_e < 1$, but *otherwise arbitrary and not specified*. Henceforth, we assume $tt_e = 0.5$, $\forall e \in E$, without loss of generality; any value of tt_e between 0 and 1 will lead to the same results in our paper.

Definition 3 (Journey). *A journey from a vertex u to a vertex v ($u \rightarrow v$ journey) is a sequence of time edges $(u, u_1, l_1), \ldots, (u_{k-1}, v, l_k)$, such that $l_i < l_{i+1}$, $\forall i = 1, \ldots, k-1$. The last label, l_k, is called the* arrival time *of the journey.*

Definition 4 (Foremost journey). *A $u \rightarrow v$ journey in a temporal graph is called* foremost journey *if its arrival time is the minimum arrival time of all $u \rightarrow v$ journeys' arrival times, under the labels assigned to the underlying graph's edges. We call this arrival time the* temporal distance, $\delta(u, v)$, *of v from u.*

Thus, no flow arrives to t (starting from s) on or before any time $l < \delta(s, t)$.

Definition 5 (Temporal Flow Network). *A* temporal flow network *$(G(L), s, t, c, B)$ is a temporal graph $G(L) = (V, E, L)$ equipped with:*

1. *a source vertex s and a sink (target) vertex t.*
2. *for each edge e, a capacity $c_e > 0$; usually the capacities are assumed to be integers.*
3. *for each node v, a buffer $B(v)$ of storage capacity $B_v > 0$; we assume $B_s = B_t = +\infty$.*

If all node capacities are infinite, we denote the network by $(G(L), s, t, c)$.

Definition 6 (Temporal Flows in Temporal Flow Networks). *Let $(G(L) = (V, E, L), s, t, c, B)$ be a temporal flow network. Denote by δ_u^+ the outgoing edges from u and by δ_u^- the incoming edges to u. Let $L_R(u)$ be the set of labels on all edges incident to u along with an extra label 0 (artificial label for initialization), i.e., $L_R(u) = \bigcup_{e \in \delta_u^+ \cup \delta_u^-} L_e \cup \{0\}$. A temporal flow on $G(L)$ consists of a non-negative real number $f(e, l)$ for each time-edge (e, l), and real numbers $b_u^-(l), b_u^\mu(l), b_u^+(l)$ for each node $u \in V$ and each "day" l, such that:*

1. *$0 \le f(e, l) \le c_e$, for every time edge (e, l),*
2. *$0 \le b_u^-(l) \le B_u$, $0 \le b_u^\mu(l) \le B_u$, $0 \le b_u^+(l) \le B_u$, for every node u and every $l \in L_R(u)$,*
3. *for every $e \in E$, $f(e, 0) = 0$,*
4. *for every $v \in V \setminus \{s\}$, $b_v^-(0) = b_v^\mu(0) = b_v^+(0) = 0$,*
5. *for every $e \in E$ and $l \notin L_e$, $f(e, l) = 0$,*
6. *at time 0 there is an infinite amount of flow "units" available at the source s,*
7. *for every $v \in V \setminus \{s\}$ and for every $l \in L$, $b_v^-(l) = b_v^+(l_{prev})$, where l_{prev} is the largest label in $L_R(v)$ that is smaller than l,*
8. *(Flow out on day l) for every $v \in V \setminus \{s\}$ and for every $l, b_v^\mu(l) = b_v^-(l) - \sum_{e \in \delta_v^+} f(e, l)$,*
9. *(Flow in on day l) for every $v \in V \setminus \{s\}$ and for every $l, b_v^+(l) = b_v^\mu(l) + \sum_{e \in \delta_v^-} f(e, l)$.*

Note 1. *One may think of $b_v^-(l), b_v^\mu(l), b_v^+(l)$ as the buffer content of liquid in v at the "morning", "noon", i.e., after the departures of flow from v, and "evening", i.e., after the arrivals of flow to v, of day l.*

Note 2. *For a temporal flow f on an acyclic $G(L)$, if one could guess the (real) numbers $f(e, l)$ for each time-edge (e, l), then the numbers $b_v^-(l), b_v^\mu(l), b_v^+(l)$, for every $v \in V$, can be computed by a single pass over an order of the vertices of $G(L)$ from s to t. This can be done by following (6) through (6) from Definition 6 from s to t.*

Definition 7 (Value of a Temporal Flow). *The value $v(f)$ of a temporal flow f is $b_t^+(l_{max})$ under f, i.e., the amount of liquid that, via f, reaches t during the lifetime of the network (l_{max} is the maximum label in L). If $b_t^+(l_{max}) > 0$ for a particular flow f, we say that f is feasible.*

Definition 8 (Mixed temporal networks). *Given a directed graph $G = (V, E)$ with a source s and a sink t in V, let $E = E_1 \cup E_2$, so that $E_1 \cap E_2 = \emptyset$, and:*

1. *the labels (availabilities) of edges in E_1 are specified, and*
2. *each of the labels of the edges in E_2 is drawn uniformly at random from the set $\{1, 2, \ldots, \alpha\}$, for some even integer α^2, independently of the others.*

We call such a network "Mixed Temporal Network $[1, \alpha]$" and denote it by $G(E_1, E_2, \alpha)$.

Note that (traditional) temporal networks as previously defined are a special case of the mixed temporal networks, in which $E_2 = \emptyset$. However, with some edges being available at random times, the value of a temporal flow (until time α) becomes a random variable and the study of relevant problems requires a different approach than the one needed for (traditional) temporal networks.

Problem 1 (Maximum Temporal Flow (MTF)). *Given a temporal flow network $\big(G(L), s, t, c, B\big)$ and a day $d \in \mathbb{N}^*$, compute the maximum $b_t^+(d)$ over all flows f in the network.*

2   LP for the MTF Problem with or Without Bounded Buffers

In the description of the MTF problem, if d is not a label in L, it is enough to compute the maximum $b_t^+(l_m)$ over all flows, where l_m is the maximum label in L that is smaller than d. Henceforth, we assume $d = l_{max}$ unless otherwise specified; the analysis does not change: if $d < l_{max}$, one can remove all time-edges with labels larger than b and solve MTF in the resulting network.

---

[2] We choose an even integer to simplify the calculations in the remainder of the paper. However, with careful adjustments, the results would still hold for an arbitrary integer.

Let Σ be the set of conditions of Definition 6. The optimization problem, Π:

$$\left\{ \begin{array}{l} \max \text{ (over all} f)\ b_t^+(d) \\ \text{subject to} \quad\quad \Sigma \end{array} \right\}$$

is a *linear program* with unknown variables $\{f(e, l), b_v^-(l), b_v^+(l)\}$, $\forall l \in L, \forall v \in V$, since each condition in Σ is either a linear equation or a linear inequality in the unknown variables. Therefore, by noticing that the number of equations and inequalities are polynomial in the size of the input of Π, we get the following:

Lemma 1. *Maximum Temporal Flow is in P, i.e., can be solved in polynomial time in the size of the input, even when the node buffers are finite, i.e., bounded.*

Note 3. *Recall that E_L is the set of time edges of a temporal graph. If $n = |V|$, $m = |E|$ and $k = |E_L| = \sum_e |L_e|$, then MTF can be solved in sequential time polynomial in $n + m + k$ when the capacities and buffer sizes can be represented with polynomial in n number of bits. In the remainder of the paper, we shall investigate more efficient approaches for MTF.*

3 Temporal Networks with Unbounded Buffers at Nodes

3.1 Basic Remarks

We consider here the MTF problem for temporal networks on underlying graphs with $B_v = +\infty$, $\forall v \in V$.

Definition 9 (Temporal Cut). *Let $\big(G(L), s, t, c\big)$ be a temporal flow network on a digraph G. A set of time-edges, S, is called a temporal cut (separating s and t) if the removal from the network of S results in a temporal flow network with no $s \rightarrow t$ journey.*

Definition 10 (Minimal Temporal Cut). *A set of time-edges, S, is called a minimal temporal cut (separating s and t) if S is a temporal cut, and no proper subset of S is a temporal cut.*

Definition 11. *Let S be a temporal cut of $\big(G(L) = (V, E, L), s, t, c\big)$. The capacity of the cut is $c(S) := \sum_{(e,l) \in S} c(e, l)$, where $c(e, l) = c_e$, $\forall l$.*

Lemma 2. *Let S be a (minimal) temporal cut in $\big(G(L) = (V, E, L), s, t, c\big)$. If we remove S from $G(L)$, no flow can ever arrive to t during the lifetime of $G(L)$.*

3.2 The Time-Extended Flow Network and Its Simplification

Let $\big(G(L) = (V, E, L), s, t, c\big)$ be a temporal flow network on a directed graph G. Let E_L be the set of time edges of $G(L)$. Following the tradition in literature [13], we construct the *time-extended* static flow network that corresponds to $G(L)$, denoted by $\text{TEG}(L) = (V^*, E^*)$. By construction, $\text{TEG}(L)$ admits the same maximum flow as $G(L)$. $\text{TEG}(L)$ is constructed as follows: for every vertex

$v \in V$ and for every time step $i = 0, 1, \ldots, l_{max}$, V^* contains a copy, v_i, of v. Also, for every time edge (x, v, l), $l \in \mathbb{N}$, $x \in V$ of $G(L)$, V^* contains a copy $v_{l+0.5}$ of v. E^* has a directed edge (called *vertical*) from a copy of vertex v to the *next* copy of v, for any $v \in V$, where the order of the copies is defined by their indices; every vertical edge has infinite capacity (as the node whose copies it connects). Furthermore, for every time edge (u, v, l) of $G(L)$, E^* has a directed edge (called *crossing*) $(u_l, v_{l+0.5})$ with capacity equal to the capacity of the edge (u, v). The source and target vertices in TEG(L) are the first copy of s and the last copy of t in V^*, respectively. Note that $|V^*| \leq |V| \cdot l_{max} + |E_L|$ and $|E^*| \leq |V| \cdot l_{max} + 2|E_L|$.

We now "simplify" TEG(L) as follows: we convert vertical edges between consecutive copies of the same vertex into a *single vertical edge (with infinite capacity)* from the first to the last copy in the sequence and we remove all intermediate copies; we only perform this simplification when no intermediate node is an endpoint of a crossing edge. We call the resulting network *simplified time-extended* network and we denote it by STEG(L) $= (V', E')$. Note that $|V'| \leq |V| + 2|E_L|$ and $|E'| \leq |V| + 3|E_L|$.

Let the first copy of any vertex $v \in V$ in the time-extended network be v_{copy_0}, the second copy v_{copy_1}, etc. An $s \to t$ flow f in $G(L)$ defines an $s \to t$ flow in the time-extended network STEG(L) as follows:

- The flow from the first copy of s to the next copy is the sum of all flow units that "leave" s in $G(L)$ throughout the time the network exists.
- The flow from the first copy of any *other* vertex to the next copy is zero.
- The flow on any crossing edge that connects some copy u_l of vertex $u \in V$ and the copy $v_{l+0.5}$ of some other vertex $v \in V$ is exactly the flow on the time edge (u, v, l).
- The flow between two *consecutive* copies v_x and v_y, for some x, y, of the same vertex $v \in V$ corresponds to the units of flow stored in v from time x up to time y and is the difference between the flow *received* at the first copy through all incoming edges and the flow *sent* from the first copy through all outgoing *crossing* edges.

Using TEG(L) and STEG(L), we can prove the following (for the proof, see [2]):

Theorem 1. *The maximum temporal flow in $(G(L) = (V, E, L), s, t, c)$ is equal to the minimum capacity (minimal) temporal cut.*

Lemma 3. *Any static flow rate algorithm A that computes the maximum flow in a static, directed, s-t network G of n vertices and m edges in time $T(n, m)$, also computes the maximum temporal flow in a $(G(L) = (V, E, L), s, t, c)$ temporal flow network in time $T(n', m')$, where $n' \leq n + 2|E_L|$ and $m' \leq n + 3|E_L|$.*

Corollary 1 (Journeys flow decomposition). *Let $(G(L) = (V, E, L), s, t, c)$ be a temporal flow network on a directed graph G. Let f be a temporal flow in $G(L)$ (f is given by the values of $f(e, l)$ for the time-edges $(e, l) \in E_L$). Then, there is a collection of $s \to t$ journeys $j_1, j_2, \ldots, j_k$ such that:*

1. $k \leq |E_L|$
2. $v(f) = v(f_1) + \ldots v(f_k)$
3. *f_i sends positive flow only on the time-edges of j_i.*

4   Mixed Temporal Networks and Their Hardness

Mixed temporal networks of the form $G(E_1, E_2, \alpha)$ (see Definition 8) can model practical cases, where some edge availabilities are exactly specified, while some other edge availabilities are randomly chosen (due to security reasons, faults, etc.); for example, in a water network, one may have planned disruptions for maintenance in some water pipes, but unplanned (random) disruptions in some others. With some edges being available at random times, the value of the maximum temporal flow (until time α) now becomes a random variable.

4.1   Temporal Networks with Random Availabilities that are Flow Cutters

We study here a special case of the mixed temporal networks $G(E_1, E_2, \alpha)$, where $E_1 = \emptyset$, i.e., *all* edges become available at random time instances, and we partially characterise such networks that eliminate the flow that arrives at t asymptotically almost surely. All missing proofs can be found in the full version of the paper [2].

Let $G = (V, E)$ be a directed graph of n vertices with a distinguished source, s, and a distinguished sink, t. Suppose that each edge $e \in E$ is available only at a *unique* moment in time (i.e., day) *selected uniformly at random from the set* $\{1, 2, \ldots, \alpha\}$, *for some even*[3] *integer* $\alpha \in \mathbb{N}$, $\alpha > 1$; suppose also that the selections of the edges' labels are independent. Let us call such a network a Temporal Network with unique random availabilities of edges, and denote it by URTN(α). Then, the following holds:

Lemma 4. *Let P_k be a directed $s \to t$ path of length k in G. Then, P_k becomes a journey in* URTN(α) *with probability at most* $\frac{1}{k!}$.

Now, consider directed graphs as described above, in which the distance from s to t is at least $c \log n$, for a constant integer $c > 2$; so any directed $s \to t$ path has at least $c \log n$ edges. Let us call such graphs "c-long $s \to t$ graphs" or simply c-long. A c-long $s \to t$ graph is called *thin* if the number of simple directed $s \to t$ paths is at most n^β, for some constant β. It can be proven that:

Lemma 5. *Consider a URTN(α) with an underlying graph G being any particular c-long and thin digraph. Then, the probability that the amount of flow from s arriving at t is positive tends to zero as n tends to $+\infty$.*

Randomly labelled c-long and thin graphs is not the only case of temporal networks that disallows flow to arrive to t asymptotically almost surely.

Definition 12. *A cut C in a (traditional) flow network G is a set of edges, the removal of which from the network leaves no directed $s \to t$ paths in G.*

---

[3] We choose an even integer to simplify the calculations. However, with careful adjustments in the calculations, the results would still hold for an arbitrary integer.

Definition 13. *A cut C_1 precedes a cut C_2 in a flow network G (denoted by $C_1 \to C_2$) if any directed $s \to t$ path that goes through an edge in C_1 must also later go through an edge in C_2.*

Definition 14 (Multiblock graphs). *A flow network is called a (c, d)-multiblock graph if it has at least $c \log n$ disjoint cuts $C_1, \ldots, C_{c \log n}$ such that $C_i \to C_{i+1}$, $i = 1, \ldots, c \log n - 1$, and for all $i = 1, \ldots, c \log n$, $|C_i| \le d$, for some constants $c > 2$, $d \ge 2$.*

Note that (c, d)-multiblocks and $(c$-long,thin$)$-graphs are two different graph classes. Figure 2 shows a $(c, 2)$-multiblock of $n = c\sqrt{k} + 2$, $k \in \mathbb{N}$, vertices which is not thin.

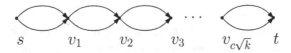

$$s \qquad v_1 \qquad v_2 \qquad v_3 \qquad \cdots \qquad v_{c\sqrt{k}} \qquad t$$

Fig. 2. A $(c, 2)$-multiblock which is not thin.

Lemma 6. *Consider a $URTN(\alpha)$ with an underlying graph G being any particular (c, d)-multiblock. Then, the probability that the amount of flow from s arriving at t is positive tends to zero as n tends to $+\infty$.*

4.2   The Complexity of Computing the Expected Maximum Temporal Flow

We consider here the following problem:

Problem 2 (Expected Maximum Temporal Flow). *What is the time complexity of computing the expected value of the maximum temporal flow, v, in $G(E_1, E_2, \alpha)$?*

Definition 15 [21, p. 441]. *Let Q be a polynomially balanced, polynomial-time decidable binary relation. The counting problem associated with Q is: Given x, how many y are there such that $(x, y) \in Q$? #P is the class of all counting problems associated with polynomially balanced polynomial-time decidable functions.*

Loosely speaking, a problem is said to be **#P**-hard if a polynomial-time algorithm for it implies that **#P** = **FP**, where **FP** is the set of functions from $\{0, 1\}^*$ to $\{0, 1\}^*$ computable by a deterministic polynomial-time Turing machine[4]. For a more formal definition, see [21]. We show the following:

Lemma 7. *Given an integer $C > 0$, it is #P-hard to compute the probability that the maximum flow value v in $G(E_1, E_2, \alpha)$ is at most C, $Pr[v \le C]$.*

---

[4] $\{0, 1\}^* = \cup_{n \ge 0}\{0, 1\}^n$, where $\{0, 1\}^n$ is the set of all strings (of bits $0, 1$) of length n.

Now, given a mixed temporal network $G(E_1, E_2, \alpha)$, let v be the random variable representing the maximum temporal flow in G.

Definition 16. *The truncated by B expected maximum temporal flow of $G(E_1, E_2, \alpha)$, denoted by $E[v, B]$, is defined as: $E[v, B] = \sum_{i=1}^{B} i Pr[v = i]$. Clearly, it is $E[v] = E[v, +\infty]$.*

The following is the main theorem of this section.

Theorem 2. *It is #P-hard to compute the expected maximum truncated Temporal Flow in a Mixed Temporal Network $G(E_1, E_2, \alpha)$.*

Open Problem 1. *Is there an FPTAS for the expected maximum flow value in mixed temporal networks?*

Open Problem 2. *What is the complexity of the maximum flow problem in periodic temporal graphs? These are graphs each edge e of which appears every x_e days ("edge period"). The maximum flow from s to t would then, in general, increase as we increase the day by which we wish to compute the flow that arrives at t. It seems that this problem requires a different approach than the one presented here, that also takes into account the different edge periods.*

References

1. Ahuja, R.K., Magnanti, T.L., Orlin, J.B.: Network Flows: Theory, Algorithms, and Applications. Prentice-Hall Inc., Upper Saddle River (1993)
2. Akrida, E.C., Czyzowicz, J., Gasieniec, L., Kuszner, L., Spirakis, P.G.: Flows in temporal networks. CoRR abs/1606.01091 (2016)
3. Akrida, E.C., Gasieniec, L., Mertzios, G.B., Spirakis, P.G.: Ephemeral networks with random availability of links: the case of fast networks. J. Parallel Distrib. Comput. **87**, 109–120 (2016)
4. Akrida, E.C., Gąsieniec, L., Mertzios, G.B., Spirakis, P.G.: On temporally connected graphs of small cost. In: Sanità, L., Skutella, M. (eds.) WAOA 2015. LNCS, vol. 9499, pp. 84–96. Springer, Cham (2015). doi:10.1007/978-3-319-28684-6_8
5. Akrida, E.C., Spirakis, P.G.: On verifying and maintaining connectivity of interval temporal networks. In: Bose, P., Gąsieniec, L.A., Römer, K., Wattenhofer, R. (eds.) ALGOSENSORS 2015. LNCS, vol. 9536, pp. 142–154. Springer, Cham (2015). doi:10.1007/978-3-319-28472-9_11
6. Aronson, J.E.: A survey of dynamic network flows. Ann. Oper. Res. **20**(1–4), 1–66 (1989)
7. Avin, C., Koucký, M., Lotker, Z.: How to explore a fast-changing world (cover time of a simple random walk on evolving graphs). In: Aceto, L., Damgård, I., Goldberg, L.A., Halldórsson, M.M., Ingólfsdóttir, A., Walukiewicz, I. (eds.) ICALP 2008. LNCS, vol. 5125, pp. 121–132. Springer, Heidelberg (2008). doi:10.1007/978-3-540-70575-8_11
8. Batra, J., Garg, N., Kumar, A., Mömke, T., Wiese, A.: New approximation schemes for unsplittable flow on a path. In: Indyk, P. (ed.) Proceedings of the Twenty-Sixth Annual ACM-SIAM Symposium on Discrete Algorithms, SODA 2015, San Diego, CA, USA, 4–6 January 2015, pp. 47–58. SIAM (2015)

9. Casteigts, A., Flocchini, P., Quattrociocchi, W., Santoro, N.: Time-varying graphs, dynamic networks. Int. J. Parallel Emerg. Distrib. Syst. (IJPEDS) **27**(5), 387–408 (2012)
10. Chaintreau, A., Mtibaa, A., Massoulié, L., Diot, C.: The diameter of opportunistic mobile networks. In: Proceedings of the 2007 ACM Conference on Emerging Network Experiment and Technology, CoNEXT 2007, New York, NY, USA, 10–13 December 2007, p. 12 (2007)
11. Clementi, A.E.F., Macci, C., Monti, A., Pasquale, F., Silvestri, R.: Flooding time of edge-Markovian evolving graphs. SIAM J. Discret. Math. (SIDMA) **24**(4), 1694–1712 (2010)
12. Erlebach, T., Hoffmann, M., Kammer, F.: On temporal graph exploration. In: Halldórsson, M.M., Iwama, K., Kobayashi, N., Speckmann, B. (eds.) ICALP 2015. LNCS, vol. 9134, pp. 444–455. Springer, Heidelberg (2015). doi:10.1007/978-3-662-47672-7_36
13. Ford, D.R., Fulkerson, D.R.: Flows in Networks. Princeton University Press, Princeton (2010)
14. Hoppe, B., Tardos, E.: The quickest transshipment problem. In: Proceedings of the Sixth Annual ACM-SIAM Symposium on Discrete Algorithms, SODA 1995, pp. 512–521. Society for Industrial and Applied Mathematics, Philadelphia (1995)
15. Hoppe, B.E.: Efficient dynamic network flow algorithms. Ph.D. thesis (1995)
16. Kamiyama, N., Katoh, N.: The universally quickest transshipment problem in a certain class of dynamic networks with uniform path-lengths. Discret. Appl. Math. **178**, 89–100 (2014)
17. Kempe, D., Kleinberg, J.M., Kumar, A.: Connectivity and inference problems for temporal networks. In: Proceedings of the 32nd Annual ACM Symposium on Theory of Computing (STOC), pp. 504–513 (2000)
18. Madry, A.: Fast approximation algorithms for cut-based problems in undirected graphs. In: 51th Annual IEEE Symposium on Foundations of Computer Science, FOCS 2010, Las Vegas, Nevada, USA, 23–26 October 2010, pp. 245–254 (2010)
19. Mertzios, G.B., Michail, O., Chatzigiannakis, I., Spirakis, P.G.: Temporal network optimization subject to connectivity constraints. In: Fomin, F.V., Freivalds, R., Kwiatkowska, M., Peleg, D. (eds.) ICALP 2013. LNCS, vol. 7966, pp. 657–668. Springer, Heidelberg (2013). doi:10.1007/978-3-642-39212-2_57
20. Orlin, J.B.: Max flows in O(nm) time, or better. In: Symposium on Theory of Computing Conference, STOC 2013, Palo Alto, CA, USA, 1–4 June 2013, pp. 765–774 (2013)
21. Papadimitriou, C.M.: Computational Complexity. Addison-Wesley, Reading (1994)
22. Powell, W.B., Jaillet, P., Odoni, A.: Stochastic and dynamic networks and routing. Handb. Oper. Res. Manag. Sci. **8**, 141–295 (1995)
23. Radzik, T.: Faster algorithms for the generalized network flow problem. Math. Oper. Res. **23**(1), 69–100 (1998)
24. Serna, M.J.: Randomized parallel approximations to max flow. In: Kao, M.-Y. (ed.) Encyclopedia of Algorithms, pp. 1750–1753. Springer, New York (2016)
25. Skutella, M.: An introduction to network flows over time. In: Cook, W., Lovász, L., Vygen, J. (eds.) Research Trends in Combinatorial Optimization, pp. 451–482. Springer, Heidelberg (2008)

Completeness Results for Counting Problems with Easy Decision

Eleni Bakali$^{(\boxtimes)}$, Aggeliki Chalki, Aris Pagourtzis(iD),
Petros Pantavos, and Stathis Zachos

School of Electrical and Computer Engineering,
National Technical University of Athens, 15780 Athens, Greece
{mpakali,achalki}@corelab.ntua.gr, {pagour,zachos}@cs.ntua.gr,
ppantavos@gmail.com

Abstract. Counting problems with easy decision are the only ones among problems in complexity class #P that are likely to be (randomly) approximable, under the assumption RP $\neq$ NP. TotP is a subclass of #P that contains many of these problems. TotP and #P share some complete problems under Cook reductions, the approximability of which does not extend to all problems in these classes (if RP $\neq$ NP); the reason is that such reductions do not preserve the function value. Therefore Cook reductions do not seem useful in obtaining (in)approximability results for counting problems in TotP and #P.

On the other hand, the existence of TotP-complete problems (apart from the generic one) under stronger reductions that preserve the function value has remained an open question thus far. In this paper we present the first such problems, the definitions of which are related to satisfiability of Boolean circuits and formulas. We also discuss implications of our results to the complexity and approximability of counting problems in general.

1 Introduction

Since Valiant introduced #P [25], the class of functions that count the number of accepting paths of a NPTM (Nondeterministic Polynomial Time Turing Machine), many counting classes arose in the literature. In [20] the class #PE was defined, as a subclass of #P that contains all functions of #P with easy decision version, that is, for a function $f \in$ #PE the problem "$f(x) \neq 0$?" is in P. #PE contains important problems such as PERMANENT [25], a special case of which is equivalent to counting perfect matchings in bipartite graphs. Another well known member of #PE is #DNF-SAT, i.e. the problem of counting satisfying assignments to DNF boolean formulas; for more such problems see [26]. Notably it was shown in [25,26] that PERMANENT, #DNF-SAT, as well as several problems presented in [26] are #P-complete, showing that counting is likely to be harder than decision (existence checking) for all these problems.

A subclass of #PE, namely TotP, was defined as the class of functions that count the total number of computation paths of the computation tree of a binary

© Springer International Publishing AG 2017
D. Fotakis et al. (Eds.): CIAC 2017, LNCS 10236, pp. 55–66, 2017.
DOI: 10.1007/978-3-319-57586-5_6

NPTM minus one [16]. Equivalently, for $f \in \text{TotP}$, $f(x)$ is the *number of branchings* of the computation tree of a binary NPTM. TotP contains PERMANENT and #DNF-SAT, as well as all self-reducible problems of #PE under a natural notion of self-reducibility for counting problems. It is intriguing that problems in TotP have varying approximability status. In particular, TotP contains well approximable problems (e.g. #DNF-SAT), not approximable (within a polynomial factor) problems unless NP = RP (e.g. #IS that is the number of independent sets of all sizes), and also problems of yet unknown approximability status, conjectured to be "intermediate" (e.g. #BIS that is the number of independent sets of all sizes of bipartite graphs).

It is known that #PE contains TotP [16] and moreover that TotP is exactly the Karp closure of self-reducible functions of #PE [21]. There is a great number of self-reducible problems with easy decision which are therefore in TotP: counting matchings, computing the determinant of a matrix, computing the partition function of several models from statistical physics, like the Ising and the hardcore model, counting colorings of a graph with a number of colors greater than the maximum degree, counting bases of a matroid, computing the volume of a convex body, counting independent sets, and many more. TotP-completeness results can shed light to the complexity and approximability of all these problems and help treat such questions in a uniform way.

Regarding completeness results, as mentioned above, there are several #P-complete problems, which belong to TotP and therefore are TotP-complete under Cook reductions. More precisely, TotP and #P are interreducible under Cook reductions [16, 17].

On the other hand, the situation is different when completeness under Karp (parsimonious) reductions is considered. In particular, there is no #P-complete problem in TotP, unless P = NP. For example, PERMANENT cannot be #P-complete under Karp reductions unless P = NP. Furthermore it also seems unlikely that PERMANENT is TotP-complete under Karp reductions. This is because such reductions preserve approximability and PERMANENT admits a FPRAS, while other problems in TotP, like #IS, are not likely to do so, as they are AP-interreducible with #SAT [9]. In this perspective completeness results in TotP and classes inside TotP may shed light on approximability of many counting problems.

In this paper, we present a first TotP-complete problem under parsimonious reductions, namely #MONOTONE-CIRCUIT-SAT: given the encoding of a monotone circuit with respect to a specific partial order (to be defined later), compute the number of inputs for which the circuit accepts. Then we reduce this to other problems, proving them to be also TotP-complete under parsimonious reductions. Finally we discuss some implications of our results in the last section.

1.1 Related Work

In recent years there has been great interest in classifying the approximation complexity of counting problems. This interest derives from the fact that very few counting problems have been proved to be in FP. At the same time the counting problems in #P with NP-complete decision version cannot have a poly-nomial time approximation, unless P = NP. Moreover in [9] it was proved that these problems are complete for #P under approximation-preserving reductions. Thus there is no FPRAS for any of them, unless NP = RP. Counting problems that could have efficient approximation algorithms (FPRAS, FPTAS) are count-ing problems with easy decision version. Such algorithms for counting problems can be found in [8,12,14,15]. Especially, the steady progress in determining the complexity of counting graph homomorphisms [10] contributed to the study of approximation counting complexity. Also the connection of counting problems with statistical physics have led to important results in this area [1–3,11].

Regarding subclasses of #P, counting classes like #L, SpanL [4], #PE [20], TotP [16], #RΣ_2 [23], #RHΠ_1 [9] have been defined. A significant open question concerns the relation between each of these classes and the problems that admit a FPRAS. We are particularly interested in the class TotP, which is related to other subclasses of #P in the following way: FP $\subseteq$ SpanL $\subseteq$ TotP $\subseteq$ #PE $\subseteq$ #P.

Furthermore, TotP is equal to IF$_t^{LN}$, the class of interval size functions defined on total p-orders with efficiently computable lexicographically nearest function [7]. This was based on [13], in which Hemaspaandra et al. defined classes of interval size functions and characterized #P in terms of such functions.

2 Preliminaries

The model of computation is the nondeterministic polynomial-time bounded Turing machine (NPTM), i.e. there is some polynomial p such that for any input x from an alphabet Σ^*, all computation paths have length at most $p(|x|)$, where $|x|$ is the length of the input. In [25] Valiant introduced the class #P:

Definition 1. *Let R be a polynomial-time decidable binary relation and p a polynomial. Let f be the function such that given $x \in \Sigma^*$, $f(x) = |\{y : |y| = p(|x|) \land R(x,y)\}|$. #P is the class of all these functions. Equivalently, #P = $\{acc_M : M$ is a NPTM$\}$, where $acc_M(x) = \#accepting$ paths of M on input x.*

The decision version of a function $f \in$ #P is the following problem: Given x, is $f(x)$ nonzero? Equivalently, is there at least one accepting path of M on input x? For each function f the related language $L_f = \{x : f(x) > 0\}$ can be defined. If a function f corresponds to the counting version of a search problem (i.e. f counts how many solutions are there for a given instance) then L_f corresponds to the existence of a solution.

Definition 2. *TotP = $\{tot_M : M$ is a NPTM$\}$, where $tot_M(x) = \#(all$ compu-tation paths of M on input $x) - 1$.*

A second class of functions with similar properties was introduced in [20]. #PE is the class that contains all the functions in #P such that their decision version is polynomial-time decidable.

Definition 3. *#PE = $\{f : f \in$ #P and L_f is polynomial time computable$\}$.*

Reductions between functions can be defined in a similar manner to the Cook/Turing and Karp/many-one reductions between languages. The latter kind of reduction is often called parsimonious, when referring to functions that count the number of solutions to NP problems. We use the terms "Cook" and "Karp", as shortcuts for "poly-time Turing" and "poly-time many-one" respectively:

Definition 4. *Polynomial-time reductions between functions:*

- *Cook (poly-time Turing) $f \leq^p_T g$: $f \in \text{FP}^g$.*
- *Karp (poly-time many-one) $f \leq^p_m g$: $\exists h \in \text{FP}, \forall x \in \Sigma^*$ $f(x) = g(h(x))$.*

The relations among #P, #PE and TotP were explored in [21]. The notion of self-reducibility is crucial for this investigation.

Definition 5. *A function $f : \Sigma^* \to \mathbb{N}$ is called poly-time self-reducible if there exist polynomials r and q, and polynomial time computable functions $h : \Sigma^* \times \mathbb{N} \to \Sigma^*$, $g : \Sigma^* \times \mathbb{N} \to \mathbb{N}$, and $t : \Sigma^* \to \mathbb{N}$ such that for all $x \in \Sigma^*$:*

(a) *$f(x) = t(x) + \sum_{i=0}^{r(|x|)} g(x,i) f(h(x,i))$, that is, f can be processed recursively by reducing x to $h(x,i)$ $(0 \leq i \leq r(|x|))$, and*
(b) *the recursion terminates after at most polynomial depth (that is,*
   $f(h(...h(h(x,i_1),i_2)...,i_{q(|x|)}))$ can be computed in polynomial time).
(c) *$|h(...h(h(x,i_1),i_2)...,i_{q(|x|)}| \in O(poly(|x|))$.*

Note that if $|h(x,i)| < |x|$ for every x and i, $0 \leq i \leq r(|x|)$, then requirement (b) holds trivially. Moreover, (c) requires that f must be computed only on inputs of polynomial length in $|x|$, which also holds if h is of decreasing length.

Theorem 1 [21]. *(a) FP $\subseteq$ TotP $\subseteq$ #PE $\subseteq$ #P. The inclusions are proper unless P = NP.*
(b) TotP is the Karp closure of self-reducible #PE functions.

Although, TotP, #PE and #P are Cook-equivalent, they are not Karp equivalent unless P = NP. This means that:

- Under Karp reductions, #P-complete, #PE-complete and TotP-complete problems constitute disjoint classes, unless P = NP.
- Under Cook reductions, TotP-complete problems are contained in #PE-complete problems which are contained in #P-complete problems.

In order to fully classify a problem, we need to prove that it is complete for a class under Karp reductions. As it can be easily observed, the fact that a problem is #P-complete under Cook reductions does not give enough information about its complexity, since it could belong in TotP. Cook reductions blur structural differences between classes.

In the rest of this section, we present definitions and observations useful for the main proof of this paper.

A tree is called (a) *binary* if every node has at most two children, (b) *full binary* if every node has either zero or two children, (c) *perfect binary* if it is binary, all interior nodes have two children and all leaves have the same depth.

Let M be a NPTM. We can modify M, without changing the total number of its paths, so that it has at most two nondeterministic choices at each step. Therefore, the computation of M on input x can be seen as a binary tree $T_{M(x)}$, i.e. a branching is created in the computation tree whenever M has to select between two choices. If there is only one choice for some state-symbol combination, we consider that the tree has no branching at this point. We conclude that we can restrict ourselves to full binary computation trees. So, it is not hard to see that $tot_M(x) = \#(\text{all paths of } M \text{ on input } x) - 1 = \#\text{branchings of } T_{M(x)}$.

Furthermore, the nondeterministic choices of the computation of M can be represented as a binary string y (a left branching corresponds to "0" and a right branching to "1"). When we write $M(x,y)$ we refer to the output of the Turing machine M on input x and nondeterministic choices y. Specifically, $M(x,y) = 1$ if M accepts x with nondeterministic choices y, and $M(x,y) = 0$ otherwise.

In the following sections we make use of two mappings from natural numbers to binary strings, as well as a special partial order on natural numbers.

Definition 6. *We define the* tree *partial order, denoted by* $\leq_{\text{tree}}$, *of* $\mathbb{N}$ *as follows. It is reflexive and transitive and, if* $y = 2x+1$ *or* $y = 2x+2$ *then* $x \leq_{\text{tree}} y$.

Note that the graph of this partial order is an infinite perfect binary tree denoted by T, the nodes of which are labeled with natural numbers, in such a way that the left to right BFS traversal of this tree yields the natural order of $\mathbb{N}$ (assuming that the left child of x is $2x + 1$ and the right one is $2x + 2$). Its root is labeled with 0, and $x \leq_{\text{tree}} y$ if and only if y is a descendant of x on this tree. The structure of T is illustrated in Fig. 1.

Using the notion of the infinite tree T we can define mappings between natural numbers and strings:

Definition 7. *1.* path $: \mathbb{N} \to \{0,1\}^*$. *It maps n to the binary string that describes the path that starts from the root of T and ends at the node with label n. For example,* path$(3) = 00$, path$(9) = 010$, path$(0) = \varepsilon$, *where ε is the empty string.*
2. num $: \{0,1\}^* \to \mathbb{N}$. *It is defined as the inverse mapping of* path.
3. bin$_k : \{0,1,\ldots,2^k - 1\} \to \{0,1\}^k$. *It maps n to its binary representation padded with leading zeros, so as to have length k. For example,* bin$_6(3) = 000011$, bin$_4(9) = 1001$, *and* bin$_3(9)$ *is not defined.*

In addition, bin$_k^{-1}$ is the inverse of bin$_k$. For simplicity, we slightly abuse notation and use bin and bin^{-1}, when the length of the binary representation is clear from the context. The functions path, num, bin$_k$ and bin$_k^{-1}$ are polynomial-time computable.

Definition 8. *If we restrict* $\leq_{\text{tree}}$ *on* $\{0,1,\ldots,2^k-1\}$ *and apply* bin$_k$, *we obtain a partial order of* $\{0,1\}^k$, *which, abusing notation, we also denote by* $\leq_{\text{tree}}$.

Let T^k denote the complete binary tree representing $\leq_{\text{tree}}$ on $\{0,1\}^k$; an illustration of T^3 is given in Fig. 2.

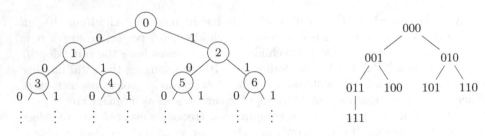

Fig. 1. The infinite perfect binary tree T.   **Fig. 2.** Tree T^3.

3   #Monotone-Circuit-Sat is TotP-complete Under Karp Reductions

In this section we define a new counting problem and we prove that it is TotP-complete. Let C_n denote a Boolean circuit (see [5]) with n input gates, and let $C_n(z)$ be the output of C_n on input $z \in \{0,1\}^n$.

Definition 9. *We call a Boolean circuit C_n non-increasing with respect to $\leq_{\text{tree}}$ if for every $x, y \in \{0,1\}^n$, $x \leq_{\text{tree}} y$ implies that $C_n(x) \geq C_n(y)$.*

Definition 10. *#MONOTONE-CIRCUIT-SAT, denoted also by $f_{\#MC}$*
Input: A Boolean circuit C_n, non-increasing with respect to $\leq_{\text{tree}}$.
Output: $f_{\#MC}(C_n) := |\{y \in \{0,1\}^n : C_n(y) = 1\}|$, i.e. the number of satisfying assignments for C_n.

3.1   #Monotone-Circuit-Sat is TotP-hard

We prove that the function $f_{\#MC}$ is TotP-hard by reducing the computation of any function $h \in \text{TotP}$ to $f_{\#MC}$.

The key observation is the following. There is a NPTM M such that for any input x, $h(x) = tot_M(x)$; let $T_{M(x)}$ denote the corresponding computation tree. Consider extending $T_{M(x)}$ to a perfect binary tree $S_{M(x)}$ with the same height, so that all leaves of the original $T_{M(x)}$ tree and all their descendants are labeled "halting". Therefore $h(x) = \#(\text{branching nodes of } T_{M(x)}) = \#(\text{non-"halting"}$ nodes of $S_{M(x)})$.

We construct a circuit C non-increasing w.r.t. $\leq_{\text{tree}}$, such that the number of accepting inputs of C equals $h(x)$. The idea is to describe a bijection between inputs of C and paths from the root to nodes of $S_{M(x)}$. C accepts an input if and only if the corresponding path ends at a non-"halting" node of $S_{M(x)}$, which in turn corresponds to a branching node of $T_{M(x)}$.

Theorem 2. *If $h \in TotP$ then $h \leq_m^p f_{\#MC}$.*

Proof. Let $h \in \text{TotP}$, and M the corresponding binary NPTM. Recall that for every input x, $h(x) = tot_M(x) = \#$branchings of $T_{M(x)}$, where $T_{M(x)}$ is the computation tree of $M(x)$. Let p be a polynomial bounding the running time of M, thus the height of $T_{M(x)}$ is at most $p(|x|)$. Given the description of M we can construct a NPTM M' such that for every input x of M:

(i) $T_{M'(x)}$ is a perfect binary tree of height $p(|x|) + 1$.
(ii) #(accepting paths of $M'(x)$) = #(branchings of $T_{M(x)}$).
(iii) For $y_1, y_2 \in \{0,1\}^{p(|x|)+1}$, if $y_1 \leq_{\text{tree}} y_2$, then $M'(x, y_1) \geq M'(x, y_2)$.

In order to describe M' we make use of the functions path and bin defined in Definition 7. The operation of M' on input x proceeds as follows:

1. Guess a binary string y of length $p(|x|) + 1$. Let $n_y = \text{bin}^{-1}(y)$.
2. Compute $z = \text{path}(n_y)$.
3. Simulate M on input x and nondeterministic choices z.
   - If the simulation reaches a halting state of M (possibly using only a prefix of z), then output 0.
   - If the simulation uses all bits of z without reaching a halting state of M, then output 1.

We now show that properties (i), (ii), (iii) hold:

(i) The computation tree of M' is a perfect binary tree of height $p(|x|) + 1$, since the only nondeterministic choices are made in Step 1 (Step 3 is deterministic).
(ii) The number of accepting paths of M' equals the number of branchings of M, since M' outputs 1 if and only if z corresponds to a computation path of M ending at a branching; recall that bin and path are bijective.
(iii) To prove the third property, it suffices to show that for all y_1, y_2 such that $y_1 \leq_{\text{tree}} y_2$ we have $M'(x, y_1) = 0 \Rightarrow M'(x, y_2) = 0$. If $y_1 \leq_{\text{tree}} y_2$, then $z_1 = \text{path}(\text{bin}^{-1}(y_1))$ is a prefix of $z_2 = \text{path}(\text{bin}^{-1}(y_2))$. This means that whenever M' simulates M with nondeterministic choices determined by z_2, it first passes through the same states as when it simulates M with nondeterministic choices determined by z_1. So, $M'(x, y_1) = 0$ means that the simulation of M reaches a halting state using (some of) the bits of z_1. Thus the remaining bits of z_2 are ignored and 0 is returned, therefore $M'(x, y_2) = 0$.

In order to complete the proof, we have to construct for each input x of h a circuit C_n^x with $n = p(|x|) + 1$ input gates, that simulates the computation of M' on input x, i.e. for all $y \in \{0,1\}^n$, $C_n^x(y) = M'(x, y)$. It is well known that such a construction can be done in polynomial time (see e.g. [22, pp. 171–172]). C_n^x is non-increasing w.r.t. $\leq_{\text{tree}}$ since M' has this property (due to (iii)). Thus, we have that $|\{y \in \{0,1\}^n : C_n^x(y) = 1\}| = \#acc_{M'}(x) = tot_M(x)$, i.e. $f_{\#MC}(C_n^x) = h(x)$ so the reduction is parsimonious.   □

3.2   #Monotone-Circuit-Sat Is in TotP

By Theorem 1(b), it suffices to prove that $f_{\#MC}$ is a self-reducible #PE function.

Proposition 1. $f_{\#MC} \in \#PE$.

Proof sketch. It is not difficult to see that $f_{\#MC} \in \#P$. Moreover, the decision version is easy since it suffices to simulate C_n on input 0^n.   □

Proposition 2. $f_{\#MC}$ *is self-reducible.*

Proof sketch. For proving that $f_{\#MC}$ is self-reducible, the intuition is that the number of satisfying assignments of a circuit C_n non-increasing w.r.t. $\leq_{\text{tree}}$, equals 0 iff 0^n is not a satisfying assignment. Otherwise it is equal to (the number of satisfying assignments that lie on the left subtree of T^n) + (the number of satisfying assignments that lie on the right subtree of T^n) +1. The proof consists of showing that we can efficiently construct two circuits non-increasing w.r.t. $\leq_{\text{tree}}$: C_{n-1}^0, with values compatible with the values of C_n on the left subtree of T^n, and C_{n-1}^1 the corresponding for the right.   □

Corollary 1. $f_{\#MC} \in TotP$.

Remark. Note that $f_{\#MC}$ is a "promise" problem, since it is not known how to check if a circuit is non-increasing w.r.t. $\leq_{\text{tree}}$. This is not an essential issue, as we can extend the function $f_{\#MC}$ on non-valid inputs to be equal to $tot_M(x)$, where M is the NPTM implied by the membership of $f_{\#MC}$ in TotP (on valid inputs).

4   More TotP-complete Problems

In this section we will show several problems to be TotP-complete. The proofs, omitted due to space limitations, will appear in the full version of the paper.

Definition 11. *Let U be a partially ordered set. A subset $V \subseteq U$ is called a lower-set (downwards closed) if for all $y, x \in U$, $(y \in V$ and $x < y) \Rightarrow x \in V$.*

Definition 12. *Let a circuit C_n with n input gates. We will call a subset V of $\{0,1\}^n$ accepting for C_n if for all $x \in V$, $C_n(x) = 1$.*

Definition 13. *We define the problem* MAX-LOWER-SET-SIZE.
Input: A circuit C_n with n input gates.
Output: The size of the maximum lower set w.r.t. $\leq_{\text{tree}}$, that is accepting for C_n.

Theorem 3. *The problem* MAX-LOWER-SET-SIZE *is TotP-complete.*

In the following we assume that each $n \in \mathbb{N}$ is encoded by path(n), and let T be the infinite perfect binary tree representing $\leq_{\text{tree}}$ on $\mathbb{N}$ (Fig. 1).

The next problem is intuitively the problem of counting the number of nodes of a subtree S of T, where S is given in a succinct way, i.e. not explicitly, but rather by a predicate that tells us whether a node v of T belongs to S.

Definition 14. SIZE-OF-SUBTREE, *denoted by f_{ss}*
Input: $(M_A, u \in \mathbb{N}, 1^k, 1^t)$ where M_A is a deterministic TM computing a predicate $A : \mathbb{N} \to \{0,1\}$ and $t \in \mathbb{N}$.
Output: The size of the maximal subtree of S with root u, where $S = \{v \in T \mid \text{distance}(u,v) \leq k, A(v) = 1$ and $A(v)$ is computed by M_A in at most t steps$\}$.

Theorem 4. *f_{ss} is TotP-complete.*

We next show another TotP-complete problem which is a special case of #SAT. Namely, the valid input formulas have the following special properties based on a clustering of the space of solutions $\{0,1\}^n$, where each cluster contains all assignments with their first k variables fixed to some values: (a) there is at most one satisfying assignment in each cluster, and it is easy to decide whether such an assignment exists and, if so, easy to find it, and·(b) if we label each cluster according to their fixed values, then there is a certain kind of monotonicity among the clusters, described below.

Definition 15. *1. For a 3-CNF formula ϕ and $k \in \mathbb{N}$ we define $f_\phi^k : \{0,1\}^k \to \mathbb{N}$ such that $f_\phi^k(a) = \#(\text{satisfying assignments of } \phi \text{ with prefix } a)$ for $a \in \{0,1\}^k$.*
2. A 3-CNF formula ϕ with n variables is called (k,n)-clustered-monotone for some $k \leq n$, if for every $a, b \in \{0,1\}^k$ such that $a \leq_{\text{tree}} b$, $f_\phi^k(a) = 0$ implies $f_\phi^k(b) = 0$.

Definition 16. *1. $Y = \{(1^k, 1^n, \phi, M, 1^t) \mid k, n, t \in \mathbb{N}, \phi \in \Phi, \text{ deterministic } TM \ M : \{0,1\}^k \times \Phi \to \mathbb{N}\}$, where Φ is the set of 3-CNF formulas on n variables.*
2. $U \subset Y$ is the set of tuples $(1^k, 1^n, \phi, M, 1^t)$ where ϕ is (k,n)-clustered monotone, and M is a deterministic TM s.t. $\forall a \in \{0,1\}^k$, $M(a, \phi) = \#(\text{satisfying assignments of } \phi \text{ with prefix } a)$, and t is an upper bound for the running time of M on every a, and on the given ϕ.

Note that in the above definition, the operation of the TM M is differentiated w.r.t. whether the instance is on U or $Y \setminus U$. In U we have the promise that ϕ is clustered monotone and that M counts the number of satisfying assignments in each clusters. In Y, both ϕ and M can be arbitrary.

Definition 17. #CLUSTERED-MONOTONE-SAT, *denoted by $f_{\#CMS}$*
Input: $y = (1^k, 1^n, \phi, M, 1^t) \in Y$
Output: $f_{\#CMS}(y) = \begin{cases} \#\text{satisfying assignments of } \phi, & \text{if } y \in U \\ \sum_{a \in S} M(a, \phi), & \text{if } y \in Y \setminus U \end{cases}$
where $S \subseteq T^k$ is the largest subtree of T^k containing 0^k s.t. $\forall a \in S$ $[M(a, \phi) > 0$ and $M(a, \phi)$ is computed within t steps$]$.

Theorem 5. *$f_{\#CMS}$ is TotP-complete.*

By introducing #CLUSTERED-MONOTONE-SAT, which is a special case of #SAT, the intuition we want to capture is the following. Every problem in TotP is reduced, as made clear from the above proof, to a 3-CNF formula that is clustered monotone, and for all formulas created in this way we have an efficient algorithm that returns the number of satisfying assignments in each cluster. So this TotP-complete special case of #SAT is much more structured than the #P-complete version. This fact, combined with other known results concerning the approximability of counting problems, may have interesting consequences, as we will discuss in the next section.

It is also worth noting that #CLUSTERED-MONOTONE-SAT is a special case of another #SAT variant which is SpanP-complete: given a formula ϕ on n variables, and a number $k \le n$, compute the number of satisfying assignments that are different in the first k variables [19].

5   Discussion on Approximability Implications

On the Approximability of TotP. It is known that there are problems in TotP, e.g. #IS (as shown in [21]), that do not admit a FPRAS unless NP = RP [9], not even a polynomial factor approximation. This follows from the fact that for self-reducible problems a polynomial factor approximation would yield a FPRAS [24].

However, it turns out that the class TotP admits some kind of polynomial time approximation: the problem SIZE-OF-SUBTREE is a special case of the backtracking-tree problem, studied in [18]; in that paper Knuth proposed a randomized algorithm. By appropriate adaptation we can use it to approximate SIZE-OF-SUBTREE. The expected output value of the algorithm is exactly the desired value, but the variance can be exponential in the worst case. Thus this algorithm would not yield a FPRAS. Approximation algorithms under other notions of approximability for SIZE-OF-SUBTREE were studied in [6].

The TotP-completeness of SIZE-OF-SUBTREE under Karp reductions implies that the above algorithmic results can be applied to every problem in TotP. Recall that TotP contains self-reducible hard counting problems with easy decision version [21]. On the other hand these simple algorithms are essentially the best we can hope for, unless NP = RP, since #IS belongs to TotP.

On the Approximability of #P and Connections to Statistical Physics. Another interesting implication comes from the TotP-completeness of the problem #CLUSTERED-MONOTONE-SAT (Definition 17), i.e. the problem of counting the number of satisfying assignments of formulas such that: (a) a solution can easily be found if one exists, and (b) their set of solutions is connected in a specific way as described before Definition 15. Combining this completeness result with the fact that #SAT can be reduced to #IS ∈ TotP (i.e. counting independent sets) by a reduction that preserves approximability [9], we get that approximating the number of satisfying assignments of an arbitrary formula is as difficult as approximating the number of satisfying assignments of a formula with the above properties.

This is particularly interesting since there is a series of papers that relate counting complexity to statistical physics [1–3], from which we know that, for the "difficult" instances of SAT, the set of satisfying assignments is widely scattered in the space of all assignments (i.e. the boolean hypercube of n dimensions), and this scattering might be responsible for the hardness of SAT. Our results show that we can reduce (with an approximation-preserving reduction) an arbitrary instance with a set of solutions that are disconnected and for which it is hard to find even one solution, to an instance with a set of solutions that are connected in a way that we have described explicitly, and for which we can easily find one solution.

This can be viewed in two ways: For an optimist it shows that approximating #SAT may be not so difficult after all (e.g. perhaps NP = RP). On the other hand, a pessimist may conclude that #SAT is not only hard in general, but also (by such "hardness amplification") even seemingly easy (e.g. structured) cases would possess the same hardness.

6   Conclusion and Open Problems

We have made an important step towards a better understanding of the complexity class TotP by presenting problems that are TotP-complete under parsimonious reductions. However, these problems are not among the well-studied problems in TotP such as #IS, PERMANENT, etc. The completeness of such problems constitutes an intriguing open question. Note that, if PERMANENT is TotP-complete under parsimonious reductions, then NP = RP.

Another interesting direction would be to explore the approximability status of the problems presented in this paper. The positive approximability of these problems would transfer to every problem in TotP.

Acknowledgments. We would like to thank Antonis Antonopoulos for many useful discussions as well as the anonymous reviewers for their observations and corrections.

References

1. Achlioptas, D.: Random Satisfiability. In: Biere, A., et al. (eds.) Handbook of Satisfiability, pp. 245–270. IOS Press, Amsterdam (2009)
2. Achlioptas, D., Coja-Oghlan, A., Ricci-Tersenghi, F.: On the solution-space geometry of random constraint satisfaction problems. Random Struct. Algorithms **38**(3), 251–268 (2011)
3. Achlioptas, D., Ricci-Tersenghi, F.: Random formulas have frozen variables. SIAM J. Comput. **39**(1), 260–280 (2009)
4. Àlvarez, C., Jenner, B.: A very hard log-space counting class. Theoret. Comput. Sci. **107**(1), 3–30 (1993)
5. Arora, S., Barak, B.: Computational Complexity: A Modern Approach. Cambridge University Press, New York (2009)
6. Bakali, E.: Self-reducible with easy decision version counting problems admit additive error approximation. Connections to counting complexity, exponential time complexity, and circuit lower bounds. CoRR abs/1611.01706 (2016)

7. Bampas, E., Gobel, A., Pagourtzis, A., Tentes, A.: On the connection between interval size functions and path counting. Comput. Complex., 1–47 (2016). doi:10.1007/s00037-016-0137-8. Springer

8. Dyer, M.: Approximate counting by dynamic programming. In: Proceedings of 35th Annual ACM Symposium on Theory of Computing (STOC), pp. 693–699 (2003)

9. Dyer, M.E., Goldberg, L.A., Greenhill, C.S., Jerrum, M.: The relative complexity of approximate counting problems. Algorithmica 38(3), 471–500 (2003)

10. Galanis, A., Goldberg, L.A., Jerrum, M.: Approximately counting H-colourings is #BIS-hard. SIAM J. Comput. 45(3), 680–711 (2016)

11. Goldberg, L.A., Jerrum, M.: The complexity of ferromagnetic ising with local fields. Comb. Probab. Comput. 16(1), 43–61 (2007)

12. Gopalan, P., Klivans, A., Meka, R., Štefankovič, D., Vempala, S., Vigoda, E.: An FPTAS for #knapsack and related counting problems. In: Proceedings of 52nd Annual Symposium on Foundations of Computer Science (FOCS), pp. 817–826 (2011)

13. Hemaspaandra, L.A., Homan, C.M., Kosub, S., Wagner, K.W.: The complexity of computing the size of an interval. SIAM J. Comput. 36(5), 1264–1300 (2007)

14. Jerrum, M., Sinclair, A.: The Markov chain Monte Carlo method: an approach to approximate counting and integration. In: Hochbaum, D. (ed.) Approximation Algorithms for NP-hard Problems, pp. 482–520. PWS, Boston (1996)

15. Karp, R.M., Luby, M., Madras, N.: Monte-Carlo approximation algorithms for enumeration problems. J. Algorithms 10(3), 429–448 (1989)

16. Kiayias, A., Pagourtzis, A., Sharma, K., Zachos, S.: Acceptor-definable counting classes. In: Manolopoulos, Y., Evripidou, S., Kakas, A.C. (eds.) PCI 2001. LNCS, vol. 2563, pp. 453–463. Springer, Heidelberg (2003). doi:10.1007/3-540-38076-0_29

17. Kiayias, A., Pagourtzis, A., Zachos, S.: Cook reductions blur structural differences between functional complexity classes. In: Proceedings of 2nd Panhellenic Logic Symposium, pp. 132–137 (1999)

18. Knuth, D.E.: Estimating the efficiency of backtrack programs. Math. Comput. 29(129), 121–136 (1975)

19. Köbler, J., Schöning, U., Toran, J.: On counting and approximation. Acta Inform. 26, 363–379 (1989)

20. Pagourtzis, A.: On the complexity of hard counting problems with easy decision version. In: Proceedings of 3rd Panhellenic Logic Symposium, Anogia, Crete (2001)

21. Pagourtzis, A., Zachos, S.: The complexity of counting functions with easy decision version. In: Královič, R., Urzyczyn, P. (eds.) MFCS 2006. LNCS, vol. 4162, pp. 741–752. Springer, Heidelberg (2006). doi:10.1007/11821069_64

22. Papadimitriou, C.H.: Computational Complexity. Addison-Wesley, New York (1994)

23. Saluja, S., Subrahmanyam, K.V., Thakur, M.: Descriptive complexity of #P functions. J. Comput. Syst. Sci. 50(3), 169–184 (1992)

24. Sinclair, A.J., Jerrum, M.R.: Approximate counting, uniform generation and rapidly mixing Markov chains. Inf. Comput. 82, 93–133 (1989)

25. Valiant, L.G.: The complexity of computing the permanent. Theoret. Comput. Sci. 8(2), 189–201 (1979)

26. Valiant, L.G.: The complexity of enumeration and reliability problems. SIAM J. Comput. 8(3), 410–421 (1979)

Tracking Paths

Aritra Banik[1(✉)], Matthew J. Katz[2], Eli Packer[3], and Marina Simakov[2]

[1] Indian Institute of Technology, Jodhpur, India
aritrabanik@gmail.com
[2] Ben-Gurion University of the Negev, Beer-Sheva, Israel
matya@cs.bgu.ac.il, simakov@bgu.ac.il
[3] Haifa Lab, IBM Research, Haifa, Israel
ELIP@il.ibm.com

Abstract. We consider several problems dealing with tracking of moving objects (e.g., vehicles) in networks. Given a graph $G = (V, E)$ and two vertices $s, t \in V$, a set of vertices $T \subseteq V$ is a *tracking set* for G (w.r.t. paths from s to t), if one can distinguish between any two paths from s to t by the order in which the vertices of T appear (or do not appear) in them. We prove that the problem of finding a minimum-cardinality tracking set w.r.t. shortest paths from s to t is NP-hard and even APX-hard. On the other hand, for the common case where G is planar, we present a 2-approximation algorithm for this problem. We also consider the following related problem: Given a graph G, two vertices s and t, and a set of forbidden vertices $V_F \subseteq V - \{s, t\}$, find a minimum-cardinality set of trackers $V^* \subset V$, such that a shortest path P from s to t passes through a forbidden vertex if and only if it passes through a vertex of V^*. We present a polynomial-time (exact) algorithm for this problem.

1 Introduction

Tracking of moving objects has received considerable attention among researchers. Much of the work on this subject has dealt with objects moving in an underlying network. One objective of a network tracking system might be to derive movement patters of the objects of interest (e.g., vehicles), or to reconstruct the exact route followed by a specific object when needed, or to detect potential flaws in the network. Tracking of moving objects (not necessarily in an underlying network) is closely related to surveillance and monitoring, whether indoor or outdoor, and to intruder detection. In a typical (physical) tracking problem, we need to place tracking devices (e.g., wireless sensors) in a way that both serves our goals and is economical. See [3] for a survey of target tracking protocols using wireless sensor network. Although this is an active area of research, most of the work in this area is heuristic-based and much of it concentrates on power management of the sensors, see e.g. [6].

In this paper we study several location-awareness problems in the physical world. More precisely, we focus on (vehicle) tracking problems in an underlying

M.J. Katz—Supported by grant 1884/16 from the Israel Science Foundation.

D. Fotakis et al. (Eds.): CIAC 2017, LNCS 10236, pp. 67–79, 2017.
DOI: 10.1007/978-3-319-57586-5_7

network. Most of our ideas apply of course to other types of objects as well as to computer networks. To the best of our knowledge, these problems have not been previously studied from a theoretical point of view.

Let $G = (V, E)$ be a network (i.e., a graph). We assume that G has a unique entry vertex s and a unique exit vertex t. (The case where the network has several entry points and several exit points can be handled by adding two special vertices and connecting them to all entry vertices and to all exit vertices, respectively.) Our goal is to place trackers at some of the vertices in a way that would allow us to reconstruct the path from s to t which has been traversed. (When a vehicle passes through a vertex with a tracker, the tracker is activated and sends a signal to the control center.) Let λ be any path and $A \subset V$ any set of vertices. We denote by T_λ^A the sequence of vertices of A obtained from λ by deleting the vertices that do not belong to A. For two given vertices s and t and a set of paths Λ between s and t, a set of vertices A is a *tracking set* w.r.t. Λ, if for any two paths λ_1 and λ_2 in Λ, $T_{\lambda_1}^A \neq T_{\lambda_2}^A$. We focus on the case where Λ is the set of all shortest paths from s to t and consider the following problems:

Problem 1 (Tracking set for shortest paths (TSSP)). *Given a graph G and two vertices s and t, find a minimum cardinality tracking set for G, w.r.t. shortest paths from s to t.*

Problem 2 (TSSP given an initial set of trackers). *Given G, two vertices s and t, and a set A of trackers, find a minimum cardinality set B of trackers, such that $A \cup B$ is a tracking set for G, w.r.t. shortest paths from s to t.*

Another variant that we consider is where there are some vertices (locations) in which one cannot place a tracker for various reasons (e.g., geographical or climate conditions, security issues, or high costs). We call such vertices *blocked vertices*, and we are interested in finding a tracking set which does not contain any of the blocked vertices (if exists).

Problem 3 (TSSP with blocked vertices). *Given G, two vertices s and t, and a set of blocked vertices $V_B \subset V$, find a minimum cardinality tracking set for G, w.r.t. shortest paths from s to t, which does not contain vertices from V_B.*

The next problem deals with the reconstruction process.

Problem 4 (Path reconstruction). *Given G, two vertices s and t, and a tracking set A for G w.r.t. shortest paths, preprocess G and A so that given a sequence T_λ^A, the path λ can be reconstructed "quickly".*

Another problem that we consider is *Catching the intruder*. In this problem, vehicles are not allowed to pass through some of the vertices (for various reasons). Our goal now is to "catch" any vehicle that has entered one or more of these forbidden vertices. More precisely, the problem is defined as follows:

Problem 5 (Catching the intruder). *Given G, two vertices s and t, and a set of forbidden vertices $V_F \subset V$, find a minimum-cardinality set of vertices V^*, such that a shortest path between s and t passes through a forbidden vertex if and only if it passes through a vertex of V^*.*

A closely related problem is the feedback vertex set problem. A *feedback vertex set* of a graph G is a subset of vertices that contains at least one vertex from every cycle in G.

Problem (Feedback vertex set [1]). Given a graph G, find a feedback vertex set of G of minimum cardinality.

Karp [8] showed that the feedback vertex set problem for directed graphs is APX-complete, which directly follows from the APX-completeness of the vertex cover problem and the existence of an approximation preserving L-reduction from the vertex cover problem to the feedback vertex set problem [5,8]. The best known approximation algorithm for undirected graphs computes a 2-approximation [1]. See remarks at the end of Sect. 3 discussing connections between our problems and the feedback vertex set problem.

Our Results. In Sect. 2, we prove that Problem 2 and Problem 1 are NP-hard. We also prove APX-hardness of Problem 1. In Sect. 3, we present a direct 2-approximation algorithm for TSSP, for the special and important case where the underlying graph is planar. In Sect. 4, we consider Problem 5 and provide a polynomial-time algorithm for solving it optimally. Finally, see remarks at the end of Sect. 3 for additional results that can be found in the full version of this paper.

2 Hardness Results

In this section we first prove that Problem 2 (TSSP given an initial set of trackers) is NP-hard. Then we prove that Problem 1 (TSSP) is NP-hard. Finally, we prove that Problem 1 does not admit a PTAS, i.e., it is APX-hard.

Lemma 1. *Problem 2 is NP-hard.*

Proof. We show a reduction from vertex cover. A *vertex cover* of a graph G is a subset of the vertices of G, such that each edge of G is incident on at least one vertex of the subset. Given a positive integer k, determining whether there exists a vertex cover of size k is an NP-complete problem [8]. Let $\{G(V,E),k\}$ be any instance of the vertex cover problem. We construct the graph $G'(V',E')$ from $G(V,E)$ as follows. We create a vertex in G' for each vertex in G and for each edge in G. We also create vertices s and t, so that $V' = V \cup E \cup \{s,t\}$. In G' there is an edge between a vertex $v \in V$ and a vertex $e \in E$ if e is incident on v in G. For each vertex $v \in V$, we create an edge between v and s in G', and for each vertex $e \in E$ we create an edge between e and t in G' (see Fig. 1, ignoring the vertices a,b,d). Finally, let $B = E$ be the given set of trackers. Consider any tracking set $A \subset V'$ for the set of all shortest paths from s and t. Observe that any shortest path between s and t is of length three. Any vertex $e \in V'$ corresponding to an edge $e \in E$ is incident on two vertices v_i and v_j such that e is the edge between v_i and v_j in G. To distinguish between the paths (s,v_i,e,t) and (s,v_j,e,t), A must contain either v_i or v_j. Hence $A \setminus B$ is a vertex cover for G.

Consider any vertex cover $V^* \subset V$ for G. We show that $V^* \cup B$ is a tracking set for all shortest paths from s to t in G'. Suppose there exists two paths (s, v_i, e_j, t) and (s, v_k, e_l, t) which are indistinguishable. If $e_j \neq e_l$, then clearly (s, v_i, e_j, t) and (s, v_k, e_l, t) are distinguishable because $e_j, e_l \in B$. Hence without loss of generality assume $e_j = e_l$. Therefore, as (s, v_i, e_j, t) and (s, v_k, e_l, t) are indistinguishable, both v_i and v_k are not in V^*. But this contradicts the fact that V^* is a vertex cover for G. Hence the result holds.   □

Next we show that Problem 1 (TSSP) is hard. Let $G(V, E)$ be any graph. We begin by constructing the graph $G'(V', E')$ from $G(V, E)$ as before. Next, we add to V' the vertices a, b, d and to E' the edges $(s, a), (s, b), (d, t)$. We also add to E' the edges (a, x), for each $x \in E \cup \{d\}$, and the edge (b, d) (see Fig. 1).

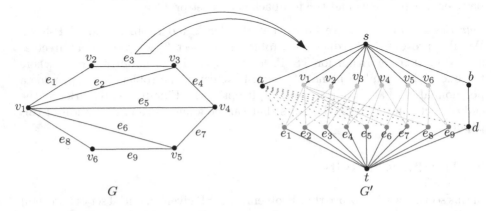

Fig. 1. Illustration of Lemma 2.

Lemma 2. *There exists a vertex cover for graph G of size k if and only if there exists a tracking set for G' of size $k + |E| + 1$.*

Proof. Let $|V^*| = k$ be any vertex cover for G. Consider the tracking set $V^* \cup E \cup \{a\}$ for G'. Observe that it is possible to track all shortest paths between s and t passing through d, because there are only two such paths, one with a single tracker at a and the other without any tracker. It is also possible to track all paths of the form (s, a, e_k, t), because each of them has two trackers, one at a and one at e_k. Suppose now that there exist two paths (s, v_i, e_k, t) and (s, v_j, e_l, t) which are indistinguishable. This implies that $e_k = e_l$ and that both v_i and v_j are not present in V^*. But this is impossible since V^* is a vertex cover for G.

Now suppose there exists a tracking set T for G' of size $k + |E| + 1$. Observe that T contains either a or b to distinguish between the paths (s, a, d, t) and (s, b, d, t). As (s, b, d, t) is the only path passing through b we can assume that there is a tracker at a.

To distinguish between the paths (s, a, x, t), where $x \in E \cup \{d\}$, all vertices in $E \cup \{d\}$ except one must contain a tracker. As a contains a tracker we can assume that each vertex in E contains a tracker (and d does not). Now $T \setminus (E \cup \{a\}) \subset V$ is of size k, and it is easy to verify that $T \setminus (E \cup \{a\})$ is a vertex cover for G.   □

We thus conclude that

Theorem 3. *Problem 1 is NP-hard.*

It is well known that minimum vertex cover is APX-hard, where the best inapproximability result is due to Dinur and Safra [5] who proved that it is NP-hard to approximate minimum vertex cover to within a factor of $\tau = 1.3606$. Next we prove that Problem 1 is APX-hard, by showing that if there exists a polynomial-time algorithm to approximate minimum tracking set to within a factor of $\tau - \delta$, for some $0 < \delta < \tau - 1$, then there also exists a polynomial-time algorithm to approximate minimum vertex cover to within a factor of τ.

Theorem 4. *Given a graph G, it is NP-hard to approximate minimum tracking set for shortest paths to within a factor of $\tau - \delta$, for any $0 < \delta < \tau - 1$.*

Proof. Let $G(V, E)$ be any graph, where $|V| = n$ and $|E| = m$. We construct the graph $G'(V', E')$ as follows. We create m sets of vertices $\{\mathcal{V}_1, \ldots, \mathcal{V}_m\}$, where $\mathcal{V}_i = \{v_{ij} : 1 \leq j \leq n\}$, and set $V' = (\cup_{1 \leq i \leq m} \mathcal{V}_i) \cup E \cup \{s_0, s_1, \ldots, s_m\} \cup \{s, t, a, b, d\}$. We add the following sets of edges to E' (see Fig. 2): (i) $\{(s, s_i) : 0 \leq i \leq m\}$, (ii) $\{(s_i, x) : x \in \mathcal{V}_i\}$, for $i = 1, \ldots, m$, (iii) $\{(s_i, b) : 0 \leq i \leq m\}$, (iv) $\{(s_0, a)\}$, (v) $\{(v_{ij}, e_k) : 1 \leq j \leq n, \ e_k \in E, \ e_k \text{ is incident on } v_j \text{ in } G\}$, for $i = 1, \ldots, m$, (vi) $\{(a, x) : x \in E \cup \{d\}\}$, (vii) $\{(b, d)\}$, and (viii) $\{(x, t) : x \in E \cup \{d\}\}$.

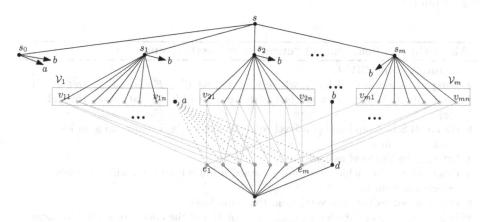

Fig. 2. The graph G'.

Let k be the size of a minimum vertex cover for G. Using arguments similar to those in Lemma 2, one can show that the size of an optimal tracking set for G' is $mk + 2m + 1$. Assume there exists a polynomial-time algorithm which produces tracking sets of size at most $(\tau - \delta_0)$ times the optimal size, where $0 < \delta_0 < \tau - 1$. Then, such an algorithm will return a tracking set T for G' of size at most $(\tau - \delta_0)(mk + 2m + 1)$. We know already that T must contain $E \cup \{a\}$, and we observe that it must also contain the set $\{s_0, \ldots, s_m\}$ except for one arbitrary member of the set. Let αk, $\alpha \geq 1$, be the number of vertices of $\mathcal{V}_i$ in T, for $i = 1, \ldots, m$. (For symmetry reasons we may assume that for any i, i' the number of vertices of $\mathcal{V}_i$ in T and the number of vertices of $\mathcal{V}_{i'}$ in T are equal.) We thus have $|T| = 2m + 1 + m\alpha k \leq (\tau - \delta_0)(mk + 2m + 1)$ or $\alpha \leq \tau - \delta_0 + (\tau - 1 - \delta_0)\frac{2m+1}{mk}$.

Now, we may assume that k is at least, say, $3(\tau - 1)/\delta_0$, since otherwise we can find a minimum vertex cover for G. But if so, we get that $\alpha < \tau$, implying that there exists an algorithm to approximate minimum vertex cover to within a factor of τ, which, as mentioned above, is impossible unless $P = NP$ [5]. $\square$

Corollary 5. *Problem 1 is APX-hard.*

3   Tracking Paths in Planar Graphs

In this section we present a 2-approximation algorithm for TSSP where the underlying graph is planar. Before presenting the algorithm, we make a few observations which are valid for any graph, not necessarily planar.

We are given a graph G (which is not necessarily planar) and two vertices s and t, and we are interested in a minimum cardinality tracking set for G, w.r.t. shortest paths from s to t. Without loss of generality, we assume that all the vertices and edges of G appear in some shortest path from s to t, otherwise we can delete all unnecessary vertices and edges in $O(|V| + |E|)$ time using Algorithm 1.

Algorithm 1. Removing all "unnecessary" vertices and edges.

---

1   **Input:** A graph $G(V, E)$.
2   **Output:** A graph $G'(V', E')$, where $V' \subset V$ and $E' \subset E$ and $v \in V'$ ($e \in E'$) if and only if there exists a shortest path from s to t in G that passes through v (e).
3   Run a BFS in G and assign a level to each vertex $v_i \in V$ depending on its distance from s.
4   Let l_{max} be the level assigned to t.
5   Delete all vertices which were not assigned a finite level (i.e., which are not reachable from s).
6   Delete all edges between vertices in the same level.
7   Delete all vertices with level $\geq l_{max}$, except t, and the edges connected to them.
8   **for** $l \leftarrow l_{max} - 1$ **down to** 1 **do**
9   Delete all vertices $v \in V$ in level l with $deg_{out}(v) = 0$.
10   Delete all edges with an endpoint at a deleted vertex.

Since we are only interested in shortest paths from s to t, we can think of G as a directed graph. The vertices of G can be divided into levels—all vertices at distance d from s are considered to be in level d (s is considered to be in level 0). Let l_{max} be the level of t. We denote the set of vertices participating in a path P or cycle C by $V(P)$ and $V(C)$, respectively.

Definition 1. Given a simple cycle C, we say that (i) $v \in V(C)$ is an *entry vertex* in C if there does not exist a vertex in $V(C)$ of level lower than that of v, i.e., $l(v) \leq l(u)$, for each $u \in V(C)$. (ii) $v \in V(C)$ is an *exit vertex* in C if there does not exist a vertex in $V(C)$ of level higher than that of v, i.e., $l(v) \geq l(u)$, for each $u \in V(C)$. (iii) If $v \in V(C)$ is not an entry nor an exit vertex, then it is a *middle vertex* in C.

Observation 1. *v is an entry vertex in C if and only if it is connected to two other vertices of C in a higher level than v, and v is an exit vertex in C if and only if it is connected to two other vertices of C in a lower level than v.*

Definition 2. A path $P = (v_1, v_2, \ldots, v_k)$ in G is an *increasing path* if $l(v_{i+1}) = l(v_i) + 1$, for $i = 1, \ldots, k - 1$. A *relevant cycle* is a simple cycle formed by two increasing paths between the same two vertices, where the paths are vertex-disjoint except of course for their common beginning and ending vertices.

Note that a relevant cycle C has a unique entry vertex and a unique exit vertex; these are the first and last vertices of the two increasing paths forming the cycle. We denote the entry vertex and the exit vertex of relevant cycle C_i by v_{in}^i and v_{out}^i, respectively.

Observation 2. *A relevant cycle C_i is composed of two (vertex-disjoint) increasing paths from v_{in}^i to v_{out}^i, both of which consisting of the same number of vertices.*

Lemma 6. *F is a tracking set for G if and only if there is at least one middle vertex from each relevant cycle of G in F.*

Proof. Assume F is a tracking set for G, we show that there is at least one middle vertex from each relevant cycle of G in F. Any relevant cycle C_i of G is composed of two vertex-disjoint paths, of equal length, from v_{in}^i to v_{out}^i, denote these paths by P_1 and P_2. In order to distinguish between P_1 and P_2 one of the vertices in $\{V(P_1) \cup V(P_2)\} \setminus \{v_{in}^i \cup v_{out}^i\}$ must contain a tracker.

We now show that if there is at least one middle vertex from each relevant cycle of G in F, then F is a tracking set for G. Let us assume that F is not a tracking set for G. Then, there must be at least two indistinguishable paths from s to t, denote these paths by P_1 and P_2. There must exist a relevant cycle C in G, such that P_1 traverses its left part, and P_2 traverses its right part (see Fig. 3) and there is no tracker on either side (otherwise we would be able to distinguish between them), but this means that there is no tracker in any of C's middle vertices—a contradiction to the definition of F.   □

Next we assume that G is planar and that a planar embedding of G is provided, and we present a 2-approximation algorithm for TSSP. Recall that we assume that all vertices and edges of G appear in some shortest path between s and t; otherwise, we can remove them by running Algorithm 1.

By Lemma 6 we know that

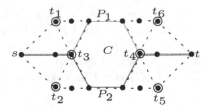

Fig. 3. A graph with a set of trackers $\{t_i\}$. The paths P_1 (red) and P_2 (blue) are indistinguishable because there is no tracker at a middle vertex of the relevant cycle C. (Color figure online)

Observation 3. *Each face F_i of G must contain a tracker in one of its middle vertices.*

Lemma 7. *The size of any tracking set for G w.r.t. shortest paths from s to t is at least $m/2$, where m is the number of faces in G.*

Proof. Each face must have a tracker in one of its middle vertices (by Observation 3). Moreover, a vertex can appear as a middle vertex in at most two faces. Therefore, any tracking set must have at least $m/2$ trackers.   □

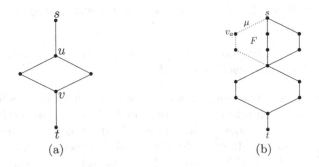

Fig. 4. Illustration of the proof of Theorem 8.

We now present a 2−approximation algorithm for TSSP, by presenting an iterative algorithm that finds a tracking set for G of size m.

Theorem 8. *There exists a tracking set of size m for G, where m is the number of faces in G.*

Proof. Denote the level of a vertex v by $l(v)$. We prove by induction the following stronger statement (from which the theorem follows immediately):
For any planar graph G with m faces, one can place m trackers, such that, for any two vertices u and v, there do not exist two different shortest paths between u and v with the same set of trackers. Notice that we view G as a directed graph, so an edge $(a, b) \in E$ can be part of a shortest path only if $l(a) < l(b)$.

Base Case: $m = 1$, i.e., the graph contains only one (bounded) face F. Assume u and v are two vertices such that there exist two shortest paths between u and v and over all such pairs of vertices $|l(u) - l(v)|$ is minimum (see Fig. 4(a)). Let these two paths be P_1 and P_2. Observe that $P_1 \cup P_2$ is the face F. Choose any middle vertex v_t of F and place a tracker at v_t. Now the two paths between u and v are distinguishable, and for any other pair of vertices u' and v' such that there exist two shortest paths between u' and v', one of the paths is a superset of P_1 and the other is a superset of P_2. Therefore one contains v_t and the other does not. Hence for any pair of vertices, if there exist two shortest paths between them, then they are distinguishable.

Induction Step: Consider any graph G with $m + 1$ faces. Let F be any face of G which shares at least one edge with the outer face F^o (see Fig. 4(b)). Observe that such a face always exists. Consider the set of vertices V_F in $F \cap F^o$. Observe that at least one of the vertices in V_F is a middle vertex of F. (Otherwise, V_F consists of the two verices v_{in} and v_{out}. But, there is another path between v_{in} and v_{out} with the same number of vertices (Observation 2), which implies that there are two edges connecting between v_{in} and v_{out}, which is impossible.) Choose one of the middle vertices v_a in V_F and place a tracker at v_a.

The outer face F^o contains two (not necessarily disjoint) paths from s to t. Denote them by λ_1 and λ_2. As v_a is a middle vertex, v_a belongs to exactly one of these paths. Without loss of generality assume $v_a \in \lambda_1$. Consider the (open) chain μ shared by F and λ_1, i.e., μ consists of all vertices and edges of the chain $F \cap \lambda_1$, except for the two extreme vertices (see Fig. 4(b)). Delete μ from G. Now the remaining graph G' has m faces, and by the induction hypothesis, there exists a set T of m trackers, such that for any two vertices u and v in G', there do not exist two different shortest paths with the same set of trackers. We claim that $T \cup \{v_a\}$ is a set of trackers, such that for any two vertices u and v in G, there do not exist two different shortest paths with the same set of trackers. Suppose there are two shortest paths P_1 and P_2 in G with the same set of trackers. There can be three cases.

Case i: P_1 and P_2 do not intersect μ. Then, P_1 and P_2 are paths in G', contradicting the induction hypothesis.

Case ii: Exactly one of P_1 and P_2 intersects μ, say P_1. Observe that if a path intersects μ it contains μ, since μ or parts of it do not "belong" to any other face. Hence P_1 contains v_a and P_2 does not. Thus they are distinguishable.

Case iii: Both P_1 and P_2 intersect μ. In this case, both paths contain μ (and v_a). But since they are different, there exist two vertices u and v in G', such that in G' there exist two different shortest paths between u and v with the same set of trackers, and these paths are contained in P_1 and P_2, respectively. But this contradicts the induction hypothesis.   □

Theorem 8 immediately implies an iterative algorithm to find a tracking set of size m for G. Using the DCEL structure for representing planar subdivisions [4], the algorithm can be implemented in linear time. By Lemma 7 we know that the optimal solution consists of at least $m/2$ trackers. Hence, we have the following result.

Theorem 9. *There exists a linear-time algorithm for computing a 2-approximation for TSSP in planar graphs.*

Remark. In the full version of this paper, we also present a linear-time heuristic for TSSP in general graphs. The heuristic is based on a variant of the feedback vertex set problem in which one is given a set of *blackout vertices*, i.e., vertices that may not appear in the solution; see Bar-Yehuda et al. [2]. The heuristic can be adjusted to serve as a heuristic for the blocked vertices variant (Problem 4).

Remark. In the full version of this paper, we also study Problem 4. Specifically, we construct a data structure of size $O(|A| \cdot |V|)$, where A is the tracking set w.r.t. shortest paths computed for G, such that given a sequence of visited trackers, one can reconstruct the traversed path P in $O(|P|)$ time.

Remark. Consider a slightly different model, where a tracker at a vertex v does not only report a vehicle that passes through v, but also reports the edges by which the vehicle entered and exited v. In this model, *any* feedback vertex set of a general graph G is necessarily a tracking set for G. Given a feedback vertex set, assuming it is not a tracking set and considering two shortest indistinguishable paths (starting and ending at the same vertices) implies a relevant cycle without a tracker (since if there existed a tracker in any of the vertices of the cycle, one would be able to distinguish between the given paths using the additional information provided by the tracker).

4   Catching the Intruder

In this section, we consider a variant of TSSP, where the goal is to "catch" trespassers: Given a graph $G = (V, E)$, two vertices s,t, and a set of *forbidden vertices* $V_F \subset V$, find a minimum-cardinality set of vertices V^*, such that a shortest path between s and t passes through at least one of the forbidden vertices if and only if it passes through at least one of the vertices of V^*. Formally, for a given shortest path P, $V_F \cap V(P) \neq \emptyset$ if and only if $V^* \cap V(P) \neq \emptyset$.

Definition 3. The set of *extended forbidden vertices* is the set of all vertices $F_V \subseteq V$, such that, if a shortest path P from s to t passes through one of them, i.e., $V(P) \cap F_V \neq \emptyset$, then P must also pass through a forbidden vertex, i.e., $V(P) \cap V_F \neq \emptyset$. (Clearly, $V_F \subset F_V$.) A *forbidden path* is then a path which passes through at least one vertex from the extended forbidden vertices set.

We first find the set of *extended forbidden vertices* by running Algorithm 2. (As before, we assume that all vertices and edges of G appear in some shortest path between s and t; otherwise, we remove all unnecessary vertices and edges by running Algorithm 1). Notice that Algorithm 2 requires $O(|V| + |E|)$ time.

Definition 4. A vertex $v \in V$ is a *first vertex*, if $v \in F_V$ and there exists a shortest path from s to v which does not pass through any other vertex of F_V. A vertex $v \in V$ is a *last vertex*, if $v \in F_V$ and there exists a shortest path from v to t which does not pass through any other vertex of F_V.

Next, we find all *first vertices* and *last vertices* by running a BFS in G. First vertices are found by running a BFS from s, each time an extended forbidden vertex v is encountered for the first time, it is added to the set of first vertices and the subtree rooted at v is not explored (since all following vertices would not be first for this specific path, however, if they are first for some other path, then they will be added at a later stage). Last vertices are found by running a similar procedure from t.

Algorithm 2. Finding the set of extended forbidden vertices.

---

1 **Input:** A directed graph $G(V, E)$ and a set of forbidden vertices $V_F \subset V$.
2 **Output:** The set of extended forbidden vertices F_V.
3 Run a BFS in G and assign a level to each vertex $v \in V$ depending on its
   distance from s.
4 $l_{max} \leftarrow l(t)$
5 $F_E \leftarrow \emptyset$
6 $F_V \leftarrow V_F$
7 **for** $l \leftarrow 1$ *to* $l_{max} - 1$ **do**
8   **foreach** v *s.t.* $l(v) = l$ **do**
9   **if** {*all edges entering* v} $\subseteq F_E$ **then**
10             $F_V \leftarrow F_V \cup \{v\}$
11   **if** $v \in F_V$ **then**
12             $F_E \leftarrow F_E \cup \{$all edges exiting $v\}$
13 **for** $l \leftarrow l_{max} - 1$ *down to* 1 **do**
14   **foreach** v *s.t.* $l(v) = l$ **do**
15   **if** {*all edges exiting* v} $\subseteq F_E$ **then**
16             $F_V \leftarrow F_V \cup \{v\}$
17   **if** $v \in F_V$ **then**
18             $F_E \leftarrow F_E \cup \{$all edges entering $v\}$
19 **return** F_V

Denote the set of first vertices by V_{first} and the set of last vertices by V_{last}. We connect all first vertices to a source v_{src} and all last vertices to a sink v_{sink}, denote all these edges by E^*. Finally, we construct the *forbidden paths graph* $G' = (V', E')$ where $V' = F_V \cup \{v_{src}, v_{sink}\}$, $E' = E_F \cup E^*$, where $E_F \subseteq E$ is the of edges for which both endpoints belong to the extended forbidden vertices set F_V. An example of the construction can be seen in Fig. 5.

Observation 4. *The desired vertex set V^* must be a subset of $F_V = V' - \{v_{src}, v_{sink}\}$. Otherwise, there is a forbidden path which is not represented in G' – which is not possible due to the way G' was constructed.*

In order to solve our problem, we must find a *minimum vertex cut* between v_{src} and v_{sink} in G', i.e., a minimum-cardinality set of vertices V_{mc} such that if we remove the vertices of V_{mc} from G' there would be no path from v_{src} to v_{sink}.

We find a minimum vertex cut in the following manner: for each vertex $v_i \in V'$ we create two new vertices v_{i1}, v_{i2} connected by the edge (v_{i1}, v_{i2}). All

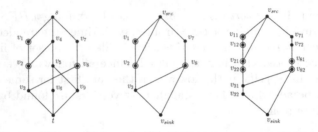

Fig. 5. The forbidden paths graph G' (middle) of a given graph G (left), and the graph constructed from G' in which we find a minimum edge cut (right). $V_F = \{v_1, v_2, v_8\}$, $F_V = \{v_1, v_2, v_3, v_7, v_8\}$, $V_{first} = \{v_1, v_2, v_7\}$, $V_{last} = \{v_3, v_8\}$.

edges originally entering v_i will now enter v_{i1}, and all edges originally exiting v_i will now exit v_{i2}. An example of the construction can be seen in Fig. 5. Notice that the size of a minimum vertex cut in G' is at least the size of a minimum edge cut in the new graph. We run a minimum edge cut algorithm in the new graph, and denote by E_{mc} the minimum cut returned. We define a minimum vertex cut V_{mc} as follows: for each $e \in E_{mc}$, if e is an edge connecting between v_{i1}, v_{i2} for some i, then v_i is added to V_{mc}. Removing v_i in this case has the same effect as removing e. Otherwise, if $e = (v_{i2}, v_{j1})$ where $i \neq j$, then v_j is added to V_{mc}. In this case, notice that one can replace e in the cut by the edge $e' = (v_{j1}, v_{j2})$ and still have a cut. The new edge corresponds to the vertex v_j which is added to the vertex cut to achieve a similar effect. Obviously we have described a polynomial-time algorithm, since computing a minimum edge cut is polynomial. Finally, set $V^* = V_{mc}$.

Lemma 10. V^* *is the desired solution.*

Proof. Assume it is not, so, if we remove the vertices in V_{mc} from G, we are still left with a path $P = (s, \ldots, v_f, \ldots, t)$ which passes through a forbidden vertex $v_f \in V_F$. By the construction of G', there must still be a path $P' = (v_{src}, \ldots, v_f, \ldots, v_{sink})$ in G', after removing the vertices from V_{mc} – a contradiction. Thus, V_{mc} must be the desired set, and its minimality follows from the construction. $\square$

Running Time. Constructing G' and splitting each vertex requires $O(|V|+|E|)$ time. On the resulting graph we run an algorithm for finding a minimum edge cut. The most efficient such algorithm is the randomized $O(n^2 \log^3 n)$-time algorithm of Karger and Stein [7]. We thus reach a total running time of $O(n^2 \log^3 n)$.

References

1. Bafna, V., Berman, P., Fujito, T.: A 2-approximation algorithm for the undirected feedback vertex set problem. SIAM J. Discret. Math. **12**(3), 289–297 (1999)

2. Bar-Yehuda, R., Geiger, D., Naor, J.S., Roth, R.M.: Approximation algorithms for the feedback vertex set problem with applications to constraint satisfaction and bayesian inference. SIAM J. Comput. **27**, 942–959 (1998)
3. Bhatti, S., Xu, J.: Survey of target tracking protocols using wireless sensor network. In: Proceedings of 5th International Conference on Wireless and Mobile Communications, pp. 110–115 (2009)
4. de Berg, M., Cheong, O., van Kreveld, M., Overmars, M.: Computational Geometry Algorithms and Applications, 3rd edn. Springer, Heidelberg (2008)
5. Dinur, I., Safra, S.: On the hardness of approximating minimum vertex cover. Ann. Math. **162**, 439–485 (2005)
6. Ganesan, D., Cristescu, R., Beferull-Lozano, B.: Power-efficient sensor placement and transmission structure for data gathering under distortion constraints. ACM Trans. Sens. Netw. **2**(2), 155–181 (2006)
7. Karger, D.R., Stein, C.: A new approach to the minimum cut problem. J. ACM **43**(4), 601–640 (1996)
8. Karp, R.M.: Reducibility among combinatorial problems. In: Proceedings of a Symposium on the Complexity of Computer Computations, IBM Thomas J. Watson Research Center, Yorktown Heights, New York, pp. 85–103. Plenum Press (1972)

On the Complexity of Finding a Potential Community

Cristina Bazgan[1,4], Thomas Pontoizeau[1(✉)], and Zsolt Tuza[2,3]

[1] Université Paris-Dauphine, PSL Research University,
CNRS, LAMSADE, 75016 Paris, France
{bazgan,thomas.pontoizeau}@lamsade.dauphine.fr
[2] Alfréd Rényi Institute of Mathematics, Hungarian Academy of Sciences,
Budapest, Hungary
[3] Department of Computer Science and Systems Technology,
University of Pannonia, Veszprém, Hungary
tuza@dcs.uni-pannon.hu
[4] Institut Universitaire de France, Paris, France

Abstract. An independent 2-clique of a graph is a subset of vertices that is an independent set and such that any two vertices inside have a common neighbor outside. In this paper, we study the complexity of finding an independent 2-clique of maximum size in several graph classes and we compare its complexity with the complexity of maximum independent set. We prove that this problem is NP-hard on apex graphs, APX-hard on line graphs, not $n^{1/2-\epsilon}$-approximable on bipartite graphs and not $n^{1-\epsilon}$-approximable on split graphs, while it is polynomial-time solvable on graphs of bounded degree and their complements, graphs of bounded treewidth, planar graphs, (C_3, C_6)-free graphs, threshold graphs, interval graphs and cographs.

Keywords: Combinatorial optimization · Complexity · Algorithm · Independent set · Inapproximability

1 Introduction

Community detection is a well established research field in the area of social networks. It can find many applications in this area with the recent development of social networks like Facebook or Linkedin. A social network can be easily modeled by a graph in which vertices represent members and edges represent relationships between those members.

There are several ways to define a community. Intuitively, a community corresponds to a dense subgraph, that is to say a subgraph with a lot of edges. If a community is defined as a group of maximum size such that all members know each other, it corresponds to the well known NP-hard problem of finding a maximum clique. However, such a condition is strong and is not always relevant to describe a community.

© Springer International Publishing AG 2017
D. Fotakis et al. (Eds.): CIAC 2017, LNCS 10236, pp. 80–91, 2017.
DOI: 10.1007/978-3-319-57586-5_8

Another way to define a community is to relax the strong condition of a clique and focus on the distance between members of a social network. Different measures have been studied to describe it. Luce introduced in [12] the notion of k-cliques while Mokken extended this notion in [13] by defining k-clubs. A k-clique (resp. a k-club) of G is a subgraph S in which any two vertices are at distance at most k in G (resp. in the subgraph induced by S). The standard term 'clique' means both a 1-clique and a 1-club.

With the recent development of social networks and particularly online dating services, it could be interesting to investigate the detection of some group of people who do not know each other, but are related by their other relationships. Such a group could be considered as a 'potential' community since it does not form a community in the first place, but could become one thanks to their proximity. This may find various applications in online dating and meet-up services in which members expect not to know the other members.

More precisely, considering a graph G, we want to define potential communities by looking at independent sets in which any two members are related within a specified distance in G. Contrary to a k-club, the distance between two vertices must be realized via vertices outside of the subgraph. We call such a subset of vertices an independent k-clique, where k is the largest distance between vertices of S in the original graph. In this paper, we study the problem of finding an independent 2-clique of maximum size.

We investigate the complexity of finding an independent 2-clique of maximum size in several graph classes. Since this problem is close to finding an independent set of maximum size, we also compare the hardness of the two problems. Figure 1 summarizes the results we prove in the paper.

The paper is structured as follows. In Sect. 2 we introduce formally some notation and definitions. In Sect. 3 we show that the complexity of MAX INDEPENDENT 2-CLIQUE jumps from polynomial-time solvable to NP-hard when the input class is extended from planar graphs to apex graphs. In Sect. 4 we present polynomial algorithms to solve MAX INDEPENDENT 2-CLIQUE in some graph classes. In Sect. 5 we show NP-hardness and non-approximability of MAX INDEPENDENT 2-CLIQUE in some other graph classes. Conclusions are given in Sect. 6. Due to space limit, some proofs will be given in a journal version.

2   Preliminaries

In this paper, all considered graphs are undirected. The *complement* $\overline{G} = (V, \overline{E})$ of a graph $G = (V, E)$ is the graph in which $uv \in E$ if and only if $uv \notin \overline{E}$, for all vertex pairs $u, v \in V$. A k-*cycle* is a cycle of length k. The maximum degree of a vertex in a graph G will be denoted by the usual notation $\Delta(G)$.

We recall that a *clique* in a graph is a set of mutually adjacent vertices. A set of vertices is called a 2-*clique* if any two vertices of the set are at distance at most 2 in G. An *independent set* in a graph is a set of vertices such that no two of them are joint by an edge. An *independent 2-clique* is a subset of vertices which is an independent set and a 2-clique at the same time.

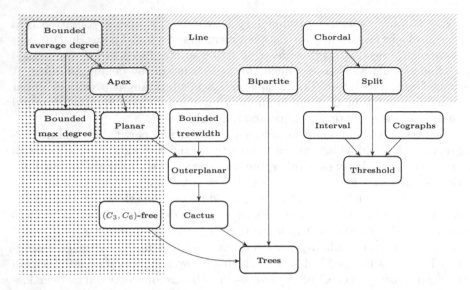

Fig. 1. Relationship among some classes of (connected) graphs, where each child is a subset of its parent. We compare the hardness of MAX INDEPENDENT 2-CLIQUE and MAX INDEPENDENT SET in studied graph classes. MAX INDEPENDENT 2-CLIQUE is NP-hard on graph classes at the top (hatched area) and is polynomial-time solvable on graph classes at the bottom (non-hatched area). MAX INDEPENDENT SET is NP-hard on graph classes on the left (dotted area) and is polynomial-time solvable on graph classes on the right (non-dotted area).

In this paper we are interested in the following optimization problem:

MAX INDEPENDENT 2-CLIQUE
Input: A graph $G = (V, E)$.
Output: A subset $S \subset V$ which is an independent 2-clique of maximum size.

The MAX INDEPENDENT 2-CLIQUE problem is closely related to another well known one:

MAX INDEPENDENT SET
Input : A graph $G = (V, E)$.
Output : A subset $S \subset V$ such that S is an independent set of maximum size.

Given a graph G, the standard notation for the maximum size of an independent set in G is $\alpha(G)$. The maximum number of vertices in an independent 2-clique of G will be denoted by $\alpha_{=2}(G)$. The subscript '=2' intends to express that the distance between any two vertices of the independent set is exactly 2.

Note that $\alpha_{=2}(G) \geq 2$ whenever at least one connected component of G is not a complete graph. (Indeed, any such component contains two vertices at distance exactly two, hence forming an independent 2-clique of size 2.) Moreover, if G is disconnected and has components $G_1, \ldots, G_k$ then

$$\alpha_{=2}(G) = \max_{1 \leq i \leq k} \alpha_{=2}(G_i)$$

For these reasons we assume throughout that G is a non-complete, connected graph (although some of the algorithms also need to handle disconnected graphs temporarily).

Some Classes of Graphs. A *cactus* is a graph in which each edge occurs in at most one cycle. A (C_3, C_6)-*free graph* is a graph containing no triangle C_3 and no induced cycle of length 6. An *interval graph* is a graph for which there exists a family of intervals on the real line and a bijection between the vertices of the graph and the intervals of the family in such a way that two vertices are joined by an edge if and only if the intersection of the two corresponding intervals is non-empty. A *threshold graph* is a graph which can be constructed from the empty graph by a sequence of two operations: insertion of an isolated vertex, and insertion of a dominating vertex (i.e., a vertex adjacent to all the other vertices). A *cograph* is a graph that can be generated from the single-vertex graph by (repeated applications of) complementation and vertex-disjoint union. A *split graph* is a graph whose vertex set can be partitioned into two subsets, one inducing an independent set S and the other one inducing a clique K. We denote by $K_{p,m}$ the complete bipartite graph with p and m vertices in its vertex parts. The *line graph* of a graph G is the graph $L(G)$ whose vertices represent the edges of G, and two vertices of $L(G)$ are adjacent if and only if the corresponding two edges of G share a vertex. A graph is *outerplanar* if it has a crossing-free embedding in the plane such that all vertices are on the same face. A graph is k-*outerplanar* if for $k = 1$, G is outerplanar and for $k > 1$ the graph has a planar embedding such that if all vertices on the exterior face are deleted, the connected components of the remaining graph are all $(k-1)$-outerplanar. A graph G is *apex* if it contains a vertex v such that $G \setminus v$ is planar. A family of graphs on n vertices is δ-*dense* if it has at least $\frac{\delta n^2}{2}$ edges. It is *everywhere-δ-dense* if the minimum degree is at least δn. A family of graphs is *dense* (resp. *everywhere-dense*) if there is a constant $\delta > 0$ such that all members of this family are δ-dense (resp. everywhere-δ-dense).

3   Complexity Jump from Planar Graphs to Apex Graphs

According to [8], MAX INDEPENDENT SET is known to be NP-hard in planar graphs, and thus also in apex graphs. On the other hand, we prove that MAX INDEPENDENT 2-CLIQUE is polynomial-time solvable on planar graphs but NP-hard on apex graphs. This shows that inserting or removing a single vertex in a graph may dramatically change the complexity of MAX INDEPENDENT 2-CLIQUE.

Theorem 1. MAX INDEPENDENT 2-CLIQUE *is NP-hard on apex graphs.*

Proof. We establish a polynomial reduction from MAX INDEPENDENT SET on cubic planar graphs, which is proved NP-hard in [8], to MAX INDEPENDENT 2-CLIQUE on apex graphs. Let $G = (V, E)$ be a cubic planar graph, an instance of MAX INDEPENDENT SET. The instance $G' = (V', E')$ of MAX INDEPENDENT 2-CLIQUE is defined by inserting an additional vertex z that is adjacent to every vertex of V. It is easy to see that $\{z\}$ itself is a one-element non-extendable independent 2-clique, while the independent 2-cliques of G' not containing z are precisely the independent sets of G.   □

This first theorem implies another interesting result:

Corollary 2. MAX INDEPENDENT 2-CLIQUE *is NP-hard on the class of graphs of average degree at most 5.*

Proof. Cubic graphs on n vertices have $3n/2$ edges, thus the graph constructed in the proof of Theorem 1 is of order $n + 1$ and has $5n/2$ edges, yielding average degree less than 5.   □

Now, in order to prove that MAX INDEPENDENT 2-CLIQUE is polynomial-time solvable on planar graphs, we use a famous theorem introduced by Courcelle in [6] which states that any problem expressible in Monadic Second-Order Logic is linear-time solvable for graphs of bounded treewidth. This allows to show first the following:

Theorem 3. MAX INDEPENDENT 2-CLIQUE *is linear-time solvable on graphs with bounded treewidth.*

Proof. We observe that the property 'Independent 2-Clique' is expressible in Monadic Second-Order Logic:

$$I2C(S) := \forall x \forall y (Sx \wedge Sy) \rightarrow (\neg edg(x, y) \wedge (\exists z, edg(x, z) \wedge edg(y, z)))$$

Since any problem expressible in Monadic Second-Order Logic is linear-time solvable for graphs of bounded treewidth (see [6]), $\alpha_{=2}$ can be determined in linear time in graphs of bounded treewidth.   □

Based on this result, we prove the following result.

Theorem 4. MAX INDEPENDENT 2-CLIQUE *is polynomial-time solvable on planar graphs.*

Proof. Let $G = (V, E)$ be a planar graph and $v \in V$ any vertex. Then all the other vertices in an independent 2-clique S containing v are at distance exactly 2 apart from v. Further, the 2-clique property for $S \setminus \{v\}$ is ensured by vertices within distance at most 3 from v. Thus, the vertices relevant for S to be an independent 2-clique induce a subgraph G' in G such that G' belongs to the class of '4-outerplanar' graphs. Graphs which are 4-outerplanar have treewidth at most 11 (more generally, all k-outerplanar graphs have treewidth at most $3k - 1$, due to [3]). Then, using Theorem 3, a polynomial-time algorithm for MAX INDEPENDENT 2-CLIQUE in planar graphs consists in solving the problem for all subgraphs G' (which have treewidth at most 11) defined from each vertex v.   □

4 Graph Classes with Polynomial-Time Algorithms

In the following we identify some graph classes on which MAX INDEPENDENT 2-CLIQUE is computable in polynomial time, while MAX INDEPENDENT SET is not always polynomial-time solvable.

First, it is interesting to notice that, according to the next propositions, MAX INDEPENDENT 2-CLIQUE is polynomial-time solvable on graphs of bounded degree and also on complements of graphs of bounded degree, while MAX INDEPENDENT SET is NP-hard on graphs of bounded degree [8] but polynomial-time solvable on their complements (using exhaustive search in the non-neighborhood of each vertex, which can be done in linear time).

Proposition 5. MAX INDEPENDENT 2-CLIQUE *is linear-time solvable on graphs with bounded maximum degree, and also on graphs of minimum degree at least* $(n-d)$, *where d is constant.*

Now, notice that a natural way to find an independent 2-clique is to take an independent set included in the neighborhood of one vertex. Then, it is interesting to investigate the properties of a graph in which an independent 2-clique is not included in the neighborhood of one vertex. We show in Lemma 6 that such a graph necessarily contains a cycle of length 3 or 6, and cannot be a cactus if such an independent 2-clique has a certain size. Such properties allow to get an easy polynomial-time algorithm for MAX INDEPENDENT 2-CLIQUE on (C_3, C_6)-free graphs, while MAX INDEPENDENT SET is NP-hard on this class of graphs (see [1]). From Theorem 4 we already know that MAX INDEPENDENT 2-CLIQUE is linear-time solvable on cactus graphs, but the property of Lemma 6 allows to give a simpler algorithm for this class of graph.

Lemma 6. *Let* $G = (V, E)$ *be a graph. Suppose that there exists an independent 2-clique S not contained in the neighborhood of a single vertex. Then G contains an induced cycle of length 3 or 6. Moreover, if* $|S| \geq 4$, *G is not a cactus.*

This lemma implies the following theorem:

Theorem 7. *Any* (C_3, C_6)*-free graph G satisfies* $\alpha_{=2}(G) = \Delta(G)$ *and* MAX INDEPENDENT 2-CLIQUE *is linear-time solvable on it.*

Recalling that a tree does not contain any cycle, for the classical class of trees we obtain the following:

Corollary 8. *Any tree T satisfies* $\alpha_{=2}(T) = \Delta(T)$ *and* MAX INDEPENDENT 2-CLIQUE *is linear-time solvable on it.*

Finally, Lemma 6 allows to give a polynomial-time algorithm for MAX INDEPENDENT 2-CLIQUE on cactus.

Proposition 9. MAX INDEPENDENT 2-CLIQUE *is linear-time solvable on cactus graphs.*

We focus in the following part of this section on classes of graphs on which both MAX INDEPENDENT 2-CLIQUE and MAX INDEPENDENT SET are polynomial-time solvable. We first investigate a subclass of split graphs, namely threshold graphs. It follows from the definitions that a threshold graph $G = (V, E)$ is a split graph with the following property: the vertices of the independent set S can be ordered as $v_1, \ldots, v_p$ such that $N_G(v_1) \subseteq N_G(v_2) \subseteq \ldots \subseteq N_G(v_p)$. We denote by $u_1, \ldots, u_q$ the vertices of the clique K, and we suppose that $d_G(u_1) \leq d_G(u_2) \leq \ldots \leq d_G(u_q)$. Without loss of generality, we assume that there is no isolated vertex in G. Note that a threshold graph can be recognized in linear time (see [10]).

Proposition 10. MAX INDEPENDENT 2-CLIQUE *is linear-time solvable on threshold graphs. Moreover, in every threshold graph G without isolated vertices we have $\alpha_{=2}(G) = \alpha(G)$.*

Proof. Let $G = (V, E)$ be a threshold graph with the previous decomposition into S and K. Let $N_G(v_p) = \{u_r, u_{r+1}, \ldots, u_q\}$, for some $r \geq 1$. Then a maximum independent 2-clique in G is S if $K \setminus N_G(v_p) = \emptyset$, and otherwise it is $S \cup \{z\}$ with any $z \in K \setminus N_G(v_p)$, since in both cases the common neighbor of all these vertices is u_q. Since MAX INDEPENDENT SET can be solved in linear time in threshold graphs [7], MAX INDEPENDENT 2-CLIQUE can be solved in linear time.   □

The previous result can be extended in two directions, for interval graphs and for cographs.

Using the results of Booth and Lueker [4] it can be tested in linear time whether a graph G is an interval graph; and if it is, then an interval representation $I_1, \ldots, I_n$ of G can also be generated.

Proposition 11. MAX INDEPENDENT 2-CLIQUE *is polynomial-time solvable on interval graphs.*

Proof. Consider any $G = (V, E)$ and let $I_1, \ldots, I_n$ be an interval representation of G. In order to determine $\alpha_{=2}(G)$, first notice that all vertices of an independent 2-clique S of G must have a common neighbor. Indeed, if I and I' are the leftmost and the rightmost intervals of S then any of their common neighbors intersects all intervals located between them, and therefore is a common neighbor of all members of S. Then, for every vertex I, we compute a maximum independent set in the subgraph induced by the neighborhood of I. An optimal solution is such an independent set with maximum size. Since MAX INDEPENDENT SET is polynomial-time solvable on interval graphs [9], the result follows.   □

We consider now the class of cographs, that contains all threshold graphs.

Proposition 12. MAX INDEPENDENT 2-CLIQUE *is polynomial-time solvable on cographs.*

Notice that since MAX INDEPENDENT SET is linear-time solvable on chordal graphs [7], it is also polynomial-time solvable on interval graphs and threshold graphs. Moreover, MAX INDEPENDENT SET is also polynomial-time solvable on cographs by bottom-up tree computation [5].

5   NP-hardness and Non-approximability

Using the reduction from the proof of Theorem 1, we can conclude:

- MAX INDEPENDENT 2-CLIQUE is NP-hard on dense (resp. everywhere dense) graphs, since MAX INDEPENDENT SET is NP-hard on dense (resp. everywhere dense) graphs. Moreover, MAX INDEPENDENT 2-CLIQUE is not $n^{1-\varepsilon}$-approximable for any $\varepsilon > 0$, if P $\neq$ NP, on everywhere dense graphs (and respectively dense graphs) since the same result holds for MAX INDEPENDENT SET on everywhere dense graphs (and respectively dense graphs). In order to get this last result, we use the same inaproximability result for MAX INDEPENDENT SET on general graphs [15] and a reduction preserving approximation from general graphs to everywhere dense graphs (that consists of adding a clique of the same size as the size of the graph and joining every vertex from the original graph to all vertices in this clique).
- MAX INDEPENDENT 2-CLIQUE is NP-hard on K_4-free graphs, since MAX INDEPENDENT SET is NP-hard on K_3-free graphs [1].

We now investigate graph classes in which MAX INDEPENDENT 2-CLIQUE is NP-hard while MAX INDEPENDENT SET is polynomial-time solvable. We first consider a graph class containing threshold graphs, namely the class of split graphs, for which MAX INDEPENDENT 2-CLIQUE becomes NP-hard (and even not $n^{1-\varepsilon}$-approximable). Since MAX INDEPENDENT SET is polynomial-time solvable on chordal graphs [7], it is also polynomial-time solvable on split graphs.

Theorem 13. *On split graphs, MAX INDEPENDENT 2-CLIQUE is NP-hard and it is not $n^{1-\varepsilon}$-approximable in polynomial time unless $P = NP$.*

We prove now that MAX INDEPENDENT 2-CLIQUE is NP-hard (and even not $n^{1/2-\varepsilon}$-approximable) on bipartite graphs while MAX INDEPENDENT SET is polynomial-time solvable since the number of vertices in a maximum independent set equals the number of edges in a minimum edge covering.

Theorem 14. *On bipartite graphs, MAX INDEPENDENT 2-CLIQUE is NP-hard and is not $n^{1/2-\varepsilon}$-approximable in polynomial time, unless $P = NP$.*

Proof. First we prove the NP-hardness. MAX INDEPENDENT SET is known to be NP-hard on 3-regular graphs [8], so MAX CLIQUE is also NP-hard on $(n-4)$-regular graphs (where n is the number of vertices), by considering its complement. We reduce MAX CLIQUE on $(n-4)$-regular graphs to MAX INDEPENDENT 2-CLIQUE on bipartite graphs. Let $G = (V, E)$ be an $(n-4)$-regular graph. We construct an instance of $G' = (V', E')$ of MAX INDEPENDENT 2-CLIQUE on bipartite graphs as follows (see Fig. 2).

Let V_1, V_2, V_3, V_4 be four copies of V. Let E_1 be a set of $|E|$ vertices corresponding to the edges in E, and define $V' := V_1 \cup V_2 \cup V_3 \cup V_4 \cup E_1$. Let there exist an edge in E' between a vertex v in V_i, $i \in \{1, 2, 3, 4\}$ and a vertex e in E_1

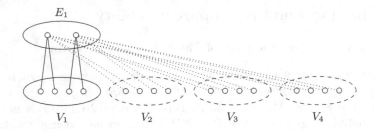

Fig. 2. The bipartite graph G', an instance of MAX INDEPENDENT 2-CLIQUE

if and only if the corresponding vertex v in V is incident with the corresponding edge e in E.

Now we show that G contains a clique of size at least k if and only if G' contains an independent 2-clique of size at least $4k$.

Given a clique $C \subseteq V$ of size at least k in G, the union of the four copies of C in G' is an independent 2-clique of size at least $4k$.

For the other direction, notice first that the value of a maximum independent set in a 3-regular graph is at least $\lceil \frac{n}{4} \rceil$. Then, the value of a maximum clique in an $(n-4)$-regular graph is also at least $\lceil \frac{n}{4} \rceil$. Thus the size of a maximum independent 2-clique in G' is at least n.

We consider now a solution C' of MAX INDEPENDENT 2-CLIQUE in G' with at least $4k \geq n$ vertices (this restriction is always possible because of the previous comment). Notice that C' cannot contain both a vertex from E_1 and a vertex from $V' \setminus E_1$ since the distance between any two vertices of C' must be 2. A solution which is a subset of E_1 would mean pairwise intersecting edges in G, hence would have size at most $\max(3, n-4) < n$. Therefore C' must be a subset of $V' \setminus E_1$. Notice that for any $i \in \{1, 2, 3, 4\}$, $C' \cap V_i$ must be a copy of a clique in G. Then C' is a union of copies of four cliques in G, and $|C'| \geq 4k$. Let C_0 be the copy of largest size, which thus has $|C_0| \geq k$. Then C_0 is the copy of a clique C of G of size at least k.

For the proof of non-approximability, we construct an E-reduction (see [11]) from MAX CLIQUE. Let $I = (V, E)$ be an instance of MAX CLIQUE. Consider a reduction similar to the one for the proof of NP-hardness, except that we now consider $\ell = |V|$ copies $V_1, \ldots, V_\ell$ instead of four copies of V; adjacencies are defined in the same way as before. We denote by $I' = (V', E')$ the corresponding instance of MAX INDEPENDENT 2-CLIQUE from the reduction. As in previous proof, starting with a clique of size $opt(I)$, we can construct an independent 2-clique of size $\ell \cdot opt(I)$ in G' and thus $opt(I') \geq \ell \cdot opt(I)$. Let S' be any independent 2-clique in I' of size at least ℓ (it always exists, take e.g. the ℓ copies of the same vertex, one copy in each V_i). As before, S' cannot contain both a vertex of E_1 and a vertex from $V \setminus E_1$ since two vertices of S' must have distance 2 in G', and S' cannot contain only vertices from E_1 since any independent 2-clique included in E_1 is of size at most $\max(3, \Delta(G)) < \ell - 1$. Moreover, each subset $V_i \cap S'$ corresponds to a clique in G. Let S be the subset $V_i \cap S'$ of largest size. We have $|S| \geq \frac{|S'|}{\ell}$

and then $opt(I) \geq |S| \geq \frac{|S'|}{\ell} = \frac{opt(I')}{\ell}$ when S' is an optimal solution. Using that $opt(I') \geq \ell \cdot opt(I)$ we get $opt(I') = \ell \cdot opt(I)$ and we obtain:

$$\epsilon(I, S) = \frac{opt(I)}{|S|} - 1 \leq \frac{\ell \cdot opt(I')}{\ell \cdot |S'|} - 1 = \epsilon(I', S')$$

Since we clearly have $opt(I') \leq p(|I|) \cdot opt(I)$ with a polynomial p, the reduction is an E-reduction. Then, since MAX CLIQUE is not $\ell^{1-\varepsilon}$-approximable unless $P = NP$ [15], the same property holds for MAX INDEPENDENT 2-CLIQUE. Thus MAX INDEPENDENT 2-CLIQUE is not $n^{1/2-\varepsilon}$ approximable where $n = |V'|$ since $n = \ell^2 + |E|$.   □

Finally we prove that MAX INDEPENDENT 2-CLIQUE is NP-hard (and even APX-hard) on line graphs, while MAX INDEPENDENT SET is polynomial-time solvable since it consists in a maximum matching in the original graph.

Theorem 15. *On line graphs,* MAX INDEPENDENT 2-CLIQUE *is NP-hard and even APX-hard.*

Proof. First we prove the NP-hardness. We establish a reduction from the MAX CLIQUE problem on graphs of minimum degree at least $n - 4$. Consider an instance $G = (V, E)$ of MAX CLIQUE with $|V| = n$. We construct a graph $G' = (V', E')$ (see Fig. 3) as follows. Let $G_0 = (V_0, E_0)$ be a copy of G. Let V' be $V_0 \cup A \cup B \cup C$ where A, B, C are three sets of n vertices. Then, let $E' = E_0 \cup E_1 \cup E_2 \cup E_3 \cup E_4$ such that E_1 is a perfect matching between V_0 and A, E_2 is the set of all possible edges (i.e., a complete bipartite graph) between the vertices of A and the vertices of B, E_3 is a perfect matching between B and C, and E_4 is the set of all possible edges between any two vertices of C (a complete subgraph). The line graph of G', denoted by $L(G')$, is an instance of MAX INDEPENDENT 2-CLIQUE. Notice that an independent 2-clique in $L(G')$ corresponds to a set of edges in G' such that, for each pair of edges $\{e_1, e_2\}$ in the set, e_1 and e_2 are not adjacent but are joined by an edge. We show that G contains a clique of size at least k if and only if $L(G')$ contains an independent 2-clique of size at least $k + n$.

Consider a clique S of size k in G, and let S_0 be its copy in G'. We define a set of edges S' of size at least $k + n$ in G' as follows. For any vertex $v \in S_0$,

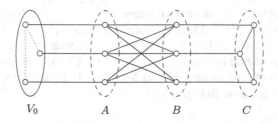

V_0 A B C

Fig. 3. The graph G' for which the corresponding line graph $L(G')$ is an instance of MAX INDEPENDENT 2-CLIQUE

add in S' its adjacent edge in E_1. Moreover add the entire E_3 to S'. We show now that any pair of edges in S' have an adjacent edge in common. Two edges of $S' \cap E_1$ have a common adjacent edge in E_0 since the subgraph induced by S_0 is a clique. Similarly, two edges of E_3 have a common adjacent edge in E_4. Moreover, an edge of $S' \cap E_1$ and an edge of E_3 have a common adjacent edge in E_2 since the subgraph induced by $A \cup B$ is $K_{n,n}$. Then, the corresponding set of vertices in $L(G')$ is an independent 2-clique of size $k + n$.

In the other direction, consider an independent 2-clique in $L(G')$ of size $k+n$. Notice that it is always possible to take the set of vertices in $L(G')$ corresponding to E_3 in G' and two edges in E_1 whose vertices in V_0 are neighbors in G', hence we can suppose that $k \geq 2$. Let S' be the set of all corresponding edges in G'. Suppose first that there is exactly one edge from E_0 in S'. Then, there are at most $n - 2$ edges from E_1 in S', and there are at most 2 edges from E_2 in S', due to the constraints of an independent 2-clique. There cannot be edges from $E_3 \cup E_4$ in S' since they would not be joined to the edge of $E_0 \cap S'$ by any edge. Then, S' contains at most $n + 1$ edges in S', which contradicts $k \geq 2$. Suppose now that there are at least two edges from E_0 in S'. Name two of them $e_{0,1}$ and $e_{0,2}$. Then, there are at most $n - 4$ edges from E_1 in S' but there is no edge from E_2 in S'. Indeed, an edge e_2 from E_2 in S' can be joined by an edge to at most one of $e_{0,1}$ and $e_{0,2}$. Then the size of S' does not exceed n, which contradicts $k \geq 2$. Thus, we can assume that there is no edge from E_0 in S'. Similarly, there is no edge from E_4 in S'. Now, notice that $|S' \cap (E_2 \cup E_3)| \leq n$ since if $S' \cap (E_2 \cup E_3)$ contained $n + 1$ edges then at least two of these edges would have a common endpoint. Consequently, $|S' \cap E_1| \geq k$. Moreover, any two edges from $S' \cap E_1$ must have a common adjacent edge in E_0 since they cannot have a common adjacent edge in E_2. Then, the subgraph of G induced by the set of vertices in V_0 which are the endpoints of the edges in $S' \cap E_1$ must be a clique whose size is at least k.

For the proof of APX-hardness, we prove that the reduction above is an L-reduction (see [14]). We proved in the NP-hardness part that any independent 2-clique in I' has a size at most $2n$. Then $opt(I') \leq 2n = 8 \cdot \frac{n}{4} \leq 8 \cdot opt(I)$ follows since $opt(I) \geq \frac{n}{4}$ in graphs of degree at least $n - 4$. Moreover, starting with a clique of size $opt(I)$, we can construct an independent 2-clique of size $opt(I) + n$ and therefore $opt(I') \geq +n + opt(I)$. Let S' be an independent 2-clique in I' of size at least $n + 2$ (we proved in the NP-hardness part that it always exists and that such a set must be included in $E_1 \cup E_2 \cup E_3$). Let S be the set of vertices in V_0 which are incident with edges in $E_1 \cap S'$. We have $|S'| - |S| \leq n$ which implies $n + |S| \geq |S'|$. Then we obtain $opt(I) - |S| \leq opt(I') - n - |S| = opt(I') - (n+|S|) \leq opt(I') - |S'|$. Since MAX INDEPENDENT SET is APX-hard on the class of graphs of maximum degree 3 [2], MAX CLIQUE is also APX-hard on the class of graphs of minimum degree at least $n - 4$. Thus, MAX INDEPENDENT 2-CLIQUE is APX-hard on line graphs. □

6   Conclusion

Even if MAX INDEPENDENT 2-CLIQUE and MAX INDEPENDENT SET are similar problems, their complexity can be very different depending on the graph class we try to solve the problem in. We showed that MAX INDEPENDENT 2-CLIQUE is NP-hard on apex, dense and everywhere dense, K_4-free, split, bipartite and line graphs while it is polynomial-time solvable on bounded treewidth, planar, bounded degree (and complement of bounded degree), (C_3, C_6)-free, interval graphs and on cographs. Many further types of graphs may be of interest, concerning separation of graph classes in which the problem is NP-hard from the ones where the problem is solvable in polynomial-time.

References

1. Alekseev, V.E.: On the local restrictions effect on the complexity of finding the graph independence number. In: Combinatorial-Algebraic Methods in Applied Mathematics, pp. 3–13 (1983)
2. Alimonti, P., Kann, V.: Hardness of approximating problems on cubic graphs. In: Bongiovanni, G., Bovet, D.P., Battista, G. (eds.) CIAC 1997. LNCS, vol. 1203, pp. 288–298. Springer, Heidelberg (1997). doi:10.1007/3-540-62592-5_80
3. Bodlaender, H.L.: A partial k-arboretum of graphs with bounded treewidth. Theoret. Comput. Sci. **209**, 1–45 (1998)
4. Booth, K.S., Lueker, G.S.: Testing for the consecutive ones property, interval graphs, and graph planarity using PQ-tree algorithms. J. Comput. Syst. Sci. **13**(3), 335–379 (1976)
5. Burlingham, L.S., Corneil, D.G., Lerchs, H.: Complement reducible graphs. Discret. Appl. Math. **3**, 163–174 (1981)
6. Courcelle, B.: The monadic second-order logic of graphs iii: tree-decompositions, minors and complexity issues. RAIRO - Informatique Théorique et Appl. **26**, 257–286 (1992)
7. Frank, A.: Some polynomial algorithms for certain graphs and hypergraphs. Congressus Numerantium No. XV, pp. 3–13 (1976)
8. Garey, M.R., Johnson, D.S., Stockmeyer, L.: Some simplified NP-complete problems. Theoret. Comput. Sci. **1**, 237–267 (1976)
9. Gupta, U.I., Lee, D.T., Leung, J.Y.-T.: Efficient algorithms for interval graphs and circular-arc graphs. Networks **12**(4), 459–467 (1982)
10. Heggernes, P., Kratsch, D.: Linear-time certifying recognition algorithms and forbidden induced subgraphs. Nord. J. Comput. **14**, 87–108 (2007)
11. Khanna, S., Motwani, R., Sudan, M., Vazirani, U.V.: On syntactic versus computational views of approximability. SIAM J. Comput. **28**(1), 164–191 (1998)
12. Luce, R.D.: Connectivity and generalized cliques in sociometric group structure. Psychometrika **15**, 169–190 (1950)
13. Mokken, R.J.: Cliques, clubs and clans. Qual. Quant. **13**(2), 161–173 (1979)
14. Papadimitriou, C.H., Yannakakis, M.: Optimization, approximation, and complexity classes. J. Comput. Syst. Sci. **43**(3), 425–440 (1991)
15. Zuckerman, D.: Linear degree extractors and the inapproximability of max clique and chromatic number. Theor. Comput. **3**(1), 103–128 (2007)

Improved Lower Bounds for Graph Embedding Problems

Hans L. Bodlaender[1,2] and Tom C. van der Zanden[1(✉)]

[1] Department of Computer Science, Utrecht University, Utrecht, The Netherlands
{H.L.Bodlaender,T.C.vanderZanden}@uu.nl
[2] Department of Mathematics and Computer Science,
Eindhoven University of Technology, Eindhoven, The Netherlands

Abstract. In this paper, we give new, tight subexponential lower bounds for a number of graph embedding problems. We introduce two related combinatorial problems, which we call STRING CRAFTING and ORTHOGONAL VECTOR CRAFTING, and show that these cannot be solved in time $2^{o(|s|/\log|s|)}$, unless the Exponential Time Hypothesis fails.

These results are used to obtain simplified hardness results for several graph embedding problems, on more restricted graph classes than previously known: assuming the Exponential Time Hypothesis, there do not exist algorithms that run in $2^{o(n/\log n)}$ time for SUBGRAPH ISOMORPHISM on graphs of pathwidth 1, INDUCED SUBGRAPH ISOMORPHISM on graphs of pathwidth 1, GRAPH MINOR on graphs of pathwidth 1, INDUCED GRAPH MINOR on graphs of pathwidth 1, INTERVALIZING 5-COLORED GRAPHS on trees, and finding a tree or path decomposition with width at most c with a minimum number of bags, for any fixed $c \geq 16$.

$2^{\Theta(n/\log n)}$ appears to be the "correct" running time for many packing and embedding problems on restricted graph classes, and we think STRING CRAFTING and ORTHOGONAL VECTOR CRAFTING form a useful framework for establishing lower bounds of this form.

1 Introduction

Many NP-complete graph problems admit faster algorithms when restricted to planar graphs. In almost all cases, these algorithms have running times that are exponential in a square root function of either the size of the instance n or some parameter k (e.g. $2^{O(\sqrt{n})}$, $n^{O(\sqrt{k})}$ or $2^{O(\sqrt{k})}n^{O(1)}$) and most of these results are tight, assuming the Exponential Time Hypothesis. This seemingly universal behaviour has been dubbed the "Square Root Phenomenon" [1]. The open question [2] of whether the Square Root Phenomenon holds for SUBGRAPH ISOMORPHISM in planar graphs, has recently been answered in the negative: assuming the Exponential Time Hypothesis, there is no $2^{o(n/\log n)}$-time algorithm for SUBGRAPH ISOMORPHISM, even when restricted to (planar) graphs

---

H.L. Bodlaender—The research of this author was partially supported by the NETWORKS project, funded by the Netherlands Organization for Scientific Research NWO.

© Springer International Publishing AG 2017
D. Fotakis et al. (Eds.): CIAC 2017, LNCS 10236, pp. 92–103, 2017.
DOI: 10.1007/978-3-319-57586-5_9

of pathwidth 2 [3]. The same lower bound holds for INDUCED SUBGRAPH and (INDUCED) MINOR and is in fact tight: the problems admit $2^{O(n/\log n)}$-time algorithms on H-minor free graphs [3].

The lower bounds in [3] follow by reductions from a problem called STRING 3-GROUPS. We introduce a new problem, STRING CRAFTING, and establish a $2^{\Omega(|s|/\log|s|)}$-time lower bound under the ETH for this problem by giving a direct reduction from 3-SATISFIABILITY. Using this result, we show that the $2^{\Omega(|s|/\log|s|)}$-time lower bounds for (Induced) Subgraph and (Induced) Minor hold even on graphs of pathwidth 1.

Alongside STRING CRAFTING, we introduce the related ORTHOGONAL VECTOR CRAFTING problem. Using this problem, we show $2^{\Omega(|n|/\log|n|)}$-time lower bounds for deciding whether a 5-coloured tree is the subgraph of an interval graph (for which the same colouring is proper) and for deciding whether a graph admits a tree (or path) decomposition of width 16 with at most a given number of bags.

For any fixed k, INTERVALIZING k-COLOURED GRAPHS can be solved in time $2^{O(n/\log n)}$ [4]. Bodlaender and Nederlof [5] conjecture a lower bound (under the Exponential Time Hypothesis) of $2^{\Omega(n/\log n)}$ time for $k \geq 6$; we settle this conjecture and show that it in fact holds for $k \geq 5$, even when restricted to trees. To complement this result for a fixed number of colours, we also show that there is no algorithm solving INTERVALIZING COLOURED GRAPHS (with an arbitrary number of colours) in time $2^{o(n)}$, even when restricted to trees.

The minimum size path and tree decomposition problems can also be solved in $2^{O(n/\log n)}$ time on graphs of bounded treewidth. This is known to be tight under the Exponential Time Hypothesis for $k \geq 39$ [5]. We improve this to $k \geq 16$; our proof is also simpler than that in [5].

Our results show that STRING CRAFTING and ORTHOGONAL VECTOR CRAFTING are a useful framework for establishing lower bounds of the form $2^{\Omega(n/\log n)}$ under the Exponential Time Hypothesis. It appears that for many packing and embedding problems on restricted graph classes, this bound is tight.

For some omitted proofs, we refer to the full version of this paper [6].

2   Preliminaries

Strings. We work with the alphabet $\{0,1\}$; i.e., strings are elements of $\{0,1\}^*$. The length of a string s is denoted by $|s|$. The ith character of a string s is denoted by $s(i)$. Given a string $s \in \{0,1\}^*$, $\overline{s}$ denotes binary complement of s, that is, each occurrence of a 0 is replaced by a 1 and vice versa; i.e., $|s| = |\overline{s}|$, and for $1 \leq i \leq |s|$, $\overline{s}(i) = 1 - s(i)$. E.g., if $s = 100$, then $\overline{s} = 011$. With s^R, we denote the string s in reverse order; e.g., if $s = 100$, then $s^R = 001$. The concatenation of strings s and t is denoted by $s \cdot t$. A string s is a *palindrome*, when $s = s^R$. By 0^n (resp. 1^n) we denote the string that consists of n 0's (resp. 1's).

Graphs. Given a graph G, we let $V(G)$ denote its vertex set and $E(G)$ its edge set. Let $Nb(v)$ denote the open neighbourhood of v, that is, the vertices adjacent to v, excluding v itself. We assume all graphs are simple.

Treewidth and Pathwidth. A *tree decomposition* of a graph $G = (V, E)$ is a tree T with vertices $t_1, \ldots, t_s$ with for each vertex t_i a *bag* $X_i \subseteq V$ such that for all $v \in V$, the set $\{t_i \in \{t_1, \ldots, t_s\} \mid v \in X_i\}$ is non-empty and induces a connected subtree of T and for all $(u, v) \in \cdot E$ there exists a bag X_i such that $\{u, v\} \in X_i$. The *width* of a tree decomposition is $\max_i\{|X_i| - 1\}$ and the *treewidth* of a graph G is the minimum width of a tree decomposition of G. A *path decomposition* is a tree decomposition where T is a path, and the pathwidth of a graph G is the minimum width of a path decomposition of G.

A graph is a *caterpillar tree* if it is connected and has pathwidth 1.

Subgraphs and Isomorphism. H is a *subgraph* of G if $V(H) \subseteq V(G)$ and $E(H) \subseteq E(G)$; we say the subgraph is *induced* if moreover $E(H) = E(G) \cap \{\{u, v\} \mid u, v \in V(H)\}$. We say a graph H is *isomorphic* to a graph G if there is a bijection $f : V(H) \to V(G)$ so that $(u, v) \in E(H) \iff (f(u), f(v)) \in E(G)$.

Contractions, Minors. We say a graph G' is obtained from G by *contracting* edge (u, v), if G' is obtained from G by replacing vertices u, v with a new vertex w which is made adjacent to all vertices in $Nb(u) \cup Nb(v)$. A graph G' is a minor of G if a graph isomorphic to G' can be obtained from G by contractions and deleting vertices and/or edges. G' is an *induced minor* if we can obtain it by contractions and deleting vertices (but not edges).

We say G' is an *r-shallow minor* if G' can be obtained as a minor of G and any subgraph of G that is contracted to form some vertex of G' has radius at most r (that is, there is a central vertex within distance at most r from any other vertex in the subgraph). Finally, G' is a *topological minor* if we can subdivide the edges of G' to obtain a graph G'' that is isomorphic to a subgraph of G (that is, we may repeatedly take an edge (u, v) and replace it by a new vertex w and edges (u, w) and (w, v)).

For each of (induced) subgraph, induced (minor), topological minor and shallow minor, we define the corresponding decision problem, that is, to decide whether a pattern graph P is isomorphic to an (induced) subgraph/(induced) minor/topological minor/shallow minor of a host graph G.

3   String Crafting and Orthogonal Vector Crafting

We now formally introduce the STRING CRAFTING problem:

STRING CRAFTING
Given: String s, and n strings $t_1, \ldots, t_n$, with $|s| = \sum_{i=1}^{n} |t_i|$.
Question: Is there a permutation $\Pi : \{1, \ldots, n\} \to \{1, \ldots, n\}$, such that the string $t^{\Pi} = t_{\Pi(1)} \cdot t_{\Pi(2)} \cdots t_{\Pi(n)}$ fulfils that for each i, $1 \le i \le |s|$, $s(i) \ge t^{\Pi}(i)$.

i.e., we permute the collection of strings $\{t_1, t_2, \ldots t_n\}$, then concatenate these, and should obtain a resulting string t^{Π} (that necessarily has the same length as s) such that on every position where t^{Π} has a 1, s also has a 1.

We also introduce the following variation of STRING CRAFTING, where, instead of requiring that whenever t^{Π} has a 1, s has a 1 as well, we require that whenever t^{Π} has a 1, s has a 0 (i.e. the strings t^{Π} and s, viewed as vectors over the reals, are orthogonal). These problems are closely related, and have the same complexity. However, sometimes one problem will be more convenient as a starting point for a reduction than the other.

ORTHOGONAL VECTOR CRAFTING

Given: String s, and n strings $t_1, \ldots, t_n$, with $|s| = \sum_{i=1}^{n} |t_i|$.
Question: Is there a permutation $\Pi : \{1, \ldots, n\} \to \{1, \ldots, n\}$, such that the string $t^{\Pi} = t_{\Pi(1)} \cdot t_{\Pi(2)} \cdots t_{\Pi(n)}$ fulfils that for each i, $1 \leq i \leq |s|$, $s(i) \cdot t^{\Pi}(i) = 0$, i.e., when viewed as vectors, s is orthogonal to t^{Π}.

Theorem 1. *Suppose the Exponential Time Hypothesis holds. Then there is no algorithm that solves the* STRING CRAFTING *problem in* $2^{o(|s|/\log|s|)}$ *time, even when all strings t_i are palindromes and start and end with a 1.*

Proof. Suppose we have an instance of 3-SATISFIABILITY with n variables and m clauses. We number the variables x_1 to x_n and for convenience, we number the clauses C_{n+1} to C_{n+m+1}.

We assume by the sparsification lemma that $m = O(n)$ [7, Corollary 2].

Let $q = \lceil \log(n+m) \rceil$, and let $r = 4q + 2$. We first assign an r-bit number to each variable and clause; more precisely, we give a mapping $id : \{1, \ldots, n+m\} \to \{0, 1\}^r$. Let $nb(i)$ be the q-bit binary representation of i. We set, for $1 \leq i \leq n+m$:

$$id(i) = 1 \cdot nb(i) \cdot \overline{nb(i)} \cdot \overline{nb(i)}^R \cdot nb(i)^R \cdot 1$$

Note that each $id(i)$ is an r-bit string that is a palindrome, ending and starting with a 1.

We first build s, by taking the concatenation of n strings, one for each variable.

Suppose the literal x_i appears c_i times in a clause, and the literal $\neg x_i$ appears d_i times in a clause. Set $f_i = c_i + d_i$. Assign the following strings to the pair of literals x_i and $\neg x_i$:

- a^{x_i} is the concatenation of the id's of all clauses in which x_i appears, followed by d_i copies of the string $1 \cdot 0^{r-2} \cdot 1$.
- $a^{\neg x_i}$ is the concatenation of the id's of all clauses in which $\neg x_i$ appears, followed by c_i copies of the string $1 \cdot 0^{r-2} \cdot 1$.
- $b^i = id(i) \cdot a^{x_i} \cdot id(i) \cdot a^{\neg x_i} \cdot id(i)$.

Now, we set $s = b^1 \cdot b^2 \cdots b^{n-1} \cdot b^n$.

We now build the collection of strings t_i. We have three different types of strings:

- *Variable selection:* For each variable x_i we have one string of length $(f_i + 2)r$ of the form $id(i) \cdot 0^{r \cdot f_i} \cdot id(i)$.
- *Clause verification:* For each clause C_i, we have a string of the form $id(i)$.

– *Filler strings:* A filler string is of the form $1 \cdot 0^{r-2} \cdot 1$. We have $n + 2m$ filler strings.

Thus, the collection of strings t_i consists of n variable selection strings (of total length $(3m + 2n)r$), m clause verification strings (of length r), and $n + 2m$ filler strings (also of length r). Notice that each of these strings is a palindrome and ends and starts with a 1.

The idea behind the reduction is that s consists of a list of variable identifiers followed by which clauses a true/false assignment to that variable would satisfy. The variable selection gadget can be placed in s in two ways: either covering all the clauses satisfied by assigning true to the variable, or covering all the clauses satisfied by assigning false to the variable. The clause verification strings then fit into s only if we have not covered all of the places where the clause can fit with variable selection strings (corresponding to that we have made some assignment that satisfies the clause).

Furthermore, note that since $\Sigma_{i=1}^{n} f_i = 3m$, the length of s is $(3n + 6m)r$, the combined length of the variable selection strings is $(2n + 3m)r$, the combined length of the clause verification strings is mr, and the filler strings have combined length $(n + 2m)r$.

In the following, we say a string t_i is mapped to a substring s' of s if s' is the substring of s corresponding to the position (and length) of t_i in t^{Π}.

Lemma 1. *The instance of* 3-SATISFIABILITY *is satisfiable, if and only if the constructed instance of* STRING CRAFTING *has a solution.*

Proof. First, we show the reverse implication. Suppose we have a satisfying assignment to the 3-SATISFIABILITY instance. Consider the substring of s formed by b^i, which is of the form $id(i) \cdot a^{x_i} \cdot id(i) \cdot a^{\neg x_i} \cdot id(i)$. If in the satisfying assignment x_i is true, we choose the permutation Π so that variable selection string $id(i) \cdot 0^{r \cdot f_i} \cdot id(i)$ corresponding to x_i is mapped to the substring $id(i) \cdot a^{\neg x_i} \cdot id(i)$; if x_i is false, we map the variable selection string onto the substring $id(i) \cdot a^{x_i} \cdot id(i)$. A filler string is mapped to the other instance of $id(i)$ in the substring.

Now, we show how the clause verification strings can be mapped. Suppose clause C_j is satisfied by a literal x_i (resp. $\neg x_i$). Since x_i is true (resp. false), the substring a^{x_i} (resp. $a^{\neg x_i}$) of s is not yet used by a variable selection gadget and contains $id(j)$ as a substring, to which we can map the clause verification string corresponding to C_j.

Note that in s now remain a number of strings of the form $1 \cdot 0^{r-2} \cdot 1$ and a number of strings corresponding to id's of clauses, together $2m$ such strings, which is exactly the number of filler strings we have left. These can thus be mapped to these strings, and we obtain a solution to the STRING CRAFTING instance. It is easy to see that with this construction, s has a 1 whenever the string constructed from the permutation does.

Next, for the forward implication, consider a solution Π to the STRING CRAFTING instance. We require the following lemma:

Lemma 2. *Suppose that $t_i = id(j)$. Then the substring w of s corresponding to the position of t_i in t^{Π} is $id(j)$.*

Proof. Because the length of each string is a multiple of r, w is either $id(k)$ for some k, or the string $1 \cdot 0^{r-2} \cdot 1$. Clearly, w can not be $1 \cdot 0^{r-2} \cdot 1$ because the construction of $id(i)$ ensures that it has more than 2 non-zero characters, so at some position w would have a 1 where w' does not. Recall that $id(i) = 1 \cdot nb(i) \cdot \overline{nb(i)} \cdot \overline{nb(i)}^R \cdot nb(i)^R \cdot 1$. If $j \neq k$, then either at some position $nb(k)$ has a 0 where $nb(j)$ has a 1 (contradicting that Π is a solution) or at some position $\overline{nb(k)}$ has a 0 where $\overline{nb(j)}$ has a 1 (again contradicting that Π is a solution). Therefore $j = k$.   □

Clearly, for any i, there are only two possible places in t^Π where the variable selection string $id(i) \cdot 0^{r \cdot f_i} \cdot id(i)$ can be mapped to: either in the place of $id(i) \cdot a^{x_i} \cdot id(i)$ in s or in the place of $id(i) \cdot a^{\neg x_i} \cdot id(i)$, since these are the only (integer multiple of r) positions where $id(i)$ occurs in s. If the former place is used we set x_i to false, otherwise we set x_i to true.

Now, consider a clause C_j, and the place where the corresponding clause verification gadget $id(j)$ is mapped to. Suppose it is mapped to some substring of $id(i) \cdot a^{x_i} \cdot id(i) \cdot a^{\neg x_i} \cdot id(i)$. If $id(j)$ is mapped to a substring of a^{x_i} then (by construction of a^{x_i}) x_i appears as a positive literal in C_j and our chosen assignment satisfies C_j (since we have set x_i to true). Otherwise, if $id(j)$ is mapped to a substring of $a^{\neg x_i}$ x_i appears negated in C_j and our chosen assignment satisfies C_j (since we have set x_i to false).

We thus obtain a satisfying assignment for the 3-SATISFIABILITY instance.

□

Since in the constructed instance, $|s| = (3n + 6m)r$ and $r = O(\log n), m = O(n)$, we have that $|s| = O(n \log n)$. A $2^{o(|s|/\log |s|)}$-time algorithm for STRING CRAFTING would give a $2^{o(n \log n / \log (n \log n))} = 2^{o(n)}$-time algorithm for deciding 3-SATISFIABILITY, violating the ETH.   □

Note that we can also restrict all strings t_i to start and end with a 0 by a slight modification of the proof.

Theorem 2. *Assuming the Exponential Time Hypothesis,* ORTHOGONAL VECTOR CRAFTING *can not be solved in* $2^{o(|s|/\log |s|)}$ *time, even when all strings* t_i *are palindromes and start and end with a 1.*

Proof. This follows from the result for STRING CRAFTING, by taking the complement of the string s.   □

Again, we can also restrict all strings t_i to start and end with a 0.

As illustrated by the following theorem, these lower bounds are tight. The algorithm is a simpler example of the techniques used in [3–5].

Theorem 3. *There exists algorithms, solving* STRING CRAFTING *and* ORTHOGONAL VECTOR CRAFTING *in* $2^{O(|s|/\log |s|)}$.

4   Lower Bounds for Graph Embedding Problems

Theorem 4. *Suppose the Exponential Time Hypothesis holds. Then there is no algorithm solving* SUBGRAPH ISOMORPHISM *in* $2^{o(n/\log n)}$ *time, even if G is a caterpillar tree of maximum degree 3 or G is connected, planar, has pathwidth 2 and has only one vertex of degree greater than 3 and P is a tree.*

Proof. By reduction from STRING CRAFTING. We first give the proof for the case that G is a caterpillar tree of maximum degree 3, We construct G from s as follows: we take a path of vertices $v_1, \ldots, v_{|s|}$ (*path vertices*). If $s(i) = 1$, we add a *hair vertex* h_i and edge (v_i, h_i) to G (obtaining a caterpillar tree). We construct P from the strings t_i by, for each string t_i repeating this construction, and taking the disjoint union of the caterpillars created in this way (resulting in a graph that is a forest of caterpillar trees, i.e., a graph of pathwidth 1). An example of this construction is depicted in Fig. 1. The constructed instance of G contains P as a subgraph, if and only if the instance of STRING CRAFTING has a solution: the order in which the caterpillars are embedded in G gives the permutation of the strings: when a caterpillar in P has a hair, G must have a hair at the specific position, which implies that a position with a 1 in the constructed string t must be a position where s also has a 1.

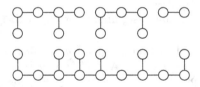

Fig. 1. Simplified example of the graphs created in the hardness reduction for Theorem 4. The bottom caterpillar represents the host graph (corresponding to string s), the top caterpillars represent the strings t_i and form the guest graph. Here $s = 101110101$ and $t_1 = 1010, t_2 = 101$ and $t_3 = 00$.

Since the constructed instance has $O(|s|)$ vertices, this establishes the first part of the lemma. For the case that G is connected, we add to the graph G constructed in the first part of the proof a vertex u and, for each path vertex v_i, an edge (v_i, u). To P we add a vertex u' that has an edge to some path vertex of each component. By virtue of their high degrees, u must be mapped to u' and the remainder of the reduction proceeds in the same way as in the first part of the proof. $\qquad\Box$

This proof can be adapted to show hardness for a number of other problems:

Theorem 5. *Suppose the Exponential Time Hypothesis holds. Then there is no algorithm solving* INDUCED SUBGRAPH, (INDUCED) MINOR, SHALLOR MINOR *or* TOPOLOGICAL MINOR *in* $2^{o(n/\log n)}$ *time, even if G is a caterpillar tree of maximum degree 3 or G is connected, planar, has pathwidth 2 and has only one vertex of degree greater than 3 and P is a tree.*

5 Tree and Path Decompositions with Few Bags

In this section, we study the following problem, and its analogously defined variant Minimum Size Path Decomposition (k-MSPD). Theorem 6 is an improvement over Theorem 3 of [5], where the same was shown for $k \geq 39$; our proof is also simpler.

MINIMUM SIZE TREE DECOMPOSITION OF WIDTH k (k-MSTD)
Given: A graph G, integer b.
Question: Does G have a tree decomposition of width at most k, that has at most b bags?

Theorem 6. *Let* $k \geq 16$. *Suppose the Exponential Time Hypothesis holds, then there is no algorithm for k-MSPD or k-MSTD using $2^{o(n/\log n)}$ time.*

Proof. By reduction from ORTHOGONAL VECTOR CRAFTING. We begin by showing the case for MSPD, but note the same reduction is used for MSTD.

For the string s, we create a connected component in the graph G as follows: for $1 \leq i \leq |s| + 1$ we create a clique C_i of size 6, and (for $1 \leq i \leq |s|$) make all vertices of C_i adjacent to all vertices of C_{i+1}. For $1 \leq i \leq |s|$, if $s(i) = 1$, we create a vertex s_i and make it adjacent to the vertices of C_i and C_{i+1}.

For each string t_i, we create a component in the same way as for s, but rather than using cliques of size 6, we use cliques of size 2: for each $1 \leq i \leq n$ and $1 \leq j \leq |t_i| + 1$ create a clique $T_{i,j}$ of size 2 and (for $1 \leq j \leq |t_i|$) make all vertices of $T_{i,j}$ adjacent to all vertices of $T_{i,j+1}$. For $1 \leq j \leq |t_i|$, if $t_i(j) = 1$, create a vertex $t_{i,j}$ and make it adjacent to the vertices of $T_{i,j}$ and $T_{i,j+1}$.

An example of the construction (for $s = 10110$ and $t_1 = 01001$) is shown in Fig. 2. We now ask whether a path decomposition of width 16 exists with at most $|s|$ bags. Due to space constraints, we omit the correctness proof of the construction.   □

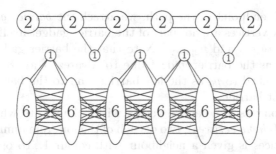

Fig. 2. Simplified example of the graph created in the hardness reduction for Theorem 6. The circles and ellipses represent cliques of various sizes. The component depicted in the top of the picture corresponds to $t_1 = 01001$, while the component at the bottom corresponds to $s = 10110$.

6   Intervalizing Coloured Graphs

In this section, we consider the following problem:

INTERVALIZING COLOURED GRAPHS
Given: A graph $G = (V, E)$ together with a proper colouring $c : V \rightarrow \{1, 2, \ldots, k\}$.
Question: Is there an interval graph G' on the vertex set V, for which c is a proper colouring, and which is a supergraph of G?

INTERVALIZING COLOURED GRAPHS is known to be NP-complete, even for 4-coloured caterpillars (with hairs of unbounded length) [8]. In contrast with this result we require five colours instead of four, and the result only holds for trees instead of caterpillars. However, we obtain a $2^{\Omega(n/\log n)}$ lower bound under the Exponential Time Hypothesis, whereas the reduction in [8] is from MULTI-PROCESSOR SCHEDULING and to our knowledge, the best lower bound obtained from it is $2^{\Omega(\sqrt[5]{n})}$ (the reduction is weakly polynomial in the length of the jobs, which is $\Theta(n^4)$, as following from the reduction from 3-PARTITION in [9]). In contrast to these hardness results, for the case with 3 colours there is an $O(n^2)$ time algorithm [10,11].

Theorem 7. INTERVALIZING COLOURED GRAPHS *does not admit a* $2^{o(n/\log n)}$- *time algorithm, even for 5-coloured trees, unless the Exponential Time Hypothesis fails.*

Proof. Let $s, t_1, \ldots, t_n$ be an instance of ORTHOGONAL VECTOR CRAFTING. We construct $G = (V, E)$ in the following way:

S-String Path. We create a path of length $2|s| - 1$ with vertices $p_0, \ldots p_{2|s|-2}$, and set $c(p_i) = 1$ if i is even and $c(p_i) = 2$ if i is odd. Furthermore, for even $0 \leq i \leq 2|s| - 2$, we create a neighbour n_i with $c(n_i) = 3$.

Barriers. To each endpoint of the path, we attach the *barrier gadget*, depicted in Fig. 3. The gray vertices are not part of the barrier gadget itself, and represent p_0 and n_0 (resp. $p_{2|s|-2}$ and $n_{2|s|-2}$). Note that the barrier gadget operates on similar principles as the barrier gadget due to Alvarez et al. [8]. We shall refer to the barrier attached to p_0 as the left barrier, and to the barrier attached to $p_{2|s|-2}$ as the right barrier.

   The barrier consists of a central vertex with colour 1, to which we connect eight neighbours (*clique vertices*), two of each of the four remaining colours. Each of the clique vertices is given a neighbour with colour 1. To one of the clique vertices with colour 2 we connect a vertex with colour 3, to which a vertex with colour 2 is connected (*blocking vertices*). This clique vertex shall be the barrier's *endpoint*. Note that the neighbour with colour 1 of this vertex is not considered part of the barrier gadget, as it is instead a path vertex. We let C_l (e_l) denote the center (endpoint) of the left barrier, and C_r (e_r) the center (endpoint) of the right barrier.

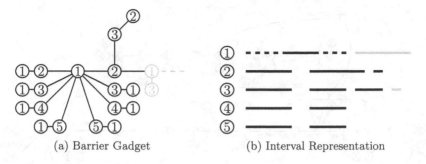

(a) Barrier Gadget   (b) Interval Representation

Fig. 3. (a) Barrier Gadget. The gray vertices are not part of the barrier gadget itself, and show how it connects to the rest of the graph. (b) How the barrier gadget may (must) be intervalized.

T-String Paths. Now, for each string t_i, we create a path of length $2|t_i| + 1$ with vertices $q_{i,0}, \ldots, q_{i,2|t_i|}$ and set $c(q_{i,j}) = 3$ if j is odd and set $c(q_{i,j}) = 2$ if j is even. We make $q_{i,1}$ adjacent to U. Furthermore, for odd $1 \leq j \leq 2|t_i| - 1$, we create a neighbour m_j with $c(m_j) = 1$. We also create two *endpoint vertices* of colour 3, one of which is adjacent to $q_{i,0}$ and the other to $q_{i,2|t_i|}$,

Connector Vertex. Next, we create a *connector vertex* of colour 5, which is made adjacent to p_1 and to $q_{i,1}$ for all $1 \leq i \leq n$. This vertex serves to make the entire graph connected.

Marking Vertices. Finally, for each $1 \leq i \leq |s|$ (resp. for each $1 \leq i \leq n$ and $1 \leq j \leq |t_i|$), if $s(i) = 1$ (resp. $t_i(j) = 1$), we give p_{2i-1} (resp. $q_{i,2j-1}$) two neighbours (called the *marking vertices*) with colour 4. For each of the marking vertices, we create a neighbour with colour 3.

This construction is depicted in Fig. 4. In this example $s = 10100$, $t_1 = 01$ and $t_2 = 001$. Note that this instance of ORTHOGONAL VECTOR CRAFTING is illustrative, and does not satisfy the restrictions required in the proof.

Informally, the construction works as follows: the barriers at the end of the path of p-vertices can not be passed by the remaining vertices, meaning we have to "weave" the shorter q-paths into the long p-path. The colours enforce that the paths are in "lockstep", that is, we have to traverse them at the same speed. We have to map every q-vertex with colour 3 to a p-vertex with colour 2, but the marking vertices prevent us from doing so if both bitstrings have a 1 at that particular position.

Due to space constraints, we omit the correctness proof of the construction.

The number of vertices of G is linear in $|s|$, and we thus obtain a $2^{o(n/\log n)}$ lower bound under the Exponential Time Hypothesis.   □

Note that the graph created in this reduction only has one vertex of super-constant degree. This is tight, since the problem is polynomial-time solvable for bounded degree graphs (for any fixed number of colours) [12].

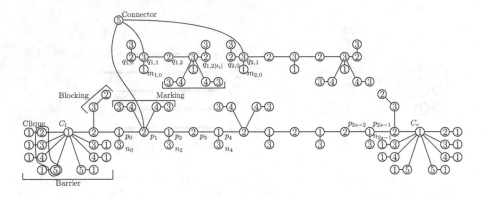

Fig. 4. Example of the graph created in the hardness reduction for Theorem 7.

To complement this result for a bounded number of colours, we also show a $2^{\Omega(n)}$-time lower bound for graphs with an unbounded number of colors, assuming the ETH. Note that this result implies that the algorithm from [4] is optimal.

Theorem 8. *Assuming the Exponential Time Hypothesis, there is no algorithm solving* INTERVALIZING COLOURED GRAPHS *in time* $2^{o(n)}$*, even when restricted to trees.*

7 Conclusions

In this paper, we have shown for several problems that, under the Exponential Time Hypothesis, $2^{\Theta(n/\log n)}$ is the best achievable running time - even when the instances are very restricted (for example in terms of pathwidth or planarity). For each of these problems, algorithms that match this lower bound are known and thus $2^{\Theta(n/\log n)}$ is (likely) the asymptotically optimal running time.

For problems where planarity or bounded treewidth of the instances (or, through bidimensionality, of the solutions) can be exploited, the optimal running time is often $2^{\Theta(\sqrt{n})}$ (or features the square root in some other way). On the other hand, each of problems studied in this paper exhibits some kind of "packing" or "embedding" behaviour. For such problems, $2^{\Theta(n/\log n)}$ is often the optimal running time. We have introduced two artificial problems, STRING CRAFTING and ORTHOGONAL VECTOR CRAFTING, that form a useful framework for proving such lower bounds.

It would be interesting to study which other problems exhibit such behaviour, or to find yet other types of running times that are "natural" under the Exponential Time Hypothesis. The loss of the $\log n$-factor in the exponent is due to the fact that $\log n$ bits or vertices are needed to "encode" n distinct elements; it would be interesting to see if there are any problems or graph classes where a more compact encoding is possible (for instance only $\log^{1-\epsilon} n$ vertices required, leading to a tighter lower bound) or where an encoding is less compact (for instance $\log^2 n$ vertices required, leading to a weaker lower bound) and whether this can be exploited algorithmically.

Acknowledgement. We thank Jesper Nederlof for helpful comments and discussions.

References

1. Marx, D.: The square root phenomenon in planar graphs. In: Fellows, M., Tan, X., Zhu, B. (eds.) AAIM/FAW -2013. LNCS, vol. 7924, p. 1. Springer, Heidelberg (2013). doi:10.1007/978-3-642-38756-2_1
2. Marx, D.: What's next? Future directions in parameterized complexity. In: Bodlaender, H.L., Downey, R., Fomin, F.V., Marx, D. (eds.) The Multivariate Algorithmic Revolution and Beyond. LNCS, vol. 7370, pp. 469–496. Springer, Heidelberg (2012). doi:10.1007/978-3-642-30891-8_20
3. Bodlaender, H.L., Nederlof, J., van der Zanden, T.C.: Subexponential time algorithms for embedding H-minor free graphs. In: Chatzigiannakis, I., Mitzenmacher, M., Rabani, Y., Sangiorgi, D. (eds.) 43rd International Colloquium on Automata, Languages, and Programming (ICALP 2016), vol. 55, Leibniz International Proceedings in Informatics (LIPIcs), Dagstuhl, Germany, Schloss Dagstuhl-Leibniz-Zentrum fuer Informatik, pp. 9: 1–9: 14 (2016)
4. Bodlaender, H.L., van Rooij, J.M.M.: Exact algorithms for Intervalizing Coloured Graphs. Theor. Comput. Syst. **58**(2), 273–286 (2016)
5. Bodlaender, H.L., Nederlof, J.: Subexponential time algorithms for finding small tree and path decompositions. In: Bansal, N., Finocchi, I. (eds.) ESA 2015. LNCS, vol. 9294, pp. 179–190. Springer, Heidelberg (2015). doi:10.1007/978-3-662-48350-3_16
6. Bodlaender, H.L., van der Zanden, T.C.: Improved lower bounds for graph embedding problems. arXiv preprint (2016). arXiv:1610.09130
7. Impagliazzo, R., Paturi, R., Zane, F.: Which problems have strongly exponential complexity? J. Comput. Syst. Sci. **63**(4), 512–530 (2001)
8. Àlvarez, C., Diáz, J., Serna, M.: The hardness of intervalizing four colored caterpillars. Discret. Math. **235**(1), 19–27 (2001)
9. Jansen, K., Land, F., Land, K.: Bounding the running time of algorithms for scheduling and packing problems. In: Dehne, F., Solis-Oba, R., Sack, J.-R. (eds.) WADS 2013. LNCS, vol. 8037, pp. 439–450. Springer, Heidelberg (2013). doi:10.1007/978-3-642-40104-6_38
10. Bodlaender, H.L., de Fluiter, B.: Intervalizing k-colored graphs. Technical report UU-CS-1995-15, Department of Information and Computing Sciences, Utrecht University (1995)
11. Bodlaender, H.L., de Fluiter, B.: On intervalizing k-colored graphs for DNA physical mapping. Discret. Appl. Math. **71**(1), 55–77 (1996)
12. Kaplan, H., Shamir, R.: Bounded degree interval sandwich problems. Algorithmica **24**(2), 96–104 (1999)

Collaboration Without Communication: Evacuating Two Robots from a Disk

Sebastian Brandt[1]([✉]), Felix Laufenberg[1], Yuezhou Lv[2], David Stolz[1], and Roger Wattenhofer[1]

[1] ETH Zürich, Zürich, Switzerland
{brandts,stolzda,wattenhofer}@ethz.ch, felix.laufenberg@gmx.de
[2] Tsinghua University, Beijing, China
lvyuezhou@gmail.com

Abstract. We consider the problem of evacuating two robots from a bounded area, through an unknown exit located on the boundary. Initially, the robots are in the center of the area and throughout the evacuation process they can only communicate with each other when they are at the same point at the same time. Having a visibility range of 0, the robots can only identify the location of the exit if they are already at the exit position. The task is to minimize the time it takes until both robots reach the exit, for a worst-case placement of the exit. For unit disks, an upper bound of 5.628 for the evacuation time is presented in [8]. Using the insight that, perhaps surprisingly, a forced meeting of the two robots as performed in the respective algorithm does not provide an exchange of any non-trivial information, we design a simpler algorithm that achieves an upper bound of 5.625. Our numerical simulations suggest that this bound is optimal for the considered natural class of algorithms. For dealing with the technical difficulties in analyzing the algorithm, we formulate a powerful new criterion that, for a given algorithm, reduces the number of possible worst-case exits radically. This criterion is of independent interest and can be applied to any area shape. Due to space restrictions, this version of the paper contains no proofs or illustrating figures; the full version can be found at http://disco.ethz.ch/publications/ciac2017-robotevac.pdf.

1 Introduction

Imagine that two robots are trapped in the middle of a room with a single door. Their goal is to evacuate both of them via the door in the shortest possible time. However, there is a problem: The position of the door is unknown to them in the beginning. Moreover, the robots have no sight and no wireless communication; they cannot see the door or the other robot except when they are right on top of it, and they can only communicate when they are at the same point at the same time. However, they do know the shape of the room and share all information before they start searching for the door. How should they divide the work of scouring the boundary for the door? Which routes should they take? Does it make sense to meet at some predefined point in order to exchange information?

© Springer International Publishing AG 2017
D. Fotakis et al. (Eds.): CIAC 2017, LNCS 10236, pp. 104–115, 2017.
DOI: 10.1007/978-3-319-57586-5_10

We consider the above problem for the most fundamental room shape possible, namely, the unit disk. More formally, the goal is to minimize the time for evacuating both robots from the room where the exit is assumed to be worst-case placed and the robots move with unit speed. The state-of-the-art algorithm for this problem due to Czyzowicz et al. [8] proceeds roughly as follows: First, both robots move to the same point on the perimeter, then they search the perimeter in different directions until they meet at the opposite point. At some predefined point in time during their search along the perimeter they leave the perimeter on symmetric non-linear routes in order to meet inside the circle upon which they return to their search. If a robot finds the exit at any point in time, then it immediately calculates the shortest route for meeting the other robot and subsequently brings the other robot to the exit. The authors of [8] show that this algorithm achieves an evacuation time of 5.628 and remark that it is possible to improve upon this result by truncating the detour to the middle slightly.

We note that the forced meeting of the robots cannot be used to exchange any non-trivial information. Moreover, the requirement to meet introduces dependencies between the parameters of the detour such as position, length, shape and angle. We prove that removing this requirement indeed allows for an improved algorithm that utilizes the independence of the aforementioned parameters while preserving the simplicity of the algorithm. In fact, we present an algorithm that simplifies the algorithm described above by omitting the forced meeting and instead using one (symmetric) detour (per robot) that is a straight line with fixed depth. An important point in omitting the forced meeting is that there is actually an implicit exchange of information between the robots even before any meeting: When a robot finds the exit, the best it can do is to meet the other robot as quickly as possible in order to communicate the location of the exit. Conversely, from not being visited by the other robot up to some point a robot can deduce that the exit does not lie in a certain part of the perimeter.

Furthermore, we show that, surprisingly, the shape of the detour and its angle to the perimeter do not affect the evacuation time if they are chosen from some reasonable range. In particular, our linear detour is optimal for the parameters chosen for the depth and the position of the detour.

Our algorithm achieves an evacuation time of 5.625, thereby slightly improving upon the previously best known upper bound. For the class of algorithms as described above with exactly one symmetric detour per robot, our numerical simulations suggest that this bound is optimal (up to numerical precision, of course). A theoretical substantiation for this optimality claim is given by the fact that for our algorithm there are three different worst-case exit placements with the same evacuation time (again, up to numerical precision). These three exit positions are characteristic for the algorithms from the mentioned class—it stands to reason that, in an optimal algorithm, they have to exhibit the same evacuation time (since otherwise the parameters of the algorithm could be changed locally in a way that improves the evacuation time for the worst of the three points) and that there is only one algorithm that has the same worst-case evacuation time for these three exit positions.

A fundamental problem regarding evacuation from a disk is that the evacuation time for a fixed algorithm and a fixed exit placement is usually the solution to some equation of the kind $x = \cos(x)$ (only more complex). For equations of this kind no closed-form solutions are known in general, which makes it difficult to find the worst-case exit placement for a fixed algorithm, not to mention to find an algorithm with an improved worst-case evacuation time. We take a substantial step in remedying this problem by proving that a very specific condition must be satisfied for an exit in order to be worst-case placed. In "reasonable" algorithms this condition is satisfied at only a few exit positions which makes it a powerful tool for determining that an exit is not worst-case placed. In fact, in order for an exit to be worst-case placed, it must satisfy one of the two following conditions: (1) The movement of one of the two robots at the exit point, resp. pick-up point[1], is not differentiable, or (2) the angles β and γ between the line connecting exit and pick-up point and the directions of movement of the robots at the exit, resp. the pick-up point, satisfy $2\cos\beta + \cos\gamma = 1$. Moreover, the presented tool is not restricted to the disk—it can be applied to any room shape. For the analysis of our aforementioned algorithm, we rely heavily on this tool. In fact, one might consider the development of this tool as the foremost contribution of this work, while its application to the disk scenario may serve as an example of its practicability.

1.1 Related Work

A thriving area in the context of problems involving mobile agents are search problems in all its diversity. Such problems include ants searching for food (cf. [11–13]), rendezvous problems (cf. [1,10,17]), pursuit-evasion games (cf. [14,20,21]) and graph exploration problems (cf. [15,16,19]). Another example are evacuation problems, where one or multiple robot(s) search for one or multiple exit(s) through which usually all of the robots have to evacuate. Evacuation problems have been studied in a centralized setting in which the robots know the search terrain and where the other robots are, and in a distributed setting where the knowledge of the robots is restricted to the area they have already explored. Very recent results concerning optimal strategies on graphs in both settings can be found in [4]. In the following, we assume that the area is known to the robots, the exit is worst-case placed and the robots move with unit speed.

Evacuation problems can be grouped into two main categories, namely, evacuation problems on graphs and geometric evacuation problems. Since our paper deals with a problem from the latter class, we will focus on the related work in this domain. Another distinction is given by the model of communication between the robots: Here, we distinguish between instantaneous wireless communication and non-wireless communication where explicit communication can only take place when the communicating entities are at the same point.

---

[1] Recall that upon finding the exit, a robot immediately takes the shortest possible tour to meet the other robot and communicate the location of the exit. We call the point where this meeting happens the *pick-up point*.

In the geometric setting, research has considered different areas from which the robots have to escape. The famous cow-path problem asks how long it takes a single robot (or cow) to evacuate through a worst-case placed exit on a line, in terms of the distance d between initial position and exit. The correct answer of $9d$ (up to lower order terms) was given by Beck and Newman [3] already in 1970, and later rediscovered by Baeza-Yates et al. [2]. In [5], Chrobak et al. show that, somewhat surprisingly, the same is true for the evacuation time of multiple robots on the line in the non-wireless communication model.

For the case of two robots in equilateral triangles and squares, Czyzowicz et al. [9] present optimal evacuation trajectories, given wireless communication.

Study of the unit disk as the confining environment was initiated by Czyzowicz et al. in [7]. The authors present upper bounds of $3 + 2\pi/k$ and $3 + \pi/k + O(k^{-4/3})$ and lower bounds of $3 + 2\pi/k - O(k^{-2})$ and $3 + \pi/k$ for the non-wireless and the wireless communication model, respectively, where k is the number of robots. Moreover, they give better upper and lower bounds for the case of 2 and 3 robots, amongst them a lower bound of approximately 5.199 and an upper bound of approximately 5.74 for the case of two robots in the non-wireless model. In [8], Czyzowicz et al. improve the latter two bounds to a lower bound of approximately 5.255 and an upper bound of 5.628. Lamprou et al. [18] present (partly matching) upper and lower bounds for two robots in the wireless model where one robot, deviating from the above, has speed of larger than 1. Finally, in [6], Czyzowicz et al. consider variations of the problem of evacuating from a disk where the two robots do not know their own initial locations.

For our paper, the algorithms from [7] and [8] that achieve the upper bounds for two robots in the non-wireless communication model are of particular interest. The algorithm from [7] proceeds as follows: Starting in the center M of the disk, both robots move to the same point A on the perimeter and start searching for the exit in opposite directions. When one of the robots finds the exit, it picks the other robot up as fast as possible and returns with it to the exit. This results in an upper bound of approximately 5.74. The authors of [8] improve on this algorithm by incorporating a *forced meeting* of the two robots before all of the perimeter is searched. For this, the robots leave the perimeter in a straight line, symmetric to each other, until they meet, and then return to their search of the perimeter if the exit has not been found yet. The authors were able to improve on this algorithm even more by moving towards the meeting point in a triangular fashion. The reasons for this further improvement are explained in Sect. 3.1.

1.2 Model

The specifics of the model for our robot evacuation problem, developed in [7], are as follows: The area from which the robots have to escape is a disk of radius 1. Somewhere on its perimeter, there is a point, called *exit*, which two robots, initially placed in the center of the disk, have to find and evacuate through. The task of the robots is to minimize the time until both robots have reached the exit, which we call the *evacuation time*. We assume that the location of the exit on the perimeter is worst-case for the algorithm the two robots perform, i.e., the exit position maximizes the evacuation time. The robots itself are point-shaped and move at unit speed. Changing direction takes no time and communication is also instantaneous, but only possible if both robots are at the same point.[2] Since this is the case in the beginning, the robots can exchange all information about each others algorithm before they start searching for the exit. The robots have no vision, and therefore can only identify the exit position when they are at the exact location of the exit. Computation also takes no time and we assume that the robots are able to actually perform all necessary computations. The robots know the shape of the area and they have the same sense of direction, i.e., we may assume that they have the same underlying coordinate system.

1.3   Notation

In the following, we give an overview of the notation and the most important terms we use.

R_1 **and** R_2 References for the two robots. We will call the robot that finds the exit first R_1, and the other robot R_2.

$\widehat{AB}$ For two points A and B on the perimeter, $\widehat{AB}$ denotes the shorter arc from A to B along the perimeter.

AB Denotes the straight line between A and B.

$|\widehat{AB}|$ **or** $|AB|$ Denote the lengths of $\widehat{AB}$, resp. AB.

Cut. A movement of a robot from the perimeter onto the disk and back to the point where the perimeter was left. Note that a cut can take any shape in general. However, in our algorithm, the term *cut* describes a linear cut, i.e., a movement from the perimeter onto the disk and back on a straight line.

Cut length. The distance traveled when moving along a cut.

Cut depth. Only used if the cut is linear, in which case the cut depth is defined as half of the cut length.

Cut position. Point where a robot leaves the perimeter to perform a cut.

Meeting protocol. A term coined in [8]. When R_1 finds the exit, the best it can do to minimize the evacuation time is to compute (and take) the shortest route to meet R_2. Since R_1 knows the algorithm R_2 follows (and therefore also that R_2 has not found the exit so far), R_1 can actually determine this shortest route.[3] Note that this route is always a straight line since otherwise there would be a shorter route, by the triangle inequality. After meeting each other, both robots travel straight to the exit. This process of picking the other

---

[2] Note that a robot can also infer information from the fact that the other robot is not at the same point as it is. For instance, it may conclude that the other robot has not already found the exit in some specific segment of the perimeter, since otherwise the other robot would have picked him up at the latest at the current position. This indirect information transfer plays an important role in our arguments that the robots cannot infer any non-trivial information from a forced meeting.

[3] We emphasize that R_1 does not calculate a shortest route to the point where R_2 is when R_1 finds the exit, but rather the shortest route for picking R_2 up, knowing that and how R_2 will move until being picked up.

robot up after finding the exit and traveling to the exit together is called the *meeting protocol.* If R_1 finds the exit at time t and the aforementioned shortest route has length x, then the evacuation time is $t + 2x$.

Pick-up point. The point where R_1 picks R_2 up, following the meeting protocol.

2   Determining the Worst-Case Placement of the Exit

If an algorithm for the two robots is fixed, it is still a challenging task to determine the worst-case exit placement and thereby the evacuation time. One reason is that already determining the pick-up point for a fixed exit placement often involves solving equations where polynomial and trigonometric functions in some variable x occur side by side. For equations of this kind no closed-form solutions are known in general. In this section, we develop a new technique to determine possible candidates for the worst-case placement of the exit. More precisely, we give a criterion that determines for a pair (exit, pick-up point) whether the exit can be excluded from the list of candidates of worst-case placed exits, by only looking at the behaviour of the algorithm in ε-neighborhoods of the exit B and the pick-up point C. The criterion is quite strong in the following sense: Let β denote the angle between the straight line from exit to pick-up point and the direction of the movement of R_1 at the exit. Let γ denote the angle between the straight line from exit to pick-up point and the direction of the movement of R_2 at the pick-up point. Then, a very specific relation between β and γ has to be satisfied in order that the exit is not excluded from the list of possible worst-case placed exits. The key result is the following theorem:

Theorem 1. *If the trajectories of the two robots are differentiable around B and C and $2\cos\beta + \cos\gamma \neq 1$, then there is an exit position that yields a larger evacuation time than placing the exit at B.*

The proof of the theorem is long and involved and can be found in the full version of the paper. According to the theorem, in order to be able to exclude an exit, the movement of the two robots at the exit and the corresponding pick-up point have to be differentiable. However, in reasonable algorithms, this property holds for almost all possible exit points. Out of these exit points, the only ones that are not excluded are those that satisfy $2\cos\beta + \cos\gamma = 1$, yielding a large reduction in the number of possible exit points. We note that our considerations are independent of the shape of the area, i.e., they hold for arbitrarily shaped areas, allowing us, for instance, to finally tackle fundamental shapes like circles.

Another interesting information can be obtained from the proof of Theorem 1: On which side of 1 the term $2\cos\beta + \cos\gamma$ lies, determines to which direction one has to move the exit in order to obtain an exit with a larger evacuation time. If $2\cos\beta + \cos\gamma < 1$, then shifting the exit at B in the direction of the movement of R_1 (if it did not find the exit at B) will provide an exit position with larger evacuation time (than the exit position at B). If $2\cos\beta + \cos\gamma > 1$, then shifting the exit at B in the reverse direction will provide an exit position with larger evacuation time. We note that it does not matter if the two robots move to the same side of the infinite line through B and C or to the same side.

3   Evacuating from a Disk

In this section we use the criterion $2\cos\beta + \cos\gamma \neq 1$, which we developed in Sect. 2, in order to improve the upper bound for evacuating two robots from a unit disk to 5.625. Like the algorithm presented in [8], which achieves the previously best known upper bound of 5.628, our algorithm consists of each robot exploring its assigned half of the perimeter, only interrupted by exactly one detour each, called cut, to the inside of the circle (and symmetric to the other's cut). In contrast, the algorithm from [8] additionally contains a forced meeting of the two robots at the far end of the cuts.

We will show that, perhaps counterintuitively, the robots cannot infer any non-trivial information from this meeting that they could not have inferred from the previous course of events. Thus, such a meeting can be omitted. We will see in more detail that all the advantages of the meeting come from the actual movement of cutting to the middle and not from an explicit exchange of information.

Without the condition that the two robots actually have to meet at the endpoint of their (symmetric) cuts, many cuts are possible candidates for improving the runtime of the evacuation algorithm. A cut is determined by four properties: the position on the perimeter where the cut starts and ends, the shape of the cut, the angle at which the cut protrudes from the perimeter and the size of the cut, which corresponds to the cut depth in the case of a linear cut. As we will show, somewhat surprisingly, the shape of the cut is optimal if it is linear, for the choices of the other three parameters made in our algorithm. Similarly, we will show that the angle does not influence the performance of the algorithm if it is chosen in a reasonable range. We will provide a choice of the remaining two parameters for linear cuts that achieves the stated bound of 5.625. Moreover, we give a rigorous proof for the evacuation time.

3.1   The Algorithm $A(y, \alpha, d)$

In this section, we describe a parameterized evacuation algorithm and provide a partitioning of R_1's half of the perimeter into segments that will be useful in the evacuation time analysis. We show that the forced meeting in the previously best algorithm from [8] does not help in exchanging non-trivial information between the robots. In Sect. 3.2, we prove that the parameters can actually be chosen in a way that improves the previously best known upper bound.

Our parameterized algorithm $A(y, \alpha, d)$ is similar to the algorithm proposed in [8]: From the center of the disk, both robots move to the same point A of the perimeter and continue on the perimeter in opposite directions. At some point C, resp. B, where $\widehat{AC} = \widehat{AB} = y$, the robots leave the perimeter at angle α in a straight line, until they reach depth d and then return straight to point C, resp. B. Then both robots continue to search along the perimeter until they meet at point D. If a robot finds the exit at any point in time, it immediately performs the meeting protocol to pick the other robot up and evacuate through the exit. Note that, for convenience, α denotes the angle between the cuts and BC.

Now we examine Algorithm $A(y, \alpha, d)$ in more detail, laying the foundation for the analysis of the evacuation time for a specific choice of the parameters y, α and d in Sect. 3.2. Since the described algorithm is symmetric, it is sufficient to analyze possible exit positions on one side of the symmetry axis, i.e., for one of the two robots. Without loss of generality, we assume that the exit lies on the arc from A to D that contains C, which implies that the robot that explores this arc is called R_1 and the other one R_2. We partition this arc into four segments by specifying the points on the arc where one segment ends and the next one begins. Note that, for simplicity, we include any of these dividing points in both its adjacent segments if not explicitly specified otherwise. The choice of the segments depends on the parameters of our algorithm.

1. **Segment $\widehat{AI_1}$:** Here, I_1 is the point with the following property: If the exit is at I_1, then R_1 will pick R_2 up at point B before R_2 performs its cut, i.e., I_1 satisfies $|\widehat{AI_1}| + |I_1B| = |AB|$. This segment contains exactly those exit positions for which the evacuation time is not influenced by the cut.
2. **Segment $\widehat{I_1I_2}$:** Here, I_2 is the point with the following property: If the exit is at I_2, then R_1 will pick R_2 up at point B after R_2 performs its cut, i.e., I_2 satisfies $|\widehat{AI_2}| + |I_2B| = |AB| + 2d$. This segment contains exactly those exit positions for which the pick-up point lies on the cut.
3. **Segment $\widehat{I_2I_3}$:** Here, $I_3 = C$. The exit positions in this segment are those for which R_1 finds the exit before performing its cut, but R_2 is picked up after performing its cut.
4. **Segment $\widehat{I_3D}$:** For this segment, we explicitly specify that I_3 itself does not belong to the segment. This segment contains exactly those exit positions that R_1 reaches after performing its cut.

One main difference of our general algorithm to the one suggested in [8] is that in the latter the robots always cut far enough to meet each other. In the following, we argue that this meeting is not necessary since no real information can be shared. Consider the following three cases for the algorithm from [8]:

Case 1: The exit is located in one of the first three segments, excluding I_3. If R_1 went now immediately to the meeting point on a straight line, then it would arrive there earlier than if it had not found the exit before performing its cut. Because of symmetry reasons, it would also arrive earlier than R_2 at the meeting point. Hence, by using the meeting protocol upon finding the exit, R_1 picks R_2 up before R_2 reaches the meeting point. Thus, R_2 never reaches the meeting point and therefore no information can be shared.

Case 2: The exit is located at I_3. In this case, there actually is an exchange of information at the meeting point, but the reason is that the predefined meeting point happens to be the pick-up point for the exit position at I_3. In other words, if R_1 (but not R_2) followed a completely different algorithm without a forced meeting but with the property that it finds the exit at I_3 at the same time as in Algorithm $A(y, \alpha, d)$, then it would still pick R_2 up at the same point at the same time and the resulting evacuation time would not change. Thus, even in this specific case, the benefit in the algorithm from [8] does not come from the forced meeting, but from the fact that R_2 cuts far enough in the direction of the exit to be picked up at the tip of its cut.

Case 3: The exit lies behind R_1's cut, in the fourth segment. At the meeting point, the only relevant information that can be shared is that neither robot has found the exit yet. However, both robots can deduce this information from the fact that they have not been picked up yet (see Case 1).

Note that in [8], two algorithms were presented, as described in Sect. 1.1: One with a linear cut and an improved one where the robots cut to the meeting point in a triangular fashion. In the latter, the exit position at I_3 is also dealt with by the explanations in the above Case 1, while Case 2 is not needed at all.

We can conclude that the meeting itself does not contribute to a better run-time of the algorithm. But it *does* limit the algorithm by introducing a dependency between cut position and cut length. At first sight it might seem as if the improvement between the two algorithms presented in [8] simply comes from the shape of the cut and therefore a shortening of the pick-up distance. However, the possibility to find parameters for the algorithm with the triangular cut that give an improved evacuation time essentially comes from a decoupling of cut position and cut length. Yet there is still some correlation between cut position and cut length which is completely nullified in our algorithm $A(y, \alpha, d)$.

3.2 The Evacuation Time for $y = 2.62843$, $\alpha = \pi/4$ and $d = 0.48793$

In this section, we show an evacuation time of 5.625 for Algorithm $A(y, \alpha, d)$ for the parameters $y = 2.62843$, $\alpha = \pi/4$ and $d = 0.48793$.[4] To do so we determine, for each of the four segments defined in Sect. 3.1, the potential candidates for the worst-case exit position and then take the maximum over the evacuation times for those exits positions. For determining these candidates we use our findings from Sect. 2. To this end, for any pair (exit position, pick-up point), let β and γ denote the same angles as in Sect. 2, i.e., β is the angle between the direction of movement of R_1 at the (potential) exit position and the line connecting exit position and pick-up point and γ the angle between this line and the direction of movement of R_2 at the (potential) pick-up point. Note that when both robots move along the perimeter, we have $\beta = \gamma$ for reasons of symmetry.

For calculating distances, observe that for any two points on the perimeter with a distance of w *along the perimeter*, their euclidian distance is $2 \sin(w/2)$. For instance, the value of $|\widehat{AI_1}|$ is equal to the solution of the equation $x +$

---

[4] These parameters are chosen in a way that for the (only) three possible global worst-case exit positions (determined in the following), the evacuation times are the same up to numerical precision. While the parameter values were determined numerically, we give a rigorous proof for the correctness of the claimed evacuation time.

$2\sin((x+y)/2) = y$, where in our case $y = 2.62843$. For the given parameters, we obtain $|\widehat{AI_1}| \approx 0.63196$, $|\widehat{AI_2}| \approx 2.5837$ and $|\widehat{AI_3}| = 2.62843$. Examining the first three segments one by one, we obtain the following lemma:

Lemma 2. *If there is a (global) worst-case exit position in the first segment, then this exit position is at I_1. If there is a (global) worst-case exit position in the second segment, then it is at I_1 or I_2. If there is a (global) worst-case exit position in the third segment, then it is at I_2 or I_3.*

For our examination of the fourth segment, we add a virtual point I_3' to the fourth segment that coincides with I_3, but has the additional property that if the exit is at I_3', then R_1 will only find the exit after performing its cut. The reason for this is that without the addition of I_3' the fourth segment is half-open which makes it possible that there is a sequence of exit positions with increasing evacuation times that converges towards I_3 and for which there is no exit position that has a larger evacuation time than all exit positions in the sequence.

Lemma 3. *If there is a (global) worst-case exit position in the fourth segment, then this exit position is at I_3'.*

Observe that the evacuation time for the exit at I_3' cannot be smaller than the evacuation time for the exit at I_3 since in the latter case R_1 could just simulate the former case which is worse than activating the meeting protocol right away. Combining this fact with Lemmas 2 and 3, we obtain the following theorem:

Theorem 4. *For $y = 2.62843$, $\alpha = \pi/4$ and $d = 0.48793$, the worst-case exit placement for Algorithm $A(y, \alpha, d)$ is at I_1, I_2 or I_3'.*

In order to determine the evacuation time for the worst-case exit, we simply take the maximum of the evacuation times for the exit placements at these three locations. We obtain evacuation times of approximately 5.6249, 5.62488 and 5.62491 for I_1, I_2 and I_3', respectively. Hence, the evacuation time for the worst-case exit is approximately 5.62491. Thus, we obtain the following corollary:

Corollary 5. *For $y = 2.62843$, $\alpha = \pi/4$ and $d = 0.48793$, the evacuation time of Algorithm $A(y, \alpha, d)$ is at most 5.625.*

Observe that, if the length of the cut is not changed, then altering the shape or angle of the cut does not affect the evacuation times for the exit positions at I_1, I_2 and I_3'. Thus, by Theorem 4, in such a case the overall evacuation time cannot decrease. We cast this insight into the following corollary:

Corollary 6. *For $y = 2.62843$, $\alpha = \pi/4$ and $d = 0.48793$, the evacuation time of Algorithm $A(y, \alpha, d)$ cannot be improved by altering shape or angle of the cut.*

Since all the inequalities in Lemmas 2 and 3 are not sharp, we can choose α in some reasonable range without compromising our evacuation time.[5] As long as the chosen α ensures that there is no worst-case exit placement such that R_2 is picked up on the cut (without the start and end points of the cut), Corollary 5 holds. The value of exactly $\pi/4$ for α is chosen for the reason of convenience.

---

[5] The same holds for the shape of the cut, by the same reason.

4   Conclusion

In this paper, we studied the evacuation of two robots from a disk using non-wireless communication. We presented a new tool for the analysis of evacuation algorithms for any area shape by showing that a strong local condition has to be satisfied in order for an exit to be worst-case placed. Using this tool and further insights, e.g., about the nature of forced meetings and the irrelevance of the chosen shape and angle in some range, we improved the state-of-the-art algorithm and gave indicators for where to look for further improvement.

However, we believe that our improved upper bound on the evacuation time is already very close to the tight bound that is the correct answer. We do not believe that our upper bound is optimal (up to numerical precision) because of the following reason: Imagine an additional second cut of very small depth ("ε-cut") close to the point opposite of the point on the perimeter where the robots start their search. If we choose the position (and the angle and depth) of this ε-cut appropriately, then the evacuation time for the exit at I_3' will be improved since R_1 will pick R_2 up at around the tip of the ε-cut which is somewhat closer to I_3' than if there was no such ε-cut. Now we can make small changes to position and depth of the first cut that result in improving the evacuation times for the exit positions at I_1 and I_2 while increasing the previously decreased evacuation time for the exit position at I_3'. By finding the parameters that again lead to equal evacuation times for these three exits, the overall evacuation time is improved. If one is careful not to let other points become worse exit positions, this approach can even be applied iteratively. However, the improvement in the evacuation time achieved by the collection of these very small cuts is negligibly small, even compared to the improvement given by our algorithm.

While the lower bound is still a long way from our upper bound, it is hard to imagine how an improvement to our algorithm apart from the ε-cuts might look like. In fact, we conjecture that, apart from these ε-cuts and numerical precision, the algorithm we presented is indeed optimal.

References

1. Alpern, S.: The rendezvous search problem. SIAM J. Control Optim. **33**(3), 673–683 (1995)
2. Baeza-Yates, R.A., Culberson, J.C., Rawlins, G.J.E.: Searching with uncertainty extended abstract. In: Karlsson, R., Lingas, A. (eds.) SWAT 1988. LNCS, vol. 318, pp. 176–189. Springer, Heidelberg (1988). doi:10.1007/3-540-19487-8_20
3. Beck, A., Newman, D.J.: Yet more on the linear search problem. Isr. J. Math. **8**(4), 419–429 (1970)
4. Borowiecki, P., Das, S., Dereniowski, D., Kuszner, Ł.: Distributed evacuation in graphs with multiple exits. In: Suomela, J. (ed.) SIROCCO 2016. LNCS, vol. 9988, pp. 228–241. Springer, Cham (2016). doi:10.1007/978-3-319-48314-6_15
5. Chrobak, M., Gasieniec, L., Gorry, T., Martin, R.: Group search on the line. In: Italiano, G.F., Margaria-Steffen, T., Pokorný, J., Quisquater, J.-J., Wattenhofer, R. (eds.) SOFSEM 2015. LNCS, vol. 8939, pp. 164–176. Springer, Heidelberg (2015). doi:10.1007/978-3-662-46078-8_14

6. Czyzowicz, J., Dobrev, S., Georgiou, K., Kranakis, E., MacQuarrie, F.: Evacuating two robots from multiple unknown exits in a circle. In: ICDCN (2016). doi:10.1145/2833312.2833318
7. Czyzowicz, J., Gasieniec, L., Gorry, T., Kranakis, E., Martin, R., Pajak, D.: Evacuating robots via unknown exit in a disk. In: Kuhn, F. (ed.) DISC 2014. LNCS, vol. 8784, pp. 122–136. Springer, Heidelberg (2014). doi:10.1007/978-3-662-45174-8_9
8. Czyzowicz, J., Georgiou, K., Kranakis, E., Narayanan, L., Opatrny, J., Vogtenhuber, B.: Evacuating robots from a disk using face-to-face communication (extended abstract). In: Paschos, V.T., Widmayer, P. (eds.) CIAC 2015. LNCS, vol. 9079, pp. 140–152. Springer, Cham (2015). doi:10.1007/978-3-319-18173-8_10
9. Czyzowicz, J., Kranakis, E., Krizanc, D., Narayanan, L., Opatrny, J., Shende, S.: Wireless autonomous robot evacuation from equilateral triangles and squares. In: Papavassiliou, S., Ruehrup, S. (eds.) ADHOC-NOW 2015. LNCS, vol. 9143, pp. 181–194. Springer, Cham (2015). doi:10.1007/978-3-319-19662-6_13
10. Dessmark, A., Fraigniaud, P., Pelc, A.: Deterministic rendezvous in graphs. In: Battista, G., Zwick, U. (eds.) ESA 2003. LNCS, vol. 2832, pp. 184–195. Springer, Heidelberg (2003). doi:10.1007/978-3-540-39658-1_19
11. Emek, Y., Langner, T., Stolz, D., Uitto, J., Wattenhofer, R.: How many ants does it take to find the food? In: Halldórsson, M.M. (ed.) SIROCCO 2014. LNCS, vol. 8576, pp. 263–278. Springer, Cham (2014). doi:10.1007/978-3-319-09620-9_21
12. Feinerman, O., Korman, A.: Memory lower bounds for randomized collaborative search and implications for biology. In: Aguilera, M.K. (ed.) DISC 2012. LNCS, vol. 7611, pp. 61–75. Springer, Heidelberg (2012). doi:10.1007/978-3-642-33651-5_5
13. Feinerman, O., Korman, A., Lotker, Z., Sereni, J.-S.: Collaborative search on the plane without communication. In: PODC (2012). doi:10.1145/2332432.2332444
14. Förster, K.-T., Nuridini, R., Uitto, J., Wattenhofer, R.: Lower bounds for the capture time: linear, quadratic, and beyond. In: Scheideler, C. (ed.) Structural Information and Communication Complexity. LNCS, vol. 9439, pp. 342–356. Springer, Cham (2015). doi:10.1007/978-3-319-25258-2_24
15. Förster, K.-T., Wattenhofer, R.: Directed graph exploration. In: Baldoni, R., Flocchini, P., Binoy, R. (eds.) OPODIS 2012. LNCS, vol. 7702, pp. 151–165. Springer, Heidelberg (2012). doi:10.1007/978-3-642-35476-2_11
16. Fraigniaud, P., Ilcinkas, D., Peer, G., Pelc, A., Peleg, D.: Graph exploration by a finite automaton. In: Fiala, J., Koubek, V., Kratochvíl, J. (eds.) MFCS 2004. LNCS, vol. 3153, pp. 451–462. Springer, Heidelberg (2004). doi:10.1007/978-3-540-28629-5_34
17. Kranakis, E., Krizanc, D., Rajsbaum, S.: Mobile agent rendezvous: a survey. In: Flocchini, P., Gasieniec, L. (eds.) SIROCCO 2006. LNCS, vol. 4056, pp. 1–9. Springer, Heidelberg (2006). doi:10.1007/11780823_1
18. Lamprou, I., Martin, R., Schewe, S.: Fast two-robot disk evacuation with wireless communication. In: Gavoille, C., Ilcinkas, D. (eds.) DISC 2016. LNCS, vol. 9888, pp. 1–15. Springer, Heidelberg (2016). doi:10.1007/978-3-662-53426-7_1
19. Megow, N., Mehlhorn, K., Schweitzer, P.: Online graph exploration: new results on old and new algorithms. In: Aceto, L., Henzinger, M., Sgall, J. (eds.) ICALP 2011. LNCS, vol. 6756, pp. 478–489. Springer, Heidelberg (2011). doi:10.1007/978-3-642-22012-8_38
20. Nowakowski, R.J., Winkler, P.: Vertex-to-vertex pursuit in a graph. Discret. Math. **43**(2–3), 235–239 (1983). doi:10.1016/0012-365X(83)90160-7
21. Parsons, T.D.: Pursuit-evasion in a graph. In: Alavi, Y., Lick, D.R. (eds.) Theory and Applications of Graphs. Lecture Notes in Mathematics, vol. 642, pp. 426–441. Springer, Heidelberg (1978)

Complexity of Single-Swap Heuristics for Metric Facility Location and Related Problems

Sascha Brauer[✉]

Department of Computer Science, Paderborn University, Paderborn, Germany
sascha.brauer@uni-paderborn.de

Abstract. Metric facility location and K-means are well-known problems of combinatorial optimization. Both admit a fairly simple heuristic called *single-swap*, which adds, drops or swaps open facilities until it reaches a local optimum. For both problems, it is known that this algorithm produces a solution that is at most a constant factor worse than the respective global optimum. In this paper, we show that single-swap applied to the weighted metric uncapacitated facility location and weighted discrete K-means problem is tightly PLS-complete and hence has exponential worst-case running time.

1 Introduction

Facility location is an important optimization problem in operations research and computational geometry. Generally speaking, the goal is to choose a set of locations, called facilities, minimizing the cost of serving a given set of clients. The service cost of a client is usually measured in some form of distance from the client to its nearest open facility. To prevent the trivial solution of opening a facility at each possible location, we usually introduce some sort of opening cost penalizing large sets of open facilities. This general framework comprises a plethora of problems using different functions to measure distance and combinations of opening and service cost. In this paper, we discuss two popular problems closely related to facility location: METRIC UNCAPACITATED FACILITY LOCATION (MUFL) and DISCRETE K-MEANS (DKM).

1.1 Problem Definitions

In an UNCAPACITATED FACILITY LOCATION (UFL) problem we are given a set of clients C, a weight function $w : C \to \mathbb{N}$ on the clients, a set of facilities F, an opening cost function $f : F \to \mathbb{R}$, and a distance function $d : C \times F \to \mathbb{R}$. The goal is to find a subset of facilities $O \subset F$ minimizing

$$\phi_{FL}(C, F, O) = \sum_{c \in C} w(c) \min_{o \in O}\{d(c, o)\} + \sum_{o \in O} f(o).$$

This problem is uncapacitated in the sense, that any open facility can serve, i.e. be the nearest open facility to, any number of clients. Simply speaking,

© Springer International Publishing AG 2017
D. Fotakis et al. (Eds.): CIAC 2017, LNCS 10236, pp. 116–127, 2017.
DOI: 10.1007/978-3-319-57586-5_11

opening a lot of facilities incurs high opening cost, but small service cost, and vice versa. MUFL is a special case of this problem, where we require the distance function d to be a metric on $C \cup F$.

DKM is a problem closely related to UFL, where we do not differentiate between clients and facilities, but are given a single set of points $C \subset \mathbb{R}^D$. We measure distance between points $p, q \in C$ as $d(p, q) = \|p - q\|^2$. Furthermore, instead of imposing an opening cost, we allow at most K locations to be opened. Hence, the goal is to find $O \subset C$ with $|O| = K$ minimizing

$$\phi_{KM}(C, O) = \sum_{c \in C} w(c) \min_{o \in O} \{\|c - o\|^2\}.$$

Notice, that we consider the *weighted* variant of both MUFL and DKM, where each client is associated with a positive weight. Such a weight can be interpreted as the importance of serving the client or as multiple clients present in the same location.

1.2   Local Search

A popular approach to solving hard problems of combinatorial optimization is local search. The general idea of a local search algorithm is to define a small *neighbourhood* for each feasible solution. Given a problem instance and an initial solution, the algorithm replaces the current solution by a better solution from its neighbourhood. This is repeated until the algorithm finds no improvement, hence has found a solution that is not worse than any solution in its neighbourhood. The runtime and the quality of the produced solutions of a local search algorithm depend heavily on its definition of neighbourhood.

Theoretical aspects of local search are captured in the definition of the complexity class PLS. There is a special type of reduction, called PLS-reduction, with respect to which PLS has complete problems [10]. Notably, there are PLS-complete problems, which exhibit two important properties. First, given an instance and an initial solution, it is PSPACE-complete to find a locally optimal solution computed by a local search started with the given initialization. Second, there is an instance and an initial solution, such that this initial solution is exponentially many local search steps away from every locally optimal solution [13]. There is a stronger version of PLS-reductions, so-called *tight* PLS-reductions which are of special interest, as they preserve both of these properties [14]. PLS-complete problems having these two properties are therefore sometimes called *tightly* PLS-complete.

In the following, we examine a local search algorithm for MUFL und DKM called the *single-swap heuristic*. For MUFL, we allow the algorithm to either close an open facility, newly open a closed facility or do both in one step (*swap* an open facility). Since feasible DKM solutions consist of exactly K open facilities, we do not allow the algorithm to solely open or close a facility, but only to swap open facilities. Formally, we define these respective neighbourhoods as

$$N_{MUFL}(O) = \{O' \subset F \mid |O \setminus O'| \leq 1 \wedge |O' \setminus O| \leq 1\} \quad \text{and}$$
$$N_{DKM}(O) = \{O' \subset C \mid |O \setminus O'| = 1 \wedge |O' \setminus O| = 1\}.$$

By MUFL/Swap and DKM/Swap we denote the respective problem as a PLS-problem associated with the described single-swap neighbourhood.

1.3   Related Work

Approximating MUFL has been subject to considerable amount of research using different algorithmic techniques. The problem can be 4-approximated using LP-rounding [17], 3-approximated using a Primal-Dual technique [9], and 1.61-approximated using a greedy algorithm [8]. However, it is known that there is no polynomial time algorithm approximating MUFL better than 1.463 unless NP $\subseteq$ DTIME($n^{\log \log n}$) [7]. Arya et al. showed that the standard local search algorithm of MUFL/Swap computes a 3-approximation for MUFL [2].

A popular generalization of DKM called K-means admits facilities to be opened anywhere in the $\mathbb{R}^D$ instead of restricting possible locations to the locations of the clients. The most popular local search algorithm for the K-means problem is called K-means algorithm, or Lloyd's algorithm [12]. It is well-known that the solutions produced by the K-means algorithm can be arbitrarily bad in comparison to an optimal solution. Furthermore, it was shown that in the worst case, the K-means algorithm requires exponentially many improvement steps to reach a local optimum, even if $D = 2$ [18]. Recently, Roughgarden and Wang proved that, given a K-means instance and an initial solution, it is PSPACE-complete to determine the local optimum computed by the K-means algorithm started on the given initial solution [15]. This is in line with several papers proving the same result for the simplex method using different pivoting rules [1,5]. Kanungo et al. proved that the standard local search algorithm of DKM/Swap computes an $\mathcal{O}(1)$-approximation for DKM and hence also for general K-means [11]. They argue that a variation of the single-swap neighbourhood, where we impose some lower bound on the improvement of a single step, yields an algorithm with polynomial runtime but a slightly worse approximation ratio. However, there is no known upper bound on the runtime of the exact single-swap heuristic, even for unweighted point sets. Another variation of single-swap is the multi-swap heuristic, where we allow the algorithm to simultaneously swap more than one facility in each iteration. For a large enough neighbourhood, i.e. swapping enough facilities in a single iteration, this heuristic yields a PTAS in Euclidean space with fixed dimension [4] and in metric spaces with bounded doubling dimension [6].

1.4   Our Contribution

In this paper, we analyze the PLS complexity of MUFL/Swap and DKM/Swap. By presenting a tight reduction from MAX 2-SAT, we show that both problems are tightly PLS-complete, hence that both local search algorithms require exponentially many steps in the worst case and that given some initial solution it is PSPACE-complete to find the solution computed by the respective algorithm started on this initial set of open facilities. Our reduction only works for the, previously introduced, weighted variants of MUFL and DKM. That is, we construct instances with a non-trivial weight for each client. Furthermore, our reduction for DKM requires the dimension of the point set to be on the order of the number of points. The performance of the single-swap heuristic is basically unaffected from using the more general variants of MUFL and DKM, since the known approximation bounds also hold for the weighted version of both problems, and since the runtime of the heuristic only depends linearly on the weights and the dimension. However, this means that our reduction is weaker than a proof of the same properties for the unweighted variants or for a constant number of dimensions would be.

Theorem 1. MUFL/Swap *and* DKM/Swap *are tightly PLS-complete.*

We prove the two parts of Theorem 1 in Sects. 3 and 4.

2   Preliminaries

We present MAX 2-SAT (SAT), a variant of the classic satisfiability problem, which is elementary in the study of PLS. An instance of SAT is a Boolean formula in conjunctive normal form, where each clause consists of exactly 2 literals and has some positive integer weight assigned to it. The cost of a truth assignment is the sum of the weights of all satisfied clauses. The PLS problem SAT/Flip consists of SAT, where the neighbourhood of an assignment is given by all assignments obtained by changing the truth value of a single variable.

Theorem 2 [16]**.** SAT/Flip *is tightly PLS-complete.*

For each clause set B and truth assignment T we denote the SAT cost of T with respect to B by $w(B, T)$. For a literal x we denote the set of all clauses in B containing x by $B(x)$. Further, we denote the set of all clauses in B satisfied by T by $B_t(T)$ and let $B_f(T) = B \setminus B_t(T)$. Finally, we set $w_{max}^B = \max_{b \in B}\{w(b)\}$.

3   The Facility Location Reduction

In the following, we formulate and prove one of our main results.

Proposition 3. SAT/Flip $\leq_{PLS}$ MUFL/Swap *and this reduction is tight.*

The following proof of Proposition 3 is divided into three parts. First, we present our construction of a PLS-reduction (Φ, Ψ), second, we argue on the correctness of this reduction and finally we show that the reduction is tight.

3.1 Construction of Φ and Ψ

First, we construct the function Φ mapping an instance $(B, w) \in$ MAX 2-SAT over the variables $\{x_n\}_{n\in[N]}$ to an instance $(C, \omega, F, f, d) \in$ METRIC UNCAPACI-TATED FACILITY LOCATION. In the following, we denote $M := |B|$. Each variable x_n appears as a facility twice, once as a positive and once as a negative literal. Formally, we set $F = \{x_n, \bar{x}_n\}_{n\in[N]}$. We further locate a client at each facility and a client corresponding to each clause, so $C = F \cup B$. We set the distance function $d : C \cup F \times C \cup F \to \mathbb{R}$ to

$$d(p, q) = d(q, p) = \begin{cases} 0 & \text{if } p = q \\ 1 & \text{if } p = x_n \wedge q = \bar{x}_n \\ \frac{4}{3} & \text{if } (p = x_n \vee p = \bar{x}_n) \wedge q = b_m \wedge p \in b_m \\ \frac{5}{3} & \text{if } (p = x_n \vee p = \bar{x}_n) \wedge q = b_m \wedge \bar{p} \in b_m \\ 2 & \text{else.} \end{cases}$$

Simply speaking, a literal has distance 1 from its negation, clauses have distance 4/3 from literals they contain, distance 5/3 from literals whose negation they contain, and all other clients/facilities have distance 2 from each other. It is easy to see that d is a metric. The weight of a client corresponding to a clause is the same as the weight of the clause. If a client corresponds to a literal, then its weight is $W = M \cdot w_{max}^B$.

$$\omega(p) = \begin{cases} w(b_m) & \text{if } p = b_m \\ W & \text{else} \end{cases}$$

The opening cost function is constant $f \equiv 2W$.

Second, we construct the function Ψ mapping solutions of $\Phi(B, w)$ back to solutions of (B, w). Given a set $O \subset F$ we let each variable x_n be true if the facility $x_n \in O$ and let it be false otherwise.

In the following, we denote $\Phi(B, w) = (C, \omega, F, 2W, d)$, $\Psi(B, w, O) = T_O$, and $d(c, O) = \min_{o\in O}\{d(c, o)\}$.

3.2 (Φ, Ψ) Is a PLS-Reduction

To prove that (Φ, Ψ) is a PLS-reduction we need to argue that $\Psi(B, w, O)$ is locally optimal for (B, w) if O is locally optimal for $\Phi(B, w)$. Observe, that Ψ is not injective, since $\Phi(B, w)$ has more feasible solutions than (B, w). We can tackle this problem by characterizing a subset of solutions for $\Phi(B, w)$ we call *reasonable* solutions.

Definition 4. *Let $O \subset F$. We call O reasonable if $|O| = N$ and*

$$\forall n \in [N] : x_n \in O \vee \bar{x}_n \in O.$$

Reasonable solutions have several useful properties, which we prove in the following. The restriction of Ψ to reasonable solutions is a bijection, the MUFL cost of a reasonable solution is closely related to the SAT cost of its image under Ψ, and all locally optimal solutions of $\Phi(B, w)$ are reasonable. This characterization of solutions is crucial to proving correctness and tightness of our reduction.

Lemma 5. *If O, O' are reasonable solutions for $\Phi(B, w)$, then*

$$w(B, T_O) < w(B, T_{O'}) \Leftrightarrow \phi_{FL}(C, F, O) > \phi_{FL}(C, F, O').$$

We obtain Lemma 5 from the fact that the MUFL cost of a reasonable solution O are a scaled variant of the SAT cost of T_O. Formally, we have that

$$\phi_{FL}(C, F, O) = \frac{4}{3} \sum_{b_m \in B_t(T_O)} w(b_m) + \frac{5}{3} \sum_{b_m \in B_f(T_O)} w(b_m) + 3WN.$$

A full proof of this claim can be found in the full version of this paper [3].

Lemma 6. *If O is a locally optimal solution for $\Phi(B, w)$, then O is reasonable.*

A detailed proof of Lemma 6 can be found in Sect. 3.3. We can combine these results to obtain the correctness of our reduction.

Corollary 7. *If O is locally optimal for $\Phi(B, w)$, then T_O is locally optimal for (B, w).*

Proof. Assume to the contrary that T_O is not locally optimal. If O is not reasonable, then it is not locally optimal by Lemma 6. Therefore, assume that O is reasonable. Since T_O is not locally optimal, we know that there exists an $n \in [N]$, such that $w(B, T_O^{\bar{n}}) > w(B, T_O)$, where $T_O^{\bar{n}}$ denotes T_O with an inverted assignment of the nth variable. Since $O^{\bar{n}} := (O \setminus \{x_n\}) \cup \{\bar{x}_n\}$ is reasonable, $\Psi(B, w, O^{\bar{n}}) = T_O^{\bar{n}}$ and by Lemma 5 we know that

$$\phi_{FL}(C, F, O^{\bar{n}}) < \phi_{FL}(C, F, O),$$

and hence can conclude that O is not locally optimal. □

3.3   Proof of Lemma 6

The following proof of Lemma 6 is presented in two steps. First, we argue in Lemma 8 that no locally optimal solution can contain both a literal and its negation. Second, we show in Lemma 9 that every locally optimal solution contains a facility corresponding to each of the variables. Combining these two results gives us Lemma 6 as a corollary. From the following results we can moreover conclude that once the single-swap algorithm has reached a reasonable solution, it will always stay at a reasonable solution. We take up on this fact in Sect. 3.4, where we argue on the tightness of our reduction.

Lemma 8. *If* $x_n, \bar{x}_n \in O$, *then* O *is not locally optimal.*

Proof. We show that closing the facility located at x_n strictly decreases the cost, and hence that O can not be locally optimal. When closing the facility x_n, we have to let all clients previously served by this facility (including the client located at x_n) be served by another facility. Choosing $\bar{x}_n$ as the replacement, we do not increase the cost by too much. More specifically, we can pay the additional cost with the cost we save from not opening x_n. Recall, that $B(x_n)$ is the set of all clauses containing the literal x_n, hence that $|B(x_n)| \leq M$. Observe, that no client in $C \setminus B(x_n)$ (except x_n) is closer to x_n than it is to $\bar{x}_n$. We obtain

$$\phi_{FL}(C, F, O) = \sum_{\substack{c \in C \setminus B(x_n) \\ c \neq x_n}} \omega(c)d(c, O) + \sum_{b_m \in B(x_n)} \omega(b_m)\frac{4}{3} + |O|\,2W$$

$$> \sum_{\substack{c \in C \setminus B(x_n) \\ c \neq x_n}} \omega(c)d(c, O) + \sum_{b_m \in B(x_n)} \omega(b_m)\frac{5}{3} + W + (|O| - 1)2W$$

$$\geq \phi_{FL}(C, F, O \setminus \{x_n\}).$$

□

Lemma 9. *If* $x_n, \bar{x}_n \notin O$, *then* O *is not locally optimal.*

Proof. Similar to before, we show that opening a facility at x_n strictly decreases the cost. When opening the facility at x_n we have to save enough service cost by serving locations from it, that we can pay for opening the facility. Connecting the clients located at x_n and $\bar{x}_n$ to the newly opened facility is sufficient. We obtain

$$\phi_{FL}(C, F, O) = \sum_{c \in C \setminus \{x_n, \bar{x}_n\}} \omega(c)d(c, O) + \underbrace{\sum_{c \in \{x_n, \bar{x}_n\}} \omega(c)d(c, O)}_{=4W} + |O|\,2W$$

$$> \sum_{c \in C \setminus \{x_n, \bar{x}_n\}} \omega(c)d(c, O) + W + (|O| + 1)2W$$

$$\geq \phi_{FL}(C, F, O \cup \{x_n\}).$$

□

3.4 (Φ, Ψ) Is a Tight Reduction

We show that (Φ, Ψ) is a tight reduction by only considering its behaviour on reasonable solutions. Lemma 6 tells us that restricted to reasonable solutions, the single-swap local search behaves on $\Phi(B, w)$ exactly the same as the flip local search behaves on (B, w). Additionally, we use the fact that once single-swap has reached a reasonable solution, it will always stay at a reasonable solution. Formally, we need to find a set of feasible solutions $\mathcal{R}$ for $(C, \omega, F, 2W, d)$, such that

1. $\mathcal{R}$ contains all local optima.
2. for every feasible solution T of (B, w), we can compute $O \in \mathcal{R}$ with $T_O = T$ in polynomial time.
3. if the transition graph $TG(C, \omega, F, 2W, d)$ contains a directed path $O \rightsquigarrow O'$, with $O, O' \in \mathcal{R}$ but all internal path vertices outside of $\mathcal{R}$, then $TG(B, w)$ contains the edge $(T_O, T_{O'})$ or $T_O = T_{O'}$.

Let $\mathcal{R}$ be the set of all reasonable solutions. $\mathcal{R}$ contains all local optima of $(C, \omega, F, 2W, d)$, by Lemma 6. The restriction of Ψ to $\mathcal{R}$ is bijective and we can obviously compute the inverse in polynomial time. To prove the final property of tight reductions, we use the following result, which is a byproduct of the proof of Lemma 6.

Corollary 10. *If $O \in \mathcal{R}$ and $O' \notin \mathcal{R}$, then $(O, O') \notin TG(C, \omega, F, 2W, d)$.*

Assume $O \rightsquigarrow O'$ is a directed path in $TG(C, \omega, F, 2W, d)$, with $O, O' \in \mathcal{R}$ but all internal path vertices outside of $\mathcal{R}$. By Corollary 10, this path consists of the single edge (O, O'). This means that $\phi_{FL}(C, F, O) > \phi_{FL}(C, F, O')$ and thus, by Lemma 5, we obtain $w(B, T_O) < w(B, T_{O'})$. Hence, we conclude the tightness proof by observing that $(T_O, T_{O'}) \in TG(B, w)$.

4   The K-means Reduction

We complement our results by showing that we can obtain tight PLS-completeness for DKM/Swap, as well.

Proposition 11. SAT/Flip $\leq_{PLS}$ DKM/Swap *and this reduction is tight.*

To prove Proposition 11, we can basically use the reduction presented in Sect. 3.1. We need to change some of the constants involved in the construction to make sure that we find a set of points in $\mathbb{R}^D$ with the required interpoint distances. However, the general approach stays the same and we obtain essentially the same intermediate results. In the following, we will point out the differences in the construction of (Φ, Ψ) and indicate which proofs require adjustments. After proving the hardness result based on the abstract definition of C, we show that there is indeed a point set in $\mathbb{R}^D$ exhibiting the required squared Euclidean distances.

4.1   Modifications to (Φ, Ψ)

As before, let (B, w) be a MAX 2-SAT instance over the variables $\{x_n\}_{n \in [N]}$. We construct an instance $(C, \omega, K) \in$ DISCRETE K-MEANS. Abstractly define the point set $C = \{x_n, \bar{x}_n\}_{n \in [N]} \cup B$. The distance function $d : C \times C \to \mathbb{R}$ is similar to before

$$d(p, q) = d(q, p) = \begin{cases} 0 & \text{if } p = q \\ 1 & \text{if } p = x_n \wedge q = \bar{x}_n \\ 1 + \epsilon & \text{if } (p = x_n \vee p = \bar{x}_n) \wedge q = b_m \wedge p \in b_m \\ 1 + c\epsilon & \text{if } (p = x_n \vee p = \bar{x}_n) \wedge q = b_m \wedge \bar{p} \in b_m \\ 1 + 2\epsilon & \text{else,} \end{cases}$$

where $1 < c < 2$ and $\epsilon = 1/(4N + 2M)$.

While the distances are scaled in comparison to the MUFL reduction, the central structure remains unchanged. The points closest to each other are literals and their negation. Clauses are closer to literals they contain, than to the literal's negation. All other point pairs have the same, even larger, distance to each other.

The weight function remains unchanged. That is, the weight of a point corresponding to a clause is the SAT weight of the clause, the weight of a point corresponding to a (negated) literal is $W = M \cdot w_{max}^B$. Finally, we choose $K = N$. Like the weight function, Ψ remains unchanged. We denote $\Phi(B, w) = (C, \omega, N)$.

4.2   Correctness of the DKM Reduction

Just as before, we have the problem that Ψ is not injective. However, we can again solve the problem using the previously introduced notion of reasonable solutions. While the first condition ($|O| = N$) is trivially fulfilled, we utilize the second property to ensure that Ψ becomes a bijection when being restricted to reasonable solutions. Moreover, we obtain results analogous to Lemmas 5 and 6.

Lemma 12. *If O, O' are reasonable solutions for $\Phi(B, w)$, then*

$$w(B, T_O) < w(B, T_{O'}) \Leftrightarrow \phi_{KM}(C, O) > \phi_{KM}(C, O').$$

The proof for Lemma 12 can be obtained by substituting the modified constants into the proof of Lemma 5. Almost all of the additional work required for the DKM correctness goes into the proof of the following Lemma 13. Here, we have to ensure that locally optimal solutions do not contain points corresponding to clauses. This was not an issue in the MUFL proof, since clauses are not available for opening in that case.

Lemma 13. *If O is a locally optimal solution for $\Phi(B, w)$, then O is reasonable.*

Using these intermediate results we can see that this is a tight reduction following the same arguments presented in Sect. 3.4.

4.3   Proof of Lemma 13

Observe, that each point $b_m \in C$ has exactly two points at distance $1 + \epsilon$ and two points at distance $1 + c\epsilon$ (the points corresponding to the literals in the clause b_m and their negations, respectively). In the following, we call these four points *adjacent* to b_m. All the other points have distance $1 + 2\epsilon$ to b_m and are hence strictly farther away. Assume to the contrary that there exists an $n \in [N]$, such that $x_n, \bar{x}_n \notin O$.

Case 1: There exists an $m \in [M] : b_m \in O$, such that $b_m = \{x_o, x_p\}$ (where one or both of these literals might be negated). One important observation is that if we exchange b_m for some other location then only its own cost and the cost of its adjacent points can increase. All other points, which might be connected to b_m, are at distance $1 + 2\epsilon$ and can hence be connected to any other location for at most the same cost.

Case 1.1: $x_o, \bar{x}_o, x_p, \bar{x}_p \notin O$. Each point adjacent to b_m has weight W and has distance at least $1 + \epsilon$ to every other points in P. Hence, we have that $\phi(\{b_m, x_o, \bar{x}_o, x_p, \bar{x}_p\}, O) \geq (4 + 4\epsilon)W$. However,

$$\phi(\{b_m, x_o, \bar{x}_o, x_p, \bar{x}_p\}, \{x_o\}) \leq (1 + c\epsilon)\omega(b_m) + W + (2 + 4\epsilon)W$$
$$< (4 + 4\epsilon)W,$$

and hence $(O \setminus \{b_m\}) \cup \{x_o\}$ is in the neighbourhood of O and has strictly smaller cost.

Case 1.2: $x_p \in O \vee \bar{x}_p \in O$ and $x_o, \bar{x}_o \notin O$. In this case, removing b_m from O does not affect the cost of x_p and $\bar{x}_p$. We obtain $\phi(\{b_m, x_o, \bar{x}_o\}, O) \geq (2 + 2\epsilon)W$. Observe, that

$$\phi(\{b_m, x_o, \bar{x}_o\}, (O \setminus \{b_m\}) \cup \{x_o\}) \leq (1 + c\epsilon)\omega(b_m) + W < 2W.$$

Case 1.3: $x_p \in O \vee \bar{x}_p \in O$ and $x_o \in O \vee \bar{x}_o \in O$. Here we have that removing b_m from O does not affect the cost of its adjacent points at all. However, similar to before we have $\phi(\{b_m, x_n, \bar{x}_n\}, O) \geq (2 + 2\epsilon)W$. Again, we obtain

$$\phi(\{b_m, x_n, \bar{x}_n\}, (C \setminus \{b_m\}) \cup \{x_n\}) \leq (1 + c\epsilon)\omega(b_m) + W < 2W.$$

Case 2: There is no $m \in [M]$, such that $b_m \in O$. Consequently, there is an $o \in [N], o \neq n : x_o, \bar{x}_o \in O$. W.l.o.g. assume that $|B(x_o)| < M$ (otherwise just exchange x_o for $\bar{x}_o$ in the following argument). Observe that

$$\phi(B(x_o) \cup \{x_o, x_n, \bar{x}_n\}, O) = (2 + 4\epsilon)W + (1 + \epsilon) \sum_{b_m \in B(x_o)} \omega(b_m).$$

The only points affected by removing x_o from O are x_o and the points corresponding to clauses in $B(x_o)$. Hence,

$$\phi(C \setminus (B(x_o) \cup \{x_o, x_n, \bar{x}_n\}), O) \geq \phi(C \setminus (B(x_o) \cup \{x_o, x_n, \bar{x}_n\}), (O \setminus \{x_o\}) \cup \{x_n\}).$$

However, recall that the points in $B(x_o)$ are at distance $(1 + c\epsilon)$ from $\bar{x}_o \in O$. We obtain

$$\phi(B(x_o) \cup \{x_o, x_n, \bar{x}_n\}, (O \setminus \{x_o\}) \cup \{x_n\})$$
$$\leq \phi(B(x_o) \cup \{x_o, \bar{x}_n\}, \{\bar{x}_o, x_n\})$$
$$= 2W + (1 + \epsilon) \sum_{b_m \in B(x_o)} \omega(b_m) + ((c - 1)\epsilon) \sum_{b_m \in B(x_o)} \omega(b_m)$$
$$< 2W + (1 + \epsilon) \sum_{b_m \in B(x_o)} \omega(b_m) + \epsilon W$$
$$< (2 + 4\epsilon)W + (1 + \epsilon) \sum_{b_m \in B(x_o)} \omega(b_m) = \phi(B(x_o) \cup \{x_o, x_n, \bar{x}_n\}, O).$$

$\square$

4.4   Embedding C into ℓ_2^2

So far, we regarded C as an abstract point set, only given by fixed pairwise interpoint distances. Due to space constraints, we only present the result that there is an isometric embedding of C into ℓ_2^2. A detailed proof of the following lemma can be found in the full version of this paper [3].

Lemma 14. *Let $\{c_1, \ldots, c_{2N+M}\}$ be some numbering of the set C as it is defined in Sect. 4.1. There is a polynomial-time algorithm that computes a matrix P whose rows form a set $\{p_n\}_{n \in [2N+M]}$, such that $\|p_i - p_j\|^2 = d(c_i, c_j)$.*

5   Related Problems and Future Work

In this work, we explore the local search complexity of the single-swap heuristic for MUFL and DKM. While we prove that the problem is tightly PLS-complete in general, our reduction requires arbitrarily many dimensions, number of clusters and a non trivial weight function on the clients. In the full version [3] we show that we can further modify our reduction and obtain the same result for the so-called DISCRETE FUZZY K-MEANS problem, a soft generalization of the DISCRETE K-MEANS problem.

One of the first follow-up question is if we can reduce the number of dimensions D down to a constant. Moreover, it is interesting to examine whether we can obtain our results for unweighted variants of these problems. The fact that the K-means method has exponential worst-case runtime even for unweighted point sets with $D = 2$ indicates that this might be possible. A potential approach to reduce the number of dimension is e.g. to embed our abstract point set using different techniques than the one presented here, since this is the only point in the proof that requires high dimensionality.

The major open result is still the conjecture of Roughgarden and Wang, that computing a local minimum of the K-means algorithm is a PLS-hard problem [15].

Acknowledgments. The author would like to thank Johannes Blömer, Jakob Juhnke and the anonymous reviewers for helpful comments which increased the quality of the paper, and Alexander Skopalik for bringing PLS to his attention.

References

1. Adler, I., Papadimitriou, C., Rubinstein, A.: On simplex pivoting rules and complexity theory. In: Lee, J., Vygen, J. (eds.) IPCO 2014. LNCS, vol. 8494, pp. 13–24. Springer, Cham (2014). doi:10.1007/978-3-319-07557-0_2
2. Arya, V., Garg, N., Khandekar, R., Meyerson, A., Munagala, K., Pandit, V.: Local search heuristics for k-median and facility location problems. SIAM J. Comput. **33**(3), 544–562 (2004)
3. Brauer, S.: Complexity of Single-Swap Heuristics for Metric Facility Location and Related Problems (Full Version). Computing Research Repository (2016). arXiv:1612.01752

4. Cohen-Addad, V., Klein, P.N., Mathieu, C.: Local search yields approximation schemes for k-means and k-median in euclidean and minor-free metrics. In: 2016 IEEE 57th Annual Symposium on Foundations of Computer Science (FOCS) (2016)
5. Fearnley, J., Savani, R.: The complexity of the simplex method. In: Proceedings of the Forty-Seventh Annual ACM on Symposium on Theory of Computing, STOC 2015, pp. 201–208 (2015)
6. Friggstad, Z., Rezapour, M., Salavatipour, M.R.: Local search yields a PTAS for k-means in doubling metrics. In: 2016 IEEE 57th Annual Symposium on Foundations of Computer Science (FOCS) (2016)
7. Guha, S., Khuller, S.: Greedy strikes back: improved facility location algorithms. J. Algorithms **31**(1), 228–248 (1999)
8. Jain, K., Mahdian, M., Saberi, A.: A new greedy approach for facility location problems. In: Proceedings of the Thirty-Fourth Annual ACM Symposium on Theory of Computing, STOC 2002, pp. 731–740 (2002)
9. Jain, K., Vazirani, V.V.: Approximation algorithms for metric facility location and k-median problems using the primal-dual schema and lagrangian relaxation. J. ACM **48**(2), 274–296 (2001)
10. Johnson, D.S., Papadimitriou, C.H., Yannakakis, M.: How easy is local search? J. Comput. Syst. Sci. **37**(1), 79–100 (1988)
11. Kanungo, T., Mount, D.M., Netanyahu, N., Piatko, C.D., Silverman, R., Wu, A.Y.: A local search approximation algorithm for k-means clustering. Comput. Geom. **28**(2), 89–112 (2004)
12. Lloyd, S.: Least squares quantization in PCM. IEEE Trans. Inf. Theor. **28**(2), 129–137 (1982)
13. Monien, B., Dumrauf, D., Tscheuschner, T.: Local search: simple, successful, but sometimes sluggish. In: Abramsky, S., Gavoille, C., Kirchner, C., Meyer auf der Heide, F., Spirakis, P.G. (eds.) ICALP 2010. LNCS, vol. 6198, pp. 1–17. Springer, Heidelberg (2010). doi:10.1007/978-3-642-14165-2_1
14. Papadimitriou, C.H., Schäffer, A.A., Yannakakis, M.: On the complexity of local search. In: Proceedings of the Twenty-Second Annual ACM Symposium on Theory of Computing, STOC 1990 (1990)
15. Roughgarden, T., Wang, J.: The complexity of the k-means method. In: 24th European Symposium on Algorithms, ESA 2016 (2016)
16. Schäffer, A.A., Yannakakis, M.: Simple local search problems that are hard to solve. SIAM J. Comput. **20**(1), 56–87 (1991)
17. Shmoys, D.B., Tardos, E., Aardal, K.: Approximation algorithms for facility location problems. In: Proceedings of the Twenty-Ninth Annual ACM Symposium on Theory of Computing STOC 1997, pp. 265–274 (1997)
18. Vattani, A.: k-means requires exponentially many iterations even in the plane. Discret. Comput. Geom. **45**(4), 596–616 (2011)

Assessing the Computational Complexity of Multi-layer Subgraph Detection

Robert Bredereck[1], Christian Komusiewicz[2], Stefan Kratsch[3],
Hendrik Molter[1(✉)], Rolf Niedermeier[1], and Manuel Sorge[1]

[1] Institut für Softwaretechnik und Theoretische Informatik, TU Berlin,
Berlin, Germany
{robert.bredereck,h.molter,rolf.niedermeier,manuel.sorge}@tu-berlin.de
[2] Institut für Informatik, Friedrich-Schiller-Universität Jena, Jena, Germany
christian.komusiewicz@uni-jena.de
[3] Institut für Informatik I, Universität Bonn, Bonn, Germany
kratsch@cs.uni-bonn.de

Abstract. Multi-layer graphs consist of several graphs (layers) over the
same vertex set. They are motivated by real-world problems where entities (vertices) are associated via multiple types of relationships (edges in
different layers). We chart the border of computational (in)tractability
for the class of subgraph detection problems on multi-layer graphs,
including fundamental problems such as maximum matching, finding
certain clique relaxations (motivated by community detection), or path
problems. Mostly encountering hardness results, sometimes even for two
or three layers, we can also spot some islands of tractability.

1 Introduction

Multi-layer graphs consist of several layers where the vertex set of all layers
is the same but each layer has an individual edge set [4,20]. They are also
known as multi-dimensional networks [3], multiplex networks [24], and edge-
colored multigraphs [1,9]. In recent years, multi-layer graphs have gained a lot
of attention in the social network analysis and data mining communities because
observational data often comes in a multimodal nature. Typical topics studied
here include clustering [5,16], detection of network communities [19,28], data
privacy [26], and general network properties [3].

In several of these applications, researchers identify vertex subsets of a multi-
layer graph that exhibit a certain structure in each of the layers. For example,
motivated by applications in genome comparison in computational biology, Gai
et al. [14] searched for maximal vertex subsets in a two-layer graph that induce
a connected graph in each of the layers. Jiang and Pei [16] and Boden et al. [5]
searched for vertex subsets that induce dense subgraphs in many layers. Such
vertex subsets model communities in a multimodal social network.

To the best of our knowledge, there is, however, no systematic work on
computational complexity classification beyond typically observing the generalization of hardness results for the one-layer case to the multi-layer one [5,16].

© Springer International Publishing AG 2017
D. Fotakis et al. (Eds.): CIAC 2017, LNCS 10236, pp. 128–139, 2017.
DOI: 10.1007/978-3-319-57586-5_12

Our aim in this article is hence to provide a general foundation for studying multi-layer subgraph problems, and to provide some initial results that pave the way for more specific complexity analyses.

We first give a general problem definition that encompasses the problems sketched above. Vaguely, they can be phrased as finding a large vertex subset that induces graphs with an interesting property in many layers. Motivated by the multitude of the desired properties, our problem definition is parameterized by a *graph property* Π, that is, any fixed set of graphs.

Π MULTI-LAYER SUBGRAPH (Π-ML-SUBGRAPH)
Input: A set of graphs $G_1, \ldots, G_t$ all on the same vertex set V and two positive integers k and ℓ.
Question: Is there a vertex set $X \subseteq V$ with $|X| \geq k$ such that for at least ℓ of the input graphs G_i it holds that $G_i[X] \in \Pi$?

We study Π-ML-SUBGRAPH mostly in the context of parameterized complexity. As parameters we use the most natural candidates: the number t of layers, the order k of the desired subgraph, and the number ℓ of layers in which we search for our subgraph, as well as their dual deletion parameters $|V| - k$ and $t - \ell$. Observe that NP-hardness and W[1]-hardness with respect to either k or $|V| - k$ in the single-layer case directly implies hardness of the multi-layer case.

Our analysis of Π-ML-SUBGRAPH starts with several easy results on hereditary graph properties Π, that is, Π is closed under taking induced subgraphs. Such properties are well-studied in the single-layer case. Using Ramsey arguments and a theorem of Khot and Raman [18], we get a trichotomy for the complexity of Π-ML-SUBGRAPH with respect to polynomial-time solvability and fixed-parameter tractability with respect to k and ℓ (Proposition 1). Second, we generalize a fixed-parameter tractability result of Cai's [8] by showing that, for graph properties Π characterized by a finite number of forbidden induced subgraphs, Π-ML-SUBGRAPH is fixed-parameter tractable with respect to combined parameter $t - \ell$ and $|V| - k$, and that it admits a polynomial-size problem kernel (Proposition 2).

Subsequently, we turn to graph properties that are not necessarily hereditary. For finding connected graphs of order at least k in ℓ of t layers, there is a simple fixed-parameter tractability result with respect to t which also gives an XP-algorithm with respect to $t - \ell$ or with respect to ℓ. This algorithm admits a generalization to each graph property that implies certain good-natured partitions of the input graphs (Proposition 3), for example c-cores and c-trusses. As a counterpart, we offer a W[1]-hardness result for Π-ML-SUBGRAPH for the combined parameter k and ℓ for a large class of graph properties Π that includes connected graphs, c-cores, and c-trusses, for example (Theorem 1).

Finally, we exhibit simple graph properties Π for which already a small number of layers leads to NP-hardness and W[1]-hardness of Π-ML-SUBGRAPH: While finding a vertex subset that induces subgraphs of order k with a perfect matching in two layers is polynomial-time solvable, it becomes NP-hard and W[1]-hard with respect to k in three layers (Theorem 2). Additionally, while finding a k-path, that is, a k-vertex graph containing a Hamiltonian path, is fixed-parameter tractable with respect to k in one layer [23], it becomes W[1]-hard in two layers (Theorem 3).

Apart from aiming to provide a broad overview on the complexity of Π-ML-SUBGRAPH, the main technical contributions are conditions on Π that make Π-ML-SUBGRAPH computationally hard (Theorem 1) and understanding the transition from tractability to hardness for perfectly matchable subgraphs (Theorem 2) and Hamiltonian subgraphs (Theorem 3). Due to the lack of space, most proofs have been deferred to a full version of this paper[1].

Related Work. As mentioned in the beginning, despite the numerous practical studies related to multi-layer networks, we are not aware of systematic work pertaining to the computational complexity of Π-ML-SUBGRAPH. The following special cases were studied from this viewpoint. Gai et al. [14] and Bui-Xuan et al. [7] studied the case where Π is the set of all connected graphs and $t = \ell = 2$. They showed that the resulting problem is polynomial-time solvable. In contrast, Cai and Ye [9] studied a modified version of this problem, where the desired vertex subset shall be of size *exactly* k instead of at least k. They showed NP-hardness and W[1]-hardness with respect to k and with respect to $|V| - k$. Agrawal et al. [1] gave a $23^{tk} \cdot \mathrm{poly}(n, t)$-time algorithm for the case in which Π is the set of all cycle free graphs and $t = \ell$.

Edge-colored graphs and multigraphs, which are equivalent to multi-layer graphs, were studied extensively. For surveys, see Bang-Jensen and Gutin [2, Chapter 16] and Kano and Li [17]. Most of the results therein, in the multi-layer terminology, pertain to paths and cycles which do not contain two consecutive edges in the same layer and to related questions like connectedness and Hamiltonicity using this notion of paths or cycles.

Preliminaries. We use the framework of parameterized complexity: Let p be a *parameter* for Π-ML-SUBGRAPH, that is, any integer depending on the input. We aim to prove *fixed-parameter tractability (FPT)* by giving an algorithm that produces a solution in $f(p) \cdot \mathrm{poly}(|V|, t)$ time, where f is a computable function, or we aim to show that such an algorithm is unlikely (*W[1]-hardness* for p). A *problem kernel* is a model for efficient data reduction. Formally, it is a polynomial-time many-one self-reduction such that the size of each resulting instance is upper-bounded by a function of the parameter p. For precise definitions and methodology we refer the reader to the literature [11].

All graphs are undirected and without self-loops or multiple edges. We use standard graph notation. A graph property Π is *hereditary* if removing any vertex from a graph in Π results again in a graph in Π. We use the following graph properties. A graph is a *c-core* if each vertex has degree at least c [27]. A graph is a *c-truss* if each edge is contained in at least $c - 2$ triangles [10]. We say that a graph is *Hamiltonian* if it contains a simple path that comprises all vertices in the graph. A *c-factor* in a graph is a subset of the edges such that each vertex is incident with exactly c edges. In sans serif font face we often denote

---

[1] A preliminary full version containing all proofs is available at arXiv:1604.07724.

graph properties. For example, c-Truss is the set of all c-trusses. By Matching we refer to the set of all graphs containing a perfect matching and by c-Factor to the set of all graphs containing a c-factor.

2 Hereditary Graph Properties

In this section we study the (parameterized) complexity of Π-ML-SUBGRAPH with respect to graph properties Π which are hereditary or whose complement is hereditary. Many natural graph properties fall into one of these categories.

First, we investigate the computational complexity of Π-ML-SUBGRAPH for *hereditary* properties Π. We give a trichotomy with regard to polynomial-time solvability, NP-hardness, and the complexity with respect to parameters k and ℓ, and we observe fixed-parameter tractability for the "deletion parameters" $|V|-k$ and $t-\ell$. The single-layer case has been studied by Lewis and Yannakakis [21] as well as Khot and Raman [18], the latter studied the parameterized complexity. We generalize the mentioned results to the multi-layer case.

Proposition 1. *Let Π be a hereditary graph property.*

1. *If Π excludes at least one complete graph and at least one edgeless graph, then Π-ML-SUBGRAPH is solvable in polynomial time.*
2. *If Π includes all complete graphs and all edgeless graphs, then Π-ML-SUBGRAPH is NP-hard and FPT when parameterized by k and ℓ combined.*
3. *If Π includes either all complete graphs or all edgeless graphs, then Π-ML-SUBGRAPH is NP-hard and W[1]-hard when parameterized by k for all ℓ.*

Note that every hereditary graph property falls into one of the three cases of Proposition 1. We remark that properties that fall into the first case are exactly those containing only a finite number of graphs. The second case is a mere classification result. In the following corollary, we give a number of hereditary properties Π and the corresponding complexity results for Π-ML-SUBGRAPH implied by Proposition 1. For their definitions we refer to the literature [6,15].

Corollary 1. *Π-ML-SUBGRAPH is NP-hard and FPT when parameterized by k and ℓ combined for $\Pi \in$ {Perfect Graph, Interval Graph, Chordal Graph, Split Graph, Asteroidal Triple Free Graph, Comparability Graph, Permutation Graph}.*

Π-ML-SUBGRAPH is NP-hard and W[1]-hard when parameterized by k for all ℓ for $\Pi \in$ {Edgeless Graph, Complete Graph, Complete Multipartite Graph, Planar Graph, c-Colorable Graph, Forest}.

Second, we consider properties Π whose complements are hereditary. For these we can observe that polynomial-time solvability transfers to the multi-layer case.

Observation 1. *Let Π be a graph property such that, if $G \in \Pi$ for some graph G and $H[X] = G$ for some graph H and vertex set X, then $H \in \Pi$. Equivalently, the complement property (containing all graphs not in Π) is hereditary. If Π can be decided in $f(n)$ time for some function f, then Π-ML-SUBGRAPH can be decided in $O(t \cdot f(n))$ for all k and ℓ.*

In the following corollary, we give two properties Π for which Π-ML-SUBGRAPH is solvable in polynomial time according to Observation 1.

Corollary 2. Π-ML-SUBGRAPH *is solvable in polynomial time for:*

- Π = "The graph has maximum degree of at least x."
- Π = "The graph has an h-index [12] of at least x."

Finally, we consider the dual parameterizations for hereditary graph properties characterized by a finite number of forbidden subgraphs. In the single-layer case, this problem has been studied by Lewis and Yannakakis [21] and Cai [8].

Proposition 2. *Let Π be a hereditary graph property that is characterized by finitely many forbidden induced subgraphs. Then Π-ML-SUBGRAPH is NP-hard and FPT when parameterized by the number $t - \ell$ of layers to delete and the number $|V| - k$ of vertices to delete combined. It also admits a polynomial-size problem kernel with respect to these parameters.*

In the following corollary, we give a number of hereditary properties Π characterizable with a finite number of forbidden subgraphs and, hence, for which Π-ML-SUBGRAPH is fixed-parameter tractable with respect to the combined parameter number $t-\ell$ of layers to delete and number $|V|-k$ of vertices to delete.

Corollary 3. *Π-ML-SUBGRAPH is NP-hard and FPT when parameterized by $t - \ell$ and $|V| - k$ combined for $\Pi \in \{$ Cluster Graph, Cograph, Line Graph, Split Graph$\}$.*

3 Non-hereditary Graph Properties

In this section, we give two results related to graph properties that are not necessarily hereditary. First, motivated by Connectivity, we give an FPT-algorithm with respect to t for graph properties in which each graph admits a certain nice vertex partitioning; this algorithm is also an XP-algorithm with respect to ℓ. Second, we give a general W[1]-hardness reduction for the combined parameter k and ℓ, capturing many classes of graph properties such as c-Core, c-Truss, Connectivity, and Matching.

Vertex-partitionable graphs. We start with investigating graph properties Π that allow for efficiently computable partitions of the graph into maximal components that each satisfy Π. It turns out that finding large Π-subgraphs in all input networks is tractable. This can be seen as a generalization of the component-detection algorithm in two layers by Gai et al. [14].

Proposition 3. *Let Π be a graph property such that for every graph $G = (V, E)$ there is a partition $\mathcal{P} := \{X_1, \ldots, X_x\}$ of V such that:*

- *$G[X_i] \in \Pi$ for all $X_i \in \mathcal{P}$,*
- *for all $X \subseteq V$ such that $G[X] \in \Pi$, we have $X \subseteq X_i$ for some $X_i \in \mathcal{P}$, and*
- *$\mathcal{P}$ can be computed in $T(|V|, |E|)$ time where T is non-decreasing in both arguments.*

Then, Π-ML-SUBGRAPH is solvable in $\binom{t}{\ell} \cdot O(|V| \cdot \ell) \cdot \max_{1 \leq i \leq t}(|E_i| + T(|V|, |E_i|))$ time.

Proof. We call a partition $\mathcal{P}$ a Π-*partition* if it fulfills the conditions of Proposition 3 with respect to Π. We describe an algorithm that outputs all maximal sets $X \subseteq V$ such that $G_i[X] \in \Pi$ for all input graphs G_i. We refer to these sets as *solutions* in the following. The algorithm maintains a partition $\mathcal{P}$ of V where, initially, $\mathcal{P} = \{V\}$.

The algorithm checks whether there is a $Y \in \mathcal{P}$ such that $G_i[Y] \notin \Pi$. If $Y \in \mathcal{P}$, then it computes in $T(|V|, |E_i|)$ time a Π-partition $\mathcal{P}_Y$ of $G_i[Y]$. The partition $\mathcal{P}$ is replaced by $(\mathcal{P} \setminus \{Y\}) \cup \mathcal{P}_Y$. If $Y \notin \mathcal{P}$, then the algorithm outputs all $Y \in \mathcal{P}$.

To see the correctness of the algorithm, first observe that for each output Y, we have $G_i[Y] \in \Pi$ for all input graphs G_i. To show maximality of each Y, we show that the algorithm maintains the invariant that each solution X is a subset of some $Y \in \mathcal{P}$. This invariant is trivially fulfilled for the initial partition $\{V\}$. Now consider a set Y that is further partitioned by the algorithm. By the invariant, any solution X that has nonempty intersection with Y is a subset of Y. Furthermore, since $\mathcal{P}_Y$ is a Π-partition of $G_i[Y]$, there is no solution X that contains vertices of two distinct sets Y_1, Y_2 of $\mathcal{P}_Y$. Thus, each solution that is a subset of Y is also a subset of some $Y' \in \mathcal{P}_Y$. Hence, each output set X is a solution since it is an element of the final partition $\mathcal{P}$ and all solutions are subsets of elements of $\mathcal{P}$.

For the running time, observe that for each $Y \in \mathcal{P}$, we can test in $O(\ell \cdot \max_{1 \leq i \leq t} T(|V|, |E_i|))$ time whether it needs to be partitioned further. At most $|V|$ partitioning steps are performed and if a set $Y \in \mathcal{P}$ does not need to be partitioned further, then it can be discarded for the remainder of the algorithm. Thus, in $O(|V|)$ applications of the "maximality test" the result is that Y is a solution and in $O(|V|)$ applications of the maximality test, Y is further partitioned. Hence, the overall number of sets Y that are elements of $\mathcal{P}$ at some point is $O(|V|)$. The overall running time now follows from the assumptions on T and from the fact that the induced subgraphs for all G_i can be computed in $O(|Y| + \ell \cdot \max_{1 \leq i \leq t} |E_i|)$ time for each Y. $\qquad\square$

Examples of graph properties covered by Proposition 3 are Connectivity and c-Edge-Connectivity. If we assume that graphs on one vertex are considered as (trivial) c-cores, then the c-Core property is covered: the nontrivial c-core of a graph is uniquely determined (it is the subgraph remaining after deleting any vertex with degree less than c). Similarly, the c-Truss property is covered by Proposition 3 if we allow one-vertex graphs to be considered as c-trusses. Observe that we can easily choose to either incorporate or disregard connectivity from the c-core and c-truss definitions.

If T is a polynomial function, as in all the examples above, then Π-ML-SUBGRAPH is fixed-parameter tractable with respect to t and polynomial time-solvable if ℓ or $t - \ell$ are constants.

Corollary 4. Π-ML-SUBGRAPH *is* FPT *when parameterized by* t *and polynomial-time solvable if* ℓ *or* $t - \ell$ *are constants for* $\Pi \in \{$ *Connectivity, c-Edge-Connectivity, c-Truss, c-Core* $\}$.

General hardness reduction. Finally, we aim to give a general description of properties Π for which Π-ML-SUBGRAPH is NP-hard and W[1]-hard when parameterized by k and ℓ combined. The next theorem is somewhat technical but covers many natural graph properties which are not covered by Corollary 1. Furthermore, it covers all graph properties from Corollary 4 and shows that for those properties Π-ML-SUBGRAPH becomes intractable when parameterized by ℓ instead of t. We list some of those properties in Corollary 5.

Intuitively, the following theorem covers graph properties Π such that for any three natural numbers x and y, and $\alpha \leq x$ it is possible to construct a graph G that has the following three different types of vertices[2] with the following property: First, there may be a fixed number $z \geq 0$ of *obligatory* vertices; second, there are x *optional* vertices; third, there are y *forbidden* vertices. Any induced subgraph of G of size at least $\alpha + z$ that has property Π has to include all obligatory vertices and may not include any forbidden vertices.

Theorem 1. *Let Π be a graph property. Π-ML-SUBGRAPH is W[1]-hard when parameterized by k and ℓ combined if there is an algorithm A that takes as input a vertex set W, a vertex set $W' \subseteq W$, and an integer α and computes a graph $G = (V, E)$ such that the following conditions hold.*

- *For each $v \in W$ there is a vertex set X_v with $|X_v| = f(\alpha)$ for some function f,*
- *$\{X_v \mid v \in W\} \cup \{Y\}$ is a partition of V for some Y with $|Y| = f'(\alpha)$ for some function f',*
- *for all $X \subseteq V$ with $|X| \geq \alpha \cdot f(\alpha) + f'(\alpha)$ we have that*

$$G[X] \in \Pi \Leftrightarrow \exists W'' \subseteq W' \text{ such that } X = \bigcup_{v \in W''} X_v \cup Y,$$

- *and A has running time $f''(\alpha) \cdot |W|^{O(1)}$ for some function f''.*

If f'' is polynomial, then we additionally get NP-hardness of Π-ML-SUBGRAPH.

The intuition is that each set X_v corresponds to one vertex $v \in W$ and every set X such that $G[X] \in \Pi$ either fully contains X_v or not. Furthermore, Y contains vertices that have to be included in X in order to have that $G[X] \in \Pi$ and all sets X_v that correspond to vertices in $v \in W \setminus W'$ have to be fully excluded from X in order to have that $G[X] \in \Pi$. For the proof, we reduce from BICLIQUE, which is W[1]-hard when parameterized by the size h of the biclique [22].

Proof. We give a parameterized reduction from BICLIQUE which, given an undirected graph H and a positive integer h, asks whether H contains a complete bipartite subgraph $K_{h,h}$. Let $(H = (U, F), h)$ be an instance of BICLIQUE and let $h \geq 2$. We construct an instance of Π-ML-SUBGRAPH in the following way.

For all $v \in U$, let $N_H(v)$ be the neighborhood of v with respect to H. Run Algorithm A on input $(U, N_H(v), h)$ to create graphs G_v for each $v \in U$. Set

---

[2] This actually describes the special case that the sets X_v from Theorem 1 all have size one.

$k = h \cdot f(h) + f'(h)$ and $\ell = h$. Now we show that $(\{G_v\}, k, \ell)$ is a yes-instance of Π-ML-SUBGRAPH if and only if (H, h) is a yes-instance of BICLIQUE.

"$\Rightarrow$": Assume that (H, h) is a yes-instance of BICLIQUE and let (C, D) with $C, D \subseteq U$ and $|C| = |D| = h$ be a biclique. Then for $X = \bigcup_{v \in C} X_v \cup Y$ and $v' \in D$, we have that all v, such that $X_v \subset X$, are neighbors of v' and hence $G_{v'}[X] \in \Pi$. Note that $|X| = h \cdot f(h) + f'(h)$ and $|\{G_v \mid v \in D\}| = h$, therefore it is a solution of Π-ML-SUBGRAPH.

"$\Leftarrow$": Assume that $(\{G_v\}, k, \ell)$ is a yes-instance of Π-ML-SUBGRAPH. There are graphs G_i, with $i \in L, L \subseteq U, |L| \geq h$, and a vertex set $X \subseteq V$ with $|X| \geq k$, such that $G_i[X] \in \Pi$ for all $i \in L$. By the construction of G_i, we know that $X = \bigcup_{v \in W'} X_v \cup Y$ for some $W' \subseteq U$ with $|W'| \geq h$. Furthermore, we know that if $i \in L$ then for all $j \in W'$ (that is $X_j \subset X$) we have that i is neighbor of j. Lastly, we have that $i \in L$ implies that $X_i \not\subset X$ and hence $i \notin W'$. Hence we have that (L, W') is a biclique in H with $|L| \geq h$ and $|W'| \geq h$. $\square$

In the following corollary, we give several properties that are polynomial-time solvable in the single-layer case but NP-hard and W[1]-hard when parameterized by k and ℓ combined in the multi-layer case.

Corollary 5. Π-ML-SUBGRAPH *is NP-hard and W[1]-hard when parameterized by k and ℓ combined for $\Pi \in \{$Connectivity, Tree, Star, c-Core, c-Connectivity, c-Truss, Matching, c-Factor$\}$.*

Proof (Sketch). We sketch Algorithm A from Theorem 1 for all properties listed above.

- Connectivity, Tree, Star, 1-Core: Let $X_v := \{v\}$ and $Y := \{u\}$. Create an edge $\{u, v\}$ for each vertex $v \in W'$.
- c-Core, c-Connectivity, $c > 1$: Let $X_v := \{v\}$ and $Y := \{u_1, \dots, u_c\}$. Create all edges $\{u, v\}$ with $u \in Y$ and $v \in W'$.
- c-Truss: Let $X_v := \{v\}$ and $Y := \{u_1, \dots, u_{c+1}\}$. Create all edges $\{u, v\}$ with $u \in Y$ and $v \in W' \cup Y$ and $u \neq v$.
- Matching: Let $X_v := \{v_1, v_2\}$ for each $v \in W$ and create edge $\{v_1, v_2\}$ if $v \in W'$.
- c-Factor: For each $v \in W'$, add a connected c-regular graph of size $f(c)$ to G, for each $v \in W \setminus W'$, add $f(c)$ vertices to V, for some function f.
- Hamiltonian: Let $X_v := \{v\}$ and create all edges $\{u, v\}$ with $u, v \in W'$. $\square$

A particular consequence of Corollary 5 is that the connected component detection algorithm for two layers by Gai et al. [14] does not generalize to Connectivity-ML-SUBGRAPH with $\ell \ll t$ without significant running time overhead.

4   Matchings, c-Factors, and Hamiltonian Subgraphs

In this section we first consider the problem of finding a set X of at least k vertices that induces in ℓ of t layers a subgraph that has a perfect matching. We also consider the more general c-factor property, asking for a subset of edges such that each vertex is incident with exactly c edges. A perfect matching is a 1-factor. Finding c-factors in single-layer graphs is polynomial-time solvable for all c (see Plummer [25] for an overview on graph factors).

Furthermore, we consider the problem of finding a set X of at least k vertices that induces in ℓ of t layers a subgraph that admits a Hamiltonian path.

Matchings and c-Factors. Corollary 5 states that Matching-ML-SUBGRAPH is W[1]-hard when parameterized by k and ℓ combined. Through closer inspection we can get a stronger result. We show that Matching-ML-SUBGRAPH is polynomial-time solvable for $\ell \leq 2$ and becomes W[1]-hard when parameterized by k already for $\ell \geq 3$. For c-Factor-ML-SUBGRAPH we show that it is already W[1]-hard when parameterized by k if $\ell \geq 2$.

For $\ell = 1$, we can simply check whether there is a c-factor in any of the layers. For $\ell = 2$ Matching-ML-SUBGRAPH can be solved by reducing it to MAXIMUM WEIGHT MATCHING. To this end, let $G_1 = (V, E_1)$ and $G_2 = (V, E_2)$ be two input graphs for which we would like to know whether there is an $X \subseteq V$ of size at least k such that both $G_1[X]$ and $G_2[X]$ have a perfect matching. We solve the problem by a simple reduction to MAXIMUM WEIGHT MATCHING, where we assume that the graph has edge weights and the task is to find a matching with maximum edge weights.

Lemma 1. *Given two graphs $G_1 = (V, E_1)$ and $G_2 = (V, E_2)$, define a graph $G' = (V', E')$ as follows:*

- $V' = \{v_1, v_2 \mid v \in V\}$ *and*
- $E' = \{\{v_1, v_2\} \mid v \in V\} \cup \{\{u_i, v_i\} \mid \{u, v\} \in E_i\}$.

Define a weight function $w\colon E' \to \mathbb{N}$ as follows; let $n := |V|$:

$$w(\{u_i, v_j\}) = \begin{cases} n & \text{if } i \neq j \text{ and } u = v, \\ n+1 & \text{if } i = j \text{ (and } u \neq v). \end{cases}$$

Let $k \in \mathbb{N}$. Then there is a set $X \subseteq V$ of size at least k such that both $G_1[X]$ and $G_2[X]$ have a perfect matching if and only if the graph G' has a matching of w-weight at least $n^2 + k$.

Due to space constraints, we omit the proof. To show that Matching-ML-SUBGRAPH remains W[1]-hard when parameterized by k for $\ell \geq 3$, we reduce from MULTICOLORED CLIQUE which is known to be W[1]-hard when parameterized by the solution size [13]. Intuitively, the reason for the computational complexity transition from two layers to three layers is as follows. Overlaying two matchings one may get cycles and paths but without connections between them. We can cope with this by finding a maximum weighted matching in an auxiliary graph. Adding a third layer, however, allows arbitrary connections between cycles and paths.

Theorem 2. *Matching-ML-SUBGRAPH can be solved in polynomial time if $\ell \leq 2$. It is NP-hard and W[1]-hard when parameterized by k for all $\ell \geq 3$ and $t \geq \ell$.*

For $c \geq 2$, c-Factor-ML-SUBGRAPH is NP-hard and W[1]-hard when parameterized by k for all $\ell \geq 2$ and $t \geq \ell$.

Proof (Sketch). The polynomial-time solvability of Matching-ML-SUBGRAPH for the case of $\ell \leq 2$ is due to Lemma 1. For the case of $\ell \geq 3$ we give a parameterized reduction from MULTICOLORED CLIQUE. Due to space constraints, we only present the contruction of the reduction. This reduction can be adapted to the c-factor case.

In MULTICOLORED CLIQUE, we are given an h-partite graph $H = (U_1 \uplus \ldots \uplus U_h, F)$ and need to determine whether it contains a clique of size h. (Such a clique necessarily contains exactly one vertex from each set U_i and cliques of more than h vertices are impossible.) Without loss of generality, we assume that the number h of colors is even. We construct an instance of Matching-ML-SUBGRAPH for $t = \ell = 3$ as follows and then argue that the construction is easily generalizable.

Vertices. First, create $h - 1$ vertices for each vertex in graph H (one vertex for each color other than his own color). Formally, for each color $1 \leq j \leq h$ and each $u_i \in U_j$ create the vertex set V_i consisting of the vertices $v_{(i,j')}$, $j' \in (\{1, \ldots, h\} \setminus \{j\})$. Second, create one *color vertex* w_j for each color $j \in \{1, \ldots, h\}$. We denote the set of color vertices as $W := \bigcup_{1 \leq j \leq h} \{w_j\}$.

Vertex selection gadget by graph G_1 and G_2. For each color $1 \leq j \leq h$ create for each $u_i \in U_j$ one cycle on $\{w_j\} \cup V_i$ in the graph $G_1 \cup G_2$ such that the edges are alternating from G_1 and from G_2. These $|U_j|$ cycles are all of length h and share only color vertex w_j. To realize this, create the following edges. For each $1 \leq z \leq h - 2$ create an edge in graph $G_{(z \bmod 2)+1}$ between $v_{(i,z)}$ and $v_{(i,z+1)}$ if $z < j - 1$, between $v_{(i,z+1)}$ and $v_{(i,z+2)}$ if $z \geq j$, and between $v_{(i,z)}$ and $v_{(i,z+2)}$ if $z = j - 1$. Create an edge between w_j and $v_{(i,1)}$ in graph G_2, between $v_{(i,h)}$ and w_j in graph G_1 if $j \neq h$, and between $v_{(i,h-1)}$ and w_j in graph G_1 if $j = h$.

Validation gadget by graph G_3. For each adjacent vertex pair u_i, $u_{i'}$ with $u_i \in U_j$ and $u_{i'} \in U_{j'}$, we create an edge between $v_{(i,j')}$ and $v_{(i',j)}$ in G_3. Furthermore, create the edge $\{w_j, w_{j+h/2}\}$ for each $1 \leq j \leq h/2$.

Finally, by setting $k = h^2$ and $t = \ell = 3$ we complete the construction, which can clearly be performed in polynomial time. It remains to show that graph H has a clique that contains each color exactly once if and only if there is a vertex set $X \subseteq V$ with $|X| \geq k$ such that graph $G_z[X]$ contains a perfect matching for each $1 \leq z \leq 3$. $\qquad\square$

Hamiltonian Subgraphs. In the following we investigate the problem variant of finding Hamiltonian subgraphs. Corollary 5 states that Hamiltonian-ML-SUBGRAPH is W[1]-hard when parameterized by k and ℓ combined. Through closer inspection we can get a stronger result. Hamiltonian-SUBGRAPH is known to be FPT when parameterized by the size of the subgraph k [23]. For the multi-layer case, we can show that it is already W[1]-hard when parameterized by k for any $\ell \geq 2$ using a reduction from MULTICOLORED BICLIQUE.

Theorem 3. *Hamiltonian*-ML-SUBGRAPH *is NP-hard and W[1]-hard when parameterized by k for all $\ell \geq 2$ and $t \geq \ell$.*

5   Conclusion

We started a systematic study of subgraph detection problems in multi-layer networks. In particular, we have shown hardness results for many multi-layer subgraph detection problems that are solvable in polynomial time in the single-layer case. We showed that Matching-ML-SUBGRAPH is solvable in polynomial time in the two-layer case, whereas it is W[1]-hard when parameterized by k for three or more layers. Considering acyclic subgraphs Agrawal et al. [1] also showed specialized algorithms for the two-layer case. Thus, it would be interesting to systematically determine which subgraph detection problems become tractable in the two-layer case and to identify problems that behave differently for two and three layers. Finally, in many applications the input graphs are directed. One of our hardness results transfers directly to this case: The reduction from MULTICOLORED BICLIQUE to Hamiltonian-ML-SUBGRAPH (Theorem 3) can be easily adapted to yield directed acyclic graphs. Hence, for directed acyclic graphs the complexity gap between the cases with one and two layers is even bigger because finding a longest path in single-layer directed acyclic graphs is polynomial-time solvable.

Acknowledgment. HM and MS were supported by the DFG, project DAPA (NI 369/12). CK was supported by the DFG, project MAGZ (KO 3669/4–1). RB was partially supported by the DFG, fellowship BR 5207/2. This work was initiated at the research retreat of the TU Berlin Algorithmics and Computational Complexity group held in Darlingerode, Harz mountains, in April 2014.

References

1. Agrawal, A., Lokshtanov, D., Mouawad, A.E., Saurabh, S.: Simultaneous feedback vertex set: a parameterized perspective. In: Proceedings of the 33rd International Symposium on Theoretical Aspects of Computer Science (STACS 2016). LIPIcs, vol. 47, pp. 7:1–7:15. Schloss Dagstuhl - Leibniz-Zentrum fuer Informatik (2016)
2. Bang-Jensen, J., Gutin, G.: Digraphs: Theory, Algorithms and Applications, 2nd edn. Springer, London (2002)
3. Berlingerio, M., Coscia, M., Giannotti, F., Monreale, A., Pedreschi, D.: Multidimensional networks: foundations of structural analysis. In: Proceedings of the 22nd International World Wide Web Conference (WWW 2013), vol. 16(5–6), pp. 567–593 (2013)
4. Boccaletti, S., Bianconi, G., Criado, R., del Genio, C.I., Gmez-Gardees, J., Romance, M., SendNadal, I., Wang, Z., Zanin, M.: The structure and dynamics of multilayer networks. Phys. Rep. **544**(1), 1–122 (2014)
5. Boden, B., Günnemann, S., Hoffmann, H., Seidl, T.: Mining coherent subgraphs in multi-layer graphs with edge labels. In: The 18th ACM SIGKDD International Conference on Knowledge Discovery and Data Mining (KDD 2012), pp. 1258–1266. ACM Press (2012)
6. Brandstädt, A., Le, V.B., Spinrad, J.P.: Graph Classes: A Survey. SIAM Monographs on Discrete Mathematics and Applications, vol. 3. SIAM, Philadelphia (1999)

7. Bui-Xuan, B., Habib, M., Paul, C.: Competitive graph searches. Theor. Comput. Sci. **393**(1–3), 72–80 (2008)
8. Cai, L.: Fixed-parameter tractability of graph modification problems for hereditary properties. Inf. Process. Lett. **58**(4), 171–176 (1996)
9. Cai, L., Ye, J.: Dual connectedness of edge-bicolored graphs and beyond. In: Csuhaj-Varjú, E., Dietzfelbinger, M., Ésik, Z. (eds.) MFCS 2014. LNCS, vol. 8635, pp. 141–152. Springer, Heidelberg (2014). doi:10.1007/978-3-662-44465-8_13
10. Cohen, J.: Trusses: cohesive subgraphs for social network analysis. Technical report, National Security Agency, p. 16 (2008)
11. Cygan, M., Fomin, F.V., Kowalik, L., Lokshtanov, D., Marx, D., Pilipczuk, M., Pilipczuk, M., Saurabh, S.: Parameterized Algorithms. Springer, Cham (2015)
12. Eppstein, D., Spiro, E.S.: The h-index of a graph and its application to dynamic subgraph statistics. J. Graph Algorithms Appl. **16**(2), 543–567 (2012)
13. Fellows, M.R., Hermelin, D., Rosamond, F.A., Vialette, S.: On the parameterized complexity of multiple-interval graph problems. Theor. Comput. Sci. **410**(1), 53–61 (2009)
14. Gai, A.T., Habib, M., Paul, C., Raffinot, M.: Identifying common connected components of graphs. Technical report, RR-LIRMM-03016, LIRMM, Université de Montpellier II (2003)
15. Golumbic, M.C.: Algorithmic Graph Theory and Perfect Graphs, Annals of Discrete Mathematics, vol. 57, 2nd edn. Elsevier B. V., Amsterdam (2004)
16. Jiang, D., Pei, J.: Mining frequent cross-graph quasi-cliques. ACM Trans. Knowl. Discov. Data. **2**(4), 16 (2009)
17. Kano, M., Li, X.: Monochromatic and heterochromatic subgraphs in edge-colored graphs–a survey. Graphs Comb. **24**(4), 237–263 (2008)
18. Khot, S., Raman, V.: Parameterized complexity of finding subgraphs with hereditary properties. Theor. Comput. Sci. **289**(2), 997–1008 (2002)
19. Kim, J., Lee, J.: Community detection in multi-layer graphs: a survey. SIGMOD Rec. **44**(3), 37–48 (2015)
20. Kivel, M., Arenas, A., Barthelemy, M., Gleeson, J.P., Moreno, Y., Porter, M.A.: Multilayer networks. J. Complex Netw. **2**(3), 203–271 (2014)
21. Lewis, J.M., Yannakakis, M.: The node-deletion problem for hereditary properties is NP-complete. J. Comput. Syst. Sci. **20**(2), 219–230 (1980)
22. Lin, B.: The parameterized complexity of k-biclique. In: Proceedings of the Twenty-Sixth Annual ACM-SIAM Symposium on Discrete Algorithms (SODA 2015), pp. 605–615. SIAM (2015)
23. Monien, B.: How to find long paths efficiently. N.-Holl. Math. Stud. **109**, 239–254 (1985)
24. Mucha, P.J., Richardson, T., Macon, K., Porter, M.A., Onnela, J.P.: Community structure in time-dependent, multiscale, and multiplex networks. Science **328**(5980), 876–878 (2010)
25. Plummer, M.D.: Graph factors and factorization: 1985–2003: a survey. Discret. Math. **307**(7–8), 791–821 (2007)
26. Rossi, L., Musolesi, M., Torsello, A.: On the k-anonymization of time-varying and multi-layer social graphs. In: Proceedings of the 9th International Conference on Web and Social Media (ICWSM 2015), pp. 377–386. AAAI Press (2015)
27. Seidman, S.B.: Network structure and minimum degree. Soc. Netw. **5**(3), 269–287 (1983)
28. Zeng, Z., Wang, J., Zhou, L., Karypis, G.: Out-of-core coherent closed quasi-clique mining from large dense graph databases. ACM Trans. Database Syst. **32**(2), 13 (2007)

Almost Optimal Cover-Free Families

Nader H. Bshouty$^{(\boxtimes)}$ and Ariel Gabizon

Department of Computer Science, Technion, 32000 Haifa, Israel
bshouty@cs.technion.ac.il

Abstract. Roughly speaking, an $(n, (r, s))$-*Cover Free Family* (CFF) is a small set of n-bit strings such that: "in any $d := r + s$ indices we see all patterns of weight r". CFFs have been of interest for a long time both in discrete mathematics as part of block design theory, and in theoretical computer science where they have found a variety of applications, for example, in parametrized algorithms where they were introduced in the recent breakthrough work of Fomin, Lokshtanov and Saurabh [16] under the name 'lopsided universal sets'.

In this paper we give the first explicit construction of cover-free families of optimal size up to lower order multiplicative terms, *for any r and s*. In fact, our construction time is almost linear in the size of the family. Before our work, such a result existed only for $r = d^{o(1)}$, and $r = \omega(d/(\log \log d \log \log \log d))$.

As a sample application, we improve the running times of parameterized algorithms from the recent work of Gabizon, Lokshtanov and Pilipczuk [18].

1 Introduction

The purpose of this paper is to give an explicit almost optimal construction of *cover free families* [20]. Before giving a formal definition, let us describe the special case of *group testing*. The problem of group testing was first presented during World War II and described as follows [10,26]: Among n soldiers, at most s carry a fatal virus. We would like to blood test the soldiers to detect the infected ones. Testing each one separately will give n tests. To minimize the number of tests we can mix the blood of several soldiers and test the mixture. If the test comes negative then none of the tested soldiers are infected. If the test comes out positive, we know that at least one of them is infected. The problem is to come up with a small number of tests.

To obtain a non-adaptive algorithm for this problem, a little thought shows that what is required is a set of tests such that for any subset T of s soldiers, and any soldier $i \notin T$, there is a test including soldier i, and precluding all soldiers in T. Let $d = s + 1$. Viewing a test as a characteristic vector $a \in \{0,1\}^n$ of the soldiers it includes, the desired property is equivalent to the following. Find a small set $\mathcal{F} \subseteq \{0,1\}^n$ such that for every $1 \leq i_1 < i_2 < \cdots < i_d \leq n$, and every $1 \leq j \leq d$, there is $a \in \mathcal{F}$ such that $a_{i_j} = 1$ and $a_{i_k} = 0$ for all $k \neq j$.

© Springer International Publishing AG 2017
D. Fotakis et al. (Eds.): CIAC 2017, LNCS 10236, pp. 140–151, 2017.
DOI: 10.1007/978-3-319-57586-5_13

1.1   Cover-Free Families

We can view $\mathcal{F}$ described above as a set of strings such that "in any d indices we see all patterns of weight one". We can generalize this property by choosing an integer $1 \leq r < d$ and requesting to see "in any d indices all patterns of weight r".

Definition 1 (Cover-Free Family). *Fix positive integers r, s, n with $r, s < n$ and let $d := r + s$. An $(n, (r, s))$-Cover Free Family (CFF) is a set $\mathcal{F} \subseteq \{0, 1\}^n$ such that for every $1 \leq i_1 < i_2 < \cdots < i_d \leq n$ and every $J \subset [d]$ of size $|J| = r$ there is a $a \in \mathcal{F}$ such that $a_{i_j} = 1$ for $j \in J$ and $a_{i_k} = 0$ for $k \notin J$.*

We will always assume $r \leq d/2$ (and therefore $r \leq s$): If not, construct an $(n, (s, r))$ -CFF and take the set of complement vectors.

We note that the definition of CFFs usually given is a different equivalent one which we now describe. Given an $(n, (r, s))$-CFF $\mathcal{F}$, denote $N = |\mathcal{F}|$ and construct the $N \times n$ boolean matrix A whose rows are the elements of $\mathcal{F}$. Now, let X be a set of N elements and think of the *columns* of A as characteristic vectors of subsets, which we will call *blocks*, $B \subseteq X$. That is, if we denote by $\mathcal{B} = \{B_1, \ldots, B_n\}$ the set of blocks corresponding to these columns, then A is the *incidence matrix* of $\mathcal{B}$, i.e. the i'th element of X is in B_j if and only if $A_{i,j} = 1$.

For this view, the CFF property of $\mathcal{F}$ implies the following: For any blocks $B_1, \ldots, B_r \in \mathcal{B}$ and any other s blocks $A_1, \ldots, A_s \in \mathcal{B}$ (distinct from the B's), there is an element of X contained in all the B's but not in any of the A's, i.e.

$$\bigcap_{i=1}^{r} B_i \nsubseteq \bigcup_{j=1}^{s} A_j.$$

This property is the usual way to define CFFs [20].

Notation: Let us denote by $N(n, (r, s))$ the minimal integer N such that there exists an $(n, (r, s))$-CFF $\mathcal{F}$ of size $|\mathcal{F}| = N$.

1.2   Previous Results

It is known that, [32], $N(n, (r, s)) \geq \Omega(N(r, s) \cdot \log n)$ where

$$N(r, s) := \frac{d\binom{d}{r}}{\log \binom{d}{r}}.$$

Using the union bound it is easy to show that for $d = r + s = o(n), r \leq s$, we have

$$N(n, (r, s)) \leq O\left(\sqrt{r} \log \binom{d}{r} \cdot N(r, s) \cdot \log n\right).$$

D'yachkov et al.'s breakthrough result, [14], implies that for $s, n \to \infty$

$$N(n, (r, s)) = \Theta\left(N(r, s) \cdot \log n\right). \tag{1}$$

The two above bounds are non-constructive.

Before proceeding to describe previous results and ours, we introduce some convenient terminology:

We will think of the parameter $d = r + s$ as going to infinity and always use the notation $o(1)$ for a term that is independent of n, and goes to 0 as $d \mapsto \infty$. We say an $(n, (r, s))$-CFF $\mathcal{F}$ is *almost optimal*, if its size $N = |\mathcal{F}|$ satisfies

$$N = N(r, s)^{1+o(1)} \cdot \log n = \begin{cases} d^{r+1+o(1)} \log n & \text{if } r = O(1) \\ \left(\frac{d}{r}\right)^{r+o(r)} \log n & \text{if } r = \omega(1), r = o(d) \cdot \\ 2^{H_2(r/d)d+o(d)} \log n & \text{if } r = O(d) \end{cases}$$

where $H_2(x)$ is the binary intopy function.

We say that such $\mathcal{F}$ can be *constructed in linear time* if it can be constructed in time $O(N(r, s)^{1+o(1)} \cdot \log n \cdot n)$. In this terminology, our goal is to obtain almost optimal CFFs that are constructible in linear time.

Let us first consider the case of constant r. It is not hard to see that in this case an $(n, (r, s))$-CFF $\mathcal{F}$ of size $d^{r+1} \log n$ is almost optimal by our definition (and in fact exceeds the optimal size in (2) only by a multiplicative $\log d$ factor). Bshouty [8] constructs $\mathcal{F}$ of such size in linear time and thus solves the case of constant r. In fact, calculation shows that for any $r = d^{o(1)}$, $\mathcal{F}$ of size

$$N = 2^{O(r)} \cdot d^{r+1} \cdot \log n$$

is almost optimal. Bshouty [7,8] constructs such $\mathcal{F}$ in linear time for any $r = o(d)$.

We proceed to the case of larger r. It follows from [31], that for an infinite sequence of integers n, an $(n, (r, s))$-CFF of size

$$M = O\left((rd)^{\log^* n} \log n\right)$$

can be constructed in polynomial time. This was the first nontrivial non-optimal construction. Fomin et al. [16] construct an $(n, (r, s))$-CFF of size

$$\binom{d}{r} 2^{O\left(\frac{d}{\log \log(d)}\right)} \log n \tag{2}$$

in linear time. This is almost optimal when $r = \omega(d/(\log \log d \log \log \log d))$. To the best of our knowledge there is no explicit construction of almost optimal $(n, (r, s))$-CFFs when $d^{o(1)} < r < \omega(d/(\log \log d \log \log \log d))$.

Note that in this range (and even for $r = \omega(1)$ and $r = o(d)$), $\mathcal{F}$ is almost optimal if and only if it has size

$$N = \binom{d}{r}^{1+o(1)} \log n = \left(\frac{d}{r}\right)^{r(1+o(1))} \cdot \log n.$$

Gabizon et al. [18] made a significant step for general r and constructed an $(n, (r, s))$-CFF of size $O((d/r)^{2 \cdot r} \cdot 2^{O(r)} \cdot \log n)$ in linear time. This is quadratically larger than optimal.

1.3   New Result

As mentioned before, there is no explicit construction of almost optimal $(n, (r, s))$-CFFs when $d^{o(1)} < r < \omega(d/(\log \log d \log \log \log d))$ and the result of [18] is quadratically larger than optimal. In this paper we close this quadratic gap and give an explicit construction of an almost optimal $(n, (r, s))$-CFF for all r and s. Our main result is

Theorem 1. *Fix any integers $r < s < d$ with $d = r + s$. There is an almost optimal $(n, (r, s))$-CFF, i.e., of size*

$$N(r, s)^{1+o(1)} \cdot \log n,$$

that can be constructed in linear time. That is, in time $O(N(r, s)^{1+o(1)} \cdot n \cdot \log n)$.

2   Applications of Result

2.1   Application to Learning Hypergraphs

Let $\mathcal{G}_{s,r}$ be a set of all labeled hypergraphs of rank at most r (the maximum size of an edge $e \subseteq V$ in the hypergraph) on the set of vertices $V = \{1, 2, \ldots, n\}$ with at most s edges. Given a hidden Sperner hypergraph[1] $G \in \mathcal{G}_{s,r}$, we need to identify it by asking *edge-detecting queries*. An edge-detecting query $Q_G(S)$, for $S \subseteq V$ is: Does S contain at least one edge of G? Our objective is to *non-adaptively* learn the hypergraph G by asking as few queries as possible.

This problem has many applications in chemical reactions, molecular biology and genome sequencing, where deterministic non-adaptive algorithms are most desirable. In chemical reactions, we are given a set of chemicals, some of which react and some which do not. When multiple chemicals are combined in one test tube, a reaction is detectable if and only if at least one set of the chemicals in the tube reacts. The goal is to identify which sets react using as few experiments as possible. The time needed to compute which experiments to do is a secondary consideration, though it is polynomial for the algorithms we present. See [3] and references within for more details and many other applications in molecular biology.

The above hypergraph $\mathcal{G}_{s,r}$ learning problem is equivalent to the problem of exact learning a monotone DNF with at most s monomials (monotone terms), where each monomial contains at most r variables (s-term r-MDNF) from membership queries [1,4]. A membership query, for an assignment $a \in \{0, 1\}^n$ returns $f(a)$ where f is the hidden s-term r-MDNF.

The non-adaptive learnability of s-term r-MDNF was studied in [9,11,17, 24,25,33]. All the algorithms are either deterministic algorithms that uses non-optimal constructions of $(n, (s, r))$-CFF or randomized algorithms that uses randomized constructions of $(n, (s, r))$-CFF. Our construction in this paper gives, for the deterministic algorithm, a better query complexity and changes the randomized algorithm to deterministic. Recently, our construction is used in [3] to give a polynomial time almost optimal algorithm for learning $\mathcal{G}_{s,r}$.

---

[1] The hypergraph is Sperner hypergraph if no edge is a subset of another. If it is not Sperner hypergraph then learning is not possible.

2.2   Application to r-Simple k-Path

Gabizon et al. [18] recently constructed deterministic algorithms for parametrized problems with 'relaxed disjointness constraints'. For example, rather than searching for a *simple* path of length k in a graph of n vertices, we can search for a path of length k where no vertex is visited more than r times, for some 'relaxation parameter' r. We call the problem of deciding whether such a path exists r-SIMPLE k-PATH. Abasi et al. [2] were the first to study r-SIMPLE k-PATH and presented a randomized algorithm running in time $O^*(r^{2k/r})$. What is perhaps surprising, is that the running time can *significantly improve* as r grows. Derandoming the result of [2, 18] obtained a deterministic algorithm for r-SIMPLE k-PATH with running time $O^*(r^{12k/r} \cdot 2^{O(k/r)})$. At the core of their derandomization is the notion of a 'multiset separator' - a small family of 'witnesses' for the fact that two multisets do not 'intersect too much' on any particular element. How small this family of witnesses can be in turn depends on how small an $(n, (2k/r, k - 2k/r))$-CFF one can construct (details on these connections are given in full paper). Plugging in our new construction into the machinery of [18], we get

Theorem 2. r-SIMPLE k-PATH *can be solved in deterministic time* $O(r^{8k/r+o(k/r)} \cdot 2^{O(k/r)} \cdot k^{O(1)} \cdot n^3 \cdot \log n)$.

For example, when both k/r and r tend to infinity, we get running time $O^*(r^{8k/r+o(k/r)})$ and [18] get $O^*(r^{12k/r+o(k/r)})$.

In a well-known work, Koutis [21] observed that practically all parametrized problems can be viewed as special cases of 'multilinear monomial detection'. [18] also studied the relaxed version of this more general problem: Given an arithmetic circuit C computing an n-variate polynomial $f \in \mathbb{Z}[X_1, \ldots, X_n]$, determine whether f contains a monomial of total degree k and individual degree at most r. We call this problem (r, k)-MONOMIAL DETECTION. [18] define such a circuit C to be *non-canceling* if it contains only variables at its leaves (i.e., no constants), and only addition and multiplication gates (i.e., no substractions). [18] showed that for non-canceling C, (r, k)-MONOMIAL DETECTION can be solved in time $O^*(|C| \cdot r^{18k/r} \cdot 2^{O(k/r)})$. We obtain

Theorem 3. *Given a non-canceling arithmetic circuit C computing $f \in \mathbb{Z}[X_1,$ $\ldots, X_n]$, (r, k)-MONOMIAL DETECTION can be solved in deterministic time* $O(|C| \cdot r^{12k/r+o(k/r)} \cdot 2^{O(k/r)} \cdot k^{O(1)} \cdot n^3 \cdot \log n)$.

Organization of Paper

In Sect. 3 we give an informal description of our CFF construction. In Sect. 4 we give a simple construction that proves Theorem 1 for any $\log^2 d \le r \le d/(\log \log d)^{\omega(1)}$. In the full paper we close the gap and give the proof for $d/(\log d)^{\omega(1)} \le r \le d/\omega(1)$.

3 Proof Overview

Our construction is essentially a generalization of [18] allowing a more flexible choices of parameters. For simplicity, we first describe the construction of [18] and then explain our improvements.

To illustrate the ideas in a simple way, the following 'adaptive' viewpoint will be convenient: We are given two disjoint subsets $C, D \subseteq [n]$ of sizes $|C| = r$ and $|D| = s$. We wish to divide $[n]$ into two separate buckets such that all elements of C fall into the first, and all elements of D fall into the second. Of course the point in CFFs is *that we do not know C and D in advance*. However, the number of different possibilites for the division that will come up in the process will be a bound on the size of an analogous $(n, (r, s))$-CFF- which will contain a vector $a \in \{0, 1\}^n$ corresponding to each way of separating $[n]$ into two buckets that came up in the adpative process.

As a first step we use a perfect hash function h to divide $[n]$ into r buckets such that each bucket contains exactly one element of C. Using a construction of Naor et al. [28], h can be chosen from a family of size $2^{O(r)} \cdot \log n$. Let us call these buckets $B_1, \ldots, B_r$. Now, suppose that we knew, for each $i \in [r]$, the number of elements s_i from D that fell into bucket B_i. In that case we could use an $(n, (1, s_i))$-CFF $\mathcal{F}_i$ to separate the element of C in B_i from the s_i elements of D, and put each in the correct final bucket.

We have such $\mathcal{F}_i$ of size $c \cdot s_i^2 \cdot \log n$ for universal constant c. Thus, the number of different choices in all buckets is $\prod_{i=1}^r c \cdot s_i^2 \cdot \log n \le c^r \cdot (s/r)^{2r} \cdot \log^r n$, as the product of the s_i's is maximized when $s_1 = \ldots s_r = s/r$. Furthermore, [18] show this can be improved to roughly $(s/r)^r \cdot \log n \le (d/r)^r \cdot \log n$ where $d = r + s$. This is done using the hitting sets for combinatorial rectangles of Linial et al. [22] (we do not go into details on this stage here). Of course, we do not know the s_i's. However, it is not too costly to simply guess them! Or rather, try all options: The number of choices for non-negative integers $s_1, \ldots, s_r$ such that $s_1 + \ldots + s_r = s$ is at most

$$\binom{d-1}{r-1} \le \binom{d}{r} \le (ed/r)^r.$$

Combining all stages, this gives us an $(n, (r, s))$-CFF of size roughly $(d/r)^{2r+O(1)} \cdot \log n$. To get an almost optimal construction, we need to get the 2 in the exponent down to a 1. We achieve this by reducing the cost of the 'guessing stage'. Instead of r buckets, we begin by dividing $[n]$ into k buckets for some $k = o(r)$, such that every bucket will contain r/k elements of C. This is done using *splitters* [28]. For concreteness, think of $k = r/\log\log d$. (In the final construction we need to choose k more delicately). Now as we only have k s_i's, there will be less possibilites to go over such that $s_1 + \ldots + s_k = s$ - specifically less than $(ed/k)^k$. On the other hand, our task in each bucket is now more costly - we need to separate r/k elements of C from s_i elements of D, rather than just *one* element of C. A careful choice of parameters show this process can be done while going over at most $(d/r)^{1+o(1)}$ options for the partition into two buckets. There are now two main technical issues left to deal with.

- The splitter construction of [28] was not analyzed as being almost-linear time, but rather, only polynomial time. We give a more careful analysis of it's runtime.
- We need to generalize a component from the construction of [18], into what we call "multi-CFFs". Roughly speaking, this is a small set of strings of length $n \cdot \ell$ that are 'simultaneously a CFF on each n-bit block'. That is, if we think of the string as divided into ℓ blocks of length n, and wish to see in each block a certain pattern of weight r_i in some subset of d_i indices of that block, there will be one string in the multi-CFF that simultaneously exhibits all patterns. We construct a small multi-CFF using a combination of "dense separating hash functions" and the hitting sets for combinatorial rectangles of [22]. See the full paper for details.

4   The First Construction

In this section we give the first construction.

4.1   Preliminary Results for the First Construction

We begin by giving some definitions and preliminary results that we will need for our first construction. The results in this subsection are from [8,28].

Let n, q and d be integers. Let $\mathcal{F}$ be a set of boolean functions $f : [q]^d \to \{0,1\}$. Let H be a family of functions $h : [n] \to [q]$. We say that H is an $(n, \mathcal{F})$-*restriction family* $((n, \mathcal{F})$-RF$)$ if for every $\{i_1, \ldots, i_d\} \subseteq [n]$, $1 \leq i_1 < i_2 < \cdots < i_d \leq n$ and every $f \in \mathcal{F}$ there is a function $h \in H$ such that $f(h(i_1), \ldots, h(i_d)) = 1$.

We say that a construction of an $(n, \mathcal{F})$-restriction family H is a *linear time construction*, if it runs in time $\tilde{O}(|H| \cdot n) = |H| \cdot n \cdot poly(\log |H|, \log n)$.

Let H be a family of functions $h : [n] \to [q]$. For $d \leq q$ we say that H is an (n, q, d)-*perfect hash family* $((n, q, d)$-PHF$)$ if for every subset $S \subseteq [n]$ of size $|S| = d$ there is a *hash function* $h \in H$ such that $h|_S$ is injective (one-to-one) on S, i.e., $|h(S)| = d$. Obviously, an (n, q, d)-PHF is an $(n, \mathcal{F})$-RF when $\mathcal{F} = \{f\}$, for some $f : [q]^d \to \{0,1\}$ satisfying $f(\sigma_1, \ldots, \sigma_d) = 1$ iff $\sigma_1, \ldots, \sigma_d$ are distinct.

In [8] Bshouty proved

Lemma 1. *Let q be a power of prime. If $q > 4(d(d-1)/2+1)$ then there is a linear time construction of an (n, q, d)-PHF of size $O\left(d^2 \log n / \log(q/d^2)\right)$.*

The following is a folklore result

Lemma 2. *Let $\mathcal{F}$ be a set of boolean functions $f : [q]^d \to \{0,1\}$. If there is a linear time construction of an $(m, \mathcal{F})$-RF where $m > 4(d(d-1)/2+1)$ of size s then there is a linear time construction of an $(n, \mathcal{F})$-RF of size $O\left(sd^2 \log n / \log(m/d^2)\right)$.*

Proof. Let H_1 be an $(m, \mathcal{F})$-RF and let H_2 be the (n, m, d)-PHF constructed in Lemma 1. Then it is easy to see that $H_1(H_2) := \{h_1(h_2) \mid h_2 \in H_2, h_1 \in H_1\}$ is an $(n, \mathcal{F})$-RF. $\qquad\square$

Another restriction family that will be used here is splitters [28]. An (n, r, k)-*splitter* is a family of functions H from $[n]$ to $[k]$ such that for all $S \subseteq [n]$ with $|S| = r$, there is $h \in H$ that splits S perfectly, i.e., for all $j \in [k]$, $|h^{-1}(j) \cap S| \in \{\lfloor r/k \rfloor, \lceil r/k \rceil\}$. Obviously, an (n, q, d)-PHF is an (n, d, q)-splitter. Define

$$\sigma(r, k) := \left(\frac{2\pi r}{k}\right)^{k/2} e^{k^2/(12r)}. \tag{3}$$

From the union bound it can be shown that there exists an (n, r, k)-splitter of size $O(\sqrt{r}\sigma(r, k) \log n)$, [28]. Naor et al. [28], use the r-wise independent probability space to construct an (m, r, k)-splitter. They show

Lemma 3. *For $k \le r$, an (m, r, k)-splitter of size $O(\sqrt{r}\sigma(r, k) \log m)$ can be constructed in time $O\left(\sqrt{r} \cdot \sigma(r, k) m^{2r} \log m\right)$.*

When $k = \omega(\sqrt{r})$, Naor et al. in [28], constructed an (n, r, k)-splitter of size $O(\sigma(r, k)^{1+o(1)} \log n)$ in polynomial time. We here show that the same construction can be done in *linear time*. They first construct an $((r/z)^2, r/z, k/z)$-splitter using Lemma 3 where $z = \Theta(r \log k/(k \log(2r/k)))$. They then use Lemma 2 to construct an $(r^2, r/z, k/z)$-splitter. Then compose z pieces of the latter to construct an (r^2, r, k)-splitter and then again use Lemma 2 to construct the final (n, r, k)-splitter.

Note here that we assume that $z|k|r$. The result can be extended to any z, k and r.

We now prove

Lemma 4. *For $k = \omega(\sqrt{r})$ and $z = 16r \log k/(k \log(4r/k))$. An (n, r, k)-splitter of size $r^{O(z)}\sigma(r, k) \log n = \sigma(r, k)^{1+o(1)} \log n$ can be constructed in time $O(\sigma(r, k)^{1+o(1)} \log n)$.*

Proof. It is easy to see that (see the full paper) z is a monotonic decreasing function in k and $16\sqrt{r} \ge z \ge 8 \log r$ for $\sqrt{r} \le k \le r$. First we construct an $((r/z)^2, r/z, k/z)$-splitter using Lemma 3. By Lemma 3, this takes time $O(\sqrt{r/z} \cdot \sigma(r/z, k/z)((r/z)^2)^{2r/z} \log(r/z)) = o(\sigma(r, k))$. By Lemma 3, the size of this splitter is $O(\sqrt{r/z} \cdot \sigma(r/z, k/z) \log(r/z))$. By Lemma 2, using the above splitter, an $(r^2, r/z, k/z)$-splitter H of size $O((r/z)^{2.5}\sigma(r/z, k/z) \log(r/z) \log r)$ can be constructed in linear time. Now, for every choice of $0 = i_0 < i_1 < i_2 < \cdots < i_{z-1} < i_z = r^2$ and $h_0, h_1, \ldots, h_{z-1} \in H$ define the function $h(j) = h_t(j) + (k/z)t$ if $i_t < j \le i_{t+1}$. It is easy to see that this gives an (r^2, r, k)-splitter. The splitter can be constructed in linear time and its size is $\binom{r^2}{z-1}\left(c_1(r/z)^{2.5}\sigma(r/z, k/z) \log(r/z) \log r\right)^z = r^{c_2 z}\sigma(r, k)$ for some constants c_1 and c_2. Now by Lemma 2, an (n, r, k)-splitter can be constructed in time $O(r^2(r^{c_2 z}\sigma(r, k)) \log n) = r^{O(z)}\sigma(r, k) \log n = \sigma(r, k)^{1+o(1)} \log n$. $\square$

The following is from [8]

Lemma 5. *There is an $(n, (r, s))$-CFF of size*

$$O\left(rs\binom{2rs}{r}\log n\right)$$

that can be constructed in linear time.

4.2 Construction I

Let $r \leq s$ be integers and $d = r + s$. Obviously, $1 \leq r \leq d/2$ and $d/2 \leq s \leq d$. We may also assume that

$$r > poly(\log d) = d^{o(1)}. \tag{4}$$

See the table in Sect. 1.2 and the discussion following it.

We first use Lemma 2 to reduce the problem to constructing a $(q, (r, s))$-CFF for $q = O(d^3)$. We then do the following. Suppose $1 \leq i_1 < i_2 < \cdots < i_d \leq q$ and let $(\xi_1, \ldots, \xi_d) \in \{0, 1\}^d$ with r ones (and s zeros) that is supposed to be assigned to $(i_1, i_2, \cdots, i_d)$. Let $i_{j_1}, \ldots, i_{j_r}$ be the entries for which $\xi_{j_1}, \ldots, \xi_{j_r}$ are equal to 1. The main idea of the construction is to first deal with entries $i_{j_1}, \ldots, i_{j_r}$ that are assigned to one and distribute them equally into k buckets, where k will be determined later. This can be done using a (q, r, k)-splitter. Each bucket will contains r/k ones and an unknown number of zeros. We do not know how many zeros, say $d_i - (r/k)$, fall in bucket i but we know that $d_1 + \cdots + d_k = d$. That is, bucket i contains d_i indices of $i_1, i_2, \cdots, i_d$ for which r/k of them are ones. We take all possible $d_1 + \cdots + d_k = d$ and for each bucket i construct $(q, d_i - (r/k), r/k)$-CFF. Taking all possible functions in each bucket for each possible $d_1 + \cdots + d_k = d$ solves the problem.

Let H_1 be an (n, q, d)-PHF such that $d^3 < q \leq 2d^3$ is a power of prime and $d = r + s$. The following follows from Lemma 2.

Lemma 6. *If H is a $(q, (r, s))$-CFF then $\{h_1(h) \mid h_1 \in H_1, h \in H\}$ is $(n, (r, s))$-CFF of size $|H| \cdot |H_1|$.*

We now construct a $(q, (r, s))$-CFF. Let H_2 be a (q, r, k)-splitter where $k < r$ will be determined later. Let $H_3'[d']$ and $H_3''[d']$ be a $(q, d' - \lfloor r/k \rfloor, \lfloor r/k \rfloor)$-CFF and $(q, d' - \lceil r/k \rceil, \lceil r/k \rceil)$-CFF respectively and define $H_3[d'] := H_3'[d'] \cup H_3''[d']$ where $d \geq d' \geq \lceil r/k \rceil$. For every $(h_1, \ldots, h_k) \in H_3[d_1] \times \cdots \times H_3[d_k]$ where $d_1 + \cdots + d_k = d$ and $g \in H_2$ define the function $H_{h_1, \ldots, h_k, g}(i) = h_{g(i)}(i)$.

We first prove

Lemma 7. *The set of all $H_{h_1, \ldots, h_k, g}$ where $(h_1, \ldots, h_k) \in H_3[d_1] \times \cdots \times H_3[d_k]$ for some $d_1 + \cdots + d_k = d$ and $g \in H_2$ is a $(q, (r, s))$-CFF.*

Proof. Consider any $1 \leq i_1 < i_2 < \cdots < i_d \leq q$ and any $(\xi_1, \ldots, \xi_d)$ of weight r. Let $S = \{i_1, \ldots, i_d\}$. Consider $I = \{i_j \mid \xi_j = 1\}$. Since H_2 is a (q, r, k)-splitter there is $g \in H_2$ such that $|g^{-1}(j) \cap I| \in \{\lfloor r/k \rfloor, \lceil r/k \rceil\}$ for all $j = 1, \ldots, k$. Let $d_j = |g^{-1}(j) \cap S|$ for $j = 1, \ldots, k$. Then $d_1 + d_2 + \cdots + d_k = d$. Since $H_3[d_j]$ is a $(q, d_j - \lfloor r/k \rfloor, \lfloor r/k \rfloor)$-CFF and $(q, d_j - \lceil r/k \rceil, \lceil r/k \rceil)$-CFF, there is $h_j \in H_3[d_j]$ such that $h_j(g^{-1}(j) \cap I) = \{1\}$ and $h_j(g^{-1}(j) \cap (S \backslash I)) = \{0\}$.

Now, if $\xi_\ell = 1$ then $i_\ell \in I$. Suppose $g(i_\ell) = j$. Then $i_\ell \in g^{-1}(j) \cap I$ and $H_{h_1, \ldots, h_k, g}(i_\ell) = h_j(i_\ell) \in h_j(g^{-1}(j) \cap I) = \{1\}$. If $\xi_\ell = 0$ then $i_\ell \in S \backslash I$. Suppose $g(i_\ell) = j$. Then $i_\ell \in g^{-1}(j) \cap (S \backslash I)$ and $H_{h_1, \ldots, h_k, g}(i_\ell) = h_j(i_\ell) \in h_j(g^{-1}(j) \cap (S \backslash I)) = \{0\}$.

4.3   Size of Construction I

We now analyze the size of the construction. We will use $c_1, c_2, \ldots$ for constants that are independent of r, s and n.

Let $d^3 < q \le 2d^3$ be a power of prime. By Lemmas 6 and 7 the size of the construction is

$$N := |H_1| \cdot |H_2| \cdot \left| \bigcup_{d_1 + \cdots + d_k = d} H_3[d_1] \times \cdots \times H_3[d_k] \right|$$

where H_1 is an (n, q, d)-PHF, H_2 is a (q, r, k)-splitter and $H_3[d']$ is a $(q, d' - \lceil r/k \rceil, \lceil r/k \rceil)$-CFF and $(q, d' - \lfloor r/k \rfloor, \lfloor r/k \rfloor)$-CFF.

Let $z = 16r \log k / (k \log(4r/k))$. By Lemmas 1, 4, 5 we have

$$N \le c_1 \frac{d^2 \log n}{\log d} \cdot r^{O(z)} \sigma(r, k)(\log d) \cdot \sum_{d_1 + \cdots + d_k = d} \prod_{i=1}^{k} c_2 \frac{d_i r}{k} \binom{2d_i \lceil r/k \rceil}{\lceil r/k \rceil} \log d$$

$$\le c_1 d^2 r^{O(z)} \left(\frac{2\pi r}{k} \right)^{k/2} e^{k^2/(12r)} (\log n) \cdot$$

$$c_3^k \left(\frac{r \log d}{k} \right)^k \sum_{d_1 + \cdots + d_k = d} \prod_{i=1}^{k} (2ed_i)^{r/k+1} d_i \tag{5}$$

$$\le c_4^k d^2 r^{O(z)} e^{k^2/(12r)} \left(\frac{r^3 \log^2 d}{k^3} \right)^{k/2} (2e)^r (\log n) \sum_{d_1 + \cdots + d_k = d} \prod_{i=1}^{k} d_i^{r/k+2}$$

$$\le c_5^k d^2 r^{O(z)} e^{k^2/(12r)} \left(\frac{r^3 \log^2 d}{k^3} \right)^{k/2} (2e)^r (\log n) \left(\frac{d}{k} \right)^k \max_{d_1 + \cdots + d_k = d} \left(\prod_{i=1}^{k} d_i \right)^{r/k+2} \tag{6}$$

$$\le c_6^k d^2 r^{O(z)} e^{k^2/(12r)} \left(\frac{r^3 \log^2 d}{k^3} \right)^{k/2} (2e)^r \left(\frac{d}{k} \right)^{r+3k} \log n \tag{7}$$

$$\le c_6^k d^2 r^{O(z)} e^{k^2/(12r)} \left(\frac{r^3 d^6 \log^2 d}{k^9} \right)^{k/2} \left(\frac{2er}{k} \right)^r \left(\frac{d}{r} \right)^r \log n$$

(5) follows from (3) and the fact that $\binom{a}{b} \le (ea/b)^b$. (6) follows from the fact that the number of k-tuples $(d_1, \ldots, d_k)$ such that $d_1 + \cdots + d_k = d$ is $\binom{d+k-1}{k-1} \le c^k (d/k)^k$ for some constant c. (7) follows from the fact that $\max_{d_1 + \cdots + d_k = d} \prod_{i=1}^{k} d_i = (d/k)^k$.

In summary, we have

$$N \le c_6^k d^2 r^{O(z)} e^{k^2/(12r)} \left(\frac{r^3 d^6 \log^2 d}{k^9} \right)^{k/2} \left(\frac{2er}{k} \right)^r \left(\frac{d}{r} \right)^r \log n.$$

Now assume $r > \log^2 d$ (see (4)) and let $k := r / \log \log d$.

Since

$$z \log r = \frac{16r \log k \log r}{k \log(4r/k)} \leq c_7 \frac{\log^2 r \log \log d}{\log \log \log d} = o(r),$$

$$\frac{k^2}{12r} = \frac{r}{12(\log \log d)^2} = o(r), \left(\frac{r^3 d^6 \log^2 d}{k^9}\right)^{k/2} = c_8^r \left(\frac{d}{r}\right)^{3k} = c_8^r \left(\frac{d}{r}\right)^{o(r)},$$

and $d/r \geq 2$, we have,

$$N \leq (c_9 \log \log d)^r (d/r)^{r(1+o(1))} \log n.$$

This is

$$\left(\frac{d}{r}\right)^{r(1+o(1))} \log n = N(r,s)^{1+o(1)} \log n$$

when $\log^2 d \leq r \leq d/(\log \log d)^{\omega(1)}$.

References

1. Angluin, D.: Queries and concept learning. Mach. Learn. **2**(4), 319–342 (1987)
2. Abasi, H., Bshouty, N.H., Gabizon, A., Haramaty, E.: On r-simple k-path. In: Csuhaj-Varjú, E., Dietzfelbinger, M., Ésik, Z. (eds.) MFCS 2014. LNCS, vol. 8635, pp. 1–12. Springer, Heidelberg (2014). doi:10.1007/978-3-662-44465-8_1
3. Abasi, H., Bshouty, N.H., Mazzawi, H.: Non-adaptive learning of a hidden hypergraph. In: Chaudhuri, K., Gentile, C., Zilles, S. (eds.) ALT 2015. LNCS (LNAI), vol. 9355, pp. 89–101. Springer, Cham (2015). doi:10.1007/978-3-319-24486-0_6
4. Angluin, D., Chen, J.: Learning a hidden graph using $O(\log n)$ queries per edge. J. Comput. Syst. Sci. **74**(4), 546–556 (2008)
5. Alon, N., Yuster, R., Zwick, U.: Color coding. In: Kao, M.Y. (ed.) Encyclopedia of Algorithms. Springer, Heidelberg (2008)
6. Boneh, D., Shaw, J.: Collusion-secure fingerprinting for digital data. IEEE Trans. Inf. Theory **44**(5), 1897–1905 (1998)
7. Bshouty, N.H.: Testers and their applications. In: ITCS 2014, pp. 327–352 (2014). Full version: Electronic Colloquium on Computational Complexity (ECCC), vol. 19, p. 11 (2012)
8. Bshouty, N.H.: Linear time constructions of some d-restriction problems. In: Paschos, V.T., Widmayer, P. (eds.) CIAC 2015. LNCS, vol. 9079, pp. 74–88. Springer, Cham (2015). doi:10.1007/978-3-319-18173-8_5
9. Chin, F.Y.L., Leung, H.C.M., Yiu, S.-M.: Non-adaptive complex group testing with multiple positive sets. Theor. Comput. Sci. **505**, 11–18 (2013)
10. Du, D.Z., Hwang, F.K.: Combinatorial group testing and its applications. Applied Mathematics, vol. 12, 2nd edn. World Scientific, New York (2000)
11. Du, D.Z., Hwang, F.: Pooling Design and Nonadaptive Group Testing: Important Tools for DNA Sequencing. World Scientific, Singapore (2006)
12. Dýachkov, A.G., Rykov, V.V.: Bounds on the length of disjunctive codes. Probl. Pereda. Inf. **18**(3), 7–13 (1982)
13. Dýachkov, A.G., Rykov, V.V., Rashad, A.M.: Superimposed distance codes. Problems Control Inform. Theory/Problemy Upravlen. Teor. Inform **18**(4), 237–250 (1989)

14. D'yachkov, A.G., Vorob'ev, I.V., Polyansky, N.A., Shchukin, VYu.: Bounds on the rate of disjunctive codes. Probl. Inf. Transm. **50**(1), 27–56 (2014)
15. Füredi, Z.: On r-cover-free families. J. Comb. Theory Ser. A **73**(1), 172–173 (1996)
16. Fomin, F.V., Lokshtanov, D., Saurabh, S.: Efficient computation of representative sets with applications in parameterized and exact algorithms. In: SODA 2014, pp. 142–151 (2014)
17. Gao, H., Hwang, F.K., Thai, M.T., Wu, W., Znati, T.: Construction of d(H)-disjunct matrix for group testing in hypergraphs. J. Comb. Optim. **12**(3), 297–301 (2006)
18. Gabizon, A., Lokshtanov, D., Pilipczuk, M.: Fast algorithms for parameterized problems with relaxed disjointness constraints. In: Bansal, N., Finocchi, I. (eds.) ESA 2015. LNCS, vol. 9294, pp. 545–556. Springer, Heidelberg (2015). doi:10.1007/978-3-662-48350-3_46
19. Indyk, P., Ngo, H.Q., Rudra, A.: Efficiently decodable non-adaptive group testing. In: 21st Annual ACM-SIAM Symposium on Discrete Algorithms (SODA 2010), pp. 1126–1142 (2010)
20. Kautz, W.H., Singleton, R.C.: Nonrandom binary superimposed codes. IEEE Trans. Inform. Theory **10**(4), 363–377 (1964)
21. Koutis, I.: Faster algebraic algorithms for path and packing problems. In: Aceto, L., Damgård, I., Goldberg, L.A., Halldórsson, M.M., Ingólfsdóttir, A., Walukiewicz, I. (eds.) ICALP 2008. LNCS, vol. 5125, pp. 575–586. Springer, Heidelberg (2008). doi:10.1007/978-3-540-70575-8_47
22. Linial, N., Luby, M., Saks, M.E., Zuckerman, D.: Efficient construction of a small hitting set for combinatorial rectangles in high dimension. Combinatorica **17**(2), 215–234 (1997)
23. Liu, L., Shen, H.: Explicit constructions of separating hash families from algebraic curves over finite fields. Des. Codes Cryptogr. **41**(2), 221–233 (2006)
24. Macula, A.J., Popyack, L.J.: A group testing method for finding patterns in data. Discret. Appl. Math. **144**, 149–157 (2004)
25. Macula, A.J., Rykov, V.V., Yekhanin, S.: Trivial two-stage group testing for complexes using almost disjunct matrices. Discret. Appl. Math. **137**(1), 97–107 (2004)
26. Ngo, H.Q., Du, D.Z.: A survey on combinatorial group testing algorithms with applications to DNA library screening. Theor. Comput. Sci. **55**, 171–182 (2000)
27. Naor, J., Naor, M.: Small-bias probability spaces: efficient constructions and applications. SIAM J. Comput. **22**(4), 838–856 (1993)
28. Naor, M., Schulman, L.J., Srinivasan, A.: Splitters and near-optimal derandomization. In: FOCS 1995, pp. 182–191 (1995)
29. Porat, E., Rothschild, A.: Explicit nonadaptive combinatorial group testing schemes. IEEE Trans. Inf. Theory **57**(12), 7982–7989 (2011)
30. Stinson, D.R., Van Trung, T., Wei, R.: Secure frameproof codes, key distribution patterns, group testing algorithms and related structures. J. Stat. Plan. Inference **86**, 595–617 (1997)
31. Stinson, D.R., Wei, R., Zhu, L.: New constructions for perfect hash families and related structures using combintorial designs and codes. J. Combin. Des. **8**(3), 189–200 (2000)
32. Stinson, D.R., Wei, R., Zhu, L.: Some new bounds for cover-free families. J. Comb. Theory Ser. A **90**(1), 224–234 (2000)
33. Torney, D.C.: Sets pooling designs. Ann. Comb. **3**, 95–101 (1999)

On the Complexity of the Star p-hub Center Problem with Parameterized Triangle Inequality

Li-Hsuan Chen[1], Sun-Yuan Hsieh[1], Ling-Ju Hung[1(✉)], Ralf Klasing[2], Chia-Wei Lee[1], and Bang Ye Wu[3]

[1] Department of Computer Science and Information Engineering, National Cheng Kung University, Tainan 701, Taiwan
{clh100p,hunglc}@cs.ccu.edu.tw, hsiehsy@mail.ncku.edu.tw, cwlee@csie.ncku.edu.tw
[2] CNRS, LaBRI, Université de Bordeaux, 351 Cours de la Libération, 33405 Talence Cedex, France
ralf.klasing@labri.fr
[3] Department of Computer Science and Information Engineering, National Chung Cheng University, Chiayi 62102, Taiwan
bangye@cs.ccu.edu.tw

Abstract. A complete weighted graph $G = (V, E, w)$ is called Δ_β-metric, for some $\beta \geq 1/2$, if G satisfies the β-triangle inequality, *i.e.*, $w(u, v) \leq \beta \cdot (w(u, x) + w(x, v))$ for all vertices $u, v, x \in V$. Given a Δ_β-metric graph $G = (V, E, w)$ and a center $c \in V$, and an integer p, the Δ_β-STAR p-HUB CENTER PROBLEM (Δ_β-SpHCP) is to find a depth-2 spanning tree T of G rooted at c such that c has exactly p children and the diameter of T is minimized. The children of c in T are called hubs. For $\beta = 1$, Δ_β-SpHCP is NP-hard. (Chen *et al.*, COCOON 2016) proved that for any $\varepsilon > 0$, it is NP-hard to approximate the Δ_β-SpHCP to within a ratio $1.5 - \varepsilon$ for $\beta = 1$. In the same paper, a $\frac{5}{3}$-approximation algorithm was given for Δ_β-SpHCP for $\beta = 1$. In this paper, we study Δ_β-SpHCP for all $\beta \geq \frac{1}{2}$. We show that for any $\varepsilon > 0$, to approximate the Δ_β-SpHCP to a ratio $g(\beta) - \varepsilon$ is NP-hard and we give $r(\beta)$-approximation algorithms for the same problem where $g(\beta)$ and $r(\beta)$ are functions of β. If $\beta \leq \frac{3-\sqrt{3}}{2}$, we have $r(\beta) = g(\beta) = 1$, *i.e.*, Δ_β-SpHCP is polynomial time solvable. If $\frac{3-\sqrt{3}}{2} < \beta \leq \frac{2}{3}$, we have $r(\beta) = g(\beta) = \frac{1+2\beta-2\beta^2}{4(1-\beta)}$. For $\frac{2}{3} \leq \beta \leq 1$, $r(\beta) = \min\{\frac{1+2\beta-2\beta^2}{4(1-\beta)}, 1 + \frac{4\beta^2}{5\beta+1}\}$. Moreover, for $\beta \geq 1$,

Parts of this research were supported by the Ministry of Science and Technology of Taiwan under grants MOST 105–2221–E–006–164–MY3, MOST 103–2218–E–006–019–MY3, and MOST 103-2221-E-006-135-MY3. L.-H. Chen, L.-J. Hung (corresponding author), and C.-W. Lee are supported by the Ministry of Science and Technology of Taiwan under grants MOST 105–2811–E–006–071, –046, and –022, respectively.
R. Klasing–Part of this work was done while Ralf Klasing was visiting the Department of Computer Science and Information Engineering at National Cheng Kung University. This study has been carried out in the frame of the "Investments for the future" Programme IdEx Bordeaux - CPU (ANR-10-IDEX-03-02). Research supported by the LaBRI under the "Projets émergents" program.

© Springer International Publishing AG 2017
D. Fotakis et al. (Eds.): CIAC 2017, LNCS 10236, pp. 152–163, 2017.
DOI: 10.1007/978-3-319-57586-5_14

we have $r(\beta) = \min\{\beta + \frac{4\beta^2 - 2\beta}{2+\beta}, 2\beta + 1\}$. For $\beta \geq 2$, the approximability of the problem (*i.e.*, upper and lower bound) is linear in β.

1 Introduction

The *hub location problems* have various applications in transportation and telecommunication systems. Variants of hub location problems have been defined and well-studied in the literatures (see the two survey papers [1,15]). Suppose that we have a set of demand nodes that want to communicate with each other through some hubs in a network. A *single allocation hub location problem* requests each demand node can only be served by exactly one hub. Conversely, if a demand node can be served by several hubs, then this kind of hub location problem is called *multi-allocation*. Classical hub location problems ask to minimize the total cost of all origin-destination pairs (see *e.g.*, [27]). However, minimizing the total routing cost would lead to the result that the poorest service quality is extremely bad. In this paper, we consider a single hub location problem with min-max criterion, called Δ_β-STAR p-HUB CENTER PROBLEM which is different from the classic hub location problems. The min-max criterion is able to avoid the drawback of minimizing the total cost.

A complete weighted graph $G = (V, E, w)$ is called Δ_β-metric, for some $\beta \geq 1/2$, if the distance function $w(\cdot, \cdot)$ satisfies $w(v, v) = 0$, $w(u, v) = w(v, u)$, and the β-triangle inequality, *i.e.*, $w(u, v) \leq \beta \cdot (w(u, x) + w(x, v))$ for all vertices $u, v, x \in V$. (If $\beta > 1$ then we speak about *relaxed triangle inequality*, and if $\beta < 1$ we speak about *sharpened triangle inequality*.) Let u, v be two vertices in a tree T. Use $d_T(u, v)$ to denote the distance between u, v in T. Define $D(T) = \max_{u,v \in T} d_T(u, v)$ called the diameter of T. We give the definition of the Δ_β-STAR p-HUB CENTER PROBLEM as follows.

Δ_β-STAR p-HUB CENTER PROBLEM (Δ_β-SpHCP).
Input: A Δ_β-metric graph $G = (V, E, w)$, a center vertex $c \in V$, and a positive integer p, $|V| \geq 2p + 1$.
Output: A depth-2 spanning tree T^* rooted at c (called the central hub) such that c has exactly p children (called hubs) and the diameter of T^*, $D(T^*)$, is minimized.

Here, we assume that the number of non-hubs is at least as many as the number of hubs, *i.e.*, $|V| \geq 2p + 1$. The assumption $|V| \geq 2p + 1$ is reasonable because in real applications, a hub could be a post office or an airport, and a non-hub could be a mail post, a customer, or a passenger.

The Δ_β-SpHCP problem is a general version of the original STAR p-HUB CENTER PROBLEM (SpHCP) since the original problem assumes the input graph to be a metric graph, *i.e.*, $\beta = 1$. We use SpHCP to denote the Δ_β-SpHCP for $\beta = 1$. Yaman and Elloumi [28] showed that SpHCP is NP-hard and gave two integer programming formulations for the same problem. Liang [24] showed that SpHCP does not admit a $(1.25 - \varepsilon)$-approximation algorithm for any $\varepsilon > 0$ unless $P = NP$ and gave a 3.5-approximation algorithm. Recently,

Chen *et al.* [17] reduced the gap between the upper and lower bounds of approximability of SpHCP. They showed that for any $\varepsilon > 0$, to approximate SpHCP to a ratio $1.5 - \varepsilon$ is NP-hard and gave 2-approximation and $\frac{5}{3}$-approximation algorithms for SpHCP.

The SINGLE ALLOCATION p-HUB CENTER PROBLEM was introduced in [14, 26] which is similar to SpHCP with min-max criterion and well-studied in [16,18, 23,25]. The difference between the two problems is that the SINGLE ALLOCATION p-HUB CENTER PROBLEM assumes that hubs are fully interconnected. Thus, for the SINGLE ALLOCATION p-HUB CENTER PROBLEM, the communication between hubs is not necessary to go through a specified central hub c.

If $\beta = 1$, Δ_β-SpHCP is NP-hard and even NP-hard to have a $(1.5 - \varepsilon)$-approximation algorithm for any $\varepsilon > 0$ [17]. In this paper, we investigate the complexity of Δ_β-SpHCP parameterized by β-triangle inequality. The motivation of this research for $\beta < 1$ is to investigate whether there exists a large subclasses of input instances of Δ_β-SpHCP that can be solved in polynomial time or admit polynomial-time approximation algorithms with a reasonable approximation ratio. For $\beta \geq 1$, it is an interesting issue to see whether there exists a polynomial-time approximation algorithm with an approximation ratio linear in β.

The well-known concept of *stability of approximation* [10,12,22] is used in our study. The idea behind this concept is to find a parameter (characteristic) of the input instances that captures the hardness of particular inputs. An approximation algorithm is called *stable* with respect to this parameter, if its approximation ratio grows with this parameter but not with the size of the input instances. A nice example is the Traveling Salesman Problem (TSP) that does not admit any polynomial-time approximation algorithm with an approximation ratio bounded by a polynomial in the size of the input instance, but is $\frac{3}{2}$-approximable for metric input instances. Here, one can characterize the input instances by their "distance" to metric instances. This can be expressed by the β-triangle inequality for any $\beta \geq \frac{1}{2}$.

In a sequence of papers [2,3,5,9–11,13], it was shown that one can partition the set of all input instances of TSP into infinitely many subclasses according to the degree of violation of the triangle inequality, and for each subclass one can guarantee upper and lower bounds on the approximation ratio. Similar studies were performed for the problem of constructing 2-connected spanning subgraphs of a given complete graph whose edge weights obey the β-triangle inequality [6], and for the problem of finding, for a given positive integer $k \geq 2$ and an edge-weighted graph G, a minimum k-edge- or k-vertex-connected spanning subgraph [7,8], demonstrating that for these problems the β-triangle inequality can serve as a measure of hardness of the input instances.

In Table 1, we list the main results of this paper. We prove that for any $\varepsilon > 0$, to approximate Δ_β-SpHCP to a ratio $g(\beta) - \varepsilon$ is NP-hard where $\beta \geq \frac{3-\sqrt{3}}{2}$ and $g(\beta)$ is a function of β. We give $r(\beta)$-approximation algorithms for Δ_β-SpHCP. If $\beta \leq \frac{3-\sqrt{3}}{2}$, we have $r(\beta) = g(\beta) = 1$, *i.e.*, Δ_β-SpHCP is polynomial time solvable. If $\frac{3-\sqrt{3}}{2} < \beta \leq \frac{2}{3}$, we have $r(\beta) = g(\beta)$. For $\frac{2}{3} \leq \beta \leq 1$, $r(\beta) =$

Table 1. The main results where Δ_β-SpHCP cannot be approximated within $g(\beta)$ and has an $r(\beta)$-approximation algorithm.

β	Lower bound $g(\beta)$	Upper bound $r(\beta)$
$[\frac{1}{2}, \frac{3-\sqrt{3}}{2}]$	1	1
$(\frac{3-\sqrt{3}}{2}, \frac{2}{3}]$	$\frac{1+2\beta-2\beta^2}{4(1-\beta)}$	$\frac{1+2\beta-2\beta^2}{4(1-\beta)}$
$[\frac{2}{3}, 0.7737\ldots]$	$\frac{5\beta+1}{4}$	$\frac{1+2\beta-2\beta^2}{4(1-\beta)}$
$[0.7737\ldots, 1]$	$\frac{5\beta+1}{4}$	$1 + \frac{4\beta^2}{5\beta+1}$
$[1, 2]$	$\beta + \frac{1}{2}$	$\beta + \frac{4\beta^2-2\beta}{2+\beta}$
$[2, \infty)$	$\beta + \frac{1}{2}$	$2\beta + 1$

$\min\{\frac{1+2\beta-2\beta^2}{4(1-\beta)}, 1 + \frac{4\beta^2}{5\beta+1}\}$ and $g(\beta) = \frac{5\beta+1}{4}$. Moreover, for $\beta \geq 1$, we have $r(\beta) = \min\{\beta + \frac{4\beta^2-2\beta}{2+\beta}, 2\beta + 1\}$ and $g(\beta) = \beta + \frac{1}{2}$. For $\beta \geq 2$, the approximability of the problem (*i.e.*, upper and lower bound) is linear in β.

For a vertex v in a tree T, we use $N_T(v)$ to denote the set of vertices adjacent to v in T and $N_T[v] = N_T(v) \cup \{v\}$. Let $f(v)$ be the parent of v in T and $f(v) = v$ if v is the root of T. Let T^* be an optimal solution of Δ_β-SpHCP in a given β-metric graph $G = (V, E, w)$. For a non-hub x in T^*, we use $f^*(x)$ to denote the hub in T^* that is adjacent to x. We use $\tilde{T}$ to denote the best solution among all solutions in $\mathcal{T}$ where $\mathcal{T}$ is the collection of all solutions satisfying that all non-hubs are adjacent to the same hub for Δ_β-SpHCP in a given β-metric graph $G = (V, E, w)$.

We close this section with the following theorem. Due to the limitation of space, we omit the proof.

Theorem 1. *Let $\beta > \frac{3-\sqrt{3}}{2}$. For any $\varepsilon > 0$, to approximate Δ_β-SpHCP to a factor $g(\beta) - \varepsilon$ is NP-hard where*

(i) $g(\beta) = \frac{1+2\beta-2\beta^2}{4(1-\beta)}$ *if* $\frac{3-\sqrt{3}}{2} < \beta \leq \frac{2}{3}$;
(ii) $g(\beta) = \frac{5\beta+1}{4}$ *if* $\frac{2}{3} \leq \beta \leq 1$;
(iii) $g(\beta) = \beta + \frac{1}{2}$ *if* $\beta \geq 1$.

2   Polynomial Time Algorithms

In this section, we show that for $\frac{1}{2} \leq \beta \leq \frac{3-\sqrt{3}}{2}$, Δ_β-SpHCP can be solved in polynomial time. Besides, we give polynomial time approximation algorithms for Δ_β-SpHCP for $\beta > \frac{3-\sqrt{3}}{2}$. For $\frac{3-\sqrt{3}}{2} < \beta \leq \frac{2}{3}$, our approximation algorithm achieves the factor that closes the gap between the upper and lower bounds of approximability for Δ_β-SpHCP.

Due to the limitation of space, we omit some proofs in this section.

Lemma 1. *Let $\frac{1}{2} \le \beta < 1$. Then the following statements hold.*

(i) *There exists a solution $\tilde{T}$ satisfying that all non-hubs are adjacent to the same hub and $D(\tilde{T}) \le \max\{1, \frac{1+2\beta-2\beta^2}{4(1-\beta)}\} \cdot D(T^*)$.*

(ii) *There exists a polynomial time algorithm to compute a solution T such that $D(T) = D(\tilde{T})$.*

According to Lemma 1, we obtain the following results.

Lemma 2. *Let $\frac{1}{2} \le \beta \le 0.7737\dots$. Then the following statements hold.*

1. *If $\beta \le \frac{3-\sqrt{3}}{2}$, then Δ_β-SpHCP can be solved in polynomial time.*
2. *If $\frac{3-\sqrt{3}}{2} < \beta \le 0.7737\dots$, there is a $\frac{1+2\beta-2\beta^2}{4(1-\beta)}$-approximation algorithm for Δ_β-SpHCP.*

Proof. Let T^* denote an optimal solution of the Δ_β-SpHCP problem. According to Lemma 1, there is a polynomial time algorithm for Δ_β-SpHCP to compute a solution T such that $D(T) \le \max\{1, \frac{1+2\beta-2\beta^2}{4(1-\beta)}\} \cdot D(T^*)$.

If $\beta \le \frac{3-\sqrt{3}}{2}$, $D(T) \le \max\{1, \frac{1+2\beta-2\beta^2}{4(1-\beta)}\} \cdot D(T^*) = D(T^*)$.

If $\frac{3-\sqrt{3}}{2} < \beta \le 0.7737\dots$,

$$D(T) \le \max\{1, \frac{1+2\beta-2\beta^2}{4(1-\beta)}\} \cdot D(T^*) = \frac{1+2\beta-2\beta^2}{4(1-\beta)} \cdot D(T^*).$$

This completes the proof.   □

Lemma 3. *Let $0.7737\dots \le \beta \le 1$. Then, there is a $(1 + \frac{4\beta^2}{5\beta+1})$-approximation algorithm for Δ_β-SpHCP.*

Proof. It is not hard to see that Algorithm 1 runs in polynomial time. Let T^* be an optimal solution of Δ_β-SpHCP. In this lemma, we show that for $0.7737\dots \le \beta \le 1$, Algorithm 1 returns a solution T such that $D(T) \le (1 + \frac{4\beta^2}{5\beta+1}) \cdot D(T^*)$. Let ℓ be the largest edge cost in T^* with one end vertex as a hub and the other end vertex as a non-hub. Note that both Algorithm APX1 and Algorithm APX2 guess all possible edges (y, z) to be the longest edge in T^* with y as a hub and z as a non-hub. Let T_1 and T_2 be the best solutions returned by Algorithm APX1 and Algorithm APX2, respectively.

CLAIM 1. $D(T_1) \le D(T^*) + 4\beta\ell$.

PROOF OF CLAIM. We first show that for any two hubs u, v in T_1, $d_{T_1}(u, v) = w(u, c) + w(v, c) \le D(T^*)$. Let T^* be an optimal solution of Δ_β-SpHCP. Let $f^*(u)$ and $f^*(v)$ be the parents of u and v in T^* respectively.

If $f^*(u) \ne f^*(v)$, there are three cases.

Algorithm 1. Approximation algorithm for Δ_β-SpHCP (G, c).

  (i) Run Algorithm APX1.
 (ii) Run Algorithm APX2.
(iii) Return the best solution found by Algorithms APX1 and APX2.

Algorithm APX1

Guess the correct edge (y, z) where $w(y, z) = \ell$ is the largest edge cost in an optimal solution T^* with y as a hub and z as a non-hub. Let $U := V \setminus \{c\}$ and $h_1 = y$. Let T_1 be the tree found by the following steps and H be the set of children of c in T_1. Initialize $H = \emptyset$.

  (i) Add edge (h_1, c) in the tree T, let $H := H \cup \{h_1\}$, and let $U := U \setminus \{h_1\}$.
 (ii) For $x \in U$, if $w(h_1, x) \leq \ell$, add edges (x, h_1) in T and let $U := U \setminus \{x\}$.
(iii) While $i = |H| + 1 \leq p$ and $U \neq \emptyset$,
       – choose $v \in U$, let $h_i = v$, add edge (h_i, c) in T, let $U := U \setminus \{v\}$, and let $H := H \cup \{h_i\}$;
       – for $x \in U$, if $w(x, h_i) \leq 2\beta\ell$, then add edge (x, h_i) in T and $U := U \setminus \{x\}$.
(iv) If $|H| < p$ and $U = \emptyset$, we change the shape of T by selecting $p - |H|$ vertices closest to c from the second layer to be the children of c, call the new tree T_1; otherwise let $T_1 := T$.

Algorithm APX2

Guess the correct edge (y, z) where $w(y, z) = \ell$ is the largest edge cost in an optimal solution T^* with y as a hub and z as a non-hub. Let T_2 be the tree found by the following steps.

  (i) Let y be the child of c in T_2.
 (ii) Pick $(p - 1)$ vertices $\{v_1, v_2, \ldots, v_{p-1}\}$ closest to c from $U \setminus \{y, z\}$. Let $N_{T_2}(c) = \{y, v_1, v_2, \ldots, v_{p-1}\}$.
(iii) Let all vertices in $U \setminus \{v_1, v_2, \ldots, v_{p-1}, y\}$ be the children of y.

– Suppose that $f^*(u) = c$ and $f^*(v) \neq u$. Then

$$d_{T_1}(u, v) = w(u, c) + w(v, c) \leq w(u, c) + w(c, f^*(v)) + w(f^*(v), v)$$
$$= d_{T^*}(u, v) \leq D(T^*).$$

– Suppose that $f^*(u) = c$ and $f^*(v) = u$. Since $w(u, v) \leq 2\beta\ell$, v is selected as a hub in Step (iv) of Algorithm APX1. Since in Step (iv), the algorithm select $(p - |H|)$ vertices closest to c from the second layer as hubs, there exists y' which is a hub in T^* and a non-hub in T_1 satisfying $w(y', c) \geq w(v, c)$. Thus,

$$d_{T_1}(u, v) = w(u, c) + w(v, c) \leq w(u, c) + w(y', c) = d_{T^*}(u, y') \leq D(T^*).$$

– Suppose that $f^*(u) \neq c$. Then

$$d_{T_1}(u, v) = w(u, c) + w(v, c)$$
$$\leq w(u, f^*(u)) + w(f^*(u), c) + w(c, f^*(v)) + w(f^*(v), v)$$
$$= d_{T^*}(u, v) \leq D(T^*).$$

If $f^*(u) = f^*(v) = c$, $d_{T_1}(u,v) = d_{T^*}(u,v) \leq D(T^*)$.

If $f^*(u) = f^*(v) \neq c$, then at most one of u,v is selected as a hub in Step (iii) of Algorithm APX1 since $w(u,v) \leq 2\beta\ell$, or both u and v are selected as hubs in Step (iv).

Suppose that u is selected as a hub in Step (iii) and and v is selected as a hub in Step (iv). We see that in Step (iv), the algorithms select $(p-|H|)$ vertices closest to c from the second layer as hubs. Thus, there exists y' which is a hub in T^* and a non-hub in T_1 satisfying $w(y',c) \geq w(v,c)$. We obtain that

$$d_{T_1}(u,v) = w(u,c) + w(v,c) \leq d_{T^*}(u,c) + w(y',c) = d_{T^*}(u,y') \leq D(T^*).$$

Suppose that both u,v are selected as hubs in Step (iv). We see that in Step (iv), the algorithm selects $(p-|H|)$ vertices closest to c from the second layer as hubs. Thus, there exist y_1, y_2 which are hubs in T^* and non-hubs in T_1 satisfying $w(y_1,c) \geq w(u,c)$ and $w(y_2,c) \geq w(v,c)$. We obtain that

$$d_{T_1}(u,v) = w(u,c) + w(v,c) \leq w(y_1,c) + w(y_2,c) = d_{T^*}(y_1,y_2) \leq D(T^*).$$

Notice that each non-hub v in T_1 is adjacent to a hub $f(v)$ in T_1 if $w(v,f(v)) \leq 2\beta\ell$.

Thus, for u,v in T_1, $d_{T_1}(u,v) \leq D(T^*) + 4\beta\ell$ and $D(T_1) \leq D(T^*) + 4\beta\ell$. This completes the proof of the claim. ∎

CLAIM 2. $D(T_2) \leq \max\{D(T^*), (D(T^*) - \ell) + \beta(D(T^*) - \ell)\}$.

PROOF OF CLAIM. Let T^* be an optimal solution. For a vertex v, use $f^*(v)$ to denote the parent of v in T^*. Notice that Algorithm APX2 guesses all possible edges (y,z) to be a longest edge in T^* with one end vertex as a hub and the other end vertex as a non-hub. In the following we assume that $w(y,z) = \ell$ is the largest edge cost in T^* with y as a hub and z as a non-hub. Since Algorithm APX2 picks $(p-1)$ vertices closest to c, y is a hub in both T^* and T_2, and $w(y,z) = \ell$, we see that for any hub v in T_2, $d_{T_2}(v,y) \leq D(T^*) - \ell$.

For two non-hubs u,v in T_2, we have the following three cases.

– $f^*(u) = f^*(v) = y$, we see that $d_{T_2}(u,v) = d_{T^*}(u,v) \leq D(T^*)$.
– $f^*(u) = y$ and $f^*(v) \neq y$, we see that

$$d_{T_2}(u,v) = w(u,y) + w(v,y) \leq \ell + \beta \cdot d_{T^*}(v,y) \leq \ell + \beta \cdot (D(T^*) - \ell) \leq D(T^*).$$

– $f^*(u) \neq y$ and $f^*(v) \neq y$, we see that

$$d_{T_2}(u,v) = w(u,y) + w(v,y) \leq \beta \cdot d_{T^*}(u,y) + \beta \cdot d_{T^*}(v,y)$$
$$\leq 2\beta(D(T^*) - \ell) \leq (D(T^*) - \ell) + \beta(D(T^*) - \ell).$$

For a non-hub u and a hub v in T_2, there are two cases.

– If $f^*(u) = y$, we see that

$$d_{T_2}(u,v) = w(u,y) + d_{T_2}(v,y) \leq \ell + D(T^*) - \ell = D(T^*).$$

– If $f^*(u) \neq y$, we see that

$$d_{T_2}(u, v) = w(u, y) + d_{T_2}(v, y) \leq \beta \cdot (D(T^*) - \ell) + (D(T^*) - \ell).$$

For two hubs u, v in T_2, $u \neq y$ and $v \neq y$, we see that $d_{T_2}(u, v) \leq D(T^*)$ since y is a hub in T^* and Algorithm APX2 picks the other $(p - 1)$ vertices closest to c as hubs.

Thus, $D(T_2) \leq \max\{D(T^*), (D(T^*) - \ell) + \beta(D(T^*) - \ell)\}$. This completes the proof of the claim. ∎

Notice that if $\frac{\ell}{D(T^*)} \geq \frac{\beta}{1+\beta}$, $D(T_2) = D(T^*)$. Thus, the worst case approximation ratio happens when $\frac{\ell}{D(T^*)} < \frac{\beta}{1+\beta}$.

If $\frac{\ell}{D(T^*)} < \frac{\beta}{1+\beta}$, $D(T_2) \leq D(T^*) - \ell + \beta(D(T^*) - \ell)$. We see that the approximation ratio of Algorithm 1 is $r(\beta) = \min\{\frac{D(T_1)}{D(T^*)}, \frac{D(T_2)}{D(T^*)}\}$. The worst case approximation ratio of Algorithm 1 happens when $D(T_1) = D(T_2)$, i.e.,

$$D(T^*) + 4\beta\ell = (D(T^*) - \ell) + \beta \cdot (D(T^*) - \ell)$$

We obtain that $\frac{\ell}{D(T^*)} = \frac{\beta}{5\beta+1}$. Thus,

$$r(\beta) = \min\{\frac{D(T_1)}{D(T^*)}, \frac{D(T_2)}{D(T^*)}\} \leq \min\{1 + \frac{4\beta^2}{5\beta+1}, 1 - \frac{\beta}{5\beta+1} + \beta(1 - \frac{\beta}{5\beta+1})\}$$
$$= 1 + \frac{4\beta^2}{5\beta+1}.$$

This completes the proof. □

In Lemma 4, we prove that if $1 \leq \beta \leq 2$, Algorithm 1 is a $(\beta + \frac{4\beta^2 - 2\beta}{2+\beta})$-approximation algorithm for Δ_β-SpHCP.

Lemma 4. *Let $1 \leq \beta \leq 2$. Then, there is a $(\beta + \frac{4\beta^2 - 2\beta}{2+\beta})$-approximation algorithm for Δ_β-SpHCP.*

If $\beta \geq 2$, we give Algorithm 2 to solve Δ_β-SpHCP and prove that Algorithm 2 is a $(2\beta + 1)$-approximation algorithm in Lemma 5.

Lemma 5. *Let $\beta \geq 2$. Then, there is a $(2\beta + 1)$-approximation algorithm for Δ_β-SpHCP.*

Proof. Let T^* be an optimal solution of Δ_β-SpHCP. Let (c, q) be the longest edge incident to c in T^*, $w(c, q) = \ell_0$, i.e., $\ell_0 = \max_{v \in N_{T^*}(c)}\{w(v, c)\}$. Let ℓ_1 and ℓ_2 be the largest and second largest edge costs in T^* with one end vertex as a hub and the other end vertex as a non-hub. Note that it is possible that $\ell_1 = \ell_2$. Our algorithm is presented as Algorithm 2. Line 1 of Algorithm 2 guesses the values of ℓ_0, ℓ_1 and ℓ_2. We certainly do not know their exact values. However, since each of them has only polynomially many possible values, we can run the algorithm for all of their possible values and take the best solution. Therefore, in the following we assume that we know ℓ_0, ℓ_1 and ℓ_2. It is easy to see that $D(T^*) \geq \ell_0 + \ell_1$ and $D(T^*) \geq \ell_1 + \ell_2$.

Algorithm 2. Approximation algorithm for Δ_β-SpHCP (G, c).

1. Guess the correct values of ℓ_0, ℓ_1 and ℓ_2. Their meanings are provided in the proof.
2. $H \leftarrow \{v \in V \setminus \{c\} \mid w(v, c) \leq \ell_0\}$.
3. Create an instance $\mathcal{J}$ of the k-center problem with forbidden centers, in which $V \setminus \{c\}$ is the set of input vertices, H is the set of allowed centers, $k = p$, and the distance function (satisfying the β-triangle inequality) is the restriction of w to $V \setminus \{c\}$.
4. Apply the greedy approximation algorithm for the k-center problem with forbidden centers (Algorithm 3), to obtain an approximate solution of $\mathcal{J}$. Assume that $H^* \subseteq H$ is the set of centers opened in the solution.
5. **return** the solution that opens H^* as the set of p hubs and assigns each vertex in $V \setminus \{c\}$ to its nearest hub in H^*.

Algorithm 3. Approximation algorithm for k-center with forbidden centers.

1. // Let C be the input vertex set, $C' \subseteq C$ be the set of allowed centers, and w be the distance function on C satisfying the β-triangle inequality. Assume w.l.o.g. that $k \leq |C'|$.
2. $R \leftarrow C; S \leftarrow \emptyset$.
3. **while** $R \neq \emptyset$ and $|S| < k$ **do**
4.   Choose an arbitrary vertex $v \in C' \cap R$.
5. $B(v) \leftarrow \{u \in R \mid w(u, v) \leq \beta(\ell_1 + \ell_2)\}$.
6. $R \leftarrow R \setminus B(v); S \leftarrow S \cup \{v\}$.
7. **end**
8. **if** $|S| < k$ and $R = \emptyset$ **then**
9.   select an arbitrary vertex set $S' \subseteq (C' \setminus S)$ of size $k - |S|; S \leftarrow S \cup S'$.
10. **return** S

Let T denote the solution returned by Algorithm 2. We next prove that Algorithm 2 is indeed a $(2\beta + 1)$-approximation algorithm for Δ_β-SpHCP by establishing an upper bound of $D(T)$. According to our choice of ℓ_0, the set H defined in line 2 contains all hub nodes in the optimal solution $N_{T^*}(c)$, *i.e.*, $N_{T^*}(c) \subseteq H$. In Line 3, we create an instance $\mathcal{J}$ of the k-center problem with forbidden centers. This problem is defined as follows: The input consists of a set C of demand points in a space satisfying the β-triangle inequality, a set $C' \subseteq C$ of allowed centers, and an integer k. The goal is to open k centers in C' such that the maximum distance between any vertex in C and its nearest center among the k opened centers is minimized. This problem is a generalization of the ordinary k-center problem (in which $C' = C$), and is a special case of the k-supplier problem (in which C' may not be a subset of C) [19–21]. There is a simple greedy approximation algorithm for this problem, which is presented in Algorithm 3. Its analysis is standard and is similar to that of the traditional k-center problem (see [19–21]), and thus is omitted here.

Hence, by applying the greedy approximation algorithm (Algorithm 3) to implement line 4 of Algorithm 2, we obtain a solution H^* of $\mathcal{J}$ with objective value at most $\beta(\ell_1 + \ell_2)$, that is,

$$\max_{v \in V \setminus \{c\}} \min_{h \in H^*} w(v, h) \le \beta(\ell_1 + \ell_2). \tag{1}$$

In Line 5 of Algorithm 2, a solution is returned that opens H^* as the set of p hubs. For each $v \in V \setminus (H^* \cup \{c\})$, let $f'(v) := \arg\min_{h \in H^*} w(v, h)$; i.e., $f'(v)$ is the hub in H^* assigned to v in the solution returned by the algorithm. Let ℓ_1' and ℓ_2' be the largest value and second-largest value in the multiset $\{w(v, f'(v)) \mid v \in V \setminus \{c\}\}$. By inequality (1), we have $\ell_1' + \ell_2' \le 2\beta(\ell_1 + \ell_2)$.

Let $x, y \in V \setminus \{c\}$ be the nodes achieving the maximum path length in T, i.e., $d_T(x, y) = D(T)$. It suffices to show that $D(T) \le (2\beta + 1) \cdot D(T^*)$.

If $f'(x) = f'(y)$, then $D(T) = w(x, f'(x)) + w(y, f'(y)) \le \ell_1' + \ell_2'$.

If $f'(x) \ne f'(y)$, then

$$D(T) = w(x, f'(x)) + w(f'(x), c) + w(f'(y), c) + w(y, f'(y)) \le \ell_1' + 2\ell_0 + \ell_2'$$

where we use $w(h, c) \le \ell_0$ for all $h \in H$ by our choice of H. Combine with the fact that $D(T^*) \ge \ell_0 + \ell_1$ and $D(T^*) \ge \ell_1 + \ell_2$, we always have

$$\begin{aligned}
D(T) &\le 2\ell_0 + \ell_1' + \ell_2' \\
&\le 2\ell_0 + 2\beta(\ell_1 + \ell_2) \\
&\le 2(\ell_0 + \ell_1) + (2\beta - 1)(\ell_1 + \ell_2) \quad \text{(using } \ell_2 \le \ell_1) \\
&\le 2 \cdot D(T^*) + (2\beta - 1) \cdot D(T^*) \\
&= (2\beta + 1) \cdot D(T^*),
\end{aligned}$$

which indicates that Algorithm 2 is a $(2\beta + 1)$-approximation algorithm for Δ_β-SpHCP. This completes the proof. □

We close this section with the following theorem.

Theorem 2. *Let $\beta \ge \frac{1}{2}$. There exists a polynomial time $r(\beta)$-approximation algorithm for Δ_β-SpHCP where*

(i) $r(\beta) = 1$ if $\beta \le \frac{3 - \sqrt{3}}{2}$;

(ii) $r(\beta) = \frac{1 + 2\beta - 2\beta^2}{4(1 - \beta)}$ if $\frac{3 - \sqrt{3}}{2} < \beta \le 0.7737\ldots$;

(iii) $r(\beta) = 1 + \frac{4\beta^2}{5\beta + 1}$ if $0.7737\ldots \le \beta \le 1$;

(iv) $r(\beta) = \beta + \frac{4\beta^2 - 2\beta}{2 + \beta}$ if $1 \le \beta \le 2$;

(v) $r(\beta) = 2\beta + 1$ if $\beta \ge 2$.

3   Conclusion

In this paper, we have studied Δ_β-SpHCP for all $\beta \ge \frac{1}{2}$. We showed that for any $\varepsilon > 0$, to approximate Δ_β-SpHCP to a ratio $g(\beta) - \varepsilon$ is NP-hard where

$g(\beta) = \frac{1+2\beta-2\beta^2}{4(1-\beta)}$ if $\frac{3-\sqrt{3}}{2} < \beta \leq \frac{2}{3}$; $g(\beta) = \frac{5\beta+1}{4}$ if $\frac{2}{3} < \beta \leq 1$; $g(\beta) = \beta + \frac{1}{2}$ if $\beta \geq 1$. Moreover, we gave $r(\beta)$-approximation algorithms for the same problem. If $\beta \leq \frac{3-\sqrt{3}}{2}$, we have $r(\beta) = g(\beta) = 1$, $i.e.$, Δ_β-SpHCP is polynomial time solvable for $\beta \leq \frac{3-\sqrt{3}}{2}$. If $\frac{3-\sqrt{3}}{2} < \beta \leq \frac{2}{3}$, we have $r(\beta) = g(\beta) = \frac{1+2\beta-2\beta^2}{4(1-\beta)}$. For $\frac{2}{3} \leq \beta \leq 1$, $r(\beta) = \min\{\frac{1+2\beta-2\beta^2}{4(1-\beta)}, 1 + \frac{4\beta^2}{5\beta+1}\}$. For $\beta \geq 1$, we have $r(\beta) = \min\{\beta + \frac{4\beta^2-2\beta}{2+\beta}, 2\beta+1\}$. In the future work, it is of interest to extend the range of β for Δ_β-SpHCP such that the gap between the upper and lower bounds of approximability can be reduced.

References

1. Alumur, S.A., Kara, B.Y.: Network hub location problems: the state of the art. Eur. J. Oper. Res. **190**, 1–21 (2008)
2. Andreae, T.: On the traveling salesman problem restricted to inputs satisfying a relaxed triangle inequality. Networks **38**, 59–67 (2001)
3. Andreae, T., Bandelt, H.-J.: Performance guarantees for approximation algorithms depending on parameterized triangle inequalities. SIAM J. Discret. Math. **8**, 1–16 (1995)
4. Ausiello, G., Crescenzi, P., Gambosi, G., Kann, V., Marchetti-Spaccamela, A., Protasi, M.: Complexity and Approximation: Combinatorial Optimization Problems and Their Approximability Properties. Springer, Heidelberg (1999)
5. Bender, M.A., Chekuri, C.: Performance guarantees for the TSP with a parameterized triangle inequality. Inf. Process. Lett. **73**, 17–21 (2000)
6. Böckenhauer, H.-J., Bongartz, D., Hromkovič, J., Klasing, R., Proietti, G., Seibert, S., Unger, W.: On the hardness of constructing minimal 2-connected spanning subgraphs in complete graphs with sharpened triangle inequality. In: Agrawal, M., Seth, A. (eds.) FSTTCS 2002. LNCS, vol. 2556, pp. 59–70. Springer, Heidelberg (2002). doi:10.1007/3-540-36206-1_7
7. Böckenhauer, H.-J., Bongartz, D., Hromkovič, J., Klasing, R., Proietti, G., Seibert, S., Unger, W.: On k-edge-connectivity problems with sharpened triangle inequality. In: Petreschi, R., Persiano, G., Silvestri, R. (eds.) CIAC 2003. LNCS, vol. 2653, pp. 189–200. Springer, Heidelberg (2003). doi:10.1007/3-540-44849-7_24
8. Böckenhauer, H.-J., Bongartz, D., Hromkovič, J., Klasing, R., Proietti, G., Seibert, S., Unger, W.: On k-connectivity problems with sharpened triangle inequality. J. Discret. Algorithms **6**(4), 605–617 (2008)
9. Böckenhauer, H.-J., Hromkovič, J., Klasing, R., Seibert, S., Unger, W.: Approximation algorithms for the TSP with sharpened triangle inequality. Inf. Process. Lett. **75**, 133–138 (2000)
10. Böckenhauer, H.-J., Hromkovič, J., Klasing, R., Seibert, S., Unger, W.: Towards the notion of stability of approximation for hard optimization tasks and the traveling salesman problem. In: Bongiovanni, G., Petreschi, R., Gambosi, G. (eds.) CIAC 2000. LNCS, vol. 1767, pp. 72–86. Springer, Heidelberg (2000). doi:10.1007/3-540-46521-9_7
11. Böckenhauer, H.-J., Hromkovič, J., Klasing, R., Seibert, S., Unger, W.: An improved lower bound on the approximability of metric TSP and approximation algorithms for the TSP with sharpened triangle inequality. In: Reichel, H., Tison, S. (eds.) STACS 2000. LNCS, vol. 1770, pp. 382–394. Springer, Heidelberg (2000). doi:10.1007/3-540-46541-3_32

12. Böckenhauer, H.-J., Hromkovič, J., Seibert, S.: Stability of approximation. In: Gonzalez, T.F. (ed.) Handbook of Approximation Algorithms and Metaheuristics. Chapman & Hall/CRC, Boca Raton (2007). Chapter 31

13. Böckenhauer, H.-J., Seibert, S.: Improved lower bounds on the approximability of the traveling salesman problem. RAIRO - Theor. Inform. Appl. **34**, 213–255 (2000)

14. Campbell, J.F.: Integer programming formulations of discrete hub location problems. Eur. J. Oper. Res. **72**, 387–405 (1994)

15. Campbell, J.F., Ernst, A.T.: Hub location problems. In: Drezner, Z., Hamacher, H.W. (eds.) Facility Location: Applications and Theory, pp. 373–407. Springer, Berlin (2002)

16. Chen, L.-H., Cheng, D.-W., Hsieh, S.-Y., Hung, L.-J., Lee, C.-W., Wu, B.Y.: Approximation algorithms for single allocation k-hub center problem. In: Proceedings of the 33rd Workshop on Combinatorial Mathematics and Computation Theory (CMCT 2016), pp. 13–18 (2016)

17. Chen, L.-H., Cheng, D.-W., Hsieh, S.-Y., Hung, L.-J., Lee, C.-W., Wu, B.Y.: Approximation algorithms for the star k-hub center problem in metric graphs. In: Dinh, T.N., Thai, M.T. (eds.) COCOON 2016. LNCS, vol. 9797, pp. 222–234. Springer, Cham (2016). doi:10.1007/978-3-319-42634-1_18

18. Ernst, A.T., Hamacher, H., Jiang, H., Krishnamoorthy, M., Woeginger, G.: Uncapacitated single and multiple allocation p-hub center problem. Comput. Oper. Res. **36**, 2230–2241 (2009)

19. Gonzalez, T.F.: Clustering to minimize the maximum intercluster distance. Theor. Comput. Sci. **38**, 293–306 (1985)

20. Hochbaum, D.S. (ed.): Approximation Algorithms for NP-hard Problems. PWS Publishing Company, Pacific Grove (1996)

21. Hochbaum, D.S., Shmoys, D.B.: A unified approach to approximation algorithms for bottleneck problems. J. ACM **33**, 533–550 (1986)

22. Hromkovič, J.: Stability of approximation algorithms for hard optimization problems. In: Pavelka, J., Tel, G., Bartošek, M. (eds.) SOFSEM 1999. LNCS, vol. 1725, pp. 29–47. Springer, Heidelberg (1999). doi:10.1007/3-540-47849-3_2

23. Kara, B.Y., Tansel, B.Ç.: On the single-assignment p-hub center problem. Eur. J. Oper. Res. **125**, 648–655 (2000)

24. Liang, H.: The hardness and approximation of the star p-hub center problem. Oper. Res. Lett. **41**, 138–141 (2013)

25. Meyer, T., Ernst, A., Krishnamoorthy, M.: A 2-phase algorithm for solving the single allocation p-hub center problem. Comput. Oper. Res. **36**, 3143–3151 (2009)

26. O'Kelly, M.E., Miller, H.J.: Solution strategies for the single facility minimax hub location problem. Pap. Reg. Sci. **70**, 376–380 (1991)

27. Todosijević, R., Urošević, D., Mladenović, N., Hanafi, S.: A general variable neighborhood search for solving the uncapacitated r-allocation p-hub median problem. Optimization Letters **1**, 1–13 (2015). doi:10.1007/s11590-015-0867-6

28. Yaman, H., Elloumi, S.: Star p-hub center problem and star p-hub median problem with bounded path lengths. Comput. Oper. Res. **39**, 2725–2732 (2012)

Parameterized Resiliency Problems via Integer Linear Programming

Jason Crampton[1], Gregory Gutin[1], Martin Koutecký[2],
and Rémi Watrigant[3(✉)]

[1] Royal Holloway, University of London, Egham, Surrey TW20 0EX, UK
`jav.crampton@gmail.com`, `greg.gutin@gmail.com`
[2] Department of Applied Mathematics, Charles University, Prague, Czech Republic
`koutecky@kam.mff.cuni.cz`
[3] Inria Sophia Antipolis Méditerranée, 06902 Sophia-Antipolis Cedex, France
`rwatriga@gmail.com`

Abstract. We introduce a framework in parameterized algorithms whose purpose is to solve resiliency versions of decision problems. In resiliency problems, the goal is to decide whether an instance remains positive after *any* (appropriately defined) perturbation has been applied to it. To tackle these kinds of problems, some of which might be of practical interest, we introduce a notion of resiliency for Integer Linear Programs (ILP) and show how to use a result of Eisenbrand and Shmonin (Math. Oper. Res., 2008) on *Parametric Linear Programming* to prove that ILP RESILIENCY is fixed-parameter tractable (FPT) under a certain parameterization.

To demonstrate the utility of our result, we consider natural resiliency version of several concrete problems, and prove that they are FPT under natural parameterizations. Our first result, for a problem which is of interest in access control, subsumes several FPT results and solves an open question from Crampton *et al.* (AAIM 2016). The second concerns the Closest String problem, for which we extend an FPT result of Gramm *et al.* (Algorithmica, 2003). We also consider problems in the fields of scheduling and social choice. We believe that many other problems can be tackled by our framework.

1 Introduction

Questions of ILP feasibility are typically answered by finding an integral assignment of variables x satisfying $Ax \leq b$. By Lenstra's theorem [18], this problem can be solved in $O^*(f(n)) := O(f(n)L^{O(1)})$ time and space, where f is a function of the number of variables n only, and L is the size of the ILP (subsequent research has obtained an algorithm of the above running time with $f(n) = n^{O(n)}$

Gutin's research was supported by Royal Society Wolfson Research Merit Award. Koutecký's research was supported by projects 14-10003S of GA ČR and 338216 of GA UK.

D. Fotakis et al. (Eds.): CIAC 2017, LNCS 10236, pp. 164–176, 2017.
DOI: 10.1007/978-3-319-57586-5_15

and using polynomial space [12,17]). In the language of parameterized complexity, this means that ILP FEASIBILITY is fixed-parameter tractable (FPT) parameterized by the number of variables. Note that there are a number of parameterized problems for which the only (known) way to prove fixed-parameter tractability is to use Lenstra's theorem[1] [7]. For more details on this topic, we refer the reader to [7,9].

The notion of *resiliency* measures the extent to which a system can tolerate modifications to its configuration and still satisfy given criteria. An organization might, for example, wish to know whether it will still be able to continue functioning, even if some of its staff become unavailable. In the language of decision problems, we would like to know whether an instance is still positive after *any* (appropriately defined) modification. Intuitively, the resiliency version of a problem is likely to be harder than the problem itself; a naive algorithm would consider every allowed modification of the input, and then see whether a solution exists.

In this paper, we introduce a framework for dealing with resiliency problems, and study their computational complexity through the lens of fixed-parameter tractability. We define resiliency for Integer Linear Programs (ILP) and show that the obtained problem can be solved in FPT time using a result of Eisenbrand and Shmonin on *Parametric Linear Programming* [10]. To illustrate the fact that our approach might be useful in different situations, we apply our framework to several concrete problems.

Crampton *et al.* analyzed the parameterized tractability of the RESILIENCY CHECKING PROBLEM (RCP) [5], which has practical applications in the context of access control [19], and can be seen as a generalization of a resiliency version of the SET COVER problem. In a nutshell, given a set of p elements R and a familly $\mathcal{U}$ of n subsets of R, the RCP asks whether for every $\mathcal{S} \subseteq \mathcal{U}$ with $|\mathcal{S}| \leq s$, one can find $\mathcal{U}_1, \ldots, \mathcal{U}_d \subseteq \mathcal{U} \setminus \mathcal{S}$, such that $\mathcal{U}_i \cap \mathcal{U}_j = \emptyset$ whenever $i \neq j$, and for every $i \in [d]$: $|\mathcal{U}_i| \leq t$, and $\bigcup_{U \in \mathcal{U}_i} U = R$. Observe that SET COVER is the particular case where $s = 0$ and $d = 1$. Thus, RCP has five natural parameters n, p, s, d and t, among which n is assumed to be large in the practical application we consider, relative to the other four parameters [19]. Using well-known tools in parameterized algorithms, Crampton *et al.* [5] were able[2] to determine the complexity of RCP (FPT, XP, W[2]-hard, para-NP-hard or para-coNP hard) for all but two combinations of p, s, d and t (these two combinations are p and p, t)[3]. In particular, in the case where $s = 0$ (when no resiliency is considered), they proved, using Lenstra's theorem, that RCP is FPT parameterized by p. However, they could not extend this result to the case of any s, and thus the

---

[1] Lenstra's theorem allows us to prove a mainly classification result, i.e. the FPT algorithm is unlikely to be efficient in practice, nevertheless Lenstra's theorem indicates that efficient FPT algorithms are a possibility, at least for subproblems of the problem under considerations.

[2] We have also established that certain sub-cases of RCP are FPT using reductions to the WORKFLOW SATISFIABILITY PROBLEM [6].

[3] By definition, a problem with several parameters $p_1, \ldots, p_\ell$ is the problem with one parameter, the sum $p_1 + \cdots + p_\ell$.

complexity of RCP parameterized by p was left open. We settle this case in this paper by showing that, in general, RCP is FPT parameterized by p. This result gives the complete picture of the parameterized complexity of RCP depending on the considered parameter.

We introduce an extension of the CLOSEST STRING problem, a problem arising in computational biology. Informally, CLOSEST STRING asks whether there exists a string that is "sufficiently close" to each member of a set of input strings. We modify the problem so that the input strings may be unreliable – due to transcription errors, for example – and show that this resiliency version of CLOSEST STRING called GLOBAL RESILIENCY CLOSEST STRING is FPT when parameterized by the number of input strings. Our resiliency result on CLOSEST STRING is a generalization of a result of Gramm *et al.* for CLOSEST STRING which was proved using Lenstra's theorem [13][4].

We introduce a resiliency version of the scheduling problem of makespan minimization on unrelated machines. We prove that this version is FPT when parameterized by the number of machines, the number of job types and the total expected downtime, generalizing a result of Mnich and Wiese [21] provided the jobs processing times are upper-bounded by a number given in unary.

Finally, we introduce a resilient swap bribery problem in the field of social choice and prove that it is FPT when parameterized by the number of candidates.

The remainder of the paper is structured in the following way. Section 2 introduces ILP resiliency and proves that it is FPT under a certain parameterization. We then apply our framework to a number of concrete problems. We establish the fixed-parameter tractability of RCP parameterized by p in Sect. 3. In Sect. 4, we introduce a resiliency version of CLOSEST STRING Problem and prove that it is FPT. We study resiliency versions of scheduling and social choice problems in Sects. 5 and 6. We conclude the paper in Sect. 7, where we discuss related literature. Due to space limits, some proofs (of results marked by a ⋆) as well as another resiliency version of the CLOSEST STRING problem were omitted. They can be found in the long version of the paper [4].

2   ILP Resiliency

Recall that questions of ILP feasibility are typically answered by finding an integral assignment of variables x satisfying $Ax \leq b$. Let us introduce resiliency for ILP as follows. We add another set of variables z, which can be seen as "resiliency variables". We then consider the following ILP[5] denoted by $\mathcal{R}$:

$$Ax \leq b \tag{1}$$

$$Cx + Dz \leq e \tag{2}$$

$$Fz \leq g \tag{3}$$

---

[4] Although not being strictly the first problem proved to be FPT using Lenstra's theorem [23], it is considered as the one which popularized this technique [7,9].

[5] To save space, we will always implicitly assume that integrality constraints are part of every ILP of this paper.

The idea is that inequalities (1) and (2) represent the intrinsic structure of the problem, among which inequalities (2) represent how the resiliency variables modify the instance. Inequalities (3), finally, represent the structure of the resiliency part. The goal of ILP RESILIENCY is to decide whether $\mathcal{R}$ is z-*resilient*, i.e. whether for *any* integral assignment of variables z satisfying inequalities (3), there exists an integral assignment of variables x satisfying (1) and (2).

In $\mathcal{R}$, we will assume that all entries of matrices in the left hand sides and vectors in the right hand sides are rational numbers. The dimensions of the vectors x and z will be denoted by n and p, respectively, and the total number of rows in A and C will be denoted by m. Let $\kappa(\mathcal{R}) := n + p + m$.

Our main result establishes that ILP RESILIENCY is FPT when parameterized by $\kappa(\mathcal{R})$, provided that part of the input is given in unary. Our method offers a generic framework to capture many situations. Firstly, it applies to ILP, a general and powerful model for representing many combinatorial problems. Secondly, the resiliency part of each problem can be represented as a whole ILP with its own variables and constraints, instead of, say, a simple additive term. Hence, we believe that our method can be applied to many other problems, as well as many different and intricate definitions of resiliency.

To prove our main result we will use the work of Eisenbrand and Shmonin [10]. For a rational polyhedron $Q \subseteq \mathbb{R}^{m+p}$, define $Q/\mathbb{Z}^p := \{h \in \mathbb{Q}^m : (h, \alpha) \in Q$ for some $\alpha \in \mathbb{Z}^p\}$. The PARAMETRIC INTEGER LINEAR PROGRAMMING (PILP) problem takes as input a rational matrix $J \in \mathbb{Q}^{m \times n}$ and a rational polyhedron $Q \subseteq \mathbb{R}^{m+p}$, and asks whether the following expression is true:

$$\forall h \in Q/\mathbb{Z}^p \quad \exists x \in \mathbb{Z}^n : \quad Jx \leq h$$

Eisenbrand and Shmonin [10, Theorem 4.2] proved that PILP is solvable in polynomial time if the number of variables $n + p$ is fixed. From this result, an interesting question is whether this running time is a uniform or non-uniform polynomial algorithm [9], and in particular for which parameters one can obtain an FPT algorithm. By looking closer at their algorithm, one can actualy obtain the following result:

Theorem 1. PILP *can be solved in time* $O^*(g(n, m, p)\varphi^{f(n,m,p)})$, *where* $\varphi \geq 2$ *is an upper bound on the encoding length of entries of* J *and* f *and* g *are some computable functions.*

Complexity Remark. Let us first mention that PILP belongs to the second level of the polynomial hierarchy, and is Π_2^P-complete [10]. Secondly, the polyhedron Q in Theorem 1 can be viewed as being defined by a system $Rh + S\alpha \leq t$, where $h \in \mathbb{R}^m$ and $\alpha \in \mathbb{Z}^p$. Then the algorithm of the theorem runs in time polynomial in the encoding lengths of R, S, and in m (the "continuous dimension"), and is FPT with respect to p (the "integer dimension").

Corollary 1. *If* n, m *and* p *are the parameters and all entries of* J *are given in unary, then* PILP *is FPT.*

Proof. We may assume that there is an upper bound $N \geq 2$ on the absolute values of entries of A and N is given in unary. Thus, the running time of the algorithm of Theorem 1 is $O^*(g(n, m, p)(\log N)^F)$, where $F = f(n, m, p)$.

It was shown in [3] that $(\log N)^F \leq (2F \log F)^F + N/2^F$, which concludes the proof. $\square$

We now prove the main result of our framework, which will be applied in the next sections to two concrete problems.

Theorem 2. ILP RESILIENCY *is FPT when parameterized by* $\kappa(\mathcal{R})$ *provided the entries of matrices* A *and* C *are given in unary.*

Proof. We will reduce ILP RESILIENCY to PARAMETRIC INTEGER LINEAR PRO-GRAMMING. Let us first define J and Q. Let $h = (h^1, h^2)$ with h^1 and h^2 being m_1 and m_2 dimensional vectors, respectively. Then the polyhedron Q is defined as follows: $h^1 = b, h^2 = e - D\alpha, F\alpha \leq g$. Furthermore, J is defined as: $Ax \leq h^1, Cx \leq h^2$.

Recall that $h^1 = b$ and $h^2 = e - D\alpha$ and α satisfies $F\alpha \leq g$, so for all $h \in Q/\mathbb{Z}^p$ there exists an integral x satisfying the above if and only if for all z satisfying $Fz \leq g$, there is an integral x satisfying (1) and (2). Moreover, the dimension of x is n, the integer dimension of Q is p and the number of inequalities of J is $m_1 + m_2 = m$, so applying Corollary 1 indeed yields the required FPT algorithm. $\square$

3   Resiliency in Access Control

Access control is an important topic in computer security and is typically achieved by enforcing a policy that specifies which users are authorized to access which resources. Authorization policies are frequently augmented by additional policies, articulating concerns such as separation of duty and resiliency. The RESILIENCY CHECKING PROBLEM (RCP) was introduced by Li *et al.* [19] and asks whether it is always possible to allocate authorized users to teams, even if some users are unavailable.

3.1   Definition of the Problem

Given a set of users U and set of resources R, an *authorization policy* is a relation $UR \subseteq U \times R$; we say u is *authorized* for resource r if $(u, r) \in UR$. For a user $u \in U$, we define $N_{UR}(u) = \{r \in R : (u, r) \in UR\}$, the *neighborhood* of u; by extension, for $V \subseteq U$, we define $N_{UR}(V) = \bigcup_{u \in V} N_{UR}(u)$, the *neighborhood* of V. Thus $N_{UR}(u)$ represents the resources for which u is authorized, and $N_{UR}(V)$ represents the resources for which the users in V are collectively authorized. We will omit the subscript UR if the authorization policy is clear from the context.

Given an authorization policy $UR \subseteq U \times R$, an instance of the RESILIENCY CHECKING PROBLEM (RCP) is defined by a resiliency policy $\mathsf{res}(P, s, d, t)$, where $P \subseteq R$, $s \geq 0$, $d \geq 1$ and $t \geq 1$. We say that UR *satisfies* $\mathsf{res}(P, s, d, t)$ if and only if for every subset $S \subseteq U$ of at most s users, there exist d pairwise disjoint subsets of users $V_1, \ldots, V_d$ such that for all $i \in \{1, \ldots, d\}$:

$$V_i \cap S = \emptyset, \tag{4}$$

$$|V_i| \leq t \text{ and } N(V_i) \supseteq P. \tag{5}$$

In other words, UR satisfies $\mathsf{res}(P, s, d, t)$ if we can find d disjoint groups of users, even if up to s users are unavailable, such that each group contains no more than t users and the users in each group are collectively authorized for the resources in P (observe that the particular case in which $s = 0$ and $d = 1$ is equivalent to the well-known SET COVER problem) Thus, we define RCP as follows:

RESILIENCY CHECKING PROBLEM (RCP)
Input: $UR \subseteq U \times R$, $P \subseteq R$, $s \geq 0$, $d \geq 1$, $t \geq 1$.
Question: Does UR satisfy $\mathsf{res}(P, s, d, t)$?

In the remainder of this section, we set $p = |P|$. Given an instance of RCP, we say that a set of d pairwise disjoint subsets of users $V = \{V_1, \ldots, V_d\}$ satisfying conditions (5) is a *set of teams*. For such a set of teams, we define $\mathcal{U}(V) = \bigcup_{i=1}^{d} V_i$. Given $U' \subseteq U$, the *restriction* of UR to U' is defined by $UR|_{U'} = UR \cap (U' \times R)$. Finally, a set of users $S \subseteq U$ is called a *blocker set* if for every set of teams $V = \{V_1, \ldots, V_d\}$, we have $\mathcal{U}(V) \cap S \neq \emptyset$. Equivalently, observe that S is a blocker set if and only if $UR|_{U \setminus S}$ does not satisfy $\mathsf{res}(P, 0, d, t)$.

3.2   Fixed-Parameter Tractability of RCP

In this section we prove that RCP is FPT parameterized by p. We first introduce some notation. In the following, $UR \subseteq U \times R$, $P \subseteq R$, $s \geq 0$, $d \geq 1$ and $t \geq 1$ will denote an input of RCP. Without loss of generality, we may assume $P = R$ and $N(u) \neq \emptyset$ for all $u \in U$. For all $N \subseteq P$, let $U_N = \{u \in U : N(u) = N\}$ (notice that we may have $U_N = \emptyset$ for some $N \subseteq P$).

Roughly speaking, the idea is that in order to construct a set of teams or a blocker set, it is sufficient to know the size of its intersection with U_N, for every $N \subseteq P$. We first define the set of *configurations*.

$$\mathcal{C} = \left\{ \{N_1, \ldots, N_b\} : b \leq t, N_i \subseteq P, i \in [b], \bigcup_{i=1}^{b} N_i = P \right\}.$$

Then, for any $N \subseteq P$, we denote the set of configurations involving N by $\mathcal{C}_N$. That is

$$\mathcal{C}_N = \{ c = \{N_1, \ldots, N_{b_c}\} \in \mathcal{C} : N = N_i \text{ for some } i \in [b_c] \}$$

Observe that since we assume $t \leq p$, we have $|\mathcal{C}| = O(2^{p^2})$. The link between sets of teams and configurations comes from the following definition: given a set of teams V, we say that a team $T \in V$ *has configuration $c \in C$* if $c = \{N(u), u \in T\}$. In other words, c represents the distinct neighborhoods of users of T in P.

We define an ILP $\mathcal{L}$ over the set of variables $x = (x_c : c \in C)$ and $z = (z_N : N \subseteq P)$, with the following inequalities:

$$\sum_{c \in C} x_c \geq d \tag{6}$$

$$\sum_{N \subseteq P} z_N \leq s \tag{7}$$

$$\sum_{c \in C_N} x_c \leq |U_N| - z_N \quad \text{for every } N \subseteq P \tag{8}$$

$$0 \leq z_N \leq |U_N| \qquad \text{for every } N \subseteq P \tag{9}$$

$$0 \leq x_c \leq d \qquad \text{for every } c \in C \tag{10}$$

Observe that $\kappa(\mathcal{L})$ is upper bounded by a function of p only. The idea behind this model is to represent a set S of at most s users by variables z (by deciding how many users to take for each set of users U_N, $N \subseteq P$), and to represent a set of teams by variables x (by deciding how many teams will have configuration $c \in C$). Then, inequalities (8) will ensure that the set of teams does not intersect with the chosen set S. However, while we would be able to solve $\mathcal{L}$ in FPT time parameterized by p by using, *e.g.*, Lenstra's ILP Theorem, the reader might realize that doing so would not solve RCP directly. Nevertheless, the following result establishes the crucial link between this system and our problem.

Lemma 1. [⋆] res(P, s, d, t) *is satisfiable if and only if $\mathcal{L}$ is z-resilient.*

Since, as we observed earlier, $\kappa(\mathcal{L})$ is bounded by a function of p only, combining Lemma 1 with Theorem 2, we obtain the following:

Theorem 3. RCP *is FPT parameterized by p.*

4   Closest String Problem

In the CLOSEST STRING problem, we are given a collection of k strings $s_1, \ldots, s_k$ of length L over a fixed alphabet Σ, and a non-negative integer d. The goal is to decide whether there exists a string s (of length L) such that $d_H(s, s_i) \leq d$ for all $i \in [k]$, where $d_H(s, s_i)$ denotes the Hamming distance between s and s_i. If such a string exists, then it will be called a *d-closest string*.

It is common to represent an instance of the problem as a matrix C with k rows and L columns (*i.e.* where each row is a string of the input); hence, in the following, the term *column* will refer to a column of this matrix. As Gramm *et al.* [13] observe, as the Hamming distance is measured column-wise, one can identify some columns sharing the same structure. Let $\Sigma = \{\varphi_1, \ldots, \varphi_{|\Sigma|}\}$. Gramm *et al.* show [13] that after a simple preprocessing of the instance, we may assume that for every column c of C, φ_i is the ith character that appears the most often (in c), for $i \in \{1, \ldots, |\Sigma|\}$ (ties broken *w.r.t.* the considered ordering of Σ). Such a preprocessed column will be called *normalized*, and by extension, a matrix consisting of normalized columns will be called *normalized*. One can observe that after this preprocessing, the number of different columns (called *column type*) is bounded by a function of k only, namely by the kth Bell number $\mathcal{B}_k = O(2^{k \log_2 k})$. The set of all column types is denoted by T. Using this observation, Gramm *et al.* [13] prove that CLOSEST STRING is FPT parameterized by k, using an ILP with a number of variables depending on k only, and then applying Lenstra's theorem.

4.1 Adding Resiliency

The motivation for studying resiliency with respect to this problem is the introduction of experimental errors, which may change the input strings [22]. While a solution of the CLOSEST STRING problem tests whether the input strings are consistent, a resiliency version asks whether these strings will remain consistent after some small changes. However, there exist several ways to define how these changes will modify the input. The version we consider here, called GLOBAL RESILIENCY CLOSEST STRING, allows at most m changes to appear anywhere in the matrix C. Another version, called COLUMN RESILIENCY CLOSEST STRING and studied in [4], we allow changes to appear column-wise. This situation might be useful if experimental errors occur more often at, say, the beginning or end of the string. It appears that both problems remain FPT parameterized by the number of input strings, generalizing in two different ways the result of Gramm *et al.* [13].

As said previously, the most natural way of defining a notion of resiliency in the context of CLOSEST STRING is to allow changes at any places of the matrix C. The only constraint is thus an upper bound on the number of total changes. To represent this, we simply use the Hamming distance between two matrices.

GLOBAL RESILIENCY CLOSEST STRING (GRCS)
Input: C, a $k \times L$ normalized matrix of elements of Σ, $d \in \mathbb{N}$, $m \leq kL$.
Question: For every C', $k \times L$ normalized matrix of elements of Σ such that the Hamming distance of C and C' is at most m, does C' admit a d-closest string?

ILP Formulation. Let $\#_t$ be the number of columns of type t in C. For two types $t, t' \in T$ let $\delta(t, t')$ be their Hamming distance (the number of different elements). Let $z_{t,t'}$, for all $t, t' \in T$, be a variable meaning "how many columns of type t in C are changed to type t' in C'" (we allow $t = t'$). Thus we have the following constraints:

$$\sum_{t' \in T} z_{t,t'} = \#_t \qquad\qquad \forall t \in T \qquad\qquad (11)$$

$$\sum_{t,t' \in T} \delta(t, t') z_{t,t'} \leq m \qquad\qquad (12)$$

These constraints clearly capture all possible scenarios of how the input strings can be modified in at most m places. Then let $\#'_t$ be a variable meaning "how many columns of C' are of type t", and let $x_{t,\varphi}$ represent the number of columns of type t in C' whose corresponding character in the solution is set to φ. Finally let $\Delta(t, \varphi)$ be the number of characters of t which are different from φ. As the remaining constraints correspond to our formulation of ILP RESILIENCY, we have:

$$\sum_{t \in T} z_{t,t'} = \#'_{t'} \qquad\qquad \forall t' \in T \qquad (13)$$

$$\sum_{\varphi \in \Sigma} x_{t,\varphi} = \#'_t \qquad\qquad \forall t \in T \qquad (14)$$

$$\sum_{t \in T} \sum_{\varphi \in \Sigma} \Delta(t, \varphi) x_{t,\varphi} \leq d \qquad\qquad (15)$$

This is the standard ILP for CLOSEST STRING [13], except that $\#'_t$ are now variables, and there exists a solution x exactly when there is a string at distance at most d from the modified strings given by the variables $\#'$. Let $\mathcal{L}$ denote the ILP composed of constraints (11), (12), (13), (14) and (15). Finally, let $\mathcal{Z}$ denote variables $z_{t,t'}$ and $\#'_t$ for every $t, t' \in T$.

Lemma 2. [⋆] *The instance is satisfiable if and only if $\mathcal{L}$ is $\mathcal{Z}$-resilient.*

It remains to observe that for the above system of constraints $\mathcal{L}$, $\kappa(\mathcal{L})$ is bounded by a function of k (since $|T| = O(2^{k \log_2 k})$). We thus get the following:

Theorem 4. GRCS *is FPT parameterized by k.*

5   Resilient Scheduling

A fundamental scheduling problem is makespan minimization on unrelated machines, where we have m machines and n jobs, and each job has a vector of processing times with respect to machines $p_j = (p_j^1, \ldots, p_j^m)$, $j \in [n]$. If the vectors p_j and $p_{j'}$ are identical for two jobs j, j', we say these jobs are of the same *type*. Here we consider the case when m and the number of types θ are parameters and the input is given as θ numbers $n_1, \ldots, n_\theta$ of job multiplicities. A *schedule* is an assignment of jobs to machines. For a particular schedule, let n_t^i be the number of jobs of type t assigned to machine i. Then, the *completion time* of machine i is $C^i = \sum_{t \in [\theta]} p_t^i n_t^i$ and the largest C^i is the *makespan* of the schedule, denoted C_{max}.

The parameterization by θ and m might seem very restrictive, but note that when m alone is a parameter, the problem is W[1]-hard even when the machines are identical and the job lengths are given in unary [16]. Also, Asahiro et al. [1] show that it is strongly NP-hard already for *restricted assignment* when there is a number p_j for each job such that for each machine i, $p_j^i \in \{p_j, \infty\}$ and all $p_j \in \{1, 2\}$ and for every job there are exactly two machines where it can run. Mnich and Wiese [21] proved that the problem is FPT with parameters θ and m.

A natural way to introduce resiliency is when we consider unexpected delays due to repairs, fixing software bugs, etc., but we have an upper bound K on the total expected downtime. We assume that the execution of jobs can be resumed after the machine becomes available again, but cannot be moved to another machine, that is, we assume preemption but not migration. Under these assumptions it does not matter when specifically the downtime happens, only the total downtime of each machine. Given m machines, n jobs and $C_{max}, K \in \mathbb{N}$, we say that a scheduling instance has a K-tolerant makespan C_{max} if, for every $d_1, \ldots, d_m \in \mathbb{N}$ such that $\sum_{i=1}^{m} d_i \leq K$, there exists a schedule where each machine $i \in [m]$ finishes by the time $C_{max} - d_i$. We obtain the following problem:

RESILIENCY MAKESPAN MINIMIZATION ON UNRELATED MACHINES
Input: m machines, θ job types $p_1, \ldots, p_\theta \in \mathbb{N}^m$, job multiplicities $n_1, \ldots, n_\theta$, and $K, C_{max} \in \mathbb{N}$.
Question: Does this instance have a K-tolerant makespan C_{max} ?

Let x_t^i be a variable expressing how many jobs of type t are scheduled to machine i. We have the following constraints, with the first constraint describing the feasible set of delays, and the subsequent constraints assuring that every job is scheduled on some machine and that every machine finishes by time $C_{max} - d_i$:

$$\sum_{i=1}^{m} d_i \leq K$$

$$\sum_{i=1}^{m} x_i^t = n_t \qquad\qquad \forall t \in [\theta]$$

$$\sum_{t=1}^{\theta} x_t^i p_t^i \leq C_{max} - d_i \qquad\qquad \forall i \in [m]$$

Theorem 2 and the system of constraints above implies the following result related to the above-mentioned result of Mnich and Wiese [21].

Theorem 5. [$\star$] RESILIENCY MAKESPAN MINIMIZATION ON UNRELATED MACHINES *is FPT when parameterized by* θ, m *and* K *and with* $\max_{t \in [\theta], i \in [m]}$ $p_i^t \leq N$ *for some number* N *given in unary.*

6 Resilient Swap Bribery

The field of computational social choice is concerned with computational problems associated with voting in elections. SWAP BRIBERY, where the goal is to find the cheapest way to bribe voters such that a preferred candidate wins, has received considerable attention. This problem models not only actual bribery, but also processes designed to influence voting (such as campaigning). It is natural to consider the case where an adversarial counterparty first performs their bribery, where we only have an estimate on their budget. The question becomes if, for each such bribery, it is possible, within a given budget, to bribe the election such that our preferred candidate still wins. The number of candidates is a well studied parameter [2,8]. In this section we will show that the resilient version of SWAP BRIBERY with unit costs (unit costs are a common setting, cf. Dorn and Schlotter [8]) is FPT using our framework. Let us now give formal definitions.

Elections. An election $E = (C, V)$ consists of a set C of m candidates $c_1, \ldots, c_m$ and a set V of voters (or votes). Each voter i is a linear order $\succ_i$ over the set C. For distinct candidates a and b, we write $a \succ_i b$ if voter i prefers a over b. We denote by $\text{rank}(c, i)$ the position of candidate $c \in C$ in the order $\succ_i$. The preferred candidate is c_1.

Swaps. Let (C, V) be an election and let $\succ_i \in V$ be a voter. A *swap* $\gamma = (a, b)_i$ in preference order $\succ_i$ means to exchange the positions of a and b in $\succ_i$; denote the resulting order by $\succ_i^\gamma$; the *cost* of $(a, b)_i$ is $\pi_i(a, b)$ (in the problem studied in this paper, we have $\pi_i(a, b) = 1$ for every voter i and candidates a, b). A swap $\gamma = (a, b)_i$ is *admissible in* $\succ_i$ if $\text{rank}(a, i) = \text{rank}(b, i) - 1$. A set Γ of swaps is *admissible in* $\succ_i$ if they can be applied sequentially in $\succ_i$, one after the other, in some order, such that each one of them is admissible. Note that the obtained vote, denoted by $\succ_i^\Gamma$, is independent from the order in which the swaps of Γ are applied. We also extend this notation for applying swaps in several votes and denote it V^Γ.

Voting Rules. A voting rule $\mathcal{R}$ is a function that maps an election to a subset of candidates, the set of winners. We will show our example for rules which are scoring protocols, but following the framework of so-called "election systems described by linear inequalities" [8] it is easily seen that the result below holds for many other voting rules. With a scoring protocol $s = (s_1, \ldots, s_m) \in \mathbb{N}^m$, a voter i gives s_1 points to his most preferred candidate, s_2 points to his second most preferred candidate and so on. The candidate with most points wins.

RESILIENCY UNIT SWAP BRIBERY
Input: An election $E = (C, V)$ with each swap of unit cost and with a scoring protocol $s \in \mathbb{N}^m$, the adversary's budget B_a, our budget B.
Question: For every adversarial bribery Γ_a of cost at most B_a, is there a bribery Γ of cost at most B such that $E = (C, (V^{\Gamma_a})^\Gamma)$ is won by c_1?

Theorem 6. [⋆] RESILIENCY UNIT SWAP BRIBERY *with a scoring protocol is FPT when parameterized by the number of candidates* m.

7   Discussion

For some time, Lenstra's theorem was the only approach in parameterized algorithms and complexity based on integer programming. Recently other tools based on integer programming have been introduced: the use of Graver bases for the n-fold integer programming problem [14], the use of ILP approaches in kernelization [11], or, conversely, kernelization results for testing ILP feasibility [15], and an integer quadratic programming analog of Lenstra's theorem [20]. Our approach is a new addition to this powerful arsenal.

References

1. Asahiro, Y., Jansson, J., Miyano, E., Ono, H., Zenmyo, K.: Approximation algorithms for the graph orientation minimizing the maximum weighted outdegree. In: Kao, M.-Y., Li, X.-Y. (eds.) AAIM 2007. LNCS, vol. 4508, pp. 167–177. Springer, Heidelberg (2007). doi:10.1007/978-3-540-72870-2_16
2. Bredereck, R., Faliszewski, P., Niedermeier, R., Skowron, P., Talmon, N.: Elections with few candidates: prices, weights, and covering problems. In: Walsh, T. (ed.) ADT 2015. LNCS (LNAI), vol. 9346, pp. 414–431. Springer, Cham (2015). doi:10.1007/978-3-319-23114-3_25
3. Chitnis, R., Cygan, M., Hajiaghayi, M., Marx, D.: Directed subset feedback vertex set is fixed-parameter tractable. ACM Trans. Algorithms **11**(4), 1–28 (2015)
4. Crampton, J., Gutin, G., Watrigant, R.: An approach to parameterized resiliency problems using integer Linear Programming. CoRR, abs/1605.08738 (2016)
5. Crampton, J., Gutin, G., Watrigant, R.: A multivariate approach for checking resiliency in access control. In: Dondi, R., Fertin, G., Mauri, G. (eds.) AAIM 2016. LNCS, vol. 9778, pp. 173–184. Springer, Cham (2016). doi:10.1007/978-3-319-41168-2_15
6. Crampton, J., Gutin, G., Watrigant, R.: Resiliency policies in access control revisited. In: Proceedings SACMAT 2016, pp. 101–111. ACM (2016)
7. Cygan, M., Fomin, F.V., Kowalik, L., Lokshtanov, D., Marx, D., Pilipczuk, M., Pilipczuk, M., Saurabh, S.: Parameterized Algorithms. Springer, Switzerland (2015)
8. Dorn, B., Schlotter, I.: Multivariate complexity analysis of swap bribery. Algorithmica **64**(1), 126–151 (2012)
9. Downey, R.G., Fellows, M.R.: Fundamentals of Parameterized Complexity: Texts in Computer Science. Springer, London (2013)
10. Eisenbrand, F., Shmonin, G.: Parametric integer programming in fixed dimension. Math. Oper. Res **33**(4), 839–850 (2008)
11. Etscheid, M., Kratsch, S., Mnich, M., Röglin, H.: Polynomial kernels for weighted problems. J. Comput. Syst. Sci. **84**, 1–10 (2017)
12. Frank, A., Tardos, É.: An application of simultaneous diophantine approximation in combinatorial optimization. Combinatorica **7**(1), 49–65 (1987)
13. Gramm, J., Niedermeier, R., Rossmanith, P.: Fixed-parameter algorithms for CLOSEST STRING and related problems. Algorithmica **37**(1), 25–42 (2003)
14. Hemmecke, R., Onn, S., Romanchuk, L.: n-fold integer programming in cubic time. Math. Prog. **137**(1), 325–341 (2013)
15. Jansen, B.M.P., Kratsch, S.: A structural approach to kernels for ILPs: treewidth and total unimodularity. In: Bansal, N., Finocchi, I. (eds.) ESA 2015. LNCS, vol. 9294, pp. 779–791. Springer, Heidelberg (2015). doi:10.1007/978-3-662-48350-3_65
16. Jansen, K., Kratsch, S., Marx, D., Schlotter, I.: Bin packing with fixed number of bins revisited. J. Comput. Syst. Sci. **79**(1), 39–49 (2013)
17. Kannan, R.: Minkowski's convex body theorem and integer programming. Math. Oper. Res. **12**(3), 415–440 (1987)
18. Lenstra, H.W.: Integer programming with a fixed number of variables. Math. Op. Res. **8**(4), 538–548 (1983)
19. Li, N., Tripunitara, M.V., Wang, Q.: Resiliency policies in access control. ACM Trans. Inf. Syst. Secur. **12**(4), 113–137 (2009)
20. Lokshtanov, D.: Parameterized integer quadratic programming: variables and coefficients. CoRR, abs/1511.00310 (2015)

21. Mnich, M., Wiese, A.: Scheduling and fixed-parameter tractability. Math. Program. **154**(1), 533–562 (2014)
22. Pevzner, P.: Computational Molecular Biology: An Algorithmic Approach. MIT Press, Cambridge (2000)
23. Sebő, A.: Integer plane multiflows with a fixed number of demands. J. Comb. Theory Ser. B **59**(2), 163–171 (1993)

Push-Pull Block Puzzles are Hard

Erik D. Demaine, Isaac Grosof, and Jayson Lynch[✉]

MIT Computer Science and Artificial Intelligence Laboratory,
32 Vassar Street, Cambridge, MA 02139, USA
{edemaine,isaacg,jaysonl}@mit.edu

Abstract. This paper proves that push-pull block puzzles in 3D are
PSPACE-complete to solve, and push-pull block puzzles in 2D with thin
walls are NP-hard to solve, settling an open question [19]. Push-pull
block puzzles are a type of recreational motion planning problem, similar
to Sokoban, that involve moving a 'robot' on a square grid with 1×1
obstacles. The obstacles cannot be traversed by the robot, but some can
be pushed and pulled by the robot into adjacent squares. Thin walls
prevent movement between two adjacent squares. This work follows in
a long line of algorithms and complexity work on similar problems [3–
9,14,16,18]. The 2D push-pull block puzzle shows up in the video games
Pukoban as well as *The Legend of Zelda: A Link to the Past*, giving
another proof of hardness for the latter [2]. This variant of block-pushing
puzzles is of particular interest because of its connections to reversibility,
since any action (e.g., push or pull) can be inverted by another valid
action (e.g., pull or push).

Keywords: Complexity · NP · PSPACE-complete · Puzzles · Motion
planning

1 Introduction

Block-pushing puzzles are a common puzzle type with one of the best known
example being *Sokoban*. Puzzles with the ability to push and pull blocks have
found their way into several popular video games including *The Legend of Zelda*
series, *Starfox Adventures*, *Half-Life* and *Tomb Raider*. Block-pushing puzzles
are also an abstraction of motion planning problems with movable obstacles.
In addition to these games, one could imagine real-world scenarios, like that
of a forklift in a warehouse, bearing similarity. Since motion planning is such
an important and computationally difficult problem, it can be useful to look at
simplified models to try to get a better understanding of the larger problem.

A significant amount of research has gone into characterizing the complexity
of block sliding puzzles. This includes PSPACE-completeness for well-known
puzzles like sliding-block puzzles [13], Sokoban [3,9], the 15-puzzle [15], 2048 [1],
Candy Crush [12] and Rush Hour [10]. Block pushing puzzles are a type of block
sliding puzzle in which the blocks are moved by a small robot within the puzzles.
This type of block sliding puzzle has gathered a significant amount of study.

© Springer International Publishing AG 2017
D. Fotakis et al. (Eds.): CIAC 2017, LNCS 10236, pp. 177–195, 2017.
DOI: 10.1007/978-3-319-57586-5_16

Table 1 gives a summary of results on block pushing puzzles. Variations include Sokoban [3,9], where blocks must reach specific targets (the *Path?* column), versions where multiple blocks can be pushed [3–5,9,14,16] (the *Push* column), versions where blocks continue to slide after being pushed [7,14] (the *Sliding* column), versions where fixed blocks are allowed [5,8] (the *Fixed?* column), and versions where the robot can pull blocks [16] (the *Pull* column).

We are particularly interested in the push-pull block model because any sequence of moves in the puzzle can be undone. Having an undirected state-space graph seems like an interesting property both mathematically and from a puzzle stand-point. This sort of player move reversibility lead to some of our gadgets being logically reversible, a notion that is fundamentally linked to quantum computation and the thermodynamics of computation.

Table 1. Summary of past and new results on block pushing and/or pulling. The *Push* and *Pull* columns describe how many blocks in a row can be moved by the robot. Here k and l are positive integers; ∗ refers to an unlimited number of blocks. The *Fixed* column notes whether fixed blocks (Yes) or thin walls (Wall) are allowed. In the problem title, F means fixed blocks are included; W means thin walls are included. The *Path* column describes whether the objective is to have the robot find a path to a target location, or to store the blocks in a specific configuration. The *Sliding* column notes whether blocks move one square or as many squares as possible before stopping.

Name	Push	Pull	Fixed?	Path?	Sliding	Complexity
Push-k	k	0	No	Path	Min	NP-hard [4]
Push-∗	∗	0	No	Path	Min	NP-hard [14]
PushPush-k	k	0	No	Path	Max	PSPACE-c. [7]
PushPush-∗	∗	0	No	Path	Max	NP-hard [14]
Push-1F	1	0	Yes	Path	Min	NP-hard [8]
Push-kF	$k \geq 2$	0	Yes	Path	Min	PSPACE-c. [5]
Push-∗F	∗	0	Yes	Path	Min	PSPACE-c. [5]
Sokoban	1	0	Yes	Storage	Min	PSPACE-c. [3]
Sokoban$(k,1)$	$k \geq 5$	1	Yes	Storage	Min	NP-hard [9]
Pull-1	0	1	No	Storage	Min	NP-hard [16]
Pull-kF	0	k	Yes	Storage	Min	NP-hard [16]
PullPull-kF	0	k	Yes	Storage	Max	NP-hard [16]
Push-k Pull-lW	k	l	Wall	Path	Min	**NP-hard** (Sect. 2)
3D Push-k Pull-lF	k	l	Yes	Path	Min	**NP-hard** (Sect. B)
3D Push-1 Pull-1W	1	1	Wall	Path	Min	**PSPACE-c.** (Sect. 3)
3D Push-k Pull-kF	$k > 1$	$k > 1$	Yes	Path	Min	**PSPACE-c.** (Sect. 3)

We add several new results showing that certain block pushing puzzles, which include the ability to push and pull blocks, are NP-hard or PSPACE-complete. The push-pull block puzzle is instantiated in the game Pukoban and heuristics for solving it have been studied [19], but its computational complexity was left as an open question.

We introduce *thin walls*, which prevent motion between two adjacent empty squares. We prove that all path planning problems in 2D with thin walls or in 3D, in which the robot can push k blocks and pull l blocks for all $k, l \in \mathbb{Z}^+$ are NP-hard. We also show that path planning problems where the robot can push and pull k blocks are PSPACE-complete, with thin walls needed only for $k = 1$. Our results are shown in the last four lines of Table 1. To prove these results, we introduce two new abstract gadgets, the set-verify and the 4-toggle, and prove hardness results for questions about their the legal state transitions.

2D Push-k Pull-j is defined as follows: There is a square lattice of cells. Each cell is connected to its orthogonal neighbors. Cells may either be empty, hold a movable block, or hold a fixed block. Additionally, in settings that allow thin walls, edges between cells may be omitted. There is also a robot on a cell. The robot may move from its current cell to an unoccupied adjacent cell. The robot may also *push* up to k movable blocks arranged in a straight line one cell forward, as long as there is an open cell with no wall in that direction. Here the robot moves into the cell occupied by the adjacent block and each subsequent block moves into the adjacent cell in the same direction. Likewise, the robot may *pull* up to j movable blocks in a straight line as long as there are no walls in the way and there is an open cell behind the robot. The robot moves into that cell, the block opposite that cell moves into the one the robot originally occupied, and subsequent blocks also move once cell toward the robot. The goal of the puzzle is for the robot to reach a specified goal cell. Given such a description, is there a legal path for the robot from its starting cell to the goal cell? The 3D problem is defined analogously on a cubic lattice.

2 Push-Pull Block Puzzles are NP-hard

In this section we show NP-hardness for Push-k Pull-l in 2D with thin walls for all positive integers k, l in Sect. 2 and Push-q Pull-r in 3D for all positive integers q, r in Sect. B.

Thin walls are a new, but natural, notion for block pushing puzzles. They prevent blocks or the robot from passing between two adjacent, empty squares, as though there were a thin wall blocking the path. We will prove hardness by a reduction from 3SAT. The 3SAT problem asks whether, given a set of variables $\{x_1, x_2, \ldots x_n\}$ and a boolean formula in conjunctive normal form with exactly three variables per clause, there exists an assignment of values to those variables that satisfies the formula [11]. To do so we will introduce an abstract gadget called the Set-Verify gadget. This gadget will then be used to construct crossover gadgets (in Appendix A), and variable and clause gadgets.

Set-Verify Gadgets. The Set-Verify gadget is an abstract gadget for motion planning problems. The gadget has four entrances/exits which have different allowable paths between them depending on the state of the gadget. There are four possible states of the Set-Verify gadget: Broken, Unset, Set, and Verified. The three relevant states are depicted in Figs. 1 and 2. Entrances to the gadget are labeled S_i, S_o, V_i, V_o and the directed arrows show the allowed passages in the shown state. The state transitions for the Set-Verify are given in Table 2. Further details are given in Appendix A.

Since the Set-Verify gadget has no hallways with length greater than 3, any capabilities the robot may have of pushing or pulling more than one block at a time are irrelevant. Thus, the following proof will apply for all positive values of j and k in Push-j Pull-k.

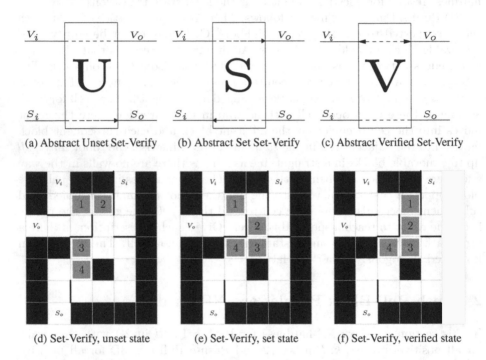

(a) Abstract Unset Set-Verify   (b) Abstract Set Set-Verify   (c) Abstract Verified Set-Verify

(d) Set-Verify, unset state   (e) Set-Verify, set state   (f) Set-Verify, verified state

Fig. 1. Diagrams of three of the states of Set-Verify gadgets along with their construction in a push-pull block puzzle. Red blocks are moveable, black blocks are fixed, thick black lines are thin walls. (Color figure online)

Table 2. State transitions of a Set-Verify gadget as seen in Fig. 1

U:

$$(U, s_i) \rightarrow (S, S_o)$$

S:

$$(S, s_o) \rightarrow (U, S_i)$$
$$(S, v_i) \rightarrow (V, v_o)$$

V:

$$(V, v_i) \rightarrow (V, v_o)$$
$$(V, v_o) \rightarrow (V, v_i)$$
$$(V, v_o) \rightarrow (S, v_i)$$

Variable and Clause Gadgets. We will be making use of the Set-Verify gadget to produce the literals in our 3SAT formula. One significant difficulty with this model is the complete reversibility of all actions. Thus we need to take care to ensure that going backward at any point does not allow the robot to cheat in solving our 3SAT instance. The directional properties of the Set-Verify allow us to create sections where we know if the robot exits, it must have either reset everything to the initial configuration or have correctly proceeded through that gadget.

Our literals will be represented by Set-Verify gadgets. They are considered true when the V_i to V_o traversal is possible, and false otherwise. Thus we can set literals to true by allowing the robot to run through the S_i to S_o passage of the gadget. This allows a simple clause gadget, shown in Fig. 5, consisting of splitting the path into three hallways, each with the corresponding verify side of our literal. We can then pass through if any of the literals are set to true, and cannot pass otherwise. Notice that the Unset and Set states do not have a backward transition. Thus the only way to go back through the clause is through the verified literal, after which the clause has been reset to the state it was in before the robot went through it.

The variables will be encoded by a series of passages which split to allow either the true or negated literals to be set, shown in Fig. 6. Once the robot has gone through at least one gadget in one hallway, there are only two possibilities remaining: either the robot can continue down the hall setting more literals to true, or the robot can go back through the gadget it has just exited, returning it to its unset state. Thus, before entering or after exiting a hallway all of the literals in that hallway will be in the same state. Additionally, unset gadgets do not allow a transition from S_o to S_i, which means at any point while setting variables, if the robot decides to go back it can only return through a hallway which has been switched to the set state. Going back through these returns them to the unset state, putting that variable gadget back in its initial configuration before the robot interacted with it.

Theorem 1. *Push-k Pull-l in 2D with thin walls is NP-hard.*

Proof. We will reduce from 3SAT. Given a 3SAT instance with variables $(x_1, x_2, \ldots x_n)$ and clauses $(x_a, x_b, \overline{x}_c), \ldots$, we will construct an equivalent Push-Pull instance as follows:

First, we will set up the clause gadgets. Each clause gadget will look like Fig. 5, with all of the Set-Verify gadgets initially in the unset state. There will be one clause gadget for each clause in the 3SAT formula. The clauses will be linked together in series, C_k out to C_{k+1} in. At the final clause gadget's exit, we will place the goal square.

Next, we will set up the variable gadgets. For each variable x_k, there will be a variable gadget X_k, consisting of a positive literal pathway, connecting to every clause where the variable is used positively, and a negative literal pathway, connecting to every clause where the variable is negated, as shown in Fig. 6. These variable gadgets will be linked together in series, X_k out to X_{k+1} in. The final variable gadget's *out* exit will be linked to the first clause gadget's *in*. Just in front of the first variable gadget's *in* entrance will be the start square.

The connections between these gadgets will consist of empty hallways, except where such hallways would cross. The hallways inside the clause and variable gadgets will also need to cross, and we will handle them similarly. We need crossovers for this reduction, rather than reducing to a PlanarSAT variant, because we need crossovers just to make the clause gadgets work.

At all crossings, we will place a Two Use Directed Crossover, from Fig. 8. The orientation of the gadget will be chosen according to a specified ordering, where the later pathway will never be used before the earlier pathway, and no pathway will every be traversed twice in the same direction. The ordering is each variable gadget's hallways, in increasing order of the variable gadgets, followed by each clause gadget's hallways, in increasing order of clause gadgets. Within the variable gadgets, the ordering will be from in to out along the positive and negative lines, with the positive lines arbitrarily placed before the negative lines. The clause gadget hallways won't cross each other.

The construction is complete. To see that it is solvable if and only if the corresponding SAT problem is satisfiable, first let us consider the case where the SAT problem is satisfiable. If the SAT problem is satisfiable, then there is an assignment of variables such that each clause is satisfied, e.g. has at least one true literal. Therefore, the PushPull construction is solvable. It can be solved by traversing each variable gadget via the side corresponding to the satisfying assignment, then traversing each clause, which is passable because it is satisfied. The crossovers do not impede traversal, since the path taken goes through each crossover at most once of each of its pathways, and strictly in the forward direction of the ordering which determined the orientation of the crossovers. Thus, the entire PushPull problem can be solved, as desired.

Next, let us consider the case where the SAT problem is not satisfiable. Consider a partial traversal of the PushPull problem, from the start cell through the variable gadgets. Regardless of any reverse transitions through a variable gadget or interactions with its clause gadget, if the robot is beyond a given variable gadget exactly one of the variable lines must be set and the other must be unset. Likewise, the interactions with the crossover gadgets do not allow any transitions other than within the variable gadgets, regardless of reversals. Moreover, interactions with the clause gadgets only change the state of Set-Verify gadgets corresponding to literals between the Set and Verified states. If a Set-Verify is Unset, its state cannot be altered via its verify line ($V_i - V_o$).

Thus, regardless of the robot's prior movements, the only literals that will be Set or Verified are at most those corresponding to a single assignment for each variable. No two literals corresponding to opposite assignments of the same variable will every be in the Set or Verified states at the same time.

Since the SAT problem is assumed to be unsatisfiable, no assignment of variables will satisfy every clause. Thus, as the robot exits the variable gadgets and enters the clause gadgets, for any prior sequence of moves, there must be some clause gadget which has all of its literals in the Unset state, corresponding to the unsatisfied clause for this setting of variables. Since all clauses must be traversed to reach the goal cell, and a clause cannot be traversed if all of its literals are Unset, the robot cannot reach the goal cell. Thus, the PushPull problem is unsolvable.

We have demonstrated that the PushPull problem is solvable if and only if the corresponding 3SAT instance is satisfiable. The reduction mentioned above is polynomial time reduction, as long as the hallways are constructed reasonably. Thus, Push-k Pull-l in 2D with thin walls is NP-hard.

3 PSPACE

In this section we show the PSPACE-completeness of 3D push-pull puzzles with equal push and pull strength. We will prove hardness by a reduction from True Quantified Boolean Formula (also known as TQBF and 3QSAT), which asks whether, given a set of variables $\{x_1, x_2, \ldots x_n, y_1, y_2, \ldots y_n\}$ and a boolean formula $\theta(x_1, \ldots x_n, y_1 \ldots y_n)$ in conjunctive normal form with exactly three variables per clause, the quantified boolean formula $\forall y_1 \exists x_1 \forall y_2 \exists x_2 \ldots$ $\theta(x_1, \ldots x_n, y_1, \ldots y_n)$ is true.

We introduce a gadget called the 4-toggle and use it to simulate 3QSAT [11]. We construct the 4-toggle gadget in 3D push-pull block puzzles, completing the reduction. In particular we prove 3D Push1-Pull1 with thin walls is PSPACE-complete and 3D Push$i$-Pull$j$, for all positive $i = j$, is PSPACE-complete. A gap between NP and PSPACE still remains for 3D puzzles with different pull and push values, as well as for 2D puzzles.

3.1 Toggles

We define an n-toggle to be a gadget which has n internal pathways and can be in one of two internal states, A or B. Each pathway has a side labeled A and another labeled B. When the toggle is in the A state, the pathways can only be traversed from A to B and similarly in the B state it can only be traversed from B to A. Whenever a pathway is traversed, the state of the toggle flips. A diagram of a 4-toggle is is given in Fig. 3 and its state transition is given in Table 3.

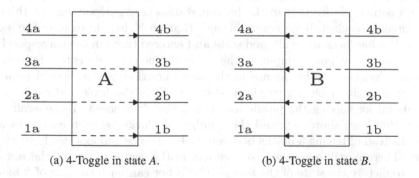

(a) 4-Toggle in state A.   (b) 4-Toggle in state B.

Fig. 2. Diagrams of the two possible states of a 4-toggle.

Figure 3a acts as a 2-toggle. The locations $1a$, $1d$, $2a$, and $2d$, are all entrances and exits to the 2-toggle, while $1b$ connects directly to $1c$, and $2b$ connects directly

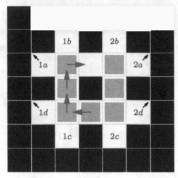
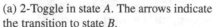
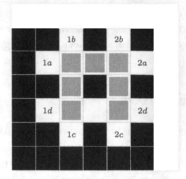

(a) 2-Toggle in state A. The arrows indicate   (b) 2-Toggle in state B.
the transition to state B.

Fig. 3. 2-toggles constructed in a push-pull block puzzle.

Table 3. State transitions of a 4-toggle as seen in Fig. 2

A:			B:		
$(A, 1a)$	$\rightarrow$	$(B, 1b)$	$(B, 1b)$	$\rightarrow$	$(A, 1a)$
$(A, 2a)$	$\rightarrow$	$(B, 2b)$	$(B, 2b)$	$\rightarrow$	$(A, 2a)$
$(A, 3a)$	$\rightarrow$	$(B, 3b)$	$(B, 3b)$	$\rightarrow$	$(A, 3a)$
$(A, 4a)$	$\rightarrow$	$(B, 4b)$	$(B, 4b)$	$\rightarrow$	$(A, 4a)$

to 2c. Notice that there is a single block missing from the ring of eight blocks. When the missing block is on top, as diagrammed, it will represent state A, and when it is on the opposite side, we call it state B. Notice that in state A, it is impossible to enter through entries $1d$ or $2d$. When we enter in the $1a$ or $2a$ sides, we can follow the moves in the series of diagrams to exit the corresponding $1d$ or $2d$ side, leaving the gadget in the B state. One can easily check that the gadget can only be left in either state A, B, or a broken state with the empty square left in a corner. Notice that in the broken state, every pathway except the one just exited is blocked. If we enter through that path, it is in exactly the same state as if it had been in an allowed state and entered through the corresponding pathway normally. For example, in the diagram one can only enter through $1a$ and after doing so the blocks are in the same position as they would be after entering in path $1a$ on a 2-toggle in state A. Thus the broken state is never more useful for solving the puzzle and can be safely ignored. To generalize to Push-k Pull-k we simply expand the number of blocks between entrances and exists. Instead of having 3 blocks between each entrance and exit, we have $2k + 1$ blocks. There is still one vacant square left in the center of one of the rows of blocks to dictate the state of the toggle. The robot can push the row of k blocks to the center or pull k blocks opening up a square in the center, giving us the same function as before.

To construct a 4-toggle we essentially take two copies of the 2-toggle, rotate them perpendicular to each other in 3D, and let them overlap on the central axis, where the block is missing. See Fig. 10a. We still interpret the lack of blocks in the same positions as in the 2-toggle as states A or B. For Push1-Pull1, this construction requires thin walls, since the exit pathways from $1b$, $2b$, $3b$ and $4b$ must pass immediately next to each other. For Push$k$-Pull$k$, with $k > 1$, thin walls are not necessary, since the exit pathways are separated from each other.

3.2 Locks

A 2-toggle and lock is a gadget consisting of a 2-toggle and a separate pathway. Traversing the separate pathway is only possible if the 2-toggle is in a specific state, and the traversal does not change the internal state of the 2-toggle. The 2-toggle functions exactly as described above.

This gadget can be implemented using a 4-toggle by connecting the $3b$ and $4b$ entrances of the 4-toggle with an additional corridor, as shown in Fig. 11. Traversing the resultant full pathway, from $3a$ to $3b$ to $4b$ to $4a$, is possible only if the initial state of the 4-toggle is A, and will leave the 4-toggle in state A. In addition, a partial traversal, such as from $3a$ to $3b$ and back to $3a$, does not change the internal state. The two unaffected pathways of the toggle, 1 and 2, continue to function as a 2-toggle.

A 2-toggle and lock can be extended to a 2-toggle with many locks. The 2-toggle with many locks is a gadget consisting of a 2-toggle and any number of separate pathways which can only be traversed when the gadget is in state B. This can be constructed using one 2-toggle and lock per separate pathway needed and attaching the toggles in series. We orient the 2-toggles so that their 2-toggles are all passable at once in one direction. When the 2-toggle is traversed, all of the internal locks' states flip, rendering the gadget passable in the opposite direction, and switching the passability and impassability of all of the external pathways.

3.3 Quantifiers

Existential Quantifier. An existential gadget is like a 2-toggle and many locks, except that instead of a 2-toggle, it has a single pathway which is always passable in both directions. Upon traversing the pathway the robot may or may not change the internal state of the 2-toggle and many locks, as it chooses. The variable is considered true if the 2-toggles and many locks is in state A and false if it is in state B. This gadget is shown in Fig. 12.

Alternating Quantifier Chain. An alternating quantifier chain, shown in Fig. 4, implements a series of alternating existential and universal variables, as well as external literal pathways, which may be traversed if and only if their corresponding variables are set to a prespecified value.

Traversing the quantifier chain repeatedly in the primary direction will cycle the universal variables through all 2^n possible settings. Upon each traversal, an

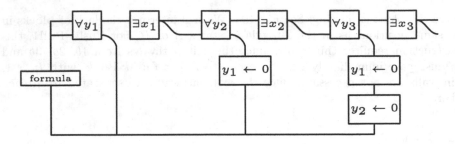

Fig. 4. A segment of the alternating quantifier chain. Each square represents the 2-toggle part of a 2-toggle and many locks.

initial sequence of the universal variables will have their values flipped. During the traversal, the robot will have the option to set a series of corresponding existential variables to whatever value it wishes. These comprise the existentials nested within the universal variables whose values were flipped. An analysis of the Quantifier Chain can be found in Appendix C.

3.4   Clause Gadget

We construct a clause gadget by putting lock pathways of three 2-toggle with many locks in parallel, as we did with Set-Verify gadgets in Fig. 5. Each of these paths can be traversed only if the corresponding variable has been set to true, or to false, depending on the orientation of that particular lock. Since they are in parallel, only one needs to be passable for the robot to be able to continue on to the next clause.

3.5   Beginning and End Conditions

The overall progression of the robot through the puzzle starts with the quantifier chain. The robot increments the universal variables and sets the appropriate existential variables arbitrarily, then traverses a series of clause gadgets to verify that the TQBF formula represented by those clauses is true under that setting of the variables. Then, the robot cycles around to the quantifier chain, and repeats.

At the beginning of this procedure, the robot must be allowed to set all of the existential variables arbitrarily. To ensure this, we will set up the quantifier gadget in the state $01 \ldots 11$, with all variables set to 1 except the highest order one. The highest order variable will be special, and will not be used in the $3CNF$ formula. The initial position of the robot will be at the entrance to the quantifier gadget. This will allow the robot to flip every universal in the quantifier gadget, from $01 \ldots 11$ to $10 \ldots 00$, and accordingly set every existential variable arbitrarily. To force the robot to go forward through the quantifier gadget instead of going backwards through the clause chain, we will add a literal onto the end of the formula gadget which is passable if and only if the highest order variable is set to 1. After this set up, the robot will progress through the loop consisting of the quantifier gadget and the formula gadget, demonstrating the appropriate existential settings for each assignment of the universal quantifiers.

At each point in this process, the robot has the option to proceed through this cycle backwards, as is guaranteed by the reversibility of the game. However, at no point does proceeding in the reverse direction give the robot the ability to access locations or set toggles to states that it could not have performed when it initially encountered the toggles or locations. Thus, any progression through the states of the alternating quantifier chain must demonstrate a TQBF solution to the formula given.

After progressing through every possible state of the universal quantifiers, the universals will be in the state $11 \ldots 11$. At this point, the robot may progress through the quantifier gadget and exit via its special pathway, the carry pathway of the highest order bit. This special pathway will lead to the goal location of the puzzle. Thus, only by traversing the quantifier - formula loop repeatedly, and demonstrating the solution to the TQBF problem, will the robot be able to reach the goal. The robot may reach the goal if and only if the corresponding quantified boolean formula is true.

Theorem 2. *Push-k Pull-k, $k > 1$ in 3D with fixed blocks is PSPACE-complete.*

Proof. By the above construction, TQBF can be reduced to Push-k Pull-k in three dimensions with fixed walls, through the intermediate step of construction a 4-toggle. This implies that Push$k$-Pull$k$ is PSPACE-hard. Since Push$k$-Pull$k$ has a polynomial-size state, the problem is in NPSPACE, and therefore in PSPACE by Savitch's Theorem [17]. So it is PSPACE-complete.

Theorem 3. *Push-1 Pull-1 in 3D with thin walls is PSPACE-complete.*

Proof. Push-1 Pull-1 in 3D with thin walls can construct a 4-toggle, and so by the same argument as in Theorem 2, it is PSPACE-complete.

4   Conclusion

In this paper, we proved hardness results about variations of block-pushing puzzles in which the robot can also pull blocks. Along the way, we analyzed the complexity of two new, simple gadgets, creating useful new toolsets with which to attack hardness of future puzzles. The results themselves are obviously of interest to game and puzzle enthusiasts, but we also hope the analysis leads to a better understanding of motion-planning problems more generally and that the techniques we developed allow us to better understand the complexity of related problems.

This work leads to many open questions to pursue in future research. For Push-Pull block puzzles, we leave several NP vs. PSPACE gaps, a feature shared with many block-pushing puzzles. One would hope to directly improve upon the results here to show tight hardness results for 2D and 3D push-pull block puzzles. One might also wonder if the gadgets used, or the introduction of thin walls, might lead to stronger results for other block-pushing puzzles. We also leave open the question of push-pull block puzzles without fixed blocks or walls. In this setting, even a single 3×3 area of clear space allows the robot to reach any point, making gadget creation challenging.

There are also interesting questions with respect to the abstract gadgets introduced in our proof. We are currently studying the complexity of smaller toggles and toggle-lock systems. It would also be interesting to know whether Set-Verify gadgets sufficient for PSPACE-hardness or if they can build full crossover gadgets. Also, there are also many variations within the framework of connected blocks with traversibility which changes with passage through the gadget. Are any other gadgets within this framework useful for capturing salient features of motion planning problems? Finally, there is the question of whether other computational complexity problems can make use of these gadgets to prove new results.

A   Additional Details on 2D Push-Pull Block Puzzles Proof

This section provides some additional figures and description to help explain the proof of Theorem 1.

Here we walk through the allowable transitions in the Set-Verify gadget and also address the potential Broken state in the Push-Pull construction which is not in the abstract gadget. In the Unset state, the $S_i \rightarrow S_o$ transition is the only possibility, changing the state to Set. In the Set state, the $S_o \rightarrow S_i$ transition is possible, changing the state back to Unset, as well as the $V_i \rightarrow V_o$ transition, which changes the state to Verified. Finally, from the Verified state, the only transitions possible are $V_o \rightarrow V_i$, changing the state back to Set, and $V_i \rightarrow V_o$, leaving the state as Verified. In the Broken state, the only possible transition is $S_o \rightarrow S_i$, changing the state to Unset. Any time we would enter the Broken state, we could instead enter the Set state, which allows strictly more transitions, and therefore will be strictly more helpful in reaching the goal. The Broken state is not helpful towards reaching the goal, so we will disregard its existence.

For the Set-Verify gadget in the Unset state, the S_i entrance is the only one which allows the robot to move any blocks. From the S_i entrance it can traverse to S_o, and it can also pull block 2 down behind them. Doing so will allow a traversal from V_i to V_o. To traverse back from S_o to S_i, the robot must first traverse back from V_o to V_i. Then, when the robot travels back from S_o to S_i, it must push block 2 back, ensuring the V_i to V_o traversal is impossible. Further, access to any sequence of entrances will not allow the robot to alter the system to allow traversals between the V_i and S_i entrances.

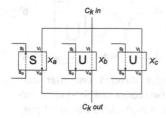

Fig. 5. Clause gadget, C_k, with variables $x_a = 1$, $x_b = 0$, $x_c = 0$.

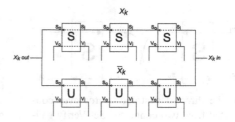

Fig. 6. A variable gadget representing X_k occurring in six clauses, three of those times negated. The value of the variable has been set to true.

Crossover Gadgets. In this section we build up the needed two use crossover gadget from a series of weaker types of crossover gadgets. One may wonder why we need crossover gadgets when Planar 3SAT is NP-complete. This only guarantees that connecting the vertices to their clauses by edges results in a planar graph, it does not ensure that we can navigate our robot between all of these gadgets in a planar manner or that our gadgets themselves are planar. The most obvious issue can be seen in the clause gadget (Fig. 5) where one of the Set Verify gadgets must lie between the other two hallways, but must also be accessible by its associated variable gadget.

Directed Destructive Crossover. This gadget, depicted in Fig. 7a, allows either a traversal from a to a' or b to b'. Once a traversal has occurred, that path may be traversed in reverse, but the other is impassable unless the original traversal is undone.

First, observe that transitions are initially only possible via the a and b entrances, since the transitions possible through a Set-Verify in the Set state can be entered through V_i and S_o, not S_i. Assume without loss of generality that the gadget is entered at a. This changes the state of the left Set-Verify to Verified. At this point, only the right S_o and left V_o transitions are passable. Taking the V_o transition either reverts all changes to the original state, or leaves the left crossover in the Verified state, which allows strictly less future transition than the original state. Therefore, we will disregard that option. Taking the S_o transition changes the right Set-Verify to Unset, and completes the crossover. At this point, the only possible transition is to undo the transition just made, from a' back to a, restoring the original state. The gadget could be entered via a, but the robot would only be able to leave via a, possibly changing the state to Set. Both options result in the robot exiting out its original entrance, and allow the same or less future transitions, so we may disregard those options. Thus, the only transition possibilities are as stated above.

In-order Directed Crossover. This gadget, depicted in Fig. 7b allows a traversal from a to a', followed by a traversal from b to b'. These traversals may also be reversed.

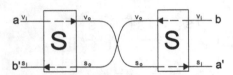

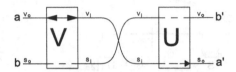

(a) The directed destructive crossover constructed from two connected Set-Verify gadgets initialized in the set position.

(b) The in-order directed crossover constructed from two connected Set-Verify gadgets initialized in the verified and unset positions.

Fig. 7. Two types of crossover gadgets

Initially, no entrance is passable except for a, since V_o is passable only in the Verified state, and S_o is passable only in the Set state. Once the left $V_o \to V_i$ transition is made, the robot has 2 options. It can either change the left Set-Verify gadget's state to Set, or leave it as Verified. In either case, the S_i entrance on that toggle is impassable, since a S_i entrance may only be traversed in the Unset state. The only transition possible on the right crossover is $S_i \to S_o$, changing the state from Unset to Set. This completes the first crossing.

Now, there are at most 2 transitions possible: from a' back to a, undoing the whole process, or entering at b. Note that entering at b is only possible if the left Set-Verify is in the Set state, so let us assume that state change occurred. In that case, the left $S_o \to S_i$ transition may be performed, changing the left Set-Verify's state to Unset. At that point, the only possible transitions are back to b, or through the right Set-Verify's $V_i \to V_o$ transition, completing the second crossover.

If the left Set-Verify was left in the Verify state, strictly less future transitions are possible compared to the case where it was changed into the set state, so we may disregard that possibility.

Two Use Directed Crossover. The Two Use Directed Crossover, depicted in Fig. 8, is the gadget needed for our proof. It allows a traversal from a to a' followed by a traversal from b to b', or from b to b' and then a to a'. These transitions may also be reversed.

It is constructed out of an In-order Directed Crossover gadget and a Destructive Directed Crossover, as

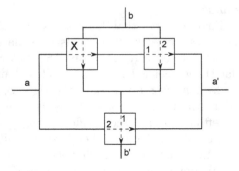

Fig. 8. The two use directed crossover is constructed from a directed destructive crossover and two in-order directed crossovers.

shown in Fig. 8. The a to a' traversal is initially passable, and goes through both gadgets, blocking the destructive crossover but leaving the in-order crossover open for the b to b' traversal. If the a to a' traversal does not occur, the b to b' traversal is possible via the destructive crossover.

Because of the behavior of the constituent crossovers which make up this gadget, no transition from a to b a to b', etc. is possible. The crossover permits reversal of each of the transitions described, but the crossings can only be reversed in queue order (last in, first out).

Two Use Crossover. Four Directed Crossovers can be combined, as shown below, to create a crossover that can be traversed in any direction [4]. This is not necessary for our proof but is shown for general interest. Unfortunately, the inability to go through this gadget multiple times in the same direction without first going back through means it likely isn't sufficient for PSPACE-completeness.

B 3D Push-Pull is NP-hard

In this section we prove that 3D Push-k Pull-l with fixed blocks is NP-hard, for all positive k and l. All of the hard work was done in Sect. 2. Here we will simply show how we can use the additional dimension to tweak the previous gadgets to build them without thin walls. We reduce from 3SAT, constructing our variables from chains of 3D Set-Verify gadgets, and our clauses from the verify side of the corresponding 3D Set-Verify gadget.

Theorem 4. *3D Push-k Pull-l with fixed blocks is NP-hard, for all positive k and l.*

Proof. We follow the proof of Theorem 1 using a modified Set-Verify gadget, shown in Fig. 9. It can be easily checked that this has the same properties as the Set-Verify given in Sect. 2. The cyclic ordering of the entrances in the 3D Set-Verify is different from that of the 2D Set-Verify, however this is not important as we no longer need to construct crossovers. Also, this construction does not use thin-walls. While this was critical in the prior construction due to the need for closely packed turns, the additional dimension allows enough freedom to keep separate hallways from being adjacent to each other. With a functional Set-Verify gadget, the remaining constructions of variables and clauses proceeded as in Sect. 2. No crossover gadgets are needed since we are working in 3D. Finally, we note that all blocks are in hallways of length at most 3, thus the gadgets still function as described for any positive push and pull values.

C Additional Details on PSPACE-Completeness Proof

This section provides some additional figures and description to help explain the proof of Theorem 2.

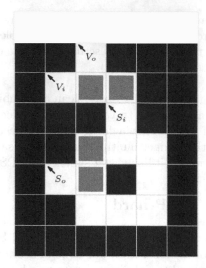

Fig. 9. A Set-Verify gadget in 3D where the entrances and exits extend upward, notated by the diagonal arrows. This gadget is in the unset state.

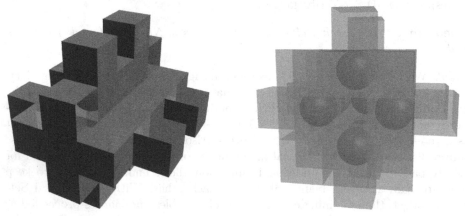

(a) Diagram of a 4-toggle showing impassible surfaces.

(b) Diagram of the internals of a 4-toggle.

Fig. 10. 3D diagrams of 4-toggles. Red spheres are blocks and blue surfaces are impassable. (Color figure online)

Binary Counter. Universal quantifiers must iterate through all possible combinations of values that they can take. In this section we construct a gadget that runs though all the states of its subcomponents as the robot progresses through the gadget. This construction will serve as the base for our universal quantifiers.

A binary counter has a fixed number of internal bits. Whenever the binary counter is traversed in the forwards direction, the binary number formed by the

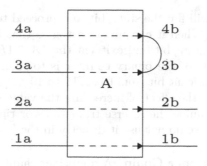

Fig. 11. Diagram of a lock. The $3a$ to $4a$ traversal is only possible in state A and returns the toggle to state A.

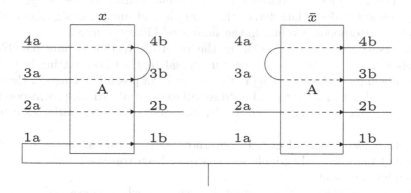

Fig. 12. An existential gadget.

internal bits increases by one and the robot leaves via one of the exits. If the binary counter is traversed in the reverse direction, the internal value is reduced by one. If the binary counter is partially traversed, but then the robot leaves via its initial entrance, the internal value does not change.

The binary counter is implemented as a series of 2-toggles, as shown in Fig. 13. To see that this produces the desired effect, identify a toggle in state A as a 0 bit, and a toggle in state B as a 1 bit. Let the entrance toggle's bit be the least significant bit, and the final toggle be the most significant. When the robot enters the binary counter in the forwards direction, it will flip the state of every toggle it passes through. When it enters a toggle that is initially in state B, and

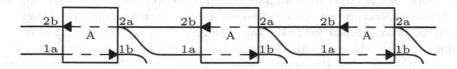

Fig. 13. The central portion of a three bit binary counter made from 2-toggles.

thus whose bit is 1, it will flip the state/bit and proceed to the next toggle, via the $2B - 2A$ pathway. When it encounters a toggle that is initially in state A / bit 0, it will flip the state/bit and exit via the $1A - 1B$ pathway. Thus, the overall effect on the bits of the binary counter is to change a sequence of bits ending at the least significant bit from $01 \ldots 11$ to $10 \ldots 00$, where the entrance is at the right. This has the effect of increasing the value of the binary counter by one. We will not examine the reverse transitions or rigorously complete the binary counter here, as we do not use it directly in the final construction.

Analysis of the Quantifier Chain. A quantifier chain is implemented much like a binary counter, with some additions. Every universal variable will be represented by a 2-toggle and many locks, where individual locks will serve as a literal. The 2-toggles are hooked up in the same manner as the 2-toggles in a binary counter gadget. This forces the 2-toggle and many locks gadgets to be set to the corresponding values in the simulated binary counter.

Traversing the quantifier chain in the reverse direction is only possible if the robot enters via the lowest order universal toggle whose setting is 1. The traversal will go back one setting in the sequence of possible settings of the universal variables, and allow the robot to set all existential variables corresponding to altered universal variables arbitrarily. No other existential variables can be changed.

There is also a special exit, the overflow exit, which can only be reached after all of the universal variable settings have been traversed. This is the goal location for the robot.

The next addition is the existential variables, which consist of existential gadgets placed just after the $2A$ exits of each universal variable, and just before the $1A$ and $2B$ entrances of the next universal variable, as shown in Fig. 4.

One portion of the apparatus which has not been analyzed thus far is the potential for the robot to re-enter the chain of existentials via a different exit pathway than the one just exited. This would be problematic if the robot re-entered via a universal gadget it had not just exited, both because the robot should not be able to take any action other than reversing its prior progress, decrementing the binary counter/universal quantifiers. Problems would also arise if the robot got access to any existential quantifiers it did not just traverse.

After a traversal, the universal quantifiers have the settings $\ldots??10 \ldots 00$, where the lowest significance 1 is on the pathway just exited. To prevent the robot from re-entering via any pathway other than the one just exited, we add a series of locks to each exit that are only passable if all lower-significance universal toggles are in state 0, as shown in Fig. 4. This does not impede the exit that the robot uses initially, since all lower-significance universal toggles are indeed 0. These locks do prevent re-entry into any higher-significance universal toggles, since the lock corresponding to the lowest-significance 1 will be closed. The robot cannot re-enter via any toggle that is in state 0, due to the arrangement of the toggle pathways. Thus, the unique re-enterable pathway is the lowest-significance toggle in state 1, as desired.

References

1. Abdelkader, A., Acharya, A., Dasler, P.: 2048 without new tiles is still hard. In: LIPIcs-Leibniz International Proceedings in Informatics, vol. 49. Schloss Dagstuhl-Leibniz-Zentrum fuer Informatik (2016)
2. Aloupis, G., Demaine, E.D., Guo, A., Viglietta, G.: Classic nintendo games are (NP-)hard. In: Proceedings of the 7th International Conference on Fun with Algorithms (FUN 2014), Lipari Island, Italy, 1–3 July 2014
3. Culberson, J.C.: Sokoban is PSPACE-complete. In: Proceedings International Conference on Fun with Algorithms (FUN 1998), pp. 65–76. Carleton Scientific, Waterloo, Ontario, Canada, June 1998
4. Demaine, E.D., Demaine, M.L., O'Rourke, J.: PushPush and Push-1 are NP-hard in 2D. In: Proceedings of the 12th Annual Canadian Conference on Computational Geometry (CCCG 2000), pp. 211–219. Fredericton, New Brunswick, Canada, 16–18 August 2000
5. Demaine, E.D., Hearn, R.A., Hoffmann, M.: Push-2-F is PSPACE-complete. In: Proceedings of the 14th Canadian Conference on Computational Geometry (CCCG 2002), pp. 31–35. Lethbridge, Alberta, Canada, 12–14 August 2002
6. Demaine, E.D., Hoffmann, M.: Pushing blocks is NP-complete for noncrossing solution paths. In: Proceedings of the 13th Canadian Conference on Computational Geometry (CCCG 2001), pp. 65–68. Waterloo, Ontario, Canada, 13–15 August 2001
7. Demaine, E.D., Hoffmann, M., Holzer, M.: PushPush-k is PSPACE-complete. In: Proceedings of the 3rd International Conference on Fun with Algorithms (FUN 2004), pp. 159–170. Isola d'Elba, Italy, 26–28 May 2004
8. Dhagat, A., O'Rourke, J.: Motion planning amidst movable square blocks. In: Proceedings of the 4th Canadian Conference on Computational Geometry (CCCG 1992) (1992)
9. Dor, D., Zwick, U.: SOKOBAN and other motion planning problems. Comput. Geom. **13**(4), 215–228 (1999)
10. Flake, G.W., Baum, E.B.: Rush hour is PSPACE-complete, or why you should generously tip parking lot attendants. Theoret. Comput. Sci. **270**(1–2), 895–911 (2002)
11. Garey, M.R., Johnson, D.S.: Computers and Intractability: A Guide to the Theory of NP-Completeness. W. H. Freeman & Company, New York (1979)
12. Guala, L., Leucci, S., Natale, E.: Bejeweled, candy crush and other match-three games are (NP-) hard. In: IEEE Conference on Computational Intelligence and Games (CIG), pp. 1–8. IEEE (2014)
13. Hearn, R.A., Demaine, E.D.: PSPACE-completeness of sliding-block puzzles and other problems through the nondeterministic constraint logic model of computation. Theoret. Comput. Sci. **343**(1), 72–96 (2005)
14. Hoffman, M.: Push-* is NP-hard. In: Proceedings of the 12th Canadian Conference on Computational Geometry (CCCG 2000), Lethbridge, Alberta, Canada (2000)
15. Ratner, D., Warmuth, M.K.: Finding a shortest solution for the n × n extension of the 15-PUZZLE is intractable. In: Proceedings of AAAI, pp. 168–172 (1986)
16. Ritt, M.: Motion planning with pull moves. arXiv:1008.2952 (2010)
17. Savitch, W.J.: Relationships between nondeterministic and deterministic tape complexities. J. Comput. Syst. Sci. **4**(2), 177–192 (1970)
18. Wilfong, G.: Motion planning in the presence of movable obstacles. Ann. Math. Artif. Intell. **3**(1), 131–150 (1991)
19. Zubaran, T., Ritt, M.: Agent motion planning with pull and push moves. In: Anais do VIII Encontro Nacional de Inteligãlncia Artificial (ENIA), Natal, July 2011

Weak Coverage of a Rectangular Barrier

Stefan Dobrev[1], Evangelos Kranakis[2], Danny Krizanc[3], Manuel Lafond[4],
Jan Maňuch[5], Lata Narayanan[6(⊠)], Jaroslav Opatrny[6], Sunil Shende[7],
and Ladislav Stacho[8]

[1] Institute of Mathematics, Slovak Academy of Sciences, Bratislava, Slovakia
[2] School of Computer Science, Carleton University, Ottawa, Canada
[3] Department of Mathematics and Computer Science, Wesleyan University,
Middletown, CT, USA
[4] Department of Mathematics and Statistics, University of Ottawa, Ottawa, Canada
[5] Department of Computer Science, University of British Columbia,
Vancouver, Canada
[6] Department of Computer Science and Software Engineering, Concordia University,
Montreal, QC, Canada
lata@cs.concordia.ca
[7] Department of Computer Science, Rutgers University, Camden, NJ, USA
[8] Department of Mathematics, Simon Fraser University, Burnaby, BC, Canada

Abstract. Assume n wireless mobile sensors are initially dispersed in an
ad hoc manner in a rectangular region. They are required to move to final
locations so that they can detect any intruder crossing the region in a
direction parallel to the sides of the rectangle, and thus provide *weak barrier coverage* of the region. We study three optimization problems related
to the movement of sensors to achieve weak barrier coverage: minimizing
the *number* of sensors moved (MinNum), minimizing the *average* distance
moved by the sensors (MinSum), and minimizing the *maximum* distance
moved by the sensors (MinMax). We give an $O(n^{3/2})$ time algorithm for
the MinNum problem for sensors of diameter 1 that are initially placed
at integer positions; in contrast we show that the problem is NP-hard
even for sensors of diameter 2 that are initially placed at integer positions. We show that the MinSum problem is solvable in $O(n \log n)$ time
for homogeneous range sensors in arbitrary initial positions for the Manhattan metric, while it is NP-hard for heterogeneous sensor ranges for
both Manhattan and Euclidean metrics. Finally, we prove that even very
restricted homogeneous versions of the MinMax problem are NP-hard.

1 Introduction

Intruder detection is an important application of wireless sensor networks. Each
sensor monitors a circular area centered at its location, and can immediately
alert a monitoring station if it detects the presence of an intruder. Collectively
the sensors can be deployed to monitor the entire region, providing so-called *area*

Research supported by NSERC, Canada.

coverage. However, for many applications, it is sufficient, and much more cost-effective, to simply monitor the boundary of the region, and provide so-called *barrier coverage*.

Barrier coverage was introduced in [16], and has been extensively studied since then [2,3,7,8,12,17,19]. The problem was posed as the deployment of sensors in a narrow belt-like rectangular region in such a way that any intruder crossing the belt would be detected. A sensor network is said to provide *strong barrier coverage* if an intruder is detected regardless of the path it follows across the given barrier (see Fig. 1(a)). In contrast, a sensor network provides *weak coverage* if an intruder is detected when it follows a straight-line path across the width of the barrier. If the location of the sensors is not known to a trespasser, weak coverage is often sufficient, and is more cost-effective.

In this paper, we consider a more general notion of weak coverage than previously considered. Given a rectangular barrier, we aim to detect intruders who cross the region in a straight-line path parallel to *either of the axes* of the rectangle (see Fig. 1(b)).

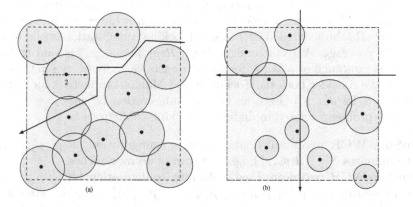

Fig. 1. (a) Strong coverage of the shaded square area by a homogeneous network, (b) Weak coverage of the shaded square area by a non-homogeneous network for paths perpendicular to the axes.

A sensor network can be deployed for the given barrier in several different ways. In *deterministic deployment*, sensors are placed in pre-defined locations that ensure intruder detection. However, when the deployment area is very large, or the terrain in the area is difficult or dangerous, a deterministic deployment might be costly or even impossible. In those instances a *random or ad hoc deployment* of sensors can be done [2]. However, this type of deployment might leave some gaps in the coverage of the area. Two approaches have been considered in order to deal with this problem. One is a *multi-round random deployment* in which the random dispersal is repeated until the coverage of the area is assured with very high probability [19]. The other approach is to use *mobile (or relocatable) sensors* [5]. After the initial dispersal, some or all sensors are instructed to *relocate to new locations* so that the desired barrier coverage is achieved. Clearly, the relocation of the sensors should be performed in the most efficient way possible. In particular, we may want to minimize the time or energy needed to perform the relocation, or the number of sensors to be relocated.

1.1   Notation and Problem Definition

We assume that n sensors are initially located in an axis-parallel rectangular area R of size $a \times b$ in the Cartesian plane. The n *sensors* $S_1, S_2, \ldots, S_n$ have sensing ranges $r_1, r_2, \ldots, r_n$ respectively. The *diameter* of a sensor is equal to twice its range. We assume that $\sum_{k=1}^{n} 2r_k \geq \max\{a, b\}$; this ensures that placed in appropriate locations, the sensors can achieve weak barrier coverage. A sensor network is called *homogeneous* if the sensing ranges of all sensors in the network is the same. Otherwise the network is called *heterogeneous*.

A *configuration* is a tuple $(R, p_1, p_2, \ldots, p_n)$ where R is the rectangle to be weakly barrier-covered and $\{p_1, p_2, \ldots, p_n\}$ are the positions of the sensors. We say a configuration is a *blocking configuration* if any straight line, perpendicular to either x or y axes, crossing the rectangle R, crosses the sensing area of at least one sensor. In other words, a blocking configuration achieves weak coverage of the rectangle R (abbreviated WCR). A non-blocking configuration is said to have *gaps* in the coverage. A given configuration $(R, p_1, p_2, \ldots, p_n)$ is said to be an *integer configuration* if $p_i = [k_i, j_i]$ for some integers k_i, j_i, for every i in the range $1 \leq i \leq n$. We consider both the Euclidean and Manhattan metrics for distance and denote it by $d(x, y)$. Given an initial configuration $(R, p_1, p_2, \ldots, p_n)$, we study three problems related to finding a blocking configuration:

- **MinSum-WCR problem:** Find a blocking configuration $\{R, p_1', p_2', \ldots, p_n'\}$ that minimizes $\sum_{k=1}^{n} d(p_k, p_k')$, i.e., the sum of all movements.
- **MinMax-WCR problem:** Find a blocking configuration $\{R, p_1', p_2', \ldots, p_n'\}$ that minimizes $\max\{d(p_1, p_1'), d(p_2, p_2'), \ldots, d(p_n, p_n')\}$, i.e., the size of the maximal move among the sensors.
- **MinNum-WCR problem:** Find a blocking configuration $\{R, p_1', p_2', \ldots, p_n'\}$ which minimizes the number of indices for which $d(p_k, p_k') \neq 0$, $1 \leq k \leq n$, i.e., minimizes the number of relocated sensors.

1.2   Our Results

For the MinNum-WCR problem, we show that the problem is NP-complete, even when the initial configuration is an integer configuration, and even when all sensors have range 1. However when the initial configuration is an integer configuration, all sensors have range 0.5, we give an $O(n^{3/2})$ algorithm for solving the MinNum-WCR problem.

When all sensors have the same range, regardless of their initial positions, we give an $O(n \log n)$ algorithm to solve the MinSum-WCR problem using the Manhattan metric. However, the problem is shown to be NP-complete for both Manhattan and Euclidean metrics when the sensors can have different ranges.

Finally, we show that the decision version of the MinMax-WCR problem is NP-complete even for a very restricted case. More specifically, given an integer configuration, with all sensor ranges equal to 0.5, the problem of deciding whether there is a blocking configuration with maximal move at most 1 (using either the Manhattan or Euclidean metric) is NP-complete. This is in sharp contrast to the one-dimensional barrier coverage case where the MinMax problem can be solved in polynomial time for arbitrary initial positions, and heterogeneous sensor ranges.

1.3 Related Work

Barrier coverage using wireless sensors was introduced as a cost-effective alternative to area coverage in [16]. The authors introduced and studied the notions of both *strong* and *weak* barrier coverage in this paper, and studied coverage of a narrow belt-like region. Since then the problem has been extensively studied, for example, see [2,3,12,17,19].

The problem of achieving barrier coverage using mobile or relocatable sensors was introduced in [7]. The authors studied a line segment barrier and gave a polynomial time algorithm for the MinMax problem when all sensors have the same range, and are initially placed on the line containing the barrier. For the same setting, the case of heterogeneous sensors was shown to be also solvable in polynomial time in [6], and the algorithm of [7] for the homogeneous case was also improved. An $O(n^2)$ algorithm for the MinSum problem with homogeneous sensors is given in [8], and an improved $O(n \log n)$ algorithm is presented in [1]. It was proved in [8] that the MinSum problem is NP-hard when sensors have heterogeneous ranges. The MinNum problem is considered in [18], and shown to be NP-hard for heterogeneous sensors and poly time for homogeneous sensors.

In [9], the complexity of the MinMax and MinSum problems when sensors are initially placed in the plane and are required to relocate to cover parallel or perpendicular barriers is studied. The authors show that while MinMax and MinSum can be solved using dynamic programming in polynomial time if sensors are required to move to the closest point on the barrier, even the feasibility of covering two perpendicular barriers is NP-hard to determine.

A stochastic optimization algorithm was considered in [14]. Distributed algorithms for the barrier coverage problem were studied in [10,11]. Further, [15] provides algorithms for deciding if a set of sensors provides k-fault tolerant protection against rectilinear attacks in both one and two dimensions. To the best of our knowledge, the problem of weak coverage of a rectangular region (in two directions) has not been studied previously.

2 MinNum-WCR Problem

For a line segment barrier, the MinNum problem was shown to be NP-complete if sensors have different ranges, but if all sensors have the same range, a polynomial-time MinNum algorithm is given in [18]. In this section, we study the MinNum-WCR problem.

2.1   Hardness Result

We show that MinNum-WCR is NP-complete, even when all sensors have sensing range 1 and the initial configuration is an integer configuration. We give a reduction from a restricted satisfiability problem, shown to be NP-complete in [4], and defined below:

3-Occ-Max-2SAT Problem:

Input: An integer t, a set of boolean variables $x_1, \ldots, x_n$ and a set of clauses $\mathcal{C} = \{C_1, \ldots, C_m\}$, each consisting of a conjunction of two literals, such that each variable appears in *exactly* 3 clauses, and no variable occurs only positively in $\mathcal{C}$, nor only negatively in $\mathcal{C}$.

Question: Does there exist an assignment of $x_1, \ldots, x_n$ that satisfies at least t clauses of $\mathcal{C}$?

Theorem 1. *MinNum-WCR is NP-complete even for integer configurations in which all sensing diameters are equal to* 2.

Proof. Given a **3-Occ-Max-2SAT** instance with variables $x_1, \ldots, x_n$ and clauses $\mathcal{C} = \{C_1, \ldots, C_m\}$, we construct a corresponding instance of the MinNum-WCR problem consisting of a set of sensors S each having radius $r = 1$, and a rectangle R to be covered. Note that $m = \frac{3}{2}n$, since there are $3n$ literals in $\mathcal{C}$ and each clause contains two literals.

R is defined to be a $(6n + 2t) \times (6n + 2t)$ square. The sensor set S contains one sensor s_{i,C_j} for each literal x_i that appears in a clause C_j. We also need two sensors α_i and β_i for each $i \in [n]$. Formally, $S = \{s_{i,C_j} : x_i \text{ occurs in clause } C_j, i \in [n], j \in [m]\} \cup \{\alpha_1, \ldots, \alpha_n, \beta_1, \ldots, \beta_n\}$. We first describe how the sensors of S are laid out on the y-axis, then on the x-axis. For a sensor $s \in S$, denote by $y(s)$ and $x(s)$ the y and x coordinate of its center, respectively. Figure 2 illustrates the y and x positioning of the sensors.

Each variable x_i has a corresponding gadget Y_i on the vertical axis, where a gadget is simply a set of sensors positioned in a particular manner. Each gadget Y_i covers the vertical range $[6(i-1)..6i]$. Let C_{j_1}, C_{j_2} and C_k be the three clauses in which x_i occurs. Choose j_1 and j_2 such that x_i occurs in C_{j_1} and C_{j_2} in the same manner (either positively in both, or negatively in both), and so that it occurs in C_k differently. Let $z = 6(i-1)$ and let $y(\alpha_i) = z+1$, $y(s_{i,C_{j_1}}) = z+2$, $y(s_{i,C_k}) = z+3$, $y(s_{i,C_{j_2}}) = z+4$ and $y(\beta_i) = z+5$. Observe that $Y_1, \ldots, Y_n$ cover the range $[0..6n]$ on the y-axis, which leaves the range $(6n..6n + 2t]$ uncovered.

Also, note that moving α_i or β_i creates a gap in Y_i. Moreover, s_{i,C_k} can be moved, or both $s_{i,C_{j_1}}$ and $s_{i,C_{j_2}}$ can be moved. However, moving both s_{i,C_k} and one of $s_{i,C_{j_1}}$ or $s_{i,C_{j_2}}$ creates a gap (see Fig. 2).

We now describe how the sensors are laid out on the x axis. Each clause $C_k \in \mathcal{C}$ has a corresponding gadget X_k that covers the range $[2(k-1)..2k]$, and is constructed as follows. Let x_i and x_j be the two variables occurring in C_k. Then X_k contains the sensors s_{i,C_k} and s_{j,C_k}, and we set $x(s_{i,C_k}) = x(s_{j,C_k}) = 2k - 1$. Thus X_i covers the $[2(k-1)..2k]$ range and $X_1, \ldots, X_m$ cover the range

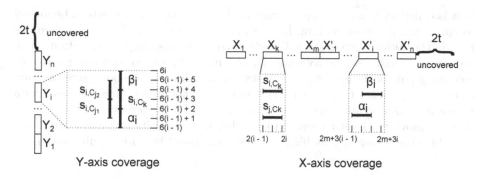

Fig. 2. An illustration of the y and x positioning of the sensors in S. The Y-axis subfigure only shows the coverage of the sensors on the Y axis, and does not depict the x coordinates of the sensors (and the X-axis subfigure does not depict the y coordinates). The depicted Y_i gadget corresponds to variable x_i occurring in clauses C_{j_1}, C_{j_2} and C_k, where x_i occurs in the same manner in the first two. An example of these clauses might be $C_{j_1} = (x_i \vee x_1), C_{j_2} = (x_i \vee x_2)$ and $C_k = (\overline{x_i} \vee x_3)$. The depicted X_k gadget corresponds to a clause C_k containing the two variables x_i and x_j. Finally, the depicted X'_i gadget contains α_i and β_i, which are partially overlapping.

$[0..2m]$. Note that so far every sensor s_{i,C_j} for $i \in [n], j \in [m]$ has been placed. We finally create one gadget X'_i for each pair (α_i, β_i). More precisely, for each $i \in [n]$, let X'_i contain the α_i, β_i sensors, and set $x(\alpha_i) = 2m + 3(i-1) + 1$ and $x(\beta_i) = 2m + 3(i-1) + 2$. Then X'_i covers the range $[2m + 3(i-1)..2m + 3i]$, and $X_1, \ldots, X_m, X'_1, \ldots, X'_n$ cover the range $[0..2m + 3n]$. Recall that $m = \frac{3}{2}n$, and so $2m + 3n = 6n$. Therefore, the x-axis also has the range $(6n..6n + 2t]$ uncovered.

It is not hard to see that this construction can be carried out in polynomial time. The proof that the `3-Occ-Max-2SAT` instance admits an assignment of $x_1, \ldots, x_n$ that satisfies t clauses of C if and only if it is possible to cover the square R by moving t sensors in the corresponding `MinNum` instance is omitted here for lack of space, and can be found in [20]. $\qquad\square$

2.2 An Efficient Algorithm for Integer Configurations

We now show that there is a polynomial algorithm to solve the MinNum-WCR problem for integer configurations when all sensor diameters are equal to 1, and the rectangle to be covered is displaced by 0.5 from the integer grid that contains sensor positions (see Fig. 3). It is not hard to see that there always exists an optimal solution to the MinNum-WCR problem which produces a final blocking configuration which is also an integer configuration.

Consider an integer configuration $(R, p_1, p_2, \ldots, p_n)$ as specified above. The position of a sensor is a pair (i, j) where i is said to be the *row* and j is the *column* in which the sensor is located. A row i (or column j) so that no sensor is located in it is called an *row gap* (resp. *column gap*), and some sensor needs to be moved to cover such a row or column. Let r and c be the number of rows and columns gaps in the initial configuration. Clearly in the final blocking configuration, there are no uncovered rows or columns. By moving a sensor, we can cover a row or column gap or both. For example, if row i and column j are both gaps, moving a sensor to position (i, j) covers both row i and column j. However, moving a sensor may also create a new row or column gap. To understand better the net effect of moving a sensor based on the other sensors in its row and column, we introduce the following classification of sensors (see Fig. 3 for an illustration).

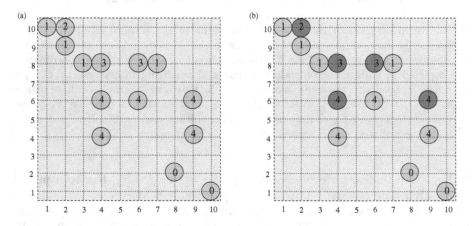

Fig. 3. (a) Classification of sensors for the MinNum algorithm. Notice that at most 2 free sensors can be removed from row 6 or column 4 without creating a new gap. (b) A maximum free set of sensors is shown in dark gray.

Definition 1. *Let S_k be a sensor in position (i, j). We say that*

1. *S_k is* free *if there is at least one other sensor located in row i and at least one other sensor in row j.*
2. *S_k is of* type 0 *if S_k is the only sensor located in row i and the only sensor in column j.*
3. *S_k is of* type 1 *if either there is another sensor located in row i but no other sensor located in column j or there is no other sensor located in row i but there is another sensor located in column j.*
4. *S_k is of* type 2 *if it is a free sensor, and there is at least one sensor of type 1 located in row i, and also at least one sensor of type 1 in column j.*
5. *S_k is of* type 3 *if it is a free sensor, and if there is at least one sensor of type 1 located in row i (column j) and only free sensors in column j (row j).*
6. *S_k is of* type 4 *if it is a free sensor, and the only sensors located in row i and column j are all free sensors.*

We call a move of a sensor a *sliding move* if the final position of the sensor is either in the same row or column as its initial position. We call a move a *jumping* move if the final position is in a different row *and* column from its initial position.

Consider a sensor of type 0. Any sliding move of such a sensor creates an additional row or column gap and can cover at most one other gap. Any jumping move of this sensor creates both a row and a column gap, and it can cover at most one previous row gap and one column gap. Thus the total number of row and column gaps cannot decrease by moving a sensor of Type 0. In what follows, we can therefore assume that sensors of type 0 and the rows and columns in which they reside are not considered any further.

A sensor of Type 1 which has another sensor in its row can make a sliding move in its column and cover a row gap. Any jumping move of this sensor can cover one row and one column gap, but it creates a column gap. Thus the total number of row and column gaps decreases by at most 1 by moving a sensor of Type 1.

However, if there is a row gap i and a column gap j, then a free sensor (of Type 2, 3, or 4) can make a jumping move to position $[i, j]$ and cover both row i and column j, without creating any new row or column gaps. Therefore, moving a free sensor can reduce the total number of row and column gaps by 2. However, moving a free sensor can change the types of sensors in its row or column, and moving *multiple* free sensors can create new empty rows or columns as for example removing all free sensors from row 6 in Fig. 3.

Denote the set of free sensors by F. We define a *maximum free set* to be a maximum-sized subset M of free sensors that can all be removed at the same time without creating new row and column gaps (see Fig. 3(b) for an example). The following theorem shows that a maximum free set can be found in polynomial time.

Theorem 2. *Given an integer configuration as input, a maximum free set M can be found in $O(n^{3/2})$ time.*

Proof. Define X to be the set of rows and columns that contain only the sensors in F, and let $B \subseteq F$ be a minimum-sized subset of F so that every row (or column) in X contains a sensor in B. We call B a minimum blocking set for X. Then clearly M is a maximum free set if and only if $F - M$ is a minimum blocking set. Therefore, to find a maximum free set, we proceed by finding a minimum blocking set.

Consider a graph $G = (V, E)$ defined as follows. The vertex set V contains a vertex corresponding to every row and column in X; we call the vertex c_i if it corresponds to column i and r_j if it corresponds to row r_j. We also introduce two extra vertices x and y. For every Type 4 sensor at position (i, j), we introduce an edge e_{ij} between the vertices r_i and c_j. For every Type 3 sensor that has only free sensors in its row i (resp. column i), we introduce an edge e_{ix} between vertex r_i (resp. c_i) and the vertex x. Finally, we add the edge e_{xy} between vertices x and y.

We claim that B is a blocking set for X if and only if $E' \cup \{e_{xy}\}$ forms an edge cover in the above graph G, where E' is the set of edges corresponding to sensors in B. To see this, observe that if a sensor of type 4 at position (i, j) is in B, then the corresponding edge e_{ij} covers both vertices r_i and c_j in the graph.

Similarly, if a type 3 sensor at position (i, j) that has free sensors only in its row (resp. column) is in B then the corresponding edge e_{ix} covers vertices r_i (resp. c_i) and x. Furthermore e_{xy} covers both x and y. Since B blocks all rows and columns in X, it follows that all vertices in G are covered by $E' \cup \{e_{xy}\}$. Conversely, consider an edge cover in G. It must include the edge e_{xy} since it is the only edge incident on y. Additionally, any set of edges that covers the remaining vertices in G must be incident on all vertices corresponding to the set X, and therefore corresponds to a set of sensors that blocks X. This completes the proof of the claim.

Since an edge cover can be found via a maximum matching in $O(\sqrt{V}E) = O(n^{3/2})$ time [13], we can find a minimum blocking set, and thereafter, a maximum free set M in $O(n^{3/2})$ time. $\square$

Assume that in the given MinNum-WCR configuration there are r row gaps and c column gaps. We can assume without loss of generality that $r \geq c$. We now give an algorithm to solve the MinNum-WCR problem:

Algorithm 1. The MinNum-WCR Algorithm

Input: I an integer configuration with r row gaps and c column gaps, with $r \geq c$

Construct the maximum free set M for I.
Recalculate the types of sensors after removing the sensors in M.
Let k be the number of free sensors in M.

Case $k \geq r$: Move c sensors from M using jumping moves to cover the first c row gaps and all c column gaps. Then $r - c$ sensors from M use sliding moves to cover the remaining row gaps. The total number of sensors moved is $c + (r - c) = r$.

Case $c \leq k < r$: Move c sensors from M using jumping moves to cover the first c row gaps and all c column gaps. Move the remaining $k - c$ sensors from M to cover $k - c$ row gaps. Finally $r - k$ sensors of Type 1 use sliding moves to cover the remaining row gaps. The total number of sensors moved is $c + (k - c) + (r - k) = r$.

Case $k < c$: Move k sensors from M using jumping moves to cover k row and column gaps. Then we use sliding moves of sensors of Type 1 to cover the remaining row and column gaps. In total $k + (c - k) + (r - k) = r + c - k$ sensors are moved.

Theorem 3. *Given an integer configuration, where all sensor ranges are 0.5, and the rectangle R to be covered is displaced by 0.5 in both axes, Algorithm 1 solves the MinNum-WCR problem in $O(n^{3/2})$ time.*

Proof. First we prove that Algorithm 1 produces a blocking configuration. Clearly, the sensors in M can be moved without creating new row or column gaps. Assume now that all sensors in M have been used by our algorithm to cover the gaps, but there still remains a row gap (or column gap). Recall that the number of sensors is assumed to be at least as large as the longer side of the rectangle. Thus, by the pigeonhole principle, there exists a row (or column, respectively) that contains more than one sensor. Such sensors must be of Type 1. One of them can make a sliding move along its column (or row) to cover the row gap, without creating any column gap or any other row gap. This shows that as long as there are gaps, there are sensors of Type 1 available to fill them as needed in the algorithm.

Next we show that Algorithm 1 moves an optimal number of sensors. Given an input configuration with r empty rows and c empty columns, assume without loss of generality that $r \geq c$. Observe that at least r sensors need to be moved to cover all empty rows, thus r is a lower bound on the number of sensors to be moved. If $k \geq c$, Algorithm 1 moves exactly r sensors, and it is optimal in the first two cases.

Suppose instead that $k < c$. At most k row and column gaps can be covered with sensors from M. It follows from the maximality of M that all remaining sensors are not free, that is, moving any of them must be done using a sliding move, which reduces either the number of row gaps or the number of column gaps by at most 1. Thus, in total we need to move $k + (c - k) + (r - k)$ sensors. Therefore Algorithm 1 moves an optimal number of sensors in this case as well.

Given a list of sensors with their coordinates, we can calculate in $O(n)$ time a list of the number of nodes in each row and a list of the number of nodes in each column. By an $O(n)$ scan of these lists we can find all nodes of Type 0 or 1, and we can mark the rows and columns containing sensors of Type 1. Now, an additional $O(n)$-time scan of the lists determines the types of all other nodes. By Theorem 2, a maximum free set of nodes can be constructed in $O(n^{3/2})$ time. After removing the max free set of sensors M, we can update the types of nodes, find row and column gaps, and calculate new positions in $O(n)$ time. Thus the total time taken by Algorithm 1 is $O(n^{3/2})$.   □

Remark 1. Notice that the algorithm described in this section assumes (a) that the sensors are initially at integer positions, and (b) that the rectangle R to be covered is displaced from the integer grid by 0.5 in each direction and (c) sensor diameters are 1. All these assumptions are necessary for the algorithm to produce a valid solution.

3   MinSum-WCR Problem

In this section, we study the MinSum-WCR problem. The MinSum problem is known to be NP-hard in the one-dimensional case of heterogeneous sensors on a line segment barrier [8]. The following theorem is therefore an immediate consequence:

Theorem 4. *For a heterogeneous sensor network, the MinSum-WCR problem is NP-complete for both Manhattan and Euclidean metrics.*

If the sensors are homogeneous and we consider the Manhattan metric, then to minimize the sum of the movements we can first minimize the sum of all horizontal movements by applying the known, one-dimensional, $O(n \log n)$ MinSum algorithm [8] to the x-coordinates of the sensors, and second we minimize the sum of all vertical movements by applying this $O(n \log n)$ MinSum algorithm to the y-coordinates of the sensors. It is easy to see that this gives an optimal solution. Thus we have the following result.

Theorem 5. *If all sensor ranges are equal, there is an $O(n \log n)$ algorithm to solve MinSum-WCR for the Manhattan metric.*

Remark 2. Unlike the algorithm for MinNum-WCR, the algorithm for MinSum-WCR works for arbitrary input configurations, and sensor ranges that are equal but of arbitrary size.

4   MinMax-WCR Problem

As seen in the previous section, the complexity of MinSum-WCR is very similar to the complexity of the MinSum barrier coverage of a line segment. However, this is not the case for the MinMax-WCR problem. For a line segment barrier, the MinMax problem can be solved using an $O(n \log n)$ algorithm even in the heterogeneous case [6]. Surprisingly, the MinMax-WCR problem is NP-hard even for a integer configuration with a very restricted possible move size, as shown in Theorem 6 below. The proof can be found in [20].

Theorem 6. *The MinMax-WCR Problem with Manhattan or Euclidean metric is NP-complete for maximum distance $D = 1$.*

5   Conclusion

In this paper we studied the complexity of establishing weak barrier coverage (WCR) in a given rectangular area using mobile sensors so that the network can detect any crossing of the area in a direction perpendicular to the sides of the rectangle. We considered the three typical optimization measures MinSum, MinMax, and MinNum for movements of sensors. For the MinNum-WCR problem, we show that the problem is NP-hard if sensors have sensing diameter 2, even if sensors are placed initially at integer locations. On the other hand, if sensors of sensing diameter 1 are placed at integer locations, we show an $O(n^{3/2})$ algorithm to solve the problem. For the MinMax-WCR problem, we show that the problem is NP-complete for both Euclidean and Manhattan metrics even when sensors are initially placed in integer locations. For the MinSum-WCR problem using the Manhattan metric, we showed that the problem is NP-complete for heterogeneous sensors, and solvable in $O(n \log n)$ time for homogeneous sensors. The complexity of MinSum-WCR for the Euclidean metric remains unknown.

References

1. Andrews, A.M., Wang, H.: Minimizing the aggregate movements for interval coverage. Algorithmica **78**(1), 47–85 (2017). doi:10.1007/s00453-016-0153-8
2. Balister, P., Bollobas, B., Sarkar, A., Kumar, S.: Reliable density estimates for coverage and connectivity in thin strips of finite length. In: ACM International Conference on Mobile Computing and Networking, pp. 75–86 (2007)
3. Ban, D., Jiang, J., Yang, W., Dou, W., Yi, H.: Strong k-barrier coverage with mobile sensors. In: Proceedings of International Wireless Communications and Mobile Computing Conference, pp. 68–72 (2010)
4. Berman, P., Karpinski, M.: On some tighter inapproximability results (extended abstract). In: Wiedermann, J., Emde Boas, P., Nielsen, M. (eds.) ICALP 1999. LNCS, vol. 1644, pp. 200–209. Springer, Heidelberg (1999). doi:10.1007/3-540-48523-6_17
5. Bhattacharya, B., Burmester, M., Hu, Y., Kranakis, E., Shi, Q., Wiese, A.: Optimal movement of mobile sensors for barrier coverage of a planar region. TCS **410**(52), 5515–5528 (2009)
6. Chen, D.Z., Gu, Y., Li, J., Wang, H.: Algorithms on minimizing the maximum sensor movement for barrier coverage of a linear domain. In: Fomin, F.V., Kaski, P. (eds.) SWAT 2012. LNCS, vol. 7357, pp. 177–188. Springer, Heidelberg (2012). doi:10.1007/978-3-642-31155-0_16
7. Czyzowicz, J., et al.: On minimizing the maximum sensor movement for barrier coverage of a line segment. In: Ruiz, P.M., Garcia-Luna-Aceves, J.J. (eds.) ADHOC-NOW 2009. LNCS, vol. 5793, pp. 194–212. Springer, Heidelberg (2009). doi:10.1007/978-3-642-04383-3_15
8. Czyzowicz, J., et al.: On minimizing the sum of sensor movements for barrier coverage of a line segment. In: Nikolaidis, I., Wu, K. (eds.) ADHOC-NOW 2010. LNCS, vol. 6288, pp. 29–42. Springer, Heidelberg (2010). doi:10.1007/978-3-642-14785-2_3
9. Dobrev, S., Durocher, S., Eftekhari, M., Georgiou, K., Kranakis, E., Krizanc, D., Narayanan, L., Opatrny, J., Shende, S., Urrutia, J.: Complexity of barrier coverage with relocatable sensors in the plane. Theoret. Comput. Sci. **579**, 64–73 (2015)
10. Eftekhari, M., Flocchini, P., Narayanan, L., Opatrny, J., Santoro, N.: Distributed barrier coverage with relocatable sensors. In: Halldórsson, M.M. (ed.) SIROCCO 2014. LNCS, vol. 8576, pp. 235–249. Springer, Cham (2014). doi:10.1007/978-3-319-09620-9_19
11. Eftekhari, M., Kranakis, E., Krizanc, D., Morales-Ponce, O., Narayanan, L., Opatrny, J., Shende, S.: Distributed algorithms for barrier coverage using relocatable sensors. Distrib. Comput. **29**(5), 361–376 (2016)
12. Eftekhari, M., Narayanan, L., Opatrny, J.: On multi-round sensor deployment for barrier coverage. In: Proceedings of 10th IEEE International Conference on Mobile Ad-hoc and Sensor Systems (IEEE MASS), pp. 310–318 (2013)
13. Garey, M.R., Johnson, D.S.: Computers and Intractability: A Guide to the Theory of NP-Completeness. W.H. Freeman & Co., New York (1979)
14. Habib, M.: Stochastic barrier coverage in wireless sensor networks based on distributed learning automata. Comput. Commun. **55**, 51–61 (2015)
15. Kranakis, E., Krizanc, D., Luccio, F.L., Smith, B.: Maintaining intruder detection capability in a rectangular domain with sensors. In: Bose, P., Gąsieniec, L.A., Römer, K., Wattenhofer, R. (eds.) ALGOSENSORS 2015. LNCS, vol. 9536, pp. 27–40. Springer, Cham (2015). doi:10.1007/978-3-319-28472-9_3

16. Kumar, S., Lai, T.H., Arora, A.: Barrier coverage with wireless sensors. In: Proceedings of the 11th Annual International Conference on Mobile Computing and Networking, pp. 284–298 (2005)
17. Li, L., Zhang, B., Shen, X., Zheng, J., Yao, Z.: A study on the weak barrier coverage problem in wireless sensor networks. Comput. Netw. **55**, 711–721 (2011)
18. Mehrandish, M., Narayanan, L., Opatrny, J.: Minimizing the number of sensors moved on line barriers. In: Proceedings of IEEE WCNC, pp. 653–658 (2011)
19. Yan, G., Qiao, D.: Multi-round sensor deployment for guaranteed barrier coverage. In: Proceedings of IEEE INFOCOM 2010, pp. 2462–2470 (2010)
20. Dobrev, S., Kranakis, E., Krizanc, D., Lafond, M., Manuch, J., Narayanan, L., Opatrny, J., Stacho, L.: Weak coverage of a rectangular barrier. arXiv 1701.07294 (2017)

Minimum Cost Perfect Matching with Delays for Two Sources

Yuval Emek[1]($\boxtimes$), Yaacov Shapiro[1]($\boxtimes$), and Yuyi Wang[2]($\boxtimes$)

[1] Technion, Haifa, Israel
yemek@technion.ac.il, shapiro.yaacov@gmail.com
[2] ETH Zürich, Zürich, Switzerland
yuwang@ethz.ch

Abstract. We study a version of the online *min-cost perfect matching with delays (MPMD)* problem recently introduced by Emek et al. (STOC 2016). In this problem, requests arrive in a continuous time online fashion and should be matched to each other. Each request emerges from one out of *n sources*, with metric inter-source distances. The algorithm is allowed to delay the matching of requests, but with a cost: when matching two requests, it pays the distance between their respective sources and the time each request has waited from its arrival until it was matched. In this paper, we consider the special case of $n = 2$ sources that captures the essence of the match-or-wait challenge (cf. rent-or-buy). It turns out that even for this degenerate metric space, the problem is far from trivial. Our results include a deterministic 3-competitive online algorithm for this problem, a proof that no deterministic online algorithm can have competitive ratio smaller than 3, and a proof that the same lower bound applies also for the restricted family of memoryless randomized algorithms.

1 Introduction

The abundance of hand held devices and the ease of application development is fueling the increasing popularity of online games. With that, the demand for head-to-head competition or players teaming up to complete a mission is growing fast. Ranging from classic games like Chess to car racing games like Asphalt 8, users wish to be matched with suitable opponents. The online gaming platform tries to find a match for each player while conflicted between two desired criteria: find a worthy opponent and find it fast. It is not hard to imagine a situation where the only available opponents are a poor match either due to a large difference between the players' skills and experience or due to the network distance between the players which may lead to significant communication delays. Should the platform wait, risking bored (and thus, dissatisfied) users? For how long?

Emek et al. [8] recently formalized this challenge in terms of the *min-cost perfect matching with delays (MPMD)* problem. In this problem requests arrive in an online fashion at the points of a finite metric space (known in advance). The online algorithm serves the requests by matching them to each other

© Springer International Publishing AG 2017
D. Fotakis et al. (Eds.): CIAC 2017, LNCS 10236, pp. 209–221, 2017.
DOI: 10.1007/978-3-319-57586-5_18

(i.e., partitioning the request set into pairs), where each match incurs a space cost equal to the metric distance between the locations of the matched requests. The crux of this problem is that it is not mandatory to serve the requests immediately; rather, the online algorithm is allowed to delay its matching commitments, but this incurs an additional time cost.

Some online gaming platforms will match a pending human player with a virtual (computer) opponent if a suitable human opponent cannot be found within a reasonable time frame. On the positive side, this allows the platform to shorten the waiting times of its users, but on the negative side, a player matched with a virtual opponent may be slightly disappointed: after all, it is more enjoyable to compete with your peers. This imposes an additional algorithmic challenge on behalf of the gaming platform: For how long should the platform wait before it matches a pending user to a virtual opponent? The challenge faced by a gaming platform that is allowed to match a human player to a virtual opponent is captured by a variant of the MPMD problem, referred to in [8] as *MPMDfp*, where the algorithm can serve a pending request without matching it to another request, paying a fixed penalty (hence the 'fp' abbreviation). Among other results, Emek et al. showed that the MPMDfp problem on an n-point metric space can be reduced to the MPMD problem on a $(2n)$-point metric space.

In this paper, we focus on a special case of the MPMD problem, referred to as *2-MPMD*, where the metric space consists of only two points with a unit distance between them, that is, the requests emerge from one of two possible *sources*. It turns out that even for this degenerate metric space, the problem is far from trivial. Moreover, the 2-MPMD problem generalizes the MPMDfp problem on a single point that corresponds to an online game in which all (user-to-user) matches are considered equally good and it is only the waiting times and penalties paid for matching users with a virtual opponent that affect the platform's total cost. This problem is interesting by its own right as it captures the essence of the wait-or-match question (cf. rent-or-buy) while abstracting away the space cost component arising from the distances in the metric space.

1.1   Model

An instance of the *2-source minimum cost perfect matching with delays (2-MPMD)* problem consists of two *sources* (denoted a and b) and a set R of *requests*. Each request $r \in R$ is characterized by its source $x(r) \in \{a, b\}$ and its *arrival time* $t(r) \in \mathbb{R}_{\geq 0}$. The request set R is provided to the algorithm in a continuous time online fashion so that request $r \in R$ is reported at time $t(r)$.

Assume throughout that $|R|$ is even. The output of the algorithm is a perfect matching of R, namely, a partition of R into unordered request pairs. Though this is computed online, the algorithm is allowed (and often required) to delay the matching of any request in R. This delay comes with a cost: the time between the arrival of a request and it being matched is added to the cost incurred by the algorithm.

Specifically, given two requests $r_1, r_2 \in R$, we denote a *match* operation $m(r_1, r_2, t)$ as the assignment of r_1 and r_2 into an unordered pair at time $t \geq max\{t(r_1), t(r_2)\}$. Match m incurs two types of costs on each request: *space cost* and *time cost*. If r_1 and r_2 have different sources (i.e., $x(r_1) \neq x(r_2)$), referred to hereafter as *matching across*, then the space cost for each of the requests is $1/2$, otherwise it is 0 (this choice of space cost convention reflects the implicit assumption that there is a unit distance between the two sources). The time costs for requests r_1 and r_2 are $t - t(r_1)$ and $t - t(r_2)$, respectively. The total cost of match operation m is denoted $cost(m)$ and is defined to be the sum of the space and time costs of the two involved requests. For algorithm ALG, we denote the set of unordered pairs that have been produced by it as M_{ALG} and the total cost incurred by the algorithm, denoted by $cost_{ALG}(R)$, is defined to be $cost_{ALG}(R) = \sum_{m \in M_{ALG}} cost(m)$.

When request r arrives, it is initially referred to as *open*; once the algorithm has matched it with another request, it becomes *matched*. Notice that this definition is only relevant in the context of a certain algorithm since two different algorithms may have matched the requests differently or at different times.

Our goal is to minimize the total cost of the match operations that our algorithm performs. Adhering to the common practice in the theory of online algorithms [4], the quality of the algorithmic solutions is measured in terms of their *competitive ratio*: Online algorithm ALG is said to be α-*competitive* if there exists a universal constant β such that $cost_{ALG}(R) \leq \alpha \cdot cost_{OPT}(R) + \beta$ for every (even size) request sequence R, where OPT is an optimal offline algorithm (the cost is taken in expectation if ALG is randomized). We notice that ALG has no a priori knowledge of R.

1.2 Related Work

Matching is a classic problem in graph theory and combinatorial optimization since the seminal work of Edmonds [6,7]. The matching problem has been studied in the context of online computation as well, starting with the classic paper of Karp et al. [11] that ignited the interest in online matching and attracted a lot of attention to the different versions of this problem [1,3,5,9,10,12–17]. In these online versions, it is usually assumed that the requests belong to one side of a bipartite graph whose other side is given in advance.

Emek et al. [8] recently introduced the MPMD problem which differs from the previously studied online matching versions in that the underlying graph (or metric space) is known in advance and the algorithmic challenge stems from the unknown locations and arrival times of the requests (whose number is unbounded). They present a randomized algorithm with competitive ratio $O(\log^2 n + \log \Delta)$, where n is the number of points in the metric space and Δ is the aspect ratio. Wang and Wattenhofer [18] show that the algorithm of [8] can be modified to treat the bipartite version of the MPMD problem (left outside the scope of the present paper), obtaining the same competitiveness. Azar et al. [2] devise another online MPMD algorithm with an improved logarithmic competitive ratio. They also prove that no (randomized) online MPMD algorithm can have competitive ratio better than $\Omega(\sqrt{\log n})$ in the all-pairs version and $\Omega(\log^{1/3} n)$ in the bipartite version.

1.3   Our Contribution

The general case of the online MPMD problem received a lot of attention in the last year, narrowing the asymptotic gap between the upper and lower bounds on the competitiveness of this problem [2,8,18]. The algorithms presented in [8] and [2] clearly imply a constant upper bound on the competitiveness of the 2-MPMD problem, but the analyses in these papers are not necessarily tailored to optimize this constant: the upper bound guaranteed by the analysis in [2] is 5; the upper bound guaranteed by the analysis in [8] is even larger, however, examining [8]'s randomized online algorithm (its 2-MPMD restriction) more carefully reveals that its competitive ratio is at most 3. Is this optimal? Can one ensure the same upper bound with a deterministic online algorithm?

In this paper, we answer these two questions on the affirmative. First, in Sect. 2, we introduce a deterministic variant of the online algorithm of [8] (restricted to the special case of the 2-MPMD problem) and establish an upper bound of 3 on its competitive ratio. Then, in Sect. 3, we prove that any deterministic online 2-MPMD algorithm must have a competitive ratio of at least 3. Finally, in Sect. 4, we investigate the competitiveness of *memoryless* online 2-MPMD algorithms—a family of randomized algorithms that include the algorithm of [8] (refer to Sect. 4 for an exact definition)—proving that it is at least 3 as well. While the upper bound of Sect. 2 clearly holds for the MPMDfp problem on a single point (recall that this is a special case of 2-MPMD), it is interesting to point out that the lower bounds of Sects. 3 and 4 also hold for that problem.

2   An Online 2-MPMD Algorithm

In this section we present a deterministic online 2-MPMD algorithm (Sect. 2.1), referred to as *delayed matching on 2 sources (DM2)*, and prove that its competitive ratio is 3 (Sect. 2.2). A matching lower bound will be established in Sect. 3.

2.1   Algorithm DM2

In this section we present our online algorithm $DM2$. While $DM2$ is designed for a continuous time environment, it is more easily understood when described as if it was operating in a discrete time environment, taking discrete time steps. The time difference dt between two consecutive steps is taken to be infinitesimally small so that we can assume without loss of generality that every request arrives in a separate time step.

The algorithm holds a counter T initialized to 0. For a given time step t and a request r_1 arriving during that time step, if there exists another open request

r_2 in the same source, then $DM2$ matches the two requests immediately. If during this time step, there are two open requests in different sources (arriving in this time step or in previous ones), then $DM2$ will increase T by dt. When T reaches τ, $DM2$ matches across and resets T back to 0. Refer to Algorithm 1 for a pseudocode.

Algorithm 1. Algorithm $DM2$ at time step t.

1: **if** there exist two open requests $r_1 \neq r_2$ with $x(r_1) = x(r_2)$ **then**
2:     $match(r_1, r_2)$
3: **else if** there exist two open requests $r_1 \neq r_2$ with $x(r_1) \neq x(r_2)$ **then**
4:     $T \leftarrow T + dt$
5:     **if** $T = \tau$ **then**
6:         $match(r_1, r_2)$
7:         $T \leftarrow 0$
8:     **end if**
9: **end if**

2.2   Analysis

Our goal in this section is to analyze the competitiveness of the online algorithm presented in Sect. 2.1, establishing the following theorem.

Theorem 1. *Algorithm $DM2$ is 3-competitive.*

We say that an (online or offline) algorithm A is *smart* if it satisfies the following property: If request r arrives at a source where there already exists an open request $r' \neq r$, i.e., $x(r') = x(r)$ and $t(r') < t(r)$, then A matches r and r' immediately, that is, at time $t(r)$. Notice that any smart algorithm will never have more than two open requests for a positive duration of time. Algorithm $DM2$ is clearly smart by definition.

Lemma 1. *There exists an online method that transforms any algorithm A into a smart algorithm $\tilde{A}$ without increasing the total cost incurred by the algorithm.*

Proof. Let R be the request sequence and consider the first occurrence of two open requests r_1, r_2 with $x(r_1) = x(r_2)$ and $t(r_1) < t(r_2)$ that algorithm A does not match immediately (i.e., at time $t(r_2)$). We construct an algorithm A' that behaves exactly like A up to time $t(r_2)$ and matches r_1 with r_2 at time $t(r_2)$ and show that the cost incurred by A' is not greater than that of A. This argument can then be repeated to turn A into the desired smart algorithm $\tilde{A}$.

Algorithm A' will match r_1 with r_2 at time $t(r_2)$ and continue as follows. If A matches r_1 with r_2 at a later time $t' > t(r_2)$, then all other matching operations of A' are identical to those of A. In this case, $\text{cost}_{A'}(R)$ is clearly smaller than $\text{cost}_A(R)$. The more interesting case is when A matches r_1 to $r'_1 \neq r_2$ at time t_1 and r_2 to $r'_2 \neq r_1$ at time t_2; in this case, A' will match r'_1 with r'_2 at time

$max\{t_1, t_2\}$. Again, all other matching operations of A' are identical to those of A.

It remains to prove that $\text{cost}_{A'}(R) \leq \text{cost}_A(R)$ also in this case. To that end, notice that

$$\begin{aligned} \text{cost}_A(R) - \text{cost}_{A'}(R) &\geq (t_1 - t(r_1) + t_1 - t(r_1') + t_2 - t(r_2) + t_2 - t(r_2')) \\ &\quad - (t(r_2) - t(r_1) + 2\max\{t_1, t_2\} - t(r_1') - t(r_2')) \\ &= 2(t_1 + t_2 - t(r_2) - \max\{t_1, t_2\}). \end{aligned}$$

The last expression is non-negative since we know that $t_1, t_2 \geq t(r_2)$.

We subsequently assume that OPT is smart. The cost incurred by a smart algorithm A (online or offline) is comprised of three *cost components*:

(C1) the space cost incurred by A for matching across;
(C2) the time cost incurred by A while there exists a single open request; and
(C3) the time cost incurred by A while there exist two open requests (one at each source).

Observation 1. *The parity of the number of open requests is the same for $DM2$ and OPT at any time t.*

From Observation 1 we conclude that cost component (C2) for $DM2$ is identical to that of OPT. We therefore ignore it in the subsequent analysis (this can only hurt the upper bound we obtain on the competitive ratio of $DM2$).

Our analysis relies on partitioning the time axis into *phases*. Each phase starts when the previous phase ends (the first phase starts at time 0) and ends when $DM2$ performs a match across (OPT might have open requests at this stage). Note also that a match across can only occur (Line 10) when T has increased from 0 to τ since the last time a match across occurred (i.e., in this phase only).

Let $M_A(P)$ denote the set of request pairs matched by algorithm A during phase P so the cost that algorithm A pays for P is $\text{cost}_A(P) = \sum_{m \in M_A(P)} \text{cost}_A(m)$. Since we ignore cost component (C2), the cost for $DM2$ of any phase P is always $\text{cost}_{DM2}(P) = 1 + 2\tau$.

Phase P is said to be *clean* if when it starts, OPT does not have any open requests. Otherwise, P is said to be *dirty*. Phase P is said to be *even numbered* if OPT performs an even number of matches across during the time interval P. Otherwise, P is said to be *odd numbered*.

Observation 2. *When a dirty phase starts, OPT has exactly two open requests.*

Proof. From Observation 1 we know OPT has an even number of open requests; it can not be larger than 2 because OPT is smart and it can not be 0 because the phase is dirty.

Observation 3. *In every phase the number of requests that appear in each source is odd.*

Proof. DM2 starts and ends each phase with no open requests, and since it also matches across exactly once during the phase, the number of requests appearing during the phase in each source must be odd.

Observation 4 follows from Observation 3 and from the fact that *DM2* matches across exactly once (and in particular an odd number of times) in each phase.

Observation 4. *Given two consecutive phases P_1 and P_2, if both are clean or both are dirty, then P_1 must be an odd numbered phase. If one of them is dirty and the other is clean, then P_1 must be even numbered.*

We now turn to define the notion of a *super phase* which consists of a clean phase followed by a maximal (possibly empty) contiguous sequence of dirty phases. This notion uniquely induces a partition of the phase sequence into super phases. We denote the cost that algorithm A pays for super phase S as $\text{cost}_A(S) = \sum_{P \in S} \text{cost}_A(P)$. By definition, when a super phase starts, both *DM2* and *OPT* have no open requests. This means that we can establish Theorem 1 by proving the following lemma.

Lemma 2. *There exists a choice of the parameter $\tau > 0$ that guarantees $\text{cost}_{DM2}(S) \leq 3 \cdot \text{cost}_{OPT}(S)$ for every super phase S.*

Proof. Consider some super phase S comprised of a single clean phase followed by $n \geq 0$ dirty phases. We know that *DM2* pays $1 + 2\tau$ for each phase, so $\text{cost}_{DM2}(S) = (n+1)(1+2\tau)$.

If S consists of exactly one phase P (i.e., $n = 0$), then by Observation 4, P is odd numbered. Therefore, *OPT* matched across at least once during super phase S, hence $\text{cost}_{OPT}(S) \geq 1$. This proves the assertion under the requirement that $\tau \leq 1$.

Assume hereafter that S contains $n \geq 1$ dirty phases. Refer to the first (clean) phase of S as P_0 and to the subsequent n dirty phases as $P_1, \ldots, P_n$ in order of appearance. By Observation 4, phase P_0 is even numbered, therefore, during this phase, *OPT* either matched across at least twice or did not match across at all. The former case implies that $\text{cost}_{OPT}(P_0) \geq 2$; in the latter case, we know by the design of *DM2* that *OPT* paid at least 2τ in time cost. This means that $\text{cost}_{OPT}(P_0) \geq min\{2, 2\tau\}$.

Since phases $P_1, \ldots, P_n$ are all dirty, Observation 4 ensures that phases $P_1, \ldots, P_{n-1}$ are all odd numbered, hence *OPT* must have matched across at least once during each one of them. It follows that $\text{cost}_{OPT}(P_i) \geq 1$ for every $1 \leq i \leq n - 1$. Therefore, the total cost of *OPT* for super phase S is $\text{cost}_{OPT}(S) \geq min\{2, 2\tau\} + (n - 1)$.

To establish the assertion, we require that $(n+1)(1+2\tau) \leq 3(min\{2, 2\tau\} + (n-1))$. To that end, we let

$$f(\tau) = (n+1)(1+2\tau) - 3(min\{2, 2\tau\} + (n-1)) = 2n\tau + 2\tau - 6min\{1, \tau\} - 2n + 4$$

and require that $f(\tau) \leq 0$. Observing that setting $\tau < 1$ implies

$$f(\tau) = 2n\tau + 2\tau - 6\tau - 2n + 4 = (1 - \tau)(4 - 2n)$$

which means that $f(\tau) > 0$ for $n = 1$, forces τ to be at least 1. Recalling the prior constraint of $\tau \leq 1$, we conclude that τ must be 1. Indeed, by setting $\tau = 1$, we obtain $f(\tau) = 0$ as required.

3   A Lower Bound for Deterministic Algorithms

We now turn to show that the algorithm presented in Sect. 2.1 is optimal by establishing a matching lower bound.

Theorem 2. *For any $\delta > 0$, no deterministic online 2-MPMD algorithm can have a competitive ratio of $3 - \delta$.*

Theorem 2 is established by proving that for every deterministic online 2-MPMD algorithm A, there exists a request sequence R with arbitrarily large $\text{cost}_{OPT}(R)$ such that $\text{cost}_A(R) \geq (3 - \delta)\text{cost}_{OPT}(R)$. The key ingredient in the construction of R is a *gadget* G satisfying $\text{cost}_A(G) \geq (3 - \delta)\text{cost}_{OPT}(G)$; the request sequence R is then constructed by repeatedly introducing instances of this gadget with sufficiently large time gaps between consecutive copies.

Gadget G is comprised of $2n$ requests, denoted by $r_1, \ldots, r_n$ and $r'_1, \ldots, r'_n$, with $x(r_i) = a$, $x(r'_i) = b$, and $t(r_i) = t(r'_i)$ for every $1 \leq i \leq n$. Requests r_1 and r'_1 are the only requests certain to appear, while the appearance of the rest of the requests r_i, r'_i depends on the behavior of A. Given that G includes requests r_i and r'_i (i.e., $n \geq i$), let t_i be the difference (in absolute value) between time $t(r_i)$ and the time when A performed $m(r_i, r'_i)$ (the time difference t_i is well defined since A must eventually match $m(r_i, r'_i)$ as otherwise its competitive ratio is unbounded). The appearance of requests r_{i+1} and r'_{i+1} then abides the following rule: if $t_i < 1$, then requests r_{i+1} and r'_{i+1} are introduced at time

$$t(r_{i+1}) = t(r'_{i+1}) = t(r_i) + t_i + \epsilon$$

for a sufficiently small $\epsilon > 0$; otherwise, the gadget ends (i.e., $n = i$). This holds until the first odd $i \geq 3$ that fulfills $\frac{i-2}{i} \geq 1 - \delta$ after which there will appear no more requests (i.e., $n = i$).

Lemma 3. *For every $\delta > 0$, the construction of G ensures that $\text{cost}_A(G) \geq (3 - \delta)\text{cost}_{OPT}(G)$.*

Proof. If $n=1$, then $\text{cost}_A(G)=1+2t_1$ whereas $\text{cost}_{OPT}(G)=1$, thus $\frac{\text{cost}_A(G)}{\text{cost}_{OPT}(G)} = \frac{1+2t_1}{1}$ which is at least 3 since $t_1 \geq 1$. If $n = 2$, then $\text{cost}_A(G) = 2 + 2t_1 + 2t_2$ whereas $\text{cost}_{OPT}(G) = 2t_1$ since OPT performs $\{m(r_1, r_2), m(r'_1, r'_2)\}$. Therefore, $\frac{\text{cost}_A(G)}{\text{cost}_{OPT}(G)} = \frac{2+2t_1+2t_2}{2t_1} = \frac{2}{2t_1} + \frac{2t_1}{2t_1} + \frac{2t_2}{2t_1}$ which is at least 3 since $t_1 < 1 \leq t_2$.

Assume hereafter that $n \geq 3$ is odd (the case of even $n \geq 3$ is similar and deferred to the full version). Notice that $\text{cost}_A(G)$ is always $n+2t_1+2t_2+\cdots+2t_n$ whereas OPT can choose between the following options (among others).

1. Perform a match across $m(r_1, r'_1, t(r_1))$ for the first two requests; match all other requests by performing $m(r_{2j}, r_{2j+1}, t(r_{2j+1}))$ and $m(r'_{2j}, r'_{2j+1}, t(r'_{2j+1}))$ for every $1 < j \leq (n-1)/2$.

   Ignoring ϵ-terms that can be made arbitrarily small, this results in $\mathrm{cost}_{OPT}(G) = 1 + 2t_2 + 2t_4 + \cdots + 2t_{n-1}$.

2. Perform a match across $m(r_n, r'_n, t(r_n))$ for the last two requests; match all other requests by performing $m(r_{2j-1}, r_{2j}, t(r_{2j}))$ and $m(r'_{2j-1}, r'_{2j}, t(r'_{2j}))$ for every $1 \leq j \leq (n-1)/2$.

   Ignoring ϵ-terms that can be made arbitrarily small, this results in $\mathrm{cost}_{OPT}(G) = 1 + 2t_1 + 2t_3 + \cdots + 2t_{n-2}$.

Denoting $T_{odd} = \sum_{j=1}^{\lfloor n/2 \rfloor} t_{2j-1}$ and $T_{even} = \sum_{j=1}^{\lceil n/2 \rceil - 1} t_{2j}$ implies that $\mathrm{cost}_{OPT}(G)$ can never exceed $\min\{1 + 2T_{even}, 1 + 2T_{odd}\}$. Since $t_i < 1$ for every $1 \leq i \leq n-1$, it follows that $\mathrm{cost}_{OPT}(G) < n$.

Consider the case where $t_i < 1$ for every $1 \leq i \leq n-1$ and $t_n \geq 1$. We examine the ratio of $\mathrm{cost}_A(G)$ and $\mathrm{cost}_{OPT}(G)$ to conclude that

$$
\frac{\mathrm{cost}_A(G)}{\mathrm{cost}_{OPT}(G)}
$$

$$
= \frac{n + 2T_{odd} + 2T_{even} + 2t_n}{1 + 2\min\{T_{even}, T_{odd}\}}
$$

$$
= \frac{1 + 2T_{odd}}{1 + 2\min\{T_{even}, T_{odd}\}} + \frac{1 + 2T_{even}}{1 + 2\min\{T_{even}, T_{odd}\}} + \frac{n - 2 + 2t_n}{1 + 2\min\{T_{even}, T_{odd}\}}
$$

$$
> \frac{1 + 2T_{odd}}{1 + 2T_{odd}} + \frac{1 + 2T_{even}}{1 + 2T_{even}} + \frac{n}{n} = 3 .
$$

It remains to consider the case where $t_i < 1$ for every $1 \leq i \leq n$ which means that n is odd and that $\frac{n-2}{n} \geq 1 - \delta$. Again, we examine the ratio of $\mathrm{cost}_A(G)$ and $\mathrm{cost}_{OPT}(G)$ to conclude that

$$
\frac{\mathrm{cost}_A(G)}{\mathrm{cost}_{OPT}(G)}
$$

$$
= \frac{n + 2T_{odd} + 2T_{even} + 2t_n}{1 + 2\min\{T_{even}, T_{odd}\}}
$$

$$
= \frac{1 + 2T_{odd}}{1 + 2\min\{T_{even}, T_{odd}\}} + \frac{1 + 2T_{even}}{1 + 2\min\{T_{even}, T_{odd}\}} + \frac{n - 2 + 2t_n}{1 + 2\min\{T_{even}, T_{odd}\}}
$$

$$
\geq \frac{1 + 2T_{odd}}{1 + 2T_{odd}} + \frac{1 + 2T_{even}}{1 + 2T_{even}} + \frac{n - 2}{n} \geq 3 - \delta .
$$

The assertion follows.

4   Memoryless Online Algorithms

We now turn our attention to randomized algorithms. As the deterministic algorithm presented in Sect. 2 is 3-competitive and the lower bound established in

Sect. 3 states that this cannot be improved, it is natural to ask whether a randomized 2-MPMD algorithm can have competitive ratio smaller than 3. While we dot know the answer to this question in the general case yet, the current section resolves it on the negative for a restricted family of randomized 2-MPMD algorithms.

Recall the notion of smart algorithms presented in Sect. 2.2. Lemma 1 guarantees that for the sake of establishing negative results, it suffices to consider the class of smart online algorithms as any algorithm can be transformed into a smart algorithm without increasing the cost (since the transformation works in an online fashion, the lemma applies to randomized algorithms as well).

Consider some randomized smart 2-MPMD algorithm ALG. By the definition of smart algorithms, ALG is fully characterized by the parameter $\lambda(t) \in \mathbb{R}_{\geq 0}$, $t \geq 0$, defined so that if there is an open request in each source throughout the infinitesimally small time interval $[t - dt, t)$, then ALG matches across at time t with probability $\lambda(t)dt$. (Strictly speaking, $\lambda(t)$ can also take the special value $1/dt$ in which case $\lambda(t)dt = 1$; in particular, the algorithm is deterministic if $\lambda(t)$ is either 0 or $1/dt$ for every $t \geq 0$.) We say that algorithm ALG is *memoryless* if there exists some $\lambda > 0$ such that $\lambda(t) = \lambda$ for every $t \geq 0$, namely, the probability that ALG matches across at time t depends only on the infinitesimally small time interval $[t - dt, t)$ and is independent of the rest of the history.

Interestingly, the restriction of the randomized online MPMD algorithm of [8] to metric spaces with 2 sources is memoryless with parameter $\lambda = 1$. Although the authors of [8] did not attempt to optimize the (constant) competitive ratio of their algorithm for that special case, a careful examination of the arguments used in the analysis of this algorithm reveals that its competitive ratio is 3. In this section, we show that this cannot be improved.

Theorem 3. *If ALG is a (randomized) memoryless 2-MPMD algorithm with parameter $\lambda \neq 1$, then its competitive ratio is greater than 3.*

Proof. Consider first the case where $\lambda < 1$. Let R be the request sequence consisting of $2n$ requests $r_1, \ldots, r_n$ and $r'_1, \ldots, r'_n$ for some arbitrarily large n so that

(1) $x(r_i) = a$ and $x(r'_i) = b$ for every $1 \leq i \leq n$;
(2) $t_1 = t(r_1) = t(r'_1) = 0$; and
(3) $t_{i+1} = t(r_{i+1}) = t(r'_{i+1}) = t_i + z$ for some sufficiently large real z for every $1 \leq i \leq n - 1$.

Given that $z > 1$, OPT will perform $m(r_i, r'_i, t_i)$ for every $1 \leq i \leq n$ for a total cost of $\text{cost}_{OPT}(R) = n$.

Assuming that all previous requests are already matched by the time r_i and r'_i arrive, ALG will perform either (i) $m(r_i, r'_i, t_i + Y_i)$ for some $0 \leq Y_i < z$; or (ii) $m(r_i, r_{i+1}, t_{i+1})$ and $m(r'_i, r'_{i+1}, t_{i+1})$, recalling that $t_{i+1} = t_i + z$. By the definition of a memoryless algorithm, the probability that (ii) occurs is

$$\mathbb{P}(\text{Exp}(\lambda) > z) = e^{-\lambda z},$$

where $\text{Exp}(\lambda)$ is an exponential random variable with rate parameter λ. Moreover, Y_i is a random variable that behaves like $\text{Exp}(\lambda)$, truncated at z, namely, $Y_i \sim \min\{\text{Exp}(\lambda), z\}$. Standard calculation (see, e.g., the proof of Lemma 4.2 in [8]) then yields that

$$\mathbb{E}(Y_i) = \frac{1}{\lambda}(1 - e^{-\lambda z}).$$

It follows that

$$\lim_{z \to \infty} \mathbb{E}(\text{cost}_{ALG}(R)) = n(1 + 2/\lambda) > 3n,$$

where the last transition follows from the assumption that $\lambda < 1$.

Suppose that $\lambda > 1$ and let R be the request sequence consisting of $4n$ requests $r_1, \ldots, r_{2n}$ and $r'_1, \ldots, r'_{2n}$ for some arbitrarily large n so that

(1) $x(r_i) = a$ and $x(r'_i) = b$ for every $1 \le i \le 2n$;
(2) $t_1 = t(r_1) = t(r'_1) = 0$;
(3) $t_{2i} = t(r_{2i}) = t(r'_{2i}) = t_{2i-1} + \epsilon$ for some sufficiently small real $\epsilon > 0$ for every $1 \le i \le n$; and
(4) $t_{2i+1} = t(r_{2i+1}) = t(r'_{2i+1}) = t_{2i} + z$ for some sufficiently large real z for every $1 \le i \le n - 1$.

Given that $\epsilon < 1$, OPT will perform $m(r_{2i-1}, r_{2i}, t_{2i})$ and $m(r'_{2i-1}, r'_{2i}, t_{2i})$ for every $1 \le i \le n$ for a total cost of $\text{cost}_{OPT}(R) = 2\epsilon n$.

Assuming that all previous requests are already matched by the time r_{2i-1} and r'_{2i-1} arrive, ALG will perform either (i) $m(r_{2i-1}, r'_{2i-1}, t_{2i-1} + y)$ for some $0 \le y < \epsilon$; or (ii) $m(r_{2i-1}, r_{2i}, t_{2i})$ and $m(r'_{2i-1}, r'_{2i}, t_{2i})$, recalling that $t_{2i} = t_{2i-1} + \epsilon$. Taking p to be the probability that (i) occurs, we conclude by the definition of a memoryless algorithm that $p = \mathbb{P}(\text{Exp}(\lambda) < \epsilon) = 1 - e^{-\lambda \epsilon}$, hence

$$\lambda \epsilon (1 - \lambda \epsilon) < \lambda \epsilon - (\lambda \epsilon)^2/2 < p < \lambda \epsilon$$

by standard approximations of the exponential function.

Condition for the time being on the event that (i) occurs and notice that ALG will now perform either (iii) $m(r_{2i}, r'_{2i}, t_{2i} + Y_i)$ for some $0 \le Y_i < z$; or (iv) $m(r_{2i}, r_{2i+1}, t_{2i+1})$ and $m(r'_{2i}, r'_{2i+1}, t_{2i+1})$, recalling that $t_{2i+1} = t_{2i} + z$. As before, the probability that (iv) occurs is $\mathbb{P}(\text{Exp}(\lambda) > z) = e^{-\lambda z}$ and $Y_i \sim \min\{\text{Exp}(\lambda), z\}$ with $\mathbb{E}(Y_i) = \frac{1}{\lambda}(1 - e^{-\lambda z})$, so by taking z to be sufficiently large with respect to n, we can assume that event (iv) never occurs. This means that every time event (i) occurs, ALG pays, on expectation, $1 + 2/\lambda$ for matching r_{2i} and r'_{2i} (and that this match occurs before time t_{2i+1}, i.e., the arrival time of the next requests).

Since every time (i) occurs, ALG pays an additional space cost of 1 for matching (across) r_{2i-1} and r'_{2i-1} and every time (ii) occurs, ALG pays a time cost of 2ϵ, it follows that

$$\mathbb{E}(\text{cost}_{ALG}(R)) \ge n(p(2 + 2/\lambda) + (1 - p)2\epsilon)$$
$$= 2n(p(1 + 1/\lambda) + (1 - p)\epsilon)$$
$$> 2n(\lambda \epsilon (1 - \lambda \epsilon)(1 + 1/\lambda) + (1 - \lambda \epsilon)\epsilon)$$
$$= (1 - \lambda \epsilon)2\epsilon n(\lambda(1 + 1/\lambda) + 1)$$
$$= (1 - \lambda \epsilon)2\epsilon n(\lambda + 2).$$

By taking $1/\epsilon$ to be sufficiently large with respect to λ (yet, much smaller than n), we conclude that $\mathbb{E}(\text{cost}_{ALG}(R)) > 3\text{cost}_{OPT}(R)$ due to the assumption that $\lambda > 1$.

Acknowledgments. We are indebted to Shay Kutten and Roger Wattenhofer for their help in making this paper happen.

References

1. Aggarwal, G., Goel, G., Karande, C., Mehta, A.: Online vertex-weighted bipartite matching and single-bid budgeted allocations. In: Proceedings of the Twenty-Second Annual ACM-SIAM Symposium on Discrete Algorithms, SODA, pp. 1253–1264 (2011)
2. Azar, Y., Chiplunkar, A., Kaplan, H.: Polylogarithmic bounds on the competitiveness of min-cost perfect matching with delays. In: Proceedings of the 28th Annual ACM-SIAM Symposium on Discrete Algorithms, SODA, pp. 1051–1061 (2017)
3. Birnbaum, B.E., Mathieu, C.: On-line bipartite matching made simple. SIGACT News **39**(1), 80–87 (2008)
4. Borodin, A., El-Yaniv, R.: Online Computation and Competitive Analysis. Cambridge University Press, Cambridge (1998). **62**(1), 1–20
5. Devanur, N.R., Jain, K., Kleinberg, R.D.: Randomized primal-dual analysis of RANKING for online bipartite matching. In: Proceedings of the Twenty-Fourth Annual ACM-SIAM Symposium on Discrete Algorithms, SODA, pp. 101–107 (2013)
6. Edmonds, J.: Maximum matching and a polyhedron with 0, 1-vertices. J. Res. Natl. Bur. Stand. B **69**, 125–130 (1965)
7. Edmonds, J.: Paths, trees, and flowers. Can. J. Math. **17**, 449–467 (1965)
8. Emek, Y., Kutten, S., Wattenhofer, R.: Online matching: haste makes waste! In: 48th Annual Symposium on Theory of Computing (STOC), June 2016. A full version can be obtained from http://ie.technion.ac.il/yemek/Publications/omhw.pdf
9. Goel, G., Mehta, A.: Online budgeted matching in random input models with applications to adwords. In: Proceedings of the Nineteenth Annual ACM-SIAM Symposium on Discrete Algorithms, SODA, pp. 982–991 (2008)
10. Kalyanasundaram, B., Pruhs, K.: Online weighted matching. J. Algorithms **14**(3), 478–488 (1993)
11. Karp, R.M., Vazirani, U.V., Vazirani, V.V.: An optimal algorithm for on-line bipartite matching. In: Proceedings of the 22nd Annual ACM Symposium on Theory of Computing, pp. 352–358 (1990)
12. Khuller, S., Mitchell, S.G., Vazirani, V.V.: On-line algorithms for weighted bipartite matching and stable marriages. Theor. Comput. Sci. **127**(2), 255–267 (1994)
13. Mehta, A.: Online matching and ad allocation. Found. Trends Theor. Comput. Sci. **8**(4), 265–368 (2013)
14. Mehta, A., Saberi, A., Vazirani, U.V., Vazirani, V.V.: AdWords and generalized on-line matching. In: 46th Annual IEEE Symposium on Foundations of Computer Science (FOCS), pp. 264–273 (2005)
15. Meyerson, A., Nanavati, A., Poplawski, L.J.: Randomized online algorithms for minimum metric bipartite matching. In: Proceedings of the Seventeenth Annual ACM-SIAM Symposium on Discrete Algorithms, SODA (2006)

16. Miyazaki, S.: On the advice complexity of online bipartite matching and online stable marriage. Inf. Process. Lett. **114**(12), 714–717 (2014)
17. Naor, J., Wajc, D.: Near-optimum online ad allocation for targeted advertising. In: Proceedings of the Sixteenth ACM Conference on Economics and Computation, EC, pp. 131–148 (2015)
18. Wang, Y., Wattenhofer, R.: Perfect Bipartite Matching with Delays (submitted)

Congestion Games with Complementarities

Matthias Feldotto$^{(\boxtimes)}$, Lennart Leder, and Alexander Skopalik

Heinz Nixdorf Institute and Department of Computer Science,
Paderborn University, Paderborn, Germany
{feldi,lleder,skopalik}@mail.upb.de

Abstract. We study a model of selfish resource allocation that seeks
to incorporate dependencies among resources as they exist in modern
networked environments. Our model is inspired by utility functions with
constant elasticity of substitution (CES) which is a well-studied model
in economics. We consider congestion games with different aggregation
functions. In particular, we study L_p norms and analyze the existence
and complexity of (approximate) pure Nash equilibria. Additionally, we
give an almost tight characterization based on monotonicity properties
to describe the set of aggregation functions that guarantee the existence
of pure Nash equilibria.

Keywords: Congestion games · Aggregation · L_p norms · Complementarities · Existence of equilibria · Approximate pure Nash equilibria

1 Introduction

Modern networked environments often lack a central authority that has the
ability or the necessary information to coordinate the allocation of resources such
as bandwidths of network links, server capacities, cloud computing resources, etc.
Hence, allocation decisions are delegated to local entities or customers. Often
they are interested in allocations that optimize for themselves rather than for
overall system performance. We study the strategic interaction that arises in
such situations using game theoretic methods.

The class of congestion games [28] is a well-known model to study scenarios
in which the players allocate shared resources. In a congestion game each player
chooses a subset of resources from a collection of allowed subsets which are called
strategies. These resources may represent links in a network, servers, switches,
etc. Each resource is equipped with a cost function that is mapping from the
number of players using it to a cost value. The cost of a player is the sum of
the costs of the resources in the chosen strategy. There are several well-known
extensions to this model. In *weighted* congestion games [17], players can have
different weights and the cost of a resource depends on the total weight of the

This work was partially supported by the German Research Foundation (DFG)
within the Collaborative Research Centre "On-The-Fly Computing" (SFB 901).
The full version of this paper is available at https://arxiv.org/abs/1701.07304.

D. Fotakis et al. (Eds.): CIAC 2017, LNCS 10236, pp. 222–233, 2017.
DOI: 10.1007/978-3-319-57586-5_19

players using it. In *player-specific* congestion games [26], the costs of resources can be different for different players.

However, all these models have in common that the cost of a player is defined as the sum of the resource costs. This is well-suited to describe latencies or delays of computer or traffic networks, for example. However, in scenarios in which bandwidth determines the costs of players this is determined by the bottleneck link. To that end, bottleneck congestion games have been introduced [5] in which the cost of a player is defined as the cost of her most expensive resource.

Both models are limited in their ability to model complementarities that naturally arise in scenarios where the performance of a resource depends to some degree on the performance of other resources. For example, a cloud-based web application may be comprised of many resources. A low performing resource negatively influences the performance of other parts and hence the overall system. Bottleneck games assume perfect complements, whereas standard congestion games assume independence. We seek to generalize both models and allow for different degrees of complementarity that may even differ between players. We are inspired by utility functions with constant elasticity of substitution (CES) [4,12], which are a well-studied and accepted model in economics. We adapt the notion to our needs and study the analogue version for cost functions that corresponds to L_p norms. Clearly, both standard congestion games and bottleneck congestion games are special cases of these games with L_1 and L_∞ norms, respectively. Using further aggregation functions instead of L_p norms even allows to model more complex dependencies. Based on natural monotonicity properties of these functions, we can characterize the existence of pure Nash equilibria.

1.1   Related Work

Congestion games were introduced by Rosenthal [28] who shows that these games are potential games. In fact, the class is isomorphic to the class of potential games as shown by Monderer and Shapley [27]. The price of anarchy in the context of network congestion games was first considered by Koutsopias and Papadimitriou [22]. The related concept of smoothness, which can be used to derive a bound on the price of anarchy, was introduced by Roughgarden [29]. Fabrikant et al. [14] show that in congestion games improvement sequences may have exponential length, and that it is in general PLS-complete to compute a pure Nash equilibrium. Chien and Sinclair [10] show that for symmetric congestion games with a mild assumption on the cost function the approximate best-response dynamics converge quickly to an approximate pure Nash equilibrium. In contrast to that, Skopalik and Vöcking [31] show that it is in general even PLS-hard to compute approximate pure Nash equilibria for any polynomially computable approximation factor. However, if the cost functions are restricted to linear or constant degree polynomials, approximate pure Nash equilibria can be computed in polynomial time as shown by Caragiannis et al. [7], even for weighted games [9] and some other variants [8,15]. Hansknecht et al. [18] use the concept of approximate potential functions to examine approximate pure Nash equilibria in weighted congestion games under different restrictions on the cost functions. For polynomial cost functions of maximal degree g they show that $(g+1)$-approximate equilibria are guaranteed to exist.

Singleton congestion games, a class of congestion games which guarantees polynomial convergence of best-response improvement sequences to pure Nash equilibria, are considered by Ieong et al. [21]. They show that the property of polynomial convergence can be generalized to so called independent-resource congestion games. It is further generalized by Ackermann et al. [2] to matroid congestion games. They show that, for non-decreasing cost functions, the matroid property is not only sufficient, but also necessary to guarantee the convergence to pure Nash equilibria in polynomial time. Milchtaich [26] studies the concept of player-specific congestion games and shows that in the singleton case these games always admit pure Nash equilibria. Ackermann et al. [1] generalize these results to matroid strategy spaces and show that the result also holds for weighted congestion games. They also examine the question of efficient computability and convergence towards these equilibria. Furthermore, they point out that in a natural sense the matroid property is maximal for the guaranteed existence of pure Nash equilibria in player-specific and weighted congestion games. Moreover, Milchtaich [26] examines congestion games in which players are both weighted and have player-specific cost functions. By constructing a game with three players he shows that these games, even in the case of singleton strategies, do not necessarily possess pure Nash equilibria. Mavronicalas et al. [25] study a special case of these games in which cost functions are not entirely player-specific. Instead, the player-specific resource costs are derived by combining the general resource cost function and a player-specific constant via a specified operation (e.g. addition or multiplication). They show that this restriction is sufficient to guarantee the existence of pure Nash-equilibria in games with three players. Dunkel and Schulz [13] show that the decision problem whether a weighted network congestion game possesses a pure Nash equilibrium is NP-hard. For player-specific network congestion games, Ackermann and Skopalik [3] show that this problem is NP-complete both in directed and in undirected graphs.

Banner and Orda [5] introduce the class of bottleneck congestion games and study their applicability in network routing scenarios. In particular, they derive bounds on the price of anarchy in network bottleneck congestion games with restricted cost functions and show that there always exists a pure Nash equilibrium which is socially optimal, but that the computation of this equilibrium is NP-hard. Harks et al. [19] give an overview on bottleneck congestion games and the complexity of computing pure Nash equilibria. Moreover, they show that in matroid bottleneck congestion games even pure strong equilibria, which are stable against coalitional deviations, can be computed efficiently. Harks et al. [20] introduce the so called Lexicographical Improvement Property, which guarantees the existence of pure Nash equilibria through a potential function argument. They show that bottleneck congestion games fulfill this property.

Feldotto et al. [16] generalize both variants and investigate the linear combination of standard and bottleneck congestion games. Kukushkin [24] introduces the concept of generalized congestion games in which players may use arbitrary monotonic aggregation functions to calculate their total cost from the costs of their single resources. He shows that, apart from monotonic mappings, additive aggregation functions that fulfill certain restrictions are the only ones for which the existence of PNE can be guaranteed. In a later paper [23], he elaborates this result by deriving properties for the players' aggregation functions which are sufficient to establish this guarantee. Another generalization of congestion games is given by Byde et al. [6] and Voice et al. [32]. They introduce the model of games with congestion-averse utility functions. They show under which properties pure Nash equilibria exist and give a polynomial time algorithm to compute them.

1.2   Our Contribution

We introduce congestion games with L_p-aggregation functions and show that pure Nash equilibria are guaranteed only if either there is one aggregation value p for all players or in the case of matroid congestion games. For games with linear cost functions in which a pure Nash equilibrium exists, we derive bounds on the price of anarchy. For general games, we show the existence of approximate pure Nash equilibria where the approximation factor scales sublinearly with the size of the largest strategy set. We show that this factor is tight and that it is NP-hard to decide whether there is an approximate equilibrium with a smaller factor. Computing an approximate PNE with that factor is PLS-hard. As a positive result, we present two different polynomial time algorithms to compute approximate equilibria in games with linear cost functions. The approximation factors of both methods have a different dependence on the parameters of the game. For matroid games, we show the existence of pure Nash equilibria not only for L_p-aggregation functions but also seek to extend it to more general aggregation functions. We can characterize the functions that guarantee existence of PNE by certain monotonicity properties.

1.3   Model/Preliminaries

A *congestion game with L_p-aggregation functions* is a tuple $\Gamma = (N, R, (\Sigma_i)_{i \in N},$ $(c_r)_{r \in R}, (p_i)_{i \in N})$. $N = \{1, \ldots, n\}$ denotes the set of players, $R = \{r_1, \ldots, r_m\}$ the set of resources. For each player $i \in N$, $\Sigma_i \subseteq 2^R$ denotes the strategy space of player i and $p_i \in \mathbb{R}$, $p_i \geq 1$ denotes the player-specific aggregation value of player i. For each resource r, $c_r : N \to \mathbb{R}$ denotes the non-decreasing cost function associated to resource r.

In a congestion game, the state $S = (S_1, \ldots, S_n)$ describes the situation that each player $i \in N$ has chosen the strategy $S_i \in \Sigma_i$. In state S, we define for each resource $r \in R$ by $n_r(S) = |\{i \in N \mid r \in S_i\}|$ the congestion of r. The cost of resource r in state S is defined as $c_r(S) = c_r(n_r(S))$. The cost of player i is defined as $c_i(S) = \left(\sum_{r \in S_i} c_r(S)^{p_i} \right)^{\frac{1}{p_i}}$. If for all $i, j \in N$ it holds that $p_i = p_j$, then we call Γ a congestion game with identical L_p-aggregation functions.

For a state $S = (S_1, \ldots, S_i, \ldots, S_n)$, we denote by (S_i', S_{-i}) the state that is reached if player i plays strategy S_i' while all other strategies remain unchanged. A state $S = (S_1, \ldots, S_n)$ is called a *pure Nash equilibrium (PNE)* if for all $i \in N$ and all $S_i' \in \Sigma_i$ it holds that $c_i(S) \leq c_i(S_i', S_{-i})$ and a *β-approximate* *pure Nash equilibrium* for a $\beta \geq 1$ if for all $i \in N$ and all $S'_i \in \Sigma_i$ it holds that $c_i(S) \leq \beta \cdot c_i(S'_i, S_{-i})$. Additionally, we define the Price of Anarchy as the worst-case ratio between the costs in any equilibria and the minimal possible costs in the game. Formally, it is given by $\frac{\max_{S \in \mathcal{PNE}} \sum_{i \in N} c_i(S)}{\min_{S \in \mathcal{S}} \sum_{i \in N} c_i(S^*)}$. A game is called (λ, μ)-smooth for $\lambda > 0$ and $\mu \leq 1$ if, for every pair of states S and S', we have $\sum_{i \in N} c_i(S'_i, S_{-i}) \leq \lambda \sum_{i \in N} c_i(S') + \mu \sum_{i \in N} c_i(S)$. In a (λ, μ)-smooth game, the Price of Anarchy is at most $\frac{\lambda}{1-\mu}$ [29].

2   Existence and Efficiency of Pure Nash Equilibria

We begin with an easy observation that congestion games in which all players use the same L_p-norm as aggregation function always possesses a PNE.

Proposition 1. *Let Γ be a congestion game with identical L_p-aggregation functions. Then Γ possesses at least one pure Nash equilibrium.*

However, if players are heterogeneous in the sense that they use different aggregation functions a PNE might not exist even for two player games.

Theorem 1. *For every $1 \leq p_1 < p_2$ there exists a 2-player congestion game with L_p-aggregation functions Γ that does not possess a pure Nash equilibrium.*

Now we will investigate the efficiency of the equilibria by analyzing the price of anarchy. We restrict ourselves to games with linear cost functions and make use of a previous result by Christodolou et al. [11]. Let $q := max_{i \in N}\ p_i$, let $d := max_{i \in N, S_i \in \Sigma_i} |S_i|$. Furthermore, let $z = \left\lfloor \frac{1}{2} \cdot \left(d^{1-\frac{1}{q}} + \sqrt{5 + 6 \cdot \left(d^{1-\frac{1}{q}} - 1 \right) + \left(d^{1-\frac{1}{q}} - 1 \right)^2} \right) \right\rfloor$ be the maximum integer such that $\frac{z^2}{z+1} \leq d^{1-\frac{1}{q}}$.

Theorem 2. *A congestion game with L_p-aggregation functions Γ in which all cost functions are linear is $\left(d^{1-\frac{1}{p}} \cdot \frac{z^2+3z+1}{2z+1}, d^{1-\frac{1}{p}} \cdot \frac{1}{2z+1} \right)$-smooth. If Γ possesses a pure Nash equilibrium, its price of anarchy is bounded by $d^{1-\frac{1}{p}} \cdot \frac{z^2+3z+1}{2z+1-d^{1-\frac{1}{p}}}$.*

3   Existence of Approximate Pure Nash Equilibria

We start by giving a bound depending on the minimal and maximal p_i-values that players use and the maximal number of resources in a strategy.

Theorem 3. *Let Γ be a congestion game with L_p-aggregation functions, let $p = \min_{i \in N} p_i$ be the minimal and $q = \max_{i \in N} p_i$ be the maximal aggregation value in the game. Furthermore, denote by $d = max_{i \in N, S_i \in \Sigma_i} |S_i|$ the size of the strategy that contains most resources. Then Γ contains a β-approximate equilibrium for $\beta = d^{\frac{1}{2} \cdot (\frac{1}{p} - \frac{1}{q})}$. Moreover, a β-approximate equilibrium will be reached from an arbitrary state after a finite number of β-improvement steps.*

In the proof we show that $\Phi(S) = \sum_{r \in R} \sum_{i=1}^{n_r(S)} c_r(i)^z$ is an approximate potential function where $z := \left(\frac{1}{2} \cdot \left(\frac{1}{p} + \frac{1}{q} \right) \right)^{-1}$. We will now complement this result by showing that this approximation quality is the best achievable: i.e., we show that for any given $p < q$ and $d \geq 2$ we can construct a game which possesses no β-approximate PNE for any $\beta < d^{\frac{1}{2}\left(\frac{1}{p}-\frac{1}{q}\right)}$.

Theorem 4. *Let $p, q, d \in \mathbb{N}$ with $p < q$ and $d \geq 2$. Then there is a congestion game with L_p-aggregation functions Γ with $N = \{1, 2\}$, $p_1 = p$, $p_2 = q$, and $d = max_{i \in N, S_i \in \Sigma_i} |S_i|$ such that Γ does not possess a β-approximate pure Nash equilibrium for any $\beta < d^{\frac{1}{2}\left(\frac{1}{p}-\frac{1}{q}\right)}$.*

We complete the discussion of approximate pure Nash equilibria by regarding the computational complexity of deciding whether an approximate PNE exists for any approximation factor smaller than β.

Theorem 5. *For any $p, q, d \in \mathbb{N}$ with $p < q$ and $d \geq 2$ it is NP-hard to decide whether a given congestion game with L_p-aggregation functions Γ, with $p \leq p_i \leq q$ for all $i \in N$, and $d = max_{i \in N, S_i \in \Sigma_i} |S_i|$ possesses a β-approximate pure Nash equilibrium for any $\beta < d^{\frac{1}{2} \cdot \left(\frac{1}{p}-\frac{1}{q}\right)}$.*

4   Computation of Approximate Equilibria

In [31] it was shown that it is PLS-hard to compute an β-approximate PNE in standard congestion games. Since these games are a special case of congestion games with L_p-aggregation functions, this negative result immediately carries over to congestion games with L_p-aggregation functions.

Proposition 2. *It is PLS-hard to compute a β-approximate pure Nash equilibrium in a congestion game Γ with L_p-aggregation functions in which all cost functions are non-negative and non-decreasing, for any β that is computable in polynomial time.*

In the light of this initial negative result, we consider games with restricted cost functions. Caragiannis et al. [7] provide an algorithm that computes approximate pure Nash equilibria for congestion games with polynomial cost functions. For linear costs, the algorithm achieves an approximation quality of $2 + \epsilon$. For polynomial functions with a maximal degree of g, the algorithm guarantees an approximation factor of $g^{O(g)}$. We will reuse the algorithmic idea in two different ways which yield to different approximation guarantees depending on the aggregation parameters p_i.

Theorem 6. *Let Γ be a congestion game with L_p-aggregation functions in which all cost functions are linear or polynomial functions of degree at most g without negative coefficients. Furthermore, let $p := min_{i \in N} p_i$, let $q := max_{i \in N} p_i$, let $d := max_{i \in N, S_i \in \Sigma_i} |S_i|$, and $z = \left(\frac{1}{2} \cdot \left(\frac{1}{p} + \frac{1}{q}\right)\right)^{-1}$.*

Then an β-approximate equilibrium of Γ can be computed in polynomial time for $\beta = min\left\{(2 + \epsilon) \cdot d^{1-\frac{1}{q}}, z^{O(1)} \cdot d^{\frac{1}{2}\left(\frac{1}{p}-\frac{1}{q}\right)}\right\}$ (linear cost functions) and for $\beta = min\left\{g^{O(g)} \cdot d^{1-\frac{1}{q}}, (g \cdot z)^{O(g)} \cdot d^{\frac{1}{2}\left(\frac{1}{p}-\frac{1}{q}\right)}\right\}$ (polynomial cost functions).

Proof. We apply the algorithm proposed by Caragiannis et al. [7] to Γ, disregarding the aggregation values. Since all cost functions are either linear or polynomial, the algorithm computes either a $(2 + \epsilon)$- or a $g^{O(g)}$-approximate equilibrium. Now we can use the proof of Theorem 3 (for $z = 1$). We get that in the state computed by the algorithm, which would be a either $(2 + \epsilon)$- or $g^{O(g)}$-approximate PNE if all players used the L_1-norm, no player i can improve her costs according to the L_{p_i}-norm by more than a factor of either $(2+\epsilon) \cdot d^{1-\frac{1}{p_i}} \le \beta$ or $g^{O(g)} \cdot d^{1-\frac{1}{p_i}} \le \beta$. Hence, the computed state is an β-approximate pure Nash equilibrium in Γ. As analyzed in [7], the running time of the algorithm is polynomial in the size of Γ and $\frac{1}{\epsilon}$. For the second approximation factor and linear costs functions we replace every $c(x)$ in Γ by a polynomial cost function $c'(x) = c(x)^z$ of degree z, where for simplicity z is assumed to be integral. For this game, the algorithm given in [7] computes a state S which is a $z^{O(z)}$-approximate equilibrium. The costs of all players are equal to the costs they would have in Γ if they accumulated their costs according to the L_z-norm without taking the z-th root. Following the argumentation of the proof of Theorem 3, we get for any player i and any strategy $S_i' \in \Sigma_i$:

$$\frac{c_i(S)}{c_i(S_i', S_{-i})} \le \left(\left(z^{O(z)} \right)^{\frac{p_i}{z}} \right)^{\frac{1}{p_i}} \cdot \left(d^{\frac{p_i}{z}-1} \right)^{\frac{1}{p_i}} = z^{O(1)} \cdot d^{\frac{1}{z}-\frac{1}{p_i}} \le z^{O(1)} \cdot d^{\frac{1}{2}(\frac{1}{p}-\frac{1}{q})}.$$

Obviously, the transformation of the cost functions can be done in polynomial time. Hence, the algorithm given in [7] computes a $z^{O(1)} \cdot d^{\frac{1}{2}(\frac{1}{p}-\frac{1}{q})}$-approximate equilibrium of Γ in polynomial time. For polynomial cost functions this will lead to a game with polynomial cost functions of a degree of at most $g \cdot z$. Hence, the algorithm from [7] computes a $(g \cdot z)^{O(g \cdot z)}$-approximate PNE. Following the reasoning of the proof, we get that the computed state is a $(g \cdot z)^{O(g)} \cdot d^{\frac{1}{2}(\frac{1}{p}-\frac{1}{q})}$-approximate PNE of the congestion game with L_p-aggregation functions. $\square$

We have derived two different upper bounds for the approximation quality of approximate equilibria that can be computed in polynomial time. Generally speaking, if d is small but players use high aggregation values, the first strategy yields the better approximation, while otherwise the second bound is better.

5   General Aggregation Functions in Matroid Games

In this section we extend our model to a more general class of aggregation functions. Instead of using the L_p norms in the cost functions of the player, they are now defined by $c_i(S) = f_i(c_{r_1}(S), c_{r_2}(S), \ldots, c_{r_m})$ with f_i being an arbitrary aggregation function for each player based on certain monotonicity properties. We now consider only matroid congestion games in which the strategy spaces of all players form the bases of a matroid on the set of resources.

Let $f : \mathbb{R}^d \to \mathbb{R}$ be a function that is defined on non-decreasingly ordered vectors. Let for all $b = (b_1, \ldots, b_d)$ and $b' = (b_1', \ldots, b_d')$ with $b_i \le b_i'$ for all $1 \le i \le d$ hold that $f(b) \le f(b')$. Then f is called **strongly monotone**. Let

$x = (x_1, \ldots, x_d)$ and $y = (y_1, \ldots, y_d)$ be vectors that differ in only one element, i.e., there are indices j and k such that $x_i = y_i$ for all $i < j$ and $i > k$, $x_j < y_k$, and $x_{i+1} = y_i$ for all $j \leq i < k$ and $f(y) < f(x)$. Furthermore, let there be a vector $z = (z_1, \ldots, z_{d-1})$ such that $f(z_1, \ldots, x_j, \ldots, z_{d-1}) < f(z_1, \ldots, y_k, \ldots, z_{d-1})$ (with x_j and y_k at their correct positions in the non-decreasingly ordered vectors). Then f is called a **strongly non-monotone** function. If f is not strongly non-monotone, then f is called a **weakly monotone** function. We remark that this definition of vectors that differ in only one element does not require these elements to be at the same position in the vectors (the case $j = k$). It is sufficient that the symmetric difference of the multisets containing all elements in x and y contains exactly two elements x_j and y_k.

Theorem 7. *Let Γ be a matroid congestion game in which each player has a personal cost aggregation function f_i according to which her costs are calculated from her single resource costs. If for all players $i \in N$ the aggregation function f_i is strongly monotone, then Γ contains a pure Nash equilibrium.*

Proof. It is sufficient to show that any strategy $S_i = \{r_1, \ldots, r_d\}$ that minimizes the sum $\sum_{j=1}^d c_{r_j}(S)$ in a state S also minimizes the cost $f_i(c_i(S))$, where $c_i(S)$ denotes the non-decreasingly ordered vector of resource costs of player i in state S. Then a pure Nash equilibrium can be computed by computing a PNE in the corresponding game in which all players use the L_1-norm. We can show that if $B = \{b_1, \ldots, b_d\}$ is a matroid basis that is minimal w.r.t. the sum $\sum_{i=1}^d b_i$, then for any other basis $B' = \{b'_1, \ldots, b'_d\}$ and all $1 \leq i \leq d$ it holds that $b_i \leq b'_i$ (w.l.o.g. assume that both B and B' are written in non-decreasing order). Hence, if B is a basis that is optimal w.r.t. sum costs, then for all other bases B' it holds that $f(B) \leq f(B')$, since f is a strongly monotone function. $\square$

Since all L_p norms are monotone functions, we can extend this result to congestion games with L_p norms:

Corollary 1. *Let Γ be a matroid congestion game with L_p norms, then Γ contains a pure Nash equilibrium.*

We have shown that strongly monotone aggregation functions are sufficient to guarantee the existence of a PNE in matroid congestion games with player-specific aggregation functions. This immediately gives rise to the question whether the monotonicity criterion is also necessary to achieve this guarantee. We investigate this question by examining if, given a non-monotone aggregation function f, we can construct a matroid congestion game in which all players use f and which does not contain a PNE. For singletons, we can immediately give a negative answer to this. Since in this case costs can be associated to single resources and all players use the same aggregation function, Rosenthal's potential function argument [28] is applicable and shows that a PNE necessarily exists. However, for matroid degrees of at least 2 the answer is positive if the aggregation function fulfills the property that we call strong non-monotonicity.

Theorem 8. *Let $f : \mathbb{R}^d \to \mathbb{R}$, $d \geq 2$ be a strongly non-monotone function. Then there is a 2-player matroid congestion game in which both players allocate matroids of degree d and use the aggregation function f, which does not contain a pure Nash equilibrium.*

As argued, it is reasonable to demand that the aggregation function f in the proof is strongly non-monotone. We will underline this by showing that the strong non-monotonicity actually is a sharp criterion: i.e., it is both sufficient and necessary to construct a game without a PNE from f.

Theorem 9. *Let f be an aggregation function that is weakly monotone and let Γ be a matroid congestion game in which all players use f as their aggregation function. Then Γ possesses a pure Nash equilibrium. Furthermore, from every state there is a sequence of best-response improvement steps that reaches a pure Nash equilibrium after a polynomial number of steps.*

Proof. Since f is weakly monotone, we have for all vectors v_x and v_y which differ in exactly one component (let v_x contain x and v_y contain y, with $x < y$) that either $f(v_x) \leq f(v_y)$ or $f(v_y) < f(v_x)$ and for any vector w_y that contains y it holds that $f(w_y) \leq f(w_x)$, where w_x results from w_y by replacing y by x. Hence, for all pairs (x, y) we have either $f(w_x) \leq f(w_y)$ for all w_x and w_y, or $f(w_y) \leq f(w_x)$ for all w_x and w_y. This means that if we replace one element by another one in an arbitrary vector, the direction in which the value of f changes (if it changes at all) depends only on the two exchanged elements, not on the rest of the vector. Based on this, we define the relation $\leq'$ on the real numbers by determining that $x \leq' y$ if and only if $f(w_x) \leq f(w_y)$ for all vectors w_x and w_y. As argued, this relation defines a total preorder on $\mathbb{R}$. Since the number of resources and players in the game Γ are finite, the number of different resource costs that can occur in the game is also finite. Hence, it is possible to enumerate all possible resource costs according to the ordering relation $\leq'$. We denote the position of the cost value $c_r(S)$ in this enumeration by $\pi(c_r(S))$: i.e., $\pi(c_r(S)) = 1$ if and only if for all $r' \in R$ and all $l \in N$ it holds that $c_r(S) \leq' c_{r'}(l)$. We have that $\pi(c_r(S)) = \pi(c_{r'}(S'))$ if and only if $c_r(S) \leq' c_{r'}(S')$ and $c_{r'}(S') \leq' c_r(S)$. We remark that $\leq'$ is not necessarily a total order. Thus the two cost values need not be equal in this case.

Using this, we define the potential function $\Phi(S) = \sum_{r \in R} \sum_{i=1}^{n_r(S)} \pi(c_r(S))$. Consider a state $S = (S_i, S_{-i})$ in which player i can improve her cost by deviating to the strategy S_i', yielding the state $S' = (S_i', S_{-i})$. Since S_i and S_i' are both bases of the same matroid M, the graph $G = (V, E)$ with $V = (S_i \setminus S_i' \cup S_i' \setminus S_i)$ and $E = \{\{r, r'\} \mid r \in S_i, r' \in S_i', S_i' \setminus \{r'\} \cup \{r\} \in M\}$ contains a perfect matching (see Corollary 39.12a in [30]). All edges in G correspond to resource pairs $\{r, r'\}$ such that $S_i' \setminus \{r'\} \cup \{r\}$ is a valid strategy for player i. Since S_i' is a best response strategy to the strategy profile S_{-i} of all other players, it must hold for all edges $\{r, r'\}$ that $f(S_i' \setminus \{r'\} \cup \{r\}, S_{-i}) \geq f(S_i', S_{-i})$. This implies that either $c_{r'}(S') \leq' c_r(S)$ or $f(S_i' \setminus \{r'\} \cup \{r\}, S_{-i}) = f(S_i', S_{-i})$ and $c_{r'}(S') > c_r(S)$. In the latter case, the strategy $S_i' \setminus \{r'\} \cup \{r\}$ is still a best-response strategy for player i. Repeating the argument yields that there must be a best-response strategy S_i'' such that in the graph defined analogously to G it holds for all edges $\{r, r'\}$ that $c_{r'}(S'') \leq' c_r(S)$, where $S'' = (S_i'', S_{-i})$.

Let $T = \{e_1, \ldots, e_k\}$ be a perfect matching in this graph. For all $\{r, r'\} \in T$ it holds that $c_{r'}(S'') \leq' c_r(S)$, and hence $\pi(c_{r'}(S'')) \leq \pi(c_r(S))$. We have to argue that T contains at least one edge $\{r, r'\}$ with $\pi(c_{r'}(S'')) < \pi(c_r(S))$. Assume that for all $\{r, r'\} \in T$ it held that $\pi(c_{r'}(S')) = \pi(c_r(S))$, i.e., $c_r(S) \leq' c_{r'}(S'')$. Then we could transform S_i into S_i'' by iteratively exchanging a single resource r for another resource r'. Since two consecutive sets S^r and $S^{r'}$ in this sequence differ only in the resources r and r', and $c_r(S) \leq' c_{r'}(S')$, it holds that $f(S^r) \leq f(S^{r'})$. Hence, none of the steps decreases the value of f, which contradicts the assumption that $f(S_i'', S_{-i}) \leq f(S_i', S_{-i}) < f(S_i, S_{-i})$. Therefore, there must be at least one edge $\{r, r'\}$ in T with $\pi(c_{r'}(S'')) < \pi(c_r(S))$, which implies $\Phi(S'') - \Phi(S) = \sum_{r' \in S_i'' \setminus S_i} \pi(c_{r'}(S'')) - \sum_{r \in S_i \setminus S_i''} \pi(c_r(S)) = \sum_{\{r, r'\} \in T} (\pi(c_r'(S'')) - \pi(c_r(S))) < 0$. By construction, the value of Φ is always integral and upper bounded by $n^2 \cdot m^2$, where n is the number of players and m the number of resources in Γ. Hence, Γ reaches a PNE from an arbitrary state after at most $n^2 \cdot m^2$ best-response improvement steps. □

The theorem states that strong non-monotonicity is necessary to construct a game without a PNE from a *single* aggregation function f. However, the technique used in the proof only requires that all aggregation functions used in the game have the same order on vectors which differ in exactly one component. It is irrelevant how these functions order vectors which differ in several components.

Corollary 2. *Let Γ be a matroid congestion game in which the costs of player i are computed according to her personal aggregation function f_i. If for all $i \in N$ the function f_i is weakly monotone and for all $i, j \in N$ and all vectors v and w that differ in exactly one component it holds that $f_i(v) \leq f_i(w) \Leftrightarrow f_j(v) \leq f_j(w)$, then Γ possesses a pure Nash equilibrium.*

This corollary is interesting mainly because it establishes an almost tight border up to which the existence of PNE can be guaranteed. If we are given two aggregation functions f and g and two vectors v and w that differ in exactly one component, with $f(v) < f(w)$ and $g(w) < g(v)$, then it is obvious that we can construct a 2-player game in which the first player uses the aggregation function f and the second g and the two players alternate between the cost vectors v and w, as we did in the proof of Theorem 8 for strongly non-monotone functions.

6 Conclusion

For congestion games with L_p-aggregation functions, we presented methods to compute approximate PNE and bound the price of anarchy which are based on previous results regarding standard congestion games. It is an open point for future work to examine if these results could be improved by specifically designing methods for congestion games with L_p-aggregation functions. Another interesting approach for further research would be to combine the application of aggregation functions with other classes of congestion games such as weighted congestion games or non-atomic congestion games, and examine the implications for the (approximate) pure Nash equilibria in these games.

References

1. Ackermann, H., Röglin, H., Vöcking, B.: Pure Nash equilibria in player-specific and weighted congestion games. Theoret. Comput. Sci. **410**(17), 1552–1563 (2009)
2. Ackermann, H., Röglin, H., Vöcking, B.: On the impact of combinatorial structure on congestion games. J. ACM **55**(6), 25:1–25:22 (2008)
3. Ackermann, H., Skopalik, A.: Complexity of pure Nash equilibria in player-specific network congestion games. Internet Math. **5**(4), 323–342 (2008)
4. Arrow, K.J., Chenery, H.B., Minhas, B.S., Solow, R.M.: Capital-labor substitution and economic efficiency. Rev. Econ. Stat. **43**(3), 225–250 (1961)
5. Banner, R., Orda, A.: Bottleneck routing games in communication networks. IEEE J. Sel. Areas Commun. **25**(6), 1173–1179 (2007)
6. Byde, A., Polukarov, M., Jennings, N.R.: Games with congestion-averse utilities. In: Mavronicolas, M., Papadopoulou, V.G. (eds.) SAGT 2009. LNCS, vol. 5814, pp. 220–232. Springer, Heidelberg (2009). doi:10.1007/978-3-642-04645-2_20
7. Caragiannis, I., Fanelli, A., Gravin, N., Skopalik, A.: Efficient computation of approximate pure Nash equilibria in congestion games. In: IEEE 52nd Annual Symposium on Foundations of Computer Science (FOCS 2011), pp. 532–541 (2011)
8. Caragiannis, I., Fanelli, A., Gravin, N.: Short sequences of improvement moves lead to approximate equilibria in constraint satisfaction games. In: Lavi, R. (ed.) SAGT 2014. LNCS, vol. 8768, pp. 49–60. Springer, Heidelberg (2014). doi:10.1007/978-3-662-44803-8_5
9. Caragiannis, I., Fanelli, A., Gravin, N., Skopalik, A.: Approximate pure Nash equilibria in weighted congestion games: existence, efficient computation, and structure. ACM Trans. Econ. Comput. **3**(1), 2:1–2:32 (2015)
10. Chien, S., Sinclair, A.: Convergence to approximate Nash equilibria in congestion games. Games Econ. Behav. **71**(2), 315–327 (2011)
11. Christodoulou, G., Koutsoupias, E., Spirakis, P.G.: On the performance of approximate equilibria in congestion games. Algorithmica **61**(1), 116–140 (2011)
12. Dixit, A.K., Stiglitz, J.E.: Monopolistic competition and optimum product diversity. Am. Econ. Rev. **67**(3), 297–308 (1977)
13. Dunkel, J., Schulz, A.S.: On the complexity of pure-strategy Nash equilibria in congestion and local-effect games. Math. Oper. Res. **33**(4), 851–868 (2008)
14. Fabrikant, A., Papadimitriou, C., Talwar, K.: The complexity of pure Nash equilibria. In: Proceedings of the Thirty-sixth Annual ACM Symposium on Theory of Computing (STOC 2004), pp. 604–612. ACM, New York (2004)
15. Feldotto, M., Gairing, M., Skopalik, A.: Bounding the potential function in congestion games and approximate pure Nash equilibria. In: Liu, T.-Y., Qi, Q., Ye, Y. (eds.) WINE 2014. LNCS, vol. 8877, pp. 30–43. Springer, Cham (2014). doi:10.1007/978-3-319-13129-0_3
16. Feldotto, M., Leder, L., Skopalik, A.: Congestion games with mixed objectives. In: Chan, T.-H.H., Li, M., Wang, L. (eds.) COCOA 2016. LNCS, vol. 10043, pp. 655–669. Springer, Cham (2016). doi:10.1007/978-3-319-48749-6_47
17. Fotakis, D., Kontogiannis, S., Spirakis, P.: Selfish unsplittable flows. Theoret. Comput. Sci. **348**(2), 226–239 (2005)

18. Hansknecht, C., Klimm, M., Skopalik, A.: Approximate pure Nash equilibria in weighted congestion games. In: Approximation, Randomization, and Combinatorial Optimization. Algorithms and Techniques (APPROX/RANDOM 2014), Dagstuhl, Germany. LIPIcs, vol. 28, pp. 242–257 (2014)
19. Harks, T., Hoefer, M., Klimm, M., Skopalik, A.: Computing pure Nash and strong equilibria in bottleneck congestion games. Math. Program. **141**(1), 193–215 (2013)
20. Harks, T., Klimm, M., Möhring, R.H.: Strong Nash equilibria in games with the lexicographical improvement property. In: Leonardi, S. (ed.) WINE 2009. LNCS, vol. 5929, pp. 463–470. Springer, Heidelberg (2009). doi:10.1007/978-3-642-10841-9_43
21. Ieong, S., McGrew, R., Nudelman, E., Shoham, Y., Sun, Q.: Fast and compact: a simple class of congestion games. In: AAAI, pp. 489–494. AAAI Press/The MIT Press (2005)
22. Koutsoupias, E., Papadimitriou, C.: Worst-case equilibria. In: Meinel, C., Tison, S. (eds.) STACS 1999. LNCS, vol. 1563, pp. 404–413. Springer, Heidelberg (1999). doi:10.1007/3-540-49116-3_38
23. Kukushkin, N.S.: Rosenthal's potential and a discrete version of the Debreu-Gorman Theorem. Autom. Remote Control **76**(6), 1101–1110 (2015)
24. Kukushkin, N.S.: Congestion games revisited. Int. J. Game Theory **36**(1), 57–83 (2007)
25. Mavronicolas, M., Milchtaich, I., Monien, B., Tiemann, K.: Congestion games with player-specific constants. In: Kučera, L., Kučera, A. (eds.) MFCS 2007. LNCS, vol. 4708, pp. 633–644. Springer, Heidelberg (2007). doi:10.1007/978-3-540-74456-6_56
26. Milchtaich, I.: Congestion games with player-specific payoff functions. Games Econ. Behav. **13**(1), 111–124 (1996)
27. Monderer, D., Shapley, L.S.: Potential games. Games Econ. Behav. **14**(1), 124–143 (1996)
28. Rosenthal, R.W.: A class of games possessing pure-strategy Nash equilibria. Int. J. Game Theory **2**(1), 65–67 (1973)
29. Roughgarden, T.: Intrinsic robustness of the price of anarchy. J. ACM **62**(5), 32:1–32:42 (2015)
30. Schrijver, A.: Combinatorial Optimization: Polyhedra and Efficiency, vol. 24. Springer Science & Business Media, Heidelberg (2002)
31. Skopalik, A., Vöcking, B.: Inapproximability of pure Nash equilibria. In: Proceedings of the Fortieth Annual ACM Symposium on Theory of Computing (STOC 2008), pp. 355–364. ACM, New York (2008)
32. Voice, T., Polukarov, M., Byde, A., Jennings, N.R.: On the impact of strategy and utility structures on congestion-averse games. In: Leonardi, S. (ed.) WINE 2009. LNCS, vol. 5929, pp. 600–607. Springer, Heidelberg (2009). doi:10.1007/978-3-642-10841-9_61

Approximating Bounded Degree Deletion
via Matroid Matching

Toshihiro Fujito[(✉)]

Department of Computer Science and Engineering,
Toyohashi University of Technology, Toyohashi 441-8580, Japan
fujito@cs.tut.ac.jp

Abstract. The *Bounded Degree Deletion* problem with degree bound b : $V \to \mathbb{Z}_+$ (denoted b-*BDD*), is that of computing a minimum cost vertex set in a graph $G = (V, E)$ such that, when it is removed from G, the degree of any remaining vertex v is no larger than $b(v)$. It will be shown that b-BDD can be approximated within $\max\{2, \bar{b}/2 + 1\}$, improving the previous best bound for $2 \leq \bar{b} \leq 5$, where $\bar{b}$ is the maximum degree bound, i.e., $\bar{b} = \max\{b(v) \mid v \in V\}$. The new bound is attained by casting b-BDD as the vertex deletion problem for such a property inducing a 2-polymatroid on the edge set of a graph, and then reducing it to the submodular set cover problem.

1 Introduction

The *Bounded Degree Deletion* problem is a well-known basic problem in graph theory. It has an application in computational biology [12] as well as in the area of property testing [28], whereas its "dual problem" of finding maximum s-plexes, introduced in 1978 [32], has applications in social network analysis [1,26]. With degree bound of $b \in \mathbb{Z}_+$, b-Bounded Degree Deletion (or b-*BDD* for short) is the problem of computing a minimum cost vertex set X in a given weighted graph $G = (V, E)$ such that the degree of any remaining vertex v is bounded by b when all the vertices in X are removed from G.

Clearly, b-BDD is a generalization of the *Vertex Cover (VC)* problem, and this can be put into a better perspective by capturing both of them as members of the same problem class called *Vertex Deletion (VD)* problems. The VD problem for a graph property π, denoted *VD(π)*, is: Given a vertex weighted graph G, find a vertex set of minimum weight whose deletion (along with all of the incident edges) from G leaves a (sub)graph satisfying the property π. Here π is *nontrivial* if infinitely many graphs satisfy π and infinitely many graphs fail to satisfy it. It is *hereditary on induced subgraphs* if, in any graph satisfying π, every vertex-induced subgraph also satisfies π. A number of well-studied graph properties are nontrivial and hereditary, including "a graph has no edge" (i.e., VD(π) = VC), and "vertex degree $\leq b$" for some fixed constant b (i.e., VD(π) = b-BDD). Concerning the computational complexity of VD(π), it is most fundamental that

T. Fujito— Supported in part by JSPS KAKENHI under Grant Number 26330010.

© Springer International Publishing AG 2017
D. Fotakis et al. (Eds.): CIAC 2017, LNCS 10236, pp. 234–246, 2017.
DOI: 10.1007/978-3-319-57586-5_20

$VD(\pi)$ is NP-hard for any π if it is nontrivial and hereditary on induced subgraphs [24].

Another member of VD has been recently introduced and actively studied for the property $\pi_k =$ "a graph has no path P_k on k vertices". The problem is called under various names such as k-path vertex cover [3–5,17,20], vertex cover P_k (VCP$_k$) [33–36], and P_k-hitting set [6] (and all of them refer to the same problem), and we denote this problem as k-Path Vertex Cover (k-PVC). A subset F of the vertex set V is called a k-path vertex cover if every path on k vertices, not necessarily induced, in G has at least one vertex from F. The k-path vertex cover number, $\psi_k(G)$, of G is the cardinality of a minimum k-path vertex cover for G. The k-PVC problem was introduced in [4] along with $\psi_k(G)$, motivated by its relation to secure connection in wireless networks [29]. The vertex weighted version of the k-PVC was considered in [36] motivated by its applications in traffic control. Clearly, $\psi_2(G)$ is the vertex cover number of G. In addition $\psi_3(G)$ corresponds to another previously studied concept of dissociation number of a graph, defined as follows. A subset of vertices in a graph G is called a dissociation set if it induces a subgraph with maximum degree at most 1. The maximum cardinality of a dissociation set in G is called the dissociation number of G and is denoted by diss(G). Clearly $\psi_3(G) = |V(G)| -$ diss(G). The problem of computing diss(G) was introduced by Yannakakis [39], who also proved it to be NP-hard in the class of bipartite graphs. See [31] for a survey on the dissociation number problem.

It must be clear by now that VC $\equiv$ 0-BDD $\equiv$ 2-PVC, and 1-BDD $\equiv$ 3-PVC (but b-BDD $\not\equiv (b+2)$-PVC for $b \geq 2$). We now summarize below algorithmic results known for these problems.

VC Approximating VC better than simple 2-approximation has been a subject of extensive research over the years, and it is now known approximable within $2 - \Theta(1/\sqrt{\log n})$ [19]. Meanwhile, VC has been shown hard to approximate within $10\sqrt{5} - 21 \approx 1.36$ unless P $=$ NP [11] (or within $2 - \epsilon$ assuming the unique games conjecture [21]), and so is any other VD problem for a nontrivial and hereditary property as VC can be reduced in approximation preserving manner to it.

b-**BDD** The first improvement over the simple $(b+2)$-approximation based on the hitting set formulation was attained in [14] using the local ratio method and b-BDD was shown approximable within max$\{2, b+1\}$. Okun and Barak considered general b-BDD where $b : V \to \mathbb{Z}_+$ is an arbitrary function, and obtained an approximation bound of $2 + \max_{v \in V} \ln b(v)$ by combination of the local ratio method and the greedy multicovering method [30].

More recently, b-BDD has been extensively studied in parameterized complexity. It has been shown that, when parameterized by the size k of the deletion set, the problem is $W[2]$-hard for unbounded b and FPT for each fixed $b \geq 0$ [12], whereas, when parameterized by treewidth tw, it is FPT with parameters k and tw, and $W[2]$-hard with only parameter tw [2]. A linear vertex kernel of b-BDD has been developed by generalizing the Nemhauser-Trotter theorem for the vertex cover problem to b-BDD [9,12,38].

Besides, 2-BDD has been recently highlighted under the name of *Co-Path/Cycle Packing* [8,9,13], mostly from the viewpoint of parameterized complexity, due to its important applications in bioinformatics.

k-**PVC** − A linear time algorithm is presented for trees and upper bounds on $\psi_k(G)$ were investigated in [4], while lower bounds for $\psi_k(G)$ in regular graphs were given in [3].

− A randomized approximation algorithm with an expected approximation ratio of 23/11 was obtained for 3-PVC [20].

− 3-PVC was shown approximable within 2 [35,36], and within 1.57 on cubic graphs [34].

− 4-PVC was shown approximable within 3 [6], and within 2 in regular graphs [10].

1.1 Our Work and Contributions

We consider general b-BDD as in [30], where b is an arbitrary function $b : V \rightarrow \mathbb{Z}_+$. Previously, a general approach of reducing VD(π) to submodular optimization was explored for a certain type of hereditary properties π [15]. We say that a graph property π is *matroidal* if, on any graph $G = (V, E)$, the edge sets of subgraphs of G satisfying π form the family of independent sets of some matroid defined on E. An archetypal example is $\pi =$ "a graph is acyclic", for which the corresponding matroid is the cycle matroid on E and the corresponding VD is the feedback vertex set problem. It was shown that VD(π) for matroidal property π can be reduced to the *submodular set cover (SSC)* problem, and thus, VD(π) for such π can be approximated by approximation algorithms for SSC. On the other hand, b-BDD is VD(π) for $\pi =$ "vertex degree of $v \leq b(v)$", and the edge sets of subgraphs satisfying π are b-*matchings* in G. A family of b-matchings does not induce a matroid in general and hence, this π is not matroidal. However, b-matchings form a family of *matroid matchings* in general, or the family of "independent sets" in a 2-*polymatroid* (more detailed description given below), and this observation allows us to make use of the above-mentioned general method even if π is not matroidal. Our major aim is to demonstrate that such a general approach could be effective for VD(π) even if π is not matroidal but if it induces a 2-polymatroid, using the case of b-BDD. More specifically, it will be shown that b-BDD can be approximated by a single generic algorithm within 2 for $\bar{b} \leq 1$ and $\bar{b}/2 + 1$ for $\bar{b} \geq 2$, where $\bar{b}$ denotes $\max_{v \in V} b(v)$, improving the previous best of $\min\{\max\{2, \bar{b} + 1\}, 2 + \ln \bar{b}\}$ [14,30] for $2 \leq \bar{b} \leq 5$, including the case of Co-Path/Cycle Packing ($\equiv$ 2-BDD).

1.2 Notations and Definitions

For a graph $G = (V, E)$, let $\delta(W)$ denote the set of edges incident to a vertex in W, i.e., $\delta(W) = \{\{u, v\} \in E \mid \{u, v\} \cap W \neq \emptyset\}$. Let $\delta(v)$ denote $\delta(\{v\})$ and $d(v) = |\delta(v)|$. To restrict edges under consideration within a certain edge set F, we use $\delta_F(W)$ and $d_F(v)$ to denote $\delta(W) \cap F$ and $|\delta(v) \cap F|$, respectively. For disjoint vertex sets X and Y, let $E(X) = \{e \in E \mid e \subseteq X\}$ and $E(X, Y) =$

$\{\{u,v\} \in E \mid u \in X, v \in Y\}$. We also use shorthand notations of $e(X) = |E(X)|$ and $e(X,Y) = |E(X,Y)|$.

2 Formulation and Submodular Optimization

2.1 BDD and Matroid Matching

Definition 1. *For a finite set N, a non-decreasing, submodular, and integer-valued function f defined on 2^N with $f(\emptyset) = 0$, is called a* polymatroid function, *and (N, f) a* polymatroid. *If f additionally satisfies $f(\{j\}) \leq k, \forall j \in N$, (N, f) is a k-polymatroid.*

Definition 2. *For any polymatroid (N, f) define another set function f^d such that*

$$f^d(S) \overset{\text{def}}{=} \sum_{j \in S} f(\{j\}) - (f(N) - f(N - S)).$$

Then f^d is a polymatroid function and (N, f^d) is called the dual polymatroid of (N, f).

Definition 3. *Let (E, f) be a 2-polymatroid.*

- *A subset $F \subseteq E$ is a matching in (E, f) if $f(F) = 2|F|$.*
- *A subset $F \subseteq E$ is spanning in (E, f) if $f(F) = f(E)$.*

Proposition 1. *Let (E, f) be a 2-polymatroid.*

- *(E, f^d) is a 2-polymatroid called the dual of (E, f).*
- *A subset $F \subseteq E$ is a matching in (E, f) iff $E - F$ is spanning in (E, f^d).*

The *Matroid Matching* problem, introduced by Lawler [22], is to compute the maximum matching in a given 2-matroid. The unweighted matroid matching problem is relatively tractable; a polynomial algorithm was obtained for linearly represented matroids [25] and an approximation scheme for general matroids [23]. It appears much harder in the weighted case, however, and, whereas randomized fully polynomial-time approximation schemes are known for linearly represented matroids [7,27], only a greedy 2-approximation is known for general matroids [18].

For a graph $G = (V, E)$ let $b : V \rightarrow \mathbb{Z}_+$. An edge set $F \subseteq E$ is a *b-matching* iff $d_F(v) \leq b(v)$, $\forall v \in V$. Thus, BDD is the problem of computing $X \subseteq V$ of minimum cost such that $V - X$ induces a *b*-matching.

Proposition 2. *Define $f : 2^E \rightarrow \mathbb{Z}_+$ such that*

$$f(F) = \sum_{v \in V} \min\{b(v), d_F(v)\}.$$

Then,

- (E, f) *is a 2-polymatroid.*
- $F \subseteq E$ *is a matching in* (E, f) *iff* F *is a b-matching in* G.

Proposition 3. *Assume* $1 \le b(v) \le d_E(v)$, $\forall v \in V$.

- $f(E) = b(V)$ *(since* $b(v) \le d_E(v)$, $\forall v \in V$*).*
- $f(e) = 2$, $\forall e \in E$ *(since* $1 \le b(v)$, $\forall v \in V$*).*

$$f^d(E) = \sum_{e \in E} f(\{e\}) - f(E) = 2|E| - b(V) = \sum_{v \in V} (d(v) - b(v)). \quad (1)$$

$$
\begin{aligned}
f^d(F) &= 2|F| - \left(\sum_{v \in V} \min\{b(v), d(v)\} - \sum_{v \in V} \min\{b(v), d_{E-F}(v)\} \right) \\
&= \sum_{v \in V} d_F(v) - \sum_{v \in V} \max\{0, \min\{b(v), d(v)\} - d_{E-F}(v)\} \\
&= \sum_{v \in V} \min\{d_F(v), d(v) - b(v)\} \quad (assuming\, b(v) \le d(v)) \quad (2)
\end{aligned}
$$

for any $F \subseteq E$.

2.2　BDD as Submodular Set Cover

In light of Proposition 1, it can be seen that $X \subseteq V$ is a b-BDD solution in $G = (V, E)$ iff $\delta(X)$ is spanning in (E, f^d) since X is a b-BDD solution iff $E - \delta(X)$ is a b-matching in G iff $E - \delta(X)$ is a matching in (E, f). Therefore, b-BDD on $G = (V, E)$ can be reduced to the problem of computing $X \subseteq V$ of minimum cost such that $\delta(X)$ is spanning in (E, f^d). More formally,

Proposition 4. *Define* $g : V \to \mathbb{Z}_+$ *such that* $g(W) = f^d(\delta(W))$. *b-BDD on* $G = (V, E)$ *can be formulated as the problem of computing* $X \subseteq V$ *of minimum cost such that* $g(X) = g(V)$.

It can be seen that g here is another non-decreasing polymatroid function, and the problem of computing minimum $X \subseteq V$ satisfying $g(X) = g(V)$ is known as the *Submodular Set Cover* problem.

Definition 4. *Let* g *be a non-decreasing submodular set function defined on the subsets of a finite ground set* N. *The* submodular set cover *problem (SSC) is to compute:*

$$\min_{S \subseteq N} \left\{ \sum_{j \in S} w_j \mid g(S) = g(N) \right\}.$$

Although the greedy algorithm, together with its performance analysis, is perhaps the most well-known heuristic for general SSC [37], another method of using a primal-dual heuristic [16] is known to deliver better solutions for some of more specific SSC problems. In this algorithm called PD, the *contraction* of g onto $N - S$ is the function g_S defined on 2^{N-S} s.t. $g_S(X) = g(X \cup S) - g(S)$ for any $S \subseteq N$. If g is non-decreasing and submodular on N, so is g_S on $N - S$, and thus, another submodular cover instance $(N - S, g_S)$ can be derived for any $S \subseteq N$. The performance of PD, when applied to general SSC, was analyzed in [16], and the following result was presented:

Proposition 5. *For an SSC instance (N, g) the performance ratio of PD is bounded by*

$$\max \left\{ \frac{\sum_{j \in X} g_S(j)}{g_S(N - S)} \right\} \tag{3}$$

where max is taken over any $S \subseteq N$ and any minimal solution X in $(N - S, g_S)$.

To apply Proposition 5 to b-BDD, consider the graph $G' = G - S$ obtained from G by removing all the vertices in S, and let us reformulate b-BDD on $G' = (V', E')$, where $V' = V - S, E' = E - \delta(S)$, as an SSC instance (E', g'). To do so, let $f' : 2^{E'} \rightarrow \mathbb{Z}_+$ be a 2-polymatroid function such that $f'(F) = \sum_{v \in V'} \min\{d'_F(v), b'(v)\}$ for $F \subseteq E'$, f'^d be the dual of f', and $g'(T) = f'^d(\delta'(T))$ for $T \subseteq V'$ (Note: Here, $\delta'(T) = \delta_{E'}(T), d'(v) = d_{E'}(v), b'(v) = \min\{b(v), d'(v)\}$ for all $T \subseteq V'$ and $v \in V'$). It can be shown then that $g_S(T) = g'(T)$ for any $S \subseteq V$ and $T \subseteq V - S$, and in particular, $g'(v) = g_S(v), \forall v \in V'$, and $g'(V') = g_S(V = S)$. Hence,

Proposition 6. *Let (E, g) and (E', g') be SSC formulations of b-BDD for $G = (V, E)$ and $G' = (V', E')$, the subgraph of G induced by V', respectively. Then,*

$$\max_{S \subseteq V} \left\{ \frac{\sum_{v \in X} g_S(v)}{g_S(V - S)} \right\} = \max \left\{ \frac{\sum_{v \in X} g'(v)}{g'(V')} \right\}$$

where max in RHS is taken over any subgraph G' of G induced by $V' \subseteq V$ and any minimal b-BDD solution X in G'.

It thus follows from Propositions 5 and 6 that the performance ratio of PD, when applied to b-BDD, can be estimated by bounding

$$\frac{\sum_{v \in X} g(v)}{g(V)}$$

for any graph $G = (V, E)$ and any minimal solution X in G.

Lemma 1. *For any minimal solution $X \subseteq V$ and $Y = V - X$,*

$$\max \left\{ 2, (\bar{b}/2 + 1) \right\} f^d(E) \geq \sum_{v \in X} f^d(\delta(v)). \tag{4}$$

Therefore, we may conclude that

Theorem 1. *The problem b-BDD can be approximated by* PD *within* $\bar{b}/2+1$ *for* $\bar{b} \geq 2$ *and within 2 when* $\bar{b} \in \{0,1\}$.

Since $\bar{b}/2 + 1$ is $<2 + \ln \bar{b}$, the bound of Okun-Barak algorithm [30], for $\bar{b} \leq 5$, it follows that, when PD is applied to b-BDD with current g, it outperforms the previous algorithms for $2 \leq \bar{b} \leq 5$.

3   Analysis

Assume $b(v) \leq d_E(v), \forall v \in V$, for the rest of paper as one can always reset $b(v)$ to $= d(v)$ w.l.o.g. if $b(v) > d(v)$. A vertex $v \in V$ is called a *tight node* in what follows if $d_E(v) = b(v)$. Let us use the following notations:

$\bar{d}(v) = d(v) - b(v)$

$\tilde{d}(v) = |\{(v,w) \in \delta(v) \mid w \text{ is not tight}\}| = \#\text{ of nodes adjacent to } v \text{ that are not tight}$

for each $v \in V$, and further classify nodes in X and $Y = V - X$ depending on whether nodes are tight or not as follows:

$$X^t = \{v \in X \mid v \text{ is tight}\}, \tilde{X} = X - X^t, Y^t = \{v \in Y \mid v \text{ is tight}\}, \tilde{Y} = Y - Y^t.$$

Observation.

– Using Eq. (1), we have

$$f^d(E) = \sum_{v \in V} \bar{d}(v) = \sum_{v \in \tilde{X} \cup \tilde{Y}} \bar{d}(v) = \bar{d}(\tilde{X}) + \bar{d}(\tilde{Y}).$$

– Because of Eq. (2), we have

$$f^d(\delta(v)) = \sum_{u \in V} \min\{d_{\delta(v)}(u), d(u) - b(u)\}$$

$$= (d(v) - b(v)) + (\#\text{ of nodes adjacent to } v \text{ that are not tight}),$$

and hence,

$$\sum_{v \in X} f^d(\delta(v)) = \sum_{v \in X}(\bar{d}(v) + \tilde{d}(v)) = \bar{d}(\tilde{X}) + \tilde{d}(X).$$

Since

$$\tilde{d}(X) = 2e(\tilde{X}) + e(\tilde{X}, X^t) + e(X, \tilde{Y}),$$

we may write

$$\sum_{v \in X} f^d(\delta(v)) = \bar{d}(\tilde{X}) + 2e(\tilde{X}) + e(\tilde{X}, X^t) + e(X, \tilde{Y}).$$

3.1 Proof of Lemma 1

It is assumed in what follows that $b \geq 2$ (the proof for the case of $b \leq 1$ is omitted due to the space limitation). Because of the preceding observations, and expanding $\bar{d}(W)$ by

$$\bar{d}(W) = 2e(W) + e(W, V - W) - b(W)$$

for any $W \subseteq V$, the proof of this lemma is reduced to showing that

$$
\left(\frac{\bar{b}}{2} + 1\right) f^d(E) - \sum_{v \in X} f^d(\delta(v))
$$

$$
= \left(\frac{\bar{b}}{2} + 1\right) \left(\bar{d}(\tilde{X}) + \bar{d}(\tilde{Y})\right) - \left(\bar{d}(\tilde{X}) + 2e(\tilde{X}) + e(\tilde{X}, X^t) + e(X, \tilde{Y})\right)
$$

$$
= \left((\bar{b} - 2)e(\tilde{X}) + \frac{\bar{b}}{2}e(\tilde{X}, V - \tilde{X}) + (\bar{b} + 2)e(\tilde{Y}) + \left(\frac{\bar{b}}{2} + 1\right) e(\tilde{Y}, V - \tilde{Y})\right)
$$

$$
- \left(e(\tilde{X}, X^t) + e(X, \tilde{Y}) + \frac{\bar{b}}{2}b(\tilde{X}) + \left(\frac{\bar{b}}{2} + 1\right) b(\tilde{Y})\right)
$$

$$
\geq 0.
$$

Observe now that

$$
\left(\frac{\bar{b}}{2}e(\tilde{X}, V - \tilde{X}) + \left(\frac{\bar{b}}{2} + 1\right) e(\tilde{Y}, V - \tilde{Y})\right) - \left(e(\tilde{X}, X^t) + e(X, \tilde{Y})\right)
$$

$$
= \left(\bar{b}e(\tilde{X}, \tilde{Y}) + \left(\frac{\bar{b}}{2} - 1\right) e(\tilde{X}, X^t) + \left(\frac{\bar{b}}{2} + 1\right) e(\tilde{Y}, Y^t) + \frac{\bar{b}}{2}\left(e(\tilde{X}, Y^t) + e(X^t, \tilde{Y})\right)\right),
$$

and hence, it amounts to showing

$$
(\bar{b} - 2)e(\tilde{X}) + (\bar{b} + 2)e(\tilde{Y}) + \bar{b}e(\tilde{X}, \tilde{Y}) + \left(\frac{\bar{b}}{2} - 1\right) e(\tilde{X}, X^t)
$$

$$
+ \left(\frac{\bar{b}}{2} + 1\right) e(\tilde{Y}, Y^t) + \frac{\bar{b}}{2}\left(e(\tilde{X}, Y^t) + e(X^t, \tilde{Y})\right) \tag{5}
$$

$$
\geq \left(\frac{\bar{b}}{2}b(\tilde{X}) + \left(\frac{\bar{b}}{2} + 1\right) b(\tilde{Y})\right)
$$

for the proof of Eq. (4).

To prove Eq. (5), values of its LHS are distributed first to edges as follows: $(\bar{b} - 2), (\bar{b} + 2), \bar{b}, (\bar{b}/2 - 1), (\bar{b}/2 + 1)$ and $\bar{b}/2$ are assigned to each edge of $E(\tilde{X}), E(\tilde{Y}), E(\tilde{X}, \tilde{Y}), E(\tilde{X}, X^t), E(\tilde{Y}, Y^t)$ and $E(\tilde{X}, Y^t) \cup E(\tilde{Y}, X^t)$, respectively. These values assigned on an edge $e = \{u, v\}$ will be redistributed to u and v, and the total value associated with each vertex will be shown to reach at least $\bar{b}b(x)/2$ for $x \in \tilde{X}$ and $(\bar{b}/2 + 1)b(y)$ for $y \in \tilde{Y}$, thus reaching the value of RHS in total.

Distributing the LHS Value to Vertices.

1. Distributing $(\bar{b} - 2)$ on $e \in E(\tilde{X})$ and $(\bar{b} + 2)$ on $e \in E(\tilde{Y})$.
   In either case, the value assigned on $e = \{u, v\}$ is evenly split and distributed to each of u and v. Thus, either u or v receives $(\bar{b}/2 - 1)$ if $\{u, v\} \in E(\tilde{X})$ and it does $(\bar{b}/2 + 1)$ if $\{u, v\} \in E(\tilde{Y})$.

2. Distributing values assigned on edges in $E(\tilde{X}, X^t) \cup E(\tilde{Y}, Y^t) \cup E(\tilde{X}, Y^t) \cup E(\tilde{Y}, X^t)$. In case $e = \{u, v\} \in E(\tilde{X}, X^t) \cup E(\tilde{X}, Y^t)$ with $u \in \tilde{X}$, distribute the whole value to u; that is, $(\bar{b}/2 - 1)$ if $e \in E(\tilde{X}, X^t)$ and $\bar{b}/2$ if $e \in E(\tilde{X}, Y^t)$, to u. Likewise, distribute the whole value on $e = \{u, v\} \in E(\tilde{Y}, Y^t) \cup E(\tilde{Y}, X^t)$ to $u \in \tilde{Y}$; that is, $(\bar{b}/2 + 1)$ if $e \in E(\tilde{Y}, Y^t)$ and $\bar{b}/2$ if $E(\tilde{Y}, X^t)$.

3. Distributing $\bar{b}$ on $e = \{x, y\} \in E(\tilde{X}, \tilde{Y})$, where $x \in \tilde{X}, y \in \tilde{Y}$.
   The way how to distribute will depend on the following cases of y:
   - Distribute whole $\bar{b}$ to x (and none to y) if $d_Y(y) = b(y)$.
   - Otherwise, i.e., if $d_Y(y) < b(y)$, then distribute

   $$\left(\frac{\bar{b}}{2} + 1\right)\left(\frac{b(y)}{b(y) + 1}\right)$$

   to y, and the rest to x, which is

   $$\bar{b} - \left(\frac{\bar{b}}{2} + 1\right)\left(\frac{b(y)}{b(y) + 1}\right) = \bar{b} - \frac{(\bar{b} + 2)b(y)}{2(b(y) + 1)} = \frac{\bar{b} - 1}{2} + \frac{\bar{b} - b(y) + 1}{2(b(y) + 1)}.$$

4. Summary. The above assignment can be summarized as follows:
   For each $e = \{u, v\} \in E$ with $u \in \tilde{X}$, u receives

   $$\begin{cases} \frac{\bar{b}}{2} - 1 & \text{if } v \in X \\ \frac{\bar{b}}{2} & \text{if } v \in Y^t \\ \bar{b} & \text{if } v \in \tilde{Y} \text{ and } d_Y(v) = b(v) \\ \frac{\bar{b}-1}{2} + \frac{\bar{b}-b(y)+1}{2(b(y)+1)} & \text{if } v \in \tilde{Y} \text{ and } d_Y(v) < b(v). \end{cases}$$

   For each $e = \{u, v\} \in E$ with $v \in \tilde{Y}$, v receives

   $$\begin{cases} \frac{\bar{b}}{2} + 1 & \text{if } u \in Y \\ \frac{\bar{b}}{2} & \text{if } v \in X^t \\ 0 & \text{if } u \in \tilde{X} \text{ and } d_Y(v) = b(v) \\ \left(\frac{\bar{b}}{2} + 1\right)\left(\frac{b(y)}{b(y)+1}\right) & \text{if } u \in \tilde{X} \text{ and } d_Y(v) < b(v). \end{cases}$$

Estimating Total Value Assigned to $y \in Y$. Let $\mathrm{val}(y)$ denote the total value assigned to $y \in \tilde{Y}$.

- Case: $d_Y(y) = b(y)$. Counting only those distributed from edges in $E(Y)$, we have

$$\mathrm{val}(y) \geq \left(\frac{\bar{b}}{2} + 1\right) d_Y(y) = \left(\frac{\bar{b}}{2} + 1\right) b(y).$$

- Case: $d_Y(y) < b(y)$. Because $d(y) = d_{X,Y}(y) + d_Y(y)$, we have

$$\text{val}(y) = d_{X,Y}(y)\left(\frac{\bar{b}}{2}+1\right)\left(\frac{b(y)}{b(y)+1}\right) + d_Y(y)\left(\frac{\bar{b}}{2}+1\right)$$

$$= \left(\frac{\bar{b}}{2}+1\right)\left(d_{X,Y}(y)\left(\frac{b(y)}{b(y)+1}\right) + d_Y(y)\right)$$

$$\geq \left(\frac{\bar{b}}{2}+1\right)(d_{X,Y}(y)+d_Y(y))\left(\frac{b(y)}{b(y)+1}\right)$$

$$\geq \left(\frac{\bar{b}}{2}+1\right)b(y),$$

where the last inequality is due to the fact that y is not tight, i.e., $d(y) \geq b(y)+1$.

It can be thus seen that total value, $\sum_{y \in \tilde{Y}} \text{val}(y)$, assigned on $\tilde{Y}$ is no less than $(\bar{b}/2+1)b(\tilde{Y})$.

Estimating Total Value Assigned to $x \in X$. Let $\text{val}(x)$ denote the total value assigned to $x \in \tilde{X}$.

- Case: $\exists\{x,y\} \in E(\tilde{X},\tilde{Y})$ with $d_Y(y) = b(y)$. While x receives b from edge $\{x,y\}$, the least amount of value distributed from any edge in $\delta(x)$ is $\left(\frac{\bar{b}}{2}-1\right)$. Hence,

$$\text{val}(x) \geq \bar{b} + \left(\frac{\bar{b}}{2}-1\right)(d(x)-1)$$

$$\geq \bar{b} + \left(\frac{\bar{b}}{2}-1\right)b(x)$$

$$= \frac{\bar{b}}{2}b(x) + \bar{b} - b(x)$$

$$\geq \frac{\bar{b}}{2}b(x).$$

- Case: $\nexists\{x,y\} \in E(\tilde{X},\tilde{Y})$ with $d_Y(y) = b(y)$. The degree of y in $G[Y]$ increases only by one when x is transfered from X to Y, and hence, the degree bound will not be violated at any node in Y even if x is removed from X. This means, since X must be a minimal solution, the degree violation must occur at x, and hence, it must be the case that $d_{X,Y}(x) > b(x)$.
Recall that least amount of value x receives from $\{x,y\} \in E(X,Y)$ is either $\bar{b}/2$ or $\frac{\bar{b}-1}{2} + \frac{\bar{b}-b(y)+1}{2(b(y)+1)}$, and observe that

$$\frac{\bar{b}-1}{2} + \frac{\bar{b}-b(y)+1}{2(b(y)+1)} \geq \frac{\bar{b}-1}{2} + \frac{\bar{b}-\bar{b}+1}{2(\bar{b}+1)} = \frac{1}{2}\left(\bar{b}-1+\frac{1}{\bar{b}+1}\right).$$

Since

$$\frac{1}{2}\left(\bar{b}-1+\frac{1}{\bar{b}+1}\right) < \bar{b}/2$$

x receives at least $\frac{1}{2}\left(\bar{b}-1+\frac{1}{b+1}\right)$ from each $\{x,y\} \in E(X,Y)$. Therefore,

$$\text{val}(x) \geq d_{X,Y}(x) \times \frac{1}{2}\left(\bar{b}-1+\frac{1}{\bar{b}+1}\right)$$

$$\geq (b(x)+1)\frac{1}{2}\left(\bar{b}-1+\frac{1}{\bar{b}+1}\right)$$

$$= \frac{1}{2}\bar{b}b(x) + \frac{1}{2}\left(\bar{b}-1-b(x)+\frac{b(x)+1}{\bar{b}+1}\right)$$

$$\geq \frac{1}{2}\bar{b}b(x)$$

since

$$\bar{b}-1-b(x)+\frac{b(x)+1}{\bar{b}+1} \geq 0.$$

Thus, $x \in \tilde{X}$ receives at $\bar{b}b(x)/2$ in either case, and total value, $\sum_{x \in \tilde{X}} \text{val}(x)$, assigned on $\tilde{X}$ is no less than $\bar{b}b(\tilde{X})/2$. Therefore, Eq. (5) must hold and this completes the proof of Lemma 1.

References

1. Balasundaram, B., Butenko, S., Hicks, I.V.: Clique relaxations in social network analysis: the maximum k-plex problem. Oper. Res. **59**(1), 133–142 (2011)
2. Betzler, N., Bredereck, R., Niedermeier, R.: On bounded-degree vertex deletion parameterized by treewidth. Discret. Appl. Math. **160**(1–2), 53–60 (2012)
3. Brešar, B., Jakovac, M., Katrenič, J., Semanišin, G., Taranenko, A.: On the vertex k-path cover. Discret. Appl. Math. **161**(13–14), 1943–1949 (2013)
4. Brešar, B., Kardoš, F., Katrenič, J., Semanišin, G.: Minimum k-path vertex cover. Discret. Appl. Math. **159**(12), 1189–1195 (2011)
5. Brešar, B., Krivos-Bellus, R., Semanišin, G., Sparl, P.: On the weighted k-path vertex cover problem. Discret. Appl. Math. **177**, 14–18 (2014)
6. Camby, E., Cardinal, J., Chapelle, M., Fiorini, S., Joret, G.: A primal-dual 3-approximation algorithm for hitting 4-vertex paths. In 9th International Colloquium on Graph Theory and Combinatorics, ICGT 2014, p. 61 (2014)
7. Camerini, P.M., Galbiati, G., Maffioli, F.: Random pseudo-polynomial algorithms for exact matroid problems. J. Algorithms **13**, 258–273 (1992)
8. Chauve, C., Tannier, E.: A methodological framework for the reconstruction of contiguous regions of ancestral genomes and its application to mammalian genome. PLoS Comput. Biol. **4**(11), e1000234 (2008)
9. Chen, Z.-Z., Fellows, M., Fu, B., Jiang, H., Liu, Y., Wang, L., Zhu, B.: A linear kernel for co-path/cycle packing. In: Chen, B. (ed.) AAIM 2010. LNCS, vol. 6124, pp. 90–102. Springer, Heidelberg (2010). doi:10.1007/978-3-642-14355-7_10
10. Devi, N.S., Mane, A.C., Mishra, S.: Computational complexity of minimum P_4 vertex cover problem for regular and $K_{1,4}$-free graphs. Discret. Appl. Math. **184**, 114–121 (2015)
11. Dinur, I., Safra, S.: On the hardness of approximating minimum vertex cover. Ann. Math. **162**(1), 439–485 (2005)

12. Fellows, M.R., Guo, J., Moser, H., Niedermeier, R.: A generalization of Nemhauser and Trotters local optimization theorem. J. Comput. Syst. Sci. **77**(6), 1141–1158 (2011)
13. Feng, Q., Wang, J., Li, S., Chen, J.: Randomized parameterized algorithms for P_2-packing and co-path packing problems. J. Comb. Optim. **29**(1), 125–140 (2015)
14. Fujito, T.: A unified approximation algorithm for node-deletion problems. Discret. Appl. Math. **86**(2–3), 213–231 (1998)
15. Fujito, T.: Approximating node-deletion problems for matroidal properties. J. Algorithms **31**(1), 211–227 (1999)
16. Fujito, T.: On approximation of the submodular set cover problem. Oper. Res. Lett. **25**(4), 169–174 (1999)
17. Jakovac, M., Taranenko, A.: On the k-path vertex cover of some graph products. Discret. Math. **313**(1), 94–100 (2013)
18. Jenkyns, T.: The efficacy of the "greedy" algorithm. In: Proceedings of the 7th Southeastern Conference on Combinatorics, Graph Theory and Computing, pp. 341–350 (1976)
19. Karakostas, G.: A better approximation ratio for the vertex cover problem. ACM Trans. Algorithms **5**(4), 41:1–41:8 (2009)
20. Kardoš, F., Katrenič, J., Schiermeyer, I.: On computing the minimum 3-path vertex cover and dissociation number of graphs. Theoret. Comput. Sci. **412**(50), 7009–7017 (2011)
21. Khot, S., Regev, O.: Vertex cover might be hard to approximate to within $2 - \epsilon$. J. Comput. Syst. Sci. **74**(3), 335–349 (2008)
22. Lawler, E.L.: Matroids with parity conditions: a new class of combinatorial optimization problems. Memo ERL-M334, Electronics Research Laboratory, College of Engineering, UC Berkeley, Berkeley, CA (1971)
23. Lee, J., Sviridenko, M., Vondrák, J.: Matroid matching: the power of local search. SIAM J. Comput. **42**(1), 357–379 (2013)
24. Lewis, J.M., Yannakakis, M.: The node-deletion problem for hereditary properties is NP-complete. J. Comput. Syst. Sci. **20**, 219–230 (1980)
25. Lovász, L.: Matroid matching and some applications. J. Combin. Theory Ser. B **28**, 208–236 (1980)
26. Moser, H., Niedermeier, R., Sorge, M.J.: Exact combinatorial algorithms and experiments for finding maximum k-plexes. J. Comb. Optim. **24**(3), 347–373 (2012)
27. Narayanan, H., Saran, H., Vazirani, V.: Randomized parallel algorithms for matroid union and intersection, with applications to arboresences and edge-disjoint spanning trees. SIAM J. Comput. **23**, 387–397 (1994)
28. Newman, I., Sohler, C.: Every property of hyperfinite graphs is testable. SIAM J. Comput. **42**(3), 1095–1112 (2013)
29. Novotný, M.: Design and analysis of a generalized canvas protocol. In: Proceedings of the 4th IFIP WG 11.2 International Conference on Information Security Theory and Practices: Security and Privacy of Pervasive Systems and Smart Devices, pp. 106–121 (2010)
30. Okun, M., Barak, A.: A new approach for approximating node deletion problems. Inform. Process. Lett. **88**(5), 231–236 (2003)
31. Orlovich, Y., Dolgui, A., Finke, G., Gordon, V., Werner, F.: The complexity of dissociation set problems in graphs. Discret. Appl. Math. **159**(13), 1352–1366 (2011)
32. Seidman, S.B., Foster, B.L.: A graph-theoretic generalization of the clique concept. J. Math. Soc. **6**(1), 139–154 (1978)
33. Tu, J.: A fixed-parameter algorithm for the vertex cover P_3 problem. Inform. Process. Lett. **115**(2), 96–99 (2015)

34. Tu, J., Yang, F.: The vertex cover P_3 problem in cubic graphs. Inform. Process. Lett. **113**(13), 481–485 (2013)
35. Tu, J., Zhou, W.: A factor 2 approximation algorithm for the vertex cover P_3 problem. Inform. Process. Lett. **111**(14), 683–686 (2011)
36. Tu, J., Zhou, W.: A primal-dual approximation algorithm for the vertex cover P_3 problem. Theoret. Comput. Sci. **412**(50), 7044–7048 (2011)
37. Wolsey, L.A.: An analysis of the greedy algorithm for the submodular set covering problem. Combinatorica **2**(4), 385–393 (1982)
38. Xiao, M.: On a generalization of Nemhauser and Trotter's local optimization theorem. J. Comput. Syst. Sci. (2016). doi:10.1016/j.jcss.2016.08.003
39. Yannakakis, M.: Node-deletion problems on bipartite graphs. SIAM J. Comput. **10**(2), 310–327 (1981)

Multi-agent Pathfinding with n Agents on Graphs with n Vertices: Combinatorial Classification and Tight Algorithmic Bounds

Klaus-Tycho Foerster[2], Linus Groner[1(✉)], Torsten Hoefler[1], Michael Koenig[1], Sascha Schmid[1], and Roger Wattenhofer[1]

[1] ETH Zurich, 8092 Zurich, Switzerland
{gronerl,mikoenig,saschmi,wattenhofer}@ethz.ch, htor@inf.ethz.ch
[2] Aalborg University, 9220 Aalborg, Denmark
ktfoerster@cs.aau.dk

Abstract. We investigate the multi-agent pathfinding (MAPF) problem with n agents on graphs with n vertices: Each agent has a unique start and goal vertex, with the objective of moving all agents in parallel movements to their goal s.t. each vertex and each edge may only be used by one agent at a time. We give a combinatorial classification of all graphs where this problem is solvable in general, including cases where the solvability depends on the initial agent placement.

Furthermore, we present an algorithm solving the MAPF problem in our setting, requiring $\mathcal{O}(n^2)$ rounds, or $\mathcal{O}(n^3)$ moves of individual agents. Complementing these results, we show that there are graphs where $\Omega(n^2)$ rounds and $\Omega(n^3)$ moves are required for any algorithm.

1 Introduction

Pathfinding for single agents on a graph is a well studied problem. *Dijkstra's algorithm* provided a solid foundation in 1959 [1] and since then, several more specialized adaptations have been conceived, such as the A^* algorithm [2] for grids and hierarchical pathfinding using the ability to pre-process maps. The applications for *multi*-agent pathfinding have grown numerous in the recent decades.

Movies such as *The Lord of the Rings* want to display huge armies clashing, but without paying an actor for each combatant [3]. Real-time strategy games incorporate larger and larger amounts of units and players expect predictable and efficient unit movement [4]. Building safety researchers can predict the movement and behaviour of human crowds during an emergency evacuation through simulation [5]. Pathfinding on graphs has also drawn attention in robotics, where it is applied to the problem of multi-robot path planning [6]. Another related field is routing in networks, where deadlock-free forwarding (pathfinding) of packets (agents) is of interest [7].

In this paper, we focus our attention on the most congested pathfinding case, where n agents are to be routed on n-vertex graphs, advancing the work

© Springer International Publishing AG 2017
D. Fotakis et al. (Eds.): CIAC 2017, LNCS 10236, pp. 247–259, 2017.
DOI: 10.1007/978-3-319-57586-5_21

of [8,9]. Motivated by real-world capacity constraints, but also following classical pathfinding research [10], we allow each edge and vertex to be used by only one agent at a time. A precise problem definition is given in Sect. 2, where we formalize the multi-agent pathfinding (MAPF) problem in the form of a labeling problem. Notwithstanding, we invite the reader to first study the background Sect. 1.1.

A main interest of this article is on classifying graphs where the MAPF problem is generally solvable with combinatorial criteria: That is, for any two initial and desired placements of agents, is there a valid sequence of moves solving the corresponding MAPF problem?

In Sect. 3, we give a clear-cut combinatorial classification of all graphs where this problem is solvable in general, including cases where the solvability depends on the initial agent placement. In the subsequent Sect. 4, we then give an algorithm[1] solving the MAPF problem in $\mathcal{O}(n^2)$ rounds and $\mathcal{O}(n^3)$ agent moves. Furthermore, we provide a class of graphs where any algorithm will require $\Omega(n^2)$ rounds and $\Omega(n^3)$ agent movements, matching our upper bounds. We conclude with a summary in Sect. 5.

1.1   Background

One of the earliest scientific works on multi-agent pathfinding on graphs is by Johnson and Story [11]: They studied the famous *15-puzzle*, where 15 agents $1, 2, \ldots, 15$ are placed on a 4×4-grid, and only one agent may move at a time to a currently unoccupied neighboring vertex. The authors showed that exactly half of the starting positions are not solvable, if the goal is to order the agents in an increasing pattern from 1 to 15, with the lower right vertex being unoccupied, and also studied larger grids – with Wilson showing the connection to alternating groups [12]. In more recent times, it was shown that finding the fastest solution for feasible problems is NP-hard already on grids, cf. [13,14].

The model of the 15-puzzle, where one agent moves at a time to an unoccupied neighboring vertex, has been studied by numerous people in various communities. One such piece of work that this article draws foundations and techniques from, in particular for lower bounds, is *Coordinating Pebble Motion On Graphs, The Diameter Of Permutation Groups, And Applications* by Kornhauser, Miller, and Spirakis. Two versions exist, one is the Master's Thesis of Kornhauser which is available as a technical report [15]. A more compact version was published at *FOCS* in 1984 [10], omitting some proofs. Even though Kornhauser uses a different model where no rotations are allowed and enforcing one unoccupied node, we arrived at the same upper and lower bounds of $\mathcal{O}(n^3)$, respectively $\Omega(n^3)$ agent moves. Our proof of the $\Omega(n^3)$ lower bound in our model is very similar to that of Kornhauser, as noted in Sect. 4.4. While their results are from the 1980's, Röger and Helmert [16] pointed out in 2012 that these findings solve some open problems in the robotics community and are still relevant in current research.

---

[1] Yu and Rus [8] also give a MAPF algorithm, cf. second to last paragraph of Sect. 1.1.

The same model as in this article was previously studied by Yu and Rus in *Pebble Motion on Graphs with Rotations: Efficient Feasibility Tests and Planning Algorithms* [8]. The authors provided an algorithm to check if a graph instance is solvable, but did not give combinatorial criteria for feasibility as provided by us. Hence, they could also not provide statements about when exactly half of the MAPF problems are solvable, as we did in Sect. 3.3. Yu and Rus also give a MAPF algorithm, differing from our methods, for which they prove an upper bound that is equivalent to our $\mathcal{O}(n^2)$ upper bound on the number of rotations. However, they did not show the lower bounds.

Lastly, Driscoll and Furst published a paper [9] in 1983 that gives a $\mathcal{O}(n^2)$ upper bound on the diameter of a class of permutation groups. While Driscoll and Furst's paper does not relate permutations to multi-agent pathfinding, our problem is in said class of permutation problems, and Driscoll and Furst's upper bound directly applies to the number of rotations in our problem. Driscoll and Furst also provide a generating set that leads to a tight lower bound, however this generating set can not be related to MAPF problems in the model discussed in this article, since it relies on two-cycles as generators.

2   Model

In this section we will first formally introduce the problem of multi-agent pathfinding, before providing some mathematical preliminaries for the concepts of permutations and permutation groups. We then use these tools to reformulate the MAPF problem as a labeling problem in Sect. 2.1. Multi-agent pathfinding (MAPF) on a graph describes a problem where k agents are distributed on vertices of a graph $G(V, E)$ with n vertices. Each agent has a destination, its goal vertex. Agents can move over edges to neighboring vertices. The problem is to find a sequence of moves, such that eventually all agents are on their goal vertex. In the problems studied here, there is always exactly one agent on each vertex, i.e., $k = n$. The movement of the agents is constrained by the following rules:

- At any given time, no more than one agent can be on any vertex.
- Any edge can only be used by one agent at a time, i.e., neighboring agents may not swap places.

The only permitted moves are thus *rotations* on *graph cycles*.

Definition 1 (rotation). *In a rotation on a graph cycle $v_1, \ldots, v_m$, the agent on a vertex v_i moves to the vertex v_{i+1} if $i \in \{1, \ldots, m-1\}$ or the vertex v_1 if $i = m$.*

To keep the terminology consistent with other works in Computer Science and Mathematics, we will be dealing with *labeled graphs* instead of *agents on graphs*:

Definition 2 (labeling). *Let $L = \{1, 2, 3, \ldots, |V|\}$ be the set of labels. A labeling of a graph $G(V, E)$ is a bijective function $l \colon V \to L$.*

Problems where objects are reordered are typically associated with the mathematical theory of *permutations* and *permutation groups*. In the following, we will give some of the basic definitions and results from those fields.

Definition 3 (permutation). *Let* $X = 1, \ldots, n$. *A permutation is a bijective function* $\pi \colon X \to X$.

There are multiple established notations for permutations. In the *two-line notation* one writes for each element x in the first row its image $\pi(x)$ in the second row:

$$\pi = \begin{pmatrix} l_1 & l_2 & l_3 & \ldots & l_{n-1} & l_n \\ \pi(l_1) & \pi(l_2) & \pi(l_3) & \ldots & \pi(l_{n-1}) & \pi(l_n) \end{pmatrix}$$

The second notation used here is the *cycle notation*: Starting from some element $x \in X$, one writes the sequence $\big(x\ \pi(x)\ \pi(\pi(x))\ \ldots\big)$ of successive images under π. The sequence is continued until x would appear again. Starting at a new element not observed yet, we do the same, and write it in a new pair of parentheses. This is repeated until every element is written down once.

Example 1. $\begin{pmatrix} 1\,2\,3\,4\,5\,6\,7 \\ 1\,5\,7\,2\,4\,3\,6 \end{pmatrix}$ could be written as $(1)\,(2\ 5\ 4)\,(3\ 7\ 6)$.

Cycles of length one are omitted, the above permutation then reads as $(2\ 5\ 4)\,(3\ 7\ 6)$. Next, a pair of labels is called an inversion, if the order of said labels is changed by the permutation.

Definition 4 (inversion). (l_i, l_j) *is an* inversion *of* π, *if* $l_i > l_j$ *and* $\pi(l_i) < \pi(l_j)$.

Definition 5 (parity of a permutation). *The* parity of a permutation *is the parity (odd or even) of the number of inversions it contains.*

Definition 6 (composition of permutations). *Two (and, iteratively, any number of) permutations can be* composed: $\pi_1 \circ \pi_2 = \pi_1 \pi_2 = \pi_2(\pi_1(x))\quad \forall x \in X$.

The set of all permutations on $1, \ldots, n$ with operation $\circ$ form the group S_n. An important subgroup of S_n is the *alternating group* A_n. It is the subgroup of S_n which contains all even permutations. It contains exactly half of the $n!$ elements of S_n. We need the following lemma to see that closure is satisfied for A_n, which is proven in numerous textbooks on the subject, such as [17]:

Lemma 1. *The composition of even permutations is even.*

We can conclude by recursion, that the composition of any number of even permutations will again result in an even permutation. The set of even permutations is thus closed under composition.

2.1   Reformulation of the MAPF Problem

The permitted operations in our model are rotations. They can be interpreted as an element of S_n. We refer to Fig. 1 for an introductory case explained in Example 2.

Example 2. If we label the bow-tie graph with labels as in Fig. 1, we can write the permutations corresponding to the rotations in cycle notation, e.g.:

- clockwise rotation in the left cycle: $\pi_{L^-} = \left(1\ 3\ 2\right)\left(4\right)\left(5\right) = \left(1\ 3\ 2\right)$,
- counterclockwise rotation in the right cycle: $\pi_{R^+} = \left(1\right)\left(2\right)\left(3\ 4\ 5\right) = \left(3\ 4\ 5\right)$,

In fact, rotations in our model always correspond to permutation cycles. (But not all permutation cycles correspond to a valid move.) It is thus justified to reformulate the MAPF problem:

Main Idea. *Let π_{goal} be the permutation that represents the goal labeling and let P_G be the set of permutations that correspond to a valid rotation.*

Find a sequence $\pi_{r_1}, \ldots, \pi_{r_m},$ where $\pi_{r_i} \in P_G$ such that

$$\pi_{r_1} \circ \pi_{r_2} \circ \ldots \circ \pi_{r_m} = \pi_{\text{goal}} \qquad (1)$$

This problem has a solution if and only if π_{goal} is an element of the group generated by P_G.

Fig. 1. Labeled bow-tie graph, consisting of two odd cycles of length three.

Lemma 2. *Rotations on graph cycles with even length correspond to odd permutations. Rotations on odd-length graph cycles correspond to even permutations.*

Proof. Rotations on graph cycles with length i correspond to permutation cycles of the same length i. It is known, cf. [17], that cyclic permutations of even length correspond to odd permutations and vice-versa. Therefore odd i give rise to even permutations, even i to odd ones.   □

3   Necessary and Sufficient Combinatorial Criteria for Solvability

We will begin this section with Theorem 1, where we specify necessary and sufficient combinatorial criteria for graphs on which the MAPF problem can be generally solved.

Definition 7. *The MAPF problem is generally solvable on a graph G, if the MAPF problem is solvable on G for any combination of an initial labeling with a goal labeling.*

Theorem 1. *The MAPF-problem on a graph G with $n \geq 2$ vertices is generally solvable, if and only if the following conditions hold:*

1. G is 2-edge-connected,
2. G contains at least two cycles,
3. G contains a cycle of even length.

In the following Sect. 3.1, we address the necessity of these criteria. Then, in Sect. 3.2, we point out that graphs fulfilling the criteria are indeed solvable, i.e., the conditions are sufficient.

Lastly in this section, we address in Sect. 3.3 that half of the MAPF problems are still solvable if the particular requirement that a graph must contain an even-length cycle is not satisfied.

3.1 The Combinatorial Conditions in Theorem 1 are Necessary

We defer the proof of conditions 1 and 2 to the full version of this article. Condition 3 is proven in the following lemma.

Lemma 3. *The MAPF-problem on a graph G is not generally solvable, if the graph does not contain an even-length cycle.*

Proof. Assume Graph G contains only odd-length cycles. Lemma 2 then implies that all π_{r_i} of Eq. 1 are even. Using Lemma 1, we see that the permutation problem can not be solved for odd π_{goal} and we will always stay in A_n.   □

In fact, as we will see in Sect. 3.3, all problems corresponding to even π_{goal} are solvable, when this last constraint is not satisfied. That is, exactly half of all problems are still solvable in that case.

3.2 The Combinatorial Conditions in Theorem 1 are Sufficient

In this section, we show that the MAPF problem on the graphs specified in Theorem 1 are indeed generally solvable. We will show that on such graphs it is possible to exchange any two labels while leaving all other labels unaffected. In terms of permutations this amounts to being able to express 2-cycles as a sequence of the permutations corresponding to the permitted rotations. (cf. our main idea). Since the set of all 2-cycles generates S_n (cf. [17]), this will conclude the proof of Theorem 1.

3.2.1 Swapping Two Labels in a Generally Solvable Graph

Lemma 4. *Let*

$$\pi_{a\times b}(x) = \begin{cases} b & if\, x = a \\ a & if\, x = b \\ x & otherwise \end{cases} \qquad \pi_{(l_1,l_2)\to(s_1,s_2)}(x) = \begin{cases} s_1 & if\, x = l_1 \\ s_2 & if\, x = l_2 \\ x' & otherwise \end{cases}$$

where x' in $\pi_{(l_1,l_2)\to(s_1,s_2)}$ is arbitrary, with the constraint that $\pi_{(l_1,l_2)\to(s_1,s_2)}$ is bijective.

Then,

$$\pi_{l_1\times l_2} = \pi_{(l_1,l_2)\to(s_1,s_2)}\pi_{s_1\times s_2}\pi^{-1}_{(l_1,l_2)\to(s_1,s_2)} \tag{2}$$

Proof. To make notation less cumbersome, we will denote $\pi_{(l_1,l_2)\to(s_1,s_2)}$ by π. The right hand side of Eq. 2 can be rewritten as $\pi\pi_{s_1\times s_2}\pi^{-1} = \pi^{-1}(\pi_{s_1\times s_2}(\pi(x)))$. We distinguish cases:

Case $x \neq l_1 \wedge x \neq l_2$: Then, $\pi(x) = x'$. Since $x' \neq s_1$ and $x' \neq s_2$ we have $\pi_{s_1\times s_2}(x') = x'$. We can then plug these values in as follows:

$$\pi^{-1}(\pi_{s_1\times s_2}(\pi(x))) = \pi^{-1}(\pi_{s_1\times s_2}(x')) = \pi^{-1}(x') = x$$

Case $x = l_1$: $\pi(l_1) = s_1$ and $\pi_{s_1\times s_2}(s_1) = s_2$:

$$\pi^{-1}(\pi_{s_1\times s_2}(\pi(x))) = \pi^{-1}(\pi_{s_1\times s_2}(s_1)) = \pi^{-1}(s_2) = l_2$$

Case $x = l_2$: analogously

In all cases, we have $\pi^{-1}(\pi_{s_1\times s_2}(\pi(x))) = \pi_{l_1\times l_2}(x)$, concluding the proof. $\square$

In other words, if we can swap a specific pair of labels (s_1 and s_2) without affecting other labels, and we are able to move any pair of labels (l_1 and l_2) to the position of the aforementioned labels, we can effectively swap any two labels by means of Eq. 2. It remains to prove that we can express some $\pi_{(l_1,l_2)\to(s_1,s_2)}$ and the suitable $\pi_{s_1\times s_2}$ for any l_1 and l_2 by means of the permitted rotations.

Lemma 5. *For any cycle c_1 in a graph that is 2-edge-connected with at least two cycles, one of the following two options holds:*

1. *There is a cycle c_2 with which it shares exactly one vertex or*
2. *There are 2 vertices in c_1 with 3 vertex-disjoint paths between them.*

The proof of this lemma is deferred to the full version of this article. According to this lemma, finding $\pi_{s_1\times s_2}$ for all 2-edge-connected graphs can be done by finding $\pi_{s_1\times s_2}$ in each of the stated cases. We will now demonstrate how swapping is possible in either case.

3.2.2 Swapping Labels in Cycles Sharing Exactly One Vertex

Let C_{n_l,n_r} denote a graph consisting of two cycles, with sizes n_l and n_r, respectively, that share exactly one vertex. As there are two cycles, four operations are permitted, namely rotations in both directions on either cycle. π_{L-} denotes the permutation associated with a clockwise rotation in the left cycle, π_{L+} the permutation associated with a counterclockwise rotation in the left cycle. π_{R+} and π_{R-} are the analogous counterparts in the right cycle. Algorithm 1 describes a procedure to swap the labels l_1 and m in C_{n_l,n_r}.

Lemma 6. *If n_l is even, Algorithm 1 terminates.*

Proof. The permutations associated with the basic rotations that we use can readily be written down in cycle notation:

$$\pi_{L-} = (l_1\ m\ l_{n_l-1}\ldots l_2) \qquad \pi_{R+} = (r_1\ r_2\ldots r_{n_r-1}\ m) \qquad \pi_{R-} = (r_1\ m\ r_{n_r-1}\ldots r_2)$$

Building on these, we can write down the composed permutations of the algorithm:

$\pi_{\text{init}} = \pi_{R-}\pi_{L-}\pi_{L-}\pi_{R+}\pi_{L-}$

$$= \begin{pmatrix} l_1 & l_2\, l_3\, l_4\, l_5\, l_6\, l_7 \ldots l_{n_l-3}\, l_{n_l-2}\, l_{n_l-1}\, m & r_1 & r_2\, r_3 \ldots r_{n_r-2}\, r_{n_r-1} \\ l_{n_l-2}\, r_1\, m\, l_1\, l_2\, l_3\, l_4 \ldots l_{n_l-6}\, l_{n_l-5}\, l_{n_l-4}\, l_{n_l-1}\, l_{n_l-3}\, r_2\, r_3 \ldots r_{n_r-2}\, r_{n_r-1} \end{pmatrix}$$

$\pi_{\text{step}} = \pi_{R-}\pi_{L-}\pi_{R+}\pi_{L-}$

$$= \begin{pmatrix} l_{n_l-2}\, r_1 & m & l_1\, l_2\, l_3\, l_4 \ldots l_{n_l-6}\, l_{n_l-5}\, l_{n_l-4}\, l_{n_l-1}\, l_{n_l-3}\, r_2\, r_3 \ldots r_{n_r-2}\, r_{n_r-1} \\ l_{n_l-4}\, l_{n_l-2}\, l_{n_l-1}\, r_1\, m\, l_1\, l_2 \ldots l_{n_l-8}\, l_{n_l-7}\, l_{n_l-6}\, l_{n_l-3}\, l_{n_l-5}\, r_2\, r_3 \ldots r_{n_r-2}\, r_{n_r-1} \end{pmatrix}$$

Assuming n_l is even, π_{step} reads in cycle notation:

$$\pi_{\text{step}} = \left(l_1\, r_1\, l_{n_l-2}\, l_{n_l-4} \ldots l_2\, m\, l_{n_l-1}\, l_{n_l-3} \ldots l_3 \right)$$

We left out $\frac{n_l}{2} - 4$ labels with each "...", namely l_i's with even i in the left case and with odd i in the right. Note that the labels $l_1, \ldots, l_{n_l-3}$ always take the place of the label with an index that is larger by 2. If n_l was odd, $n_l - 1$ would be even and l_{n_l-1} would be in the cycle much earlier, such that not all labels would be in the same cycle.

Applying a cyclic permutation k-fold has step the labels k steps forward in the order of the cycle. For each label x in the permutation cycle we can count k positions to the right in the cyclic representation of π_{step} to find $\pi_{\text{step}}^k(x)$. In this way, we find $\pi_{\text{step}}^{\frac{n_l}{2}-1}$:

$$\pi_{\text{step}}^{\frac{n_l}{2}-1} = \begin{pmatrix} l_{n_l-2}\, r_1\, m\, l_1\, l_2\, l_3\, l_4 \ldots l_{n_l-6}\, l_{n_l-5}\, l_{n_l-4}\, l_{n_l-1}\, l_{n_l-3}\, r_2\, r_3 \ldots r_{n_r-2}\, r_{n_r-1} \\ m & l_2\, l_3\, l_4\, l_5\, l_6\, l_7 \ldots l_{n_l-3}\, l_{n_l-2}\, l_{n_l-1}\, l_1 & r_1 & r_2\, r_3 \ldots r_{n_r-2}\, r_{n_r-1} \end{pmatrix}$$

We've written down $\pi_{\text{step}}^{\frac{n_l}{2}-1}$ such that it is easy to see that

$$\pi_{\text{init}}\pi_{\text{step}}^{\frac{n_l}{2}-1} = \begin{pmatrix} l_1\, l_2\, l_3\, l_4\, l_5\, l_6\, l_7 \ldots l_{n_l-3}\, l_{n_l-2}\, l_{n_l-1}\, m\, r_1\, r_2\, r_3 \ldots r_{n_r-2}\, r_{n_r-1} \\ m\, l_2\, l_3\, l_4\, l_5\, l_6\, l_7 \ldots l_{n_l-3}\, l_{n_l-2}\, l_{n_l-1}\, l_1\, r_1\, r_2\, r_3 \ldots r_{n_r-2}\, r_{n_r-1} \end{pmatrix}$$

Which is our goal permutation. That is, after $\frac{n_l}{2} - 1$ repetitions of the loop in Algorithm 1, we are at the desired configuration, and the algorithm terminates.

□

Algorithm 1. Swapping Two Labels in C_{n_l,n_r}

$\pi := \pi_{R-}\pi_{L-}\pi_{L-}\pi_{R+}\pi_{L-}$
$\pi_{\text{step}} := \pi_{R-}\pi_{L-}\pi_{R+}\pi_{L-}$
while $\pi \neq \pi_{\text{goal}}$ **do**
$\quad\lfloor\ \pi := \pi\pi_{\text{step}}$

3.2.3 Swapping Labels in a Cycle Containing 2 Vertices with Three Paths Between Them

In the case when there are two vertices with three vertex-disjoint paths between them, swapping two labels is simpler, and possible with just 3 rotations. One possibility of performing such a swap is illustrated in Fig. 2.

3.2.4   Travelling to Swapspot

It remains to express $\pi_{(l_1,l_2)\to(s_1,s_2)}$ for any l_1, l_2, s_1 and s_2, where the initial vertices of s_1 and s_2 are neighbors. We will do this in two phases. First, we move l_1 and l_2 such that they are neighbors. Then, these neighbors are moved to the place where they can be swapped. For details, we refer to the full version.

3.3   Solvable Problems on Not Generally Solvable Graphs

We have now specified the class of graphs on which the MAPF problem is generally solvable. On those that are not generally solvable, some problems are still solvable. In the cases where a graph is not 2-edge connected, one can consider each 2-edge connected component separately, as no label can cross bridges. The solvable problems are then those where the labels only travel within subgraphs that fulfill the constraints of Theorem 1. Another case is when there is only one cycle present, where the solvable problems are exactly those obtained by rotations on this cycle.

However, if a graph is still 2-edge connected and contains at least two cycles, but only contains cycles of odd length, a more interesting observation can be made. In fact, exactly half of the problems can still be solved. In Sect. 3.2 we presented a method to express 2-cycles as a sequence of the permitted rotations. Without the presence of even cycles, it is possible to express 3-cycles with a very similar method. Recall that 3-cycles are a generating set of the alternating group A_n, which contains half of the elements of S_n. The details of 3-cycling are deferred to the full version.

4   Algorithms, Lower and Upper Bounds

In this section, we use the mechanisms studied so far to construct an algorithm that solves the MAPF problem in $\mathcal{O}(n^3)$ label movements and $\mathcal{O}(n^2)$ rotations. We will also present a class of graphs, on which the MAPF problem cannot be solved with less than $\Omega(n^3)$ label movements and $\Omega(n^2)$ rotations, meaning that our algorithm is optimal in terms of the asymptotic number of operations in the worst case.

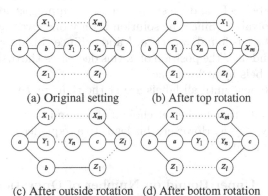

(a) Original setting   (b) After top rotation

(c) After outside rotation   (d) After bottom rotation

Fig. 2. Swapping in graphs with two vertices with three vertex-disjoint paths between them.

4.1 Complexity Measures

The complexity of a solution to the problem can be described in different ways. In this chapter we will investigate the complexity with respect to three related measures. An upper bound on the length of the sequence of permutations found by the algorithm is given in all three measures, and a class of graphs is given on which these upper bounds for all three measures are tight in an asymptotic sense.

One way of describing the complexity of a solution is that of the total number of *rotations*. In this case, every rotation increases the complexity by one. This is effectively the length of the sequence found in our main idea. For some problem instances, rotations can be performed in parallel. On these problems, measuring the complexity with the number of rotations might not give a good representation of the running time. The number of *rounds* thus can be used as a second measure. However, the algorithms used in this paper never use the possibility of parallel rotations. Therefore here, the number of rotations equals the number of rounds. As we will see, our lower bounds are tight regardless. Lastly, the number of *label movements* is studied. That is, the number of rotations a label was involved in, summed up over all labels.

4.2 The Algorithm

We have established the notions of swapping and 3-cycling labels. Using these mechanisms we can directly build an algorithm:

1. As long as there are labels in the wrong place, pick one wrongly placed label, say a. Then, pick the label $b := \pi_{\text{goal}}(a)$.
2. Set c to be an arbitrary incorrectly placed label such that $a \neq c$ and $\pi_{\text{goal}}(a) \neq c$. If no such c can be found, a and b are the only wrongly placed labels left, and we swap them. If swapping is not possible, the problem is not solvable. (Since the solution is only one swap away, π_{goal} is not in A_n.)
3. If a c is found, 3-cycle a, b and c. Since this only moves wrongly placed labels, and fixes the position of a, this decreases the overall number of wrongly placed labels by at least one.
4. Repeat until all labels are at the right place.

Note that by better choices of a, b and c, we can fix at least two labels with every 3-cycle. However, this leads to a sequence of operations of the same asymptotic length.

4.3 Upper Bound on Number of Operations

Lemma 7. *Swapping two labels and 3-cycling three labels without affecting any other labels both take $\mathcal{O}(n)$ rotations.*

For a full proof, we refer to the full article. The gist is that there is a constant number of steps involved, each with a complexity in $\mathcal{O}(n)$ rotations. These complexities are mainly determined by the length of paths and cycles in the graph.

Theorem 2. *The Algorithm described in Sect. 4.2 terminates in $\mathcal{O}(n^2)$ rotations, $\mathcal{O}(n^2)$ rounds and $\mathcal{O}(n^3)$ label movements.*

Proof. Since on n labels and n vertices, there can be at most n wrongly placed labels, and we fix at least one with every 3-cycle and every swap, we will need at most n such operations. In other words, the added number of 3-cycles and swaps performed is in $\mathcal{O}(n)$.

We have seen in Lemma 7 that both swapping and cycling take $\mathcal{O}(n)$ rotations. Having $\mathcal{O}(n)$ swaps or cycling operations costing $\mathcal{O}(n)$ rotations each, we get the claimed overall bounds of $\mathcal{O}(n^2)$ rotations. Clearly, each rotation moves at most n labels, which directly implies the upper bound of $\mathcal{O}(n^3)$ label movements.

The worst case in terms of number of rounds, is when all rotations are done sequentially. Therefore, an upper bound on the number of rotations is also an upper bound on the number of rounds. I.e., the upper bound of $\mathcal{O}(n^2)$ rotations directly implies an upper bound of $\mathcal{O}(n^2)$ rounds.   □

4.4 Lower Bound on Number of Operations

We will now give a class of graphs and a MAPF problem on which any algorithm takes at least $\Omega(n^2)$ rotations, $\Omega(n^2)$ rounds and $\Omega(n^3)$ label movements, providing lower bounds that are asymptotically tight. The class of graphs is the same as Kornhauser et al. [10] used for their model.

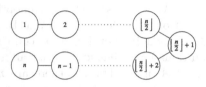

Fig. 3. Graph LB_n for the proof of the lower bound

Consider the graph of Fig. 3, that is the cyclic graph on n vertices with an added edge between the $\lfloor \frac{n}{2} \rfloor$-th and the $\lfloor \frac{n}{2} + 2 \rfloor$-th vertex. We denote this graph by LB_n.

4.4.1 Rotations and Rounds

Lemma 8. *There is a MAPF problem on LB_n for which any solution requires $\Omega(n^2)$ rotations.*

Proof. Assume n to be odd. We define d_i to be the *semi-circular distance* between label i and label $i + 1$. The semi-circular distance is the shortest path between the labels on the cyclic graph, that does not use the added edge. The d_i are maximal for $d_i = \lfloor \frac{n}{2} \rfloor$, and are at least 1.

Following Kornhauser et al. [10], we define the notion of *entropy* as $E = \sum_{i=1}^{\lfloor \frac{n}{2} \rfloor} d_i$. We chose an initial labeling, for which $E = \lfloor \frac{n}{2} \rfloor^2$, with our goal configuration having $E = \lfloor \frac{n}{2} \rfloor$. There are six permitted operations on LB_n: A rotation on the outer cycle, denoted by A, a rotation on the cycle $(\lfloor \frac{n}{2} \rfloor \ \lfloor \frac{n}{2} \rfloor + 1 \ \lfloor \frac{n}{2} \rfloor + 2)$, denoted by B and a rotation on the cycle not including $\lfloor \frac{n}{2} \rfloor + 1$, denoted by C, as well as their respective inverses A^{-1}, B^{-1} and C^{-1}. We can study the effect of the three operations on the entropy. Clearly, A and A^{-1} do not change the entropy. Rotating B or C can only change the d_i that include the labels on vertices $\lfloor\frac{n}{2}\rfloor$, $\lfloor\frac{n}{2}\rfloor + 1$ and $\lfloor\frac{n}{2}\rfloor + 2$, and by at most 2 each. Each rotation thus decreases E by at most 12. Having $E = \lfloor\frac{n}{2}\rfloor^2$ at the beginning and $E = \lfloor\frac{n}{2}\rfloor$ at the goal configuration, we can say that we need at least $\frac{\lfloor\frac{n}{2}\rfloor^2 - \lfloor\frac{n}{2}\rfloor}{12} \in \Omega(n^2)$ operations. $\qquad\square$

Lemma 9. *There is a MAPF problem on LB_n for which any solution requires $\Omega(n^2)$ rounds.*

Proof. Since all three cycles in LB_n pairwise share vertices, only one rotation can be performed at a time. Therefore, the lower bound on the number of rotations from Lemma 8 is also a lower bound for the number of rounds. $\qquad\square$

4.4.2 Label Movements

We now look at the number of label movements.

Lemma 10. *There is a MAPF problem on LB_n where any solution requires $\Omega(n^3)$ label movements.*

Proof. We can assume that in an optimal solution, no more than one consecutive operation is performed on cycle B. (Since, e.g., BB can be replaced by B^{-1}, $B^{-1}B$ by doing nothing at all, consecutive operations on B indicate non-optimal solutions.) We thus know, that after each operation on B, there will be one on either A or C. Thus, if there are m operations, at least $\lfloor\frac{m}{2}\rfloor$ operations are performed on A and C. Those require at least $n - 1$ label movements. (Namely, if C is moved.) As any solution will use at least $\Omega(n^2)$ rotations, so will a solution that is optimal with respect to label movements. Hence, $(n-1)\lfloor\frac{\Omega(n^2)}{2}\rfloor \in \Omega(n^3)$ is a lower bound for the number of label movements. $\qquad\square$

5   Conclusion

We studied combinatorial classifications and algorithms for the multi-agent pathfinding (MAPF) problem on graphs G with n agents. We proved that the MAPF problem is only generally solvable, if the graphs G are 2-edge-connected, contain at least two cycles, and contain at least one cycle of even length. Should the last of these three combinatorial conditions be violated, we showed that exactly half of the MAPF problems on these graphs are solvable.

Furthermore, we specified an algorithm that solves feasible MAPF problems in $\mathcal{O}(n^2)$ operations or $\mathcal{O}(n^3)$ agent-movements. We also specified a class of graphs, where at least $\Omega(n^2)$ operations or $\Omega(n^3)$ agent-movements are required, meaning that on general graphs, our algorithms are asymptotically optimal.

Acknowledgements. We would like to thank the anonymous reviewers for their helpful comments. Klaus-Tycho Foerster is supported by the Danish Villum Foundation.

References

1. Dijkstra, E.W.: A note on two problems in connexion with graphs. Numer. Math. **1**(1), 269–271 (1959)
2. Hart, P.E., Nilsson, N.J., Raphael, B.: A formal basis for the heuristic determination of minimum cost paths. IEEE Trans. Syst. Sci. Cybern. **4**(2), 100–107 (1968)
3. Scott, R.: Sparking life: notes on the performance capture sessions for the lord of the rings: the two towers. SIGGRAPH Comput. Graph. **37**(4), 17–21 (2003)
4. Silver, D.: Cooperative pathfinding. In: Artificial Intelligence and Interactive Digital Entertainment Conference (2005)
5. Pelechano, N., Malkawi, A.: Evacuation simulation models: challenges in modeling high rise building evacuation with cellular automata approaches. Autom. Constr. **17**(4), 377–385 (2008)
6. Svestka, P., Overmars, M.H.: Coordinated path planning for multiple robots. Robot. Auton. Syst. **23**(3), 125–152 (1998)
7. Domke, J., Hoefler, T., Matsuoka, S.: Routing on the dependency graph: a new approach to deadlock-free high-performance routing. In: Symposium on High-Performance Parallel and Distributed Computing (2016)
8. Yu, J., Rus, D.: Pebble motion on graphs with rotations: efficient feasibility tests and planning algorithms. In: Akin, H.L., Amato, N.M., Isler, V., Stappen, A.F. (eds.) Algorithmic Foundations of Robotics XI. STAR, vol. 107, pp. 729–746. Springer, Cham (2015). doi:10.1007/978-3-319-16595-0_42
9. Driscoll, J.R., Furst, M.L.: On the diameter of permutation groups. In: Symposium on Theory of Computing (1983)
10. Kornhauser, D., Miller, G., Spirakis, P.: Coordinating pebble motion on graphs, the diameter of permutation groups, and applications. In: Symposium on Foundations of Computer Science (1984)
11. Johnson, W.W.: Notes on the "15" puzzle. Am. J. Math. **2**(4), 397–404 (1879)
12. Wilson, R.M.: Graph puzzles, homotopy, and the alternating group. J. Comb. Theory, Ser. B **16**(1), 86–96 (1974)
13. Ratner, D., Warmuth, M.: The $n^2 - 1$ puzzle and related relocation problems. J. Symb. Comput. **10**(2), 111–137 (1990)
14. Goldreich, O.: Finding the shortest move-sequence in the graph-generalized 15-puzzle is NP-hard. In: Goldreich, O. (ed.) Studies in Complexity and Cryptography. Miscellanea on the Interplay Between Randomness and Computation. LNCS, vol. 6650, pp. 1–5. Springer, Heidelberg (2011). doi:10.1007/978-3-642-22670-0_1
15. Kornhauser, D.: Coordinating pebble motion on graphs, the diameter of permutation groups, and applications. Master's thesis MIT/LCS/TR-320, Massachusetts Institute of Technology (1984)
16. Röger, G., Helmert, M.: Non-optimal multi-agent pathfinding is solved (since 1984). In: Symposium on Combinatorial Search (2012)
17. Jacobson, N.: Basic Algebra. Freeman, San Francisco (1974)

On the Combinatorial Power
of the Weisfeiler-Lehman Algorithm

Martin Fürer[(⊠)]

Department of Computer Science and Engineering,
Pennsylvania State University, University Park, PA, USA
furer@cse.psu.edu

Abstract. The classical Weisfeiler-Lehman method WL[2] uses edge colors to produce a powerful graph invariant. It is at least as powerful in its ability to distinguish non-isomorphic graphs as the most prominent algebraic graph invariants. It determines not only the spectrum of a graph, and the angles between standard basis vectors and the eigenspaces, but even the angles between projections of standard basis vectors into the eigenspaces. Here, we investigate the combinatorial power of WL[2]. For sufficiently large k, WL[k] determines all combinatorial properties of a graph. Many traditionally used combinatorial invariants are determined by WL[k] for small k. We focus on two fundamental invariants, the number of cycles C_p of length p, and the number of cliques K_p of size p. We show that WL[2] determines the number of cycles of lengths up to 6, but not those of length 8. Also, WL[2] does not determine the number of 4-cliques.

Keywords: Weisfeiler-Lehman algorithm · Graph invariants · Counting cycles · Graph isomorphism

1 Introduction

1.1 Weisfeiler-Lehman Method

Two graphs are isomorphic, if there is a bijection of their vertices mapping edges to edges and non-edges to non-edges. An automorphism of a graph is an isomorphism from the graph to itself. The graph isomorphism problem is closely connected to the graph automorphism problem. Two connected graphs G and G' with disjoint vertex sets are isomorphic, iff their union has an automorphism mapping one vertex of G into a vertex of G'. Obviously in this case, all vertices of G are mapped to vertices of G'.

The most natural and most practical way to detect that two graphs are not isomorphic is vertex classification [17]. The idea is to give different colors to two

---

M. Fürer—This work was partially supported by NSF Grant CCF-1320814. Part of this work has been done while visiting Theoretical Computer Science, ETH Zürich, Switzerland.

© Springer International Publishing AG 2017
D. Fotakis et al. (Eds.): CIAC 2017, LNCS 10236, pp. 260–271, 2017.
DOI: 10.1007/978-3-319-57586-5_22

vertices whenever it is obvious that neither of them can be mapped to the other one by an isomorphism. Thus vertex classification could start by coloring the vertices by their degree. One can easily go further. If one vertex u has more neighbors of a certain degree than another vertex v, then obviously u and v should also be colored differently.

A simple way to capture these observations, is to start with all vertices of $G = (V, E)$ having the same color, and then refining the coloring in rounds. In round $i + 1$, vertices u and v receive different colors, if they already had different colors in round i, or if the multisets of colors of neighbors of u and v in round i are different.[1] During each round some color classes are split into two or more classes, until this process stops after at most $n = |V|$ rounds. Nowadays, this method of vertex classification is also known as WL[1]. It is at the heart of all software tools for graph isomorphism testing.

The classical Weisfeiler-Lehman method WL[2] [18], classifies edges in a similar way. Still, it is a bit more involved. In fact, all ordered pairs of vertices are classified, not just the edges. In other words, we can think of handling a complete directed graph with colored edges, including self-loops in all vertices.

At the start, the edges of the complete graph are partitioned into 3 color classes: the previous edges, the previous non-edges, and the self-loops. In round $i + 1$ every directed edge (u, v) is colored with a pair whose first component is its previous color, and whose second component is the multiset of all pairs of previous colors on paths of length 2 from u to v. In each round, the actually occurring colors are lexicographically ordered and replaced by an initial segment of the natural numbers. This time, after $O(n^2)$ rounds, the algorithm stops, because a *stable coloring* is reached, i.e., no color class of edges is further divided.

Sometimes, it is useful to keep for each round the mapping assigning to each detailed color (pair of old color and some multiset) a simplified color (small integer). We refer to this information as the definition of colors.

It has been noticed that WL[2], has a natural k-dimensional extension WL[k] by various researchers, including some authors of [8] who tried to prove that WL[k] solves the graph isomorphism problem for graphs of degree at most k. It seems that the first published definition of WL[k] has been in [7]. The CFI algorithm [8] has introduced and popularized the term WL[k] at the suggestion of Babai as an editor to honor the influence of Weisfeiler and Lehman [18] towards the development of this algorithm.

Weisfeiler and Lehman did not use the WL[k] algorithm, but extended WL[2] by individualizing a sequence of vertices. A sequence $v_1, \ldots, v_\ell$ is individualized by giving a unique color to each vertex of the sequence before the WL[k] algorithm starts. Note that WL[$k + \ell$] is at least as powerful as doing WL[k] for every possible individualization of ℓ vertices.

WL[k] is defined as follows. The initial color $W^0(v_1, \ldots, v_k)$ is according to the isomorphism type of $(v_1, \ldots, v_k)$. To be precise, $(u_1, \ldots, u_k)$ is isomorphic to $(v_1, \ldots, v_k)$ if

---

[1] A multiset differs from a set by assigning a positive integer multiplicity to each element.

- for all i, j, $u_i = u_j$, iff $v_i = v_j$, and
- for all i, j, $\{u_i, u_j\} \in E$, iff $\{v_i, v_j\} \in E$.

For each coloring $f : V^k \to C$ and each $w \in V$, define the operation

$$\text{sift}(f, (u_1, u_2, \ldots, u_k), w))$$
$$= \langle f(w, u_2, \ldots, u_k), f(u_1, w, \ldots, u_k), \ldots, f(u_1, u_2, \ldots, w) \rangle.$$

Hence, $\text{sift}(W^i, (u_1, u_2, \ldots, u_k), w))$ is the k-tuple of W^i colors of the k-tuples arising from substituting w in turn for each of the k positions in $(u_1, u_2, \ldots, u_k)$. Thus, intuitively in each round of WL[3], triangles T are colored by the multiset of the triples of colors used on the triangular faces of the tetrahedra with one face being T. To be precise, actually ordered triples of vertices are used instead of triangles. Now the next color of $(u_1, u_2, \ldots, u_k)$ is

$$(f(u_1, u_2, \ldots, u_k), \text{multiset}\{\text{sift}(f, (u_1, u_2, \ldots, u_k), w) \mid w \in V\}).$$

It should be noticed that for every k, WL[$k + 1$] is at least as powerful as WL[k], because every stable coloring $\mathcal{C}$ of the $k+1$-tuples defines a stable coloring $\mathcal{C}'$ of the k-tuples by $\mathcal{C}'(v_1, \ldots, v_k) = \mathcal{C}(v_1, \ldots, v_k, v_k)$, i.e., by just repeating the last component. Thus for example WL[2] does not only color the edges, but also the vertices. The color of a vertex v shows up as the color of the self-loop at v. The stable partition of the k-tuples of vertices produced by WL[$k + 1$] is at least as fine as that produced by WL[k], because WL[k] produces the coarsest stable partition of the k-tuples.

1.2   Graph Invariants

A *graph invariant* is any function defined on graphs whose value is constant on classes of isomorphic graphs. In particular, the value does not depend on the enumeration of the vertices. In other words, a graph invariant is a function defined on adjacency matrices whose value does not change, when the same permutation is applied to the rows and columns of an adjacency matrix.

Many simple combinatorial graph invariants are often used to quickly conclude that two graphs are non-isomorphic. Some such invariants are, the number of vertices n, the number of edges m, the number of triangles, the degree (maximum number of neighbors of any vertex), the multiset of degrees of vertices. Graph invariants can also be just boolean properties like being bipartite, being connected, being acyclic, or containing a given graph as a subgraph or induced subgraph.

Some more complicated invariants are obtained by counting cliques and cycles. We use the words *path* and *cycle* to refer to a simple path or simple cycle respectively (i.e., an open or closed vertex disjoint path). More precisely, we refer to the set of their edges. Thus, e.g., a K_3 consists of 1 cycle.

Let C_k^v be the number of k-cycles with one vertex being v. Now the multiset of all C_k^v for fixed k and varying over all $v \in V$ is a nice invariant. Similar invariants are obtained by varying over all edges instead of vertices, and by considering

k-cliques instead of k-cycles. We will mainly focus on the graph invariants #k-cliques, the total number of subsets of k vertices forming a complete graph, and #k-cycles, the number of cycles of length k which are occurring in the given graph.

The ultimate combinatorial invariant is obtained by the WL[k] method. Its strength increases with k, and it determines the isomorphism type for $k = n$. We call the invariant WL[k] too. The invariant consists of the multiset of colors of k-tuples in the stable refinement, together with all the definitions of colors occurring during the coloring rounds.

A graph invariant identifies a graph G in a class of graphs, if all graphs in the class with the same invariant as G are isomorphic to G. In other words, up to isomorphisms, G is the only graph in the class with this invariant. A graph invariant identifies a graph G, if it identifies G in the class of all graphs. An invariant identifies a class of graphs, if it identifies all graphs G of this class in the class of all graphs. For example, the spectrum does not identify the trees, while the lexicographically first adjacency matrix (varying over all enumerations of the vertices) identifies all graphs. Of course, no fast algorithm is known to compute the lexicographically first adjacency matrix of a graph.

WL[n] trivially identifies all graphs of size at most n. On the other hand, even WL[1] is sufficient to identify almost all graphs [5]. In fact, for almost all graphs, WL[1] stops after the second round with all vertices receiving distinct colors. The remaining graphs can be handled sufficiently fast to obtain an $O(n^2)$ expected time algorithm [6] (linear in the input size of a random graph). Even almost all regular graphs can be identified by WL[2], resulting in a linear expected time algorithm for identifying the regular graphs [15]. In general, it is difficult to find instances of graphs that are not easily identified. One source of such graphs are strongly regular graphs, which are the graphs where WL[2] stops immediately after assigning the initial colors without doing any refinements.

Algebraic graph invariants are among the most widely studied invariants. Examples of algebraic invariants are the spectrum (the multiset of eigenvalues of the adjacency matrix), the Laplacian spectrum, the multiset of angles of the standard basis vectors with the eigenspace for a given eigenvalue. The standard basis vectors are those with a component 1 in one vertex and components 0 in all other vertices. Note that multisets rather than n-tuples have to be used here, because in general no ordering of the vertices can be defined in an invariant way.

The standard algebraic graph invariants have a low distinguishing power, compared to strong combinatorial invariants. Already WL[2] determines the spectrum. The WL[2] color of a vertex determines the lengths of the projections of its standard basis vectors into the eigenspaces, and the WL[2] color of an edge determines the angle between the projections of its endpoints [11] (see also [12]). The spectrum of the k-th power of a graph G is more powerful than the spectrum of G itself, but not as powerful as WL[2k] [2].

1.3   The Graph Isomorphism Problem

The graph isomorphism problem, i.e., testing whether two graphs are isomorphic is not known to be in P, but not believed to be NP-complete, as this would have strange consequences like the collapse of the polynomial hierarchy. Babai [3,4] has recently shown the graph isomorphism problem to be in pseudo-polynomial time (i.e., in time $2^{(\log n)^{O(1)}}$). This result builds on the milestone work of Luks [16], who proved that graphs of bounded degree can be tested in polynomial time. These results rely heavily on group theoretical methods. Since the early eighties the author was involved in an oral debate, whether combinatorial methods could solve the bounded degree case too. In particular, it was open whether WL[k], with k being the degree of the graph, could solve the bounded degree graph isomorphism problem. This would be a very natural algorithm, running in polynomial time, or more precisely, in time $O(n^{k+1} \log n)$. It was even not clear whether a constant k would be sufficient for all graphs. Some support for this possibility was provided by the result that WL[5] always makes at least some progress [9,13,14] except for some known trivial cases.

These questions have been answered by the CFI result [8]. It shows that WL[k] requires $k = \Omega(n)$ in order to identify all graphs of size n. We now introduce this construction, since we use it for our proofs later. It starts with an arbitrary graph H called the global graph. For the $\Omega(n)$ result, H has to be an expander graph, but any low degree graph can be used for the construction. Here we only describe the interesting case of H being regular of degree 3. We show how to produce two similar graphs G and $\widetilde{G}$ from H. The graphs G and $\widetilde{G}$ are not isomorphic, but WL[2] uses edge colors with the same multiplicities.

1. Every vertex v of H is replaced by 4 vertices v_0, v_1, v_2, v_3 of G arranged counterclockwise in the corners of a square, but without the edges of the square. Note, that there are 3 partitions of $\{v_0, v_1, v_2, v_3\}$ into two subsets of vertices of size two:
   (a) Bottom $\{v_0, v_1\}$, Top $\{v_2, v_3\}$,
   (b) Left $\{v_0, v_3\}$, Right $\{v_1, v_2\}$,
   (c) Slash $\{v_0, v_2\}$, Backslash $\{v_1, v_3\}$,
2. Consider every edge $\{u, v\}$ of H to consist of 2 directed edges (u, v) and (v, u). For every vertex u of H label the 3 outgoing edges in an arbitrary way with the 3 partitions a, b, c, from above.
3. Now introduce 8 edges of G to replace every edge $\{u, v\}$ of H. For example, if (u, v) is labeled **a**, and (v, u) is labeled **b**, then the bottom u-nodes are connected to the left v-nodes, and the top u-nodes are connected to the right v-nodes. In other words, the edge $\{u_i, v_j\}$ is introduced, if either $i \in \{0, 1\}$ and $j \in \{0, 3\}$, or $i \in \{2, 3\}$ and $j \in \{1, 2\}$.

Finally, $\widetilde{G}$ is constructed from G by picking an arbitrary edge of H and flipping the corresponding connections in G. In the previous example, the bottom u-nodes would be connected to the right v-nodes, and the top u-nodes would be connected to the left v-nodes.

Fact 1. *The location of a flip is undefined. It can easily be moved from an edge incident to a vertex v of H to any of the other edges incident on v by doing a Bottom-Top ($\{v_0, v_1\} \leftrightarrow \{v_2, v_3\}$) exchange and/or a Left-Right ($\{v_0, v_3\} \leftrightarrow \{v_1, v_2\}$) exchange. In several steps, the flip can be moved to any edge in the same connected component.*

Fact 2. *Only the parity of the flips matter. If G is manipulated by introducing an even number of flips, we obtain a graph isomorphic to G. If G is manipulated by introducing an odd number of flips, we obtain a graph isomorphic to $\bar{G}$.*

1.4 Summary of Main Results

In this paper, we study the power of WL[2] in comparison with the graph invariants #k-cliques and #k-cycles for different values of k. In the next section we study the positive results. For cliques we only have the trivial result that 3-cliques are identified by WL[2]. For cycles, astonishingly WL[2] is much more powerful. Of course, 3-cycles and 4-cycles are identified, but surprisingly also 5-cycles and 6-cycles are identified. Section 3 contains the negative result for 4-cliques, and Sect. 4 is devoted to the negative result for 8-cycles.

2 Positive Results

Recall that we use the words *path* and *cycle* to refer to the set of edges of a simple path or cycle. Walks (not necessarily simple paths) and closed walks are not so interesting in our context. For example, for regular graphs, their numbers are determined by the graph size and the degree. It is not hard to see that WL[2] can easily count walks and closed walks of any length. More interesting is the task of counting (simple) paths and cycles.

We say that WL[k] counts the number of j-cycles or solves the problem #j-cycles, if it produces a multiset of colors (including their definitions) that is only produced for graphs that have exactly the same number of j-cycles. In the same way, we define WL[k] counting the number of j-cliques or solving the #j-clique problem. Similarly, we say that an edge $\{u, v\}$ knows a certain number, if the color of (u, v) and its definition determines that number.

Theorem 1. *WL[2] counts the number of triangles.*

Proof. Obviously, WL[2] trivially counts the number of triangles. After 1 round, every edge knows the number of triangles it is involved in. Therefore, the multiset of all colors of edges determines the total number of triangles. If c_j edges are involved in j triangles, then the total number of triangles is $\frac{1}{3} \sum_{j=1}^{n-2} c_j j$.   □

For #k-cliques, this trivial positive result is all we get. For #k-cycles we can do much better. But first we look at the problem of counting the number of paths of length 4 between a given pair of vertices. This could easily be used to show that WL[2] also counts the total number of paths of length 4. We don't treat counting the paths of length 5, as it can be done along the lines of the #6-cycles problem. Counting the paths of length $k < 5$ is easy.

We say that a coloring algorithm WL[k] computes a function or decides a property of graphs, if the multiset of stable colors of the k-tuples determines the function value or the property respectively. This means that whenever two graphs have the same multiset of colors, then they agree in the function value or the property respectively.

Here, having the same multiset of colors of k-tuples can be defined in two equivalent ways.

- The two graphs are colored simultaneously, i.e., when the names of the colors are reduced to small integers, a small integer abbreviates the same long name in both graphs. In this scenario, there is no need to retain the color definitions.
- Each graph is colored separately, but the definitions of the colors are included. The two graphs must have the same number of equally defined colors. Here, it is important to include an additional color definition, when the color partition is already stable. A key example consists of two strongly regular graphs with the same number of vertices and edges, but with different parameters λ (number of common neighbors of adjacent vertices) and μ (number of common neighbors of non-adjacent vertices). In each graph the edge coloring is stable from the beginning, as even the first refinement round has no effect. But the new colors in the two graphs have different definitions.

Similarly, we define what it means for a k-tuple *to know* a function value or a property. It means that WL[k] colors the k-tuple with a color (including its definition) that only shows up when the function has this value or the graph has this property respectively.

Lemma 1. *WL[2] can count the number of paths of length 4 between any pair of vertices.*

Proof. We show that every edge (u, v) knows the number of paths of length 4 from u to v. Let p_{uv}^ℓ be the number of paths of length ℓ from u to v. Every vertex pair (u, v) knows p_{uv}^1 from the start and p_{uv}^2 after 1 round. For all $\ell_1, \ell_1' \in \{0, 1\}$ and $\ell_2, \ell_2' \in \mathbb{N}$, after 2 rounds, (u, v) knows

$$n_{\ell_1 \ell_2 \ell_1' \ell_2'} := \#\{w \mid w \notin \{u, v\} \wedge p_{uw}^i = \ell_i \wedge p_{wv}^i = \ell_i' \text{ for } i \in \{1, 2\}\}.$$

Then (u, v) knows the number of paths of length 4 from u to v, which is

$$\sum_{\ell_1, \ell_2, \ell_1', \ell_2'} n_{\ell_1 \ell_2 \ell_1' \ell_2'} (\ell_2 - p_{uv}^1 \, \ell_1')(\ell_2' - p_{uv}^1 \, \ell_1) - \sum_{x \in V \setminus \{u, v\}} p_{ux}^1 (d(x) - 2) p_{xv}^1,$$

where $d(w)$ is the degree of vertex w. Of course, (u, w) knows $d(w)$ after 1 round.

For the correctness of this formula, notice when combining all paths of length 2 from u to w with all paths of length 2 from w to v, we also encounter two kinds of undesirable walks. First, we don't allow the paths of length 2 from u to w through v and from w to v through u. Finally, we subtract all walks u, x, w, x, v for any vertex x. □

Theorem 2. *WL[2] solves #k-cycles for $k \leq 6$.*

Proof. For $k = 4$ the result is easy. Every edge e can count the number of quadrangles of which it is a diagonal. In one round the edge $e = (u, v)$ knows the number of common neighbors $n(e) = p_{uv}^2$. Then the over all number of quadrangles is $\frac{1}{2}\sum_{e \in E} \binom{n(e)}{2}$.

For $k = 5$ the result follows from the lemma. Every pair (u, v) knows the number p_{uv}^4 of paths of length 4 from u to v, and it knows whether u and v are adjacent. Thus (u, v) knows in how many 5-cycles it is involved.

$k = 6$ is the interesting case. Any closed path $v_0, v_1, \ldots, v_5, v_0$ of length 6 can be broken down into a path of length 4 and a path of length 2 from v_0 to v_4.

In order to count the number of 6-cycles, we count for every fixed pair (v_0, v_4) with $v_0 \neq v_4$ the number of 6-cycles $H = v_0, v_1, v_2, v_3, v_4, v_5, v_0$ which are subgraphs of the given graph G for variable v_1, v_2, v_3, v_5. From now on the pair (v_0, v_4) is fixed. Let $\#H$ be the number of such subgraphs H. Let $\#H(*)$ be the number of subgraphs consisting of a path v_0, v_1, v_2, v_3, v_4 of length 4 and a path v_0, v_5, v_4 of length 2 where v_5 might possibly be identified with v_1, v_2, or v_3. Let $H[u_1 = w_1, \ldots, u_p = w_p]$ be any graph obtained from a graph of type H by identifying u_i with w_i for $1 \leq i \leq p$. Then

$$\#H = \#H(*) - \#H(v_1 = v_5) - \#H(v_2 = v_5) - \#H(v_3 = v_5).$$

After 1 round, the pairs (v_0, v_2), (v_2, v_4), and (v_0, v_4) all know the number of paths of length 2 between them. After 2 rounds, the pair (v_0, v_4), also knows the number $p_{v_0 v_4}^4$ of paths v_0, v_1, v_2, v_3, v_4 of length 4 from v_0 to v_4 by Lemma 1.

Thus after 2 rounds, the pair (v_0, v_4) knows $\#H(*) = p_{v_0 v_4}^2 p_{v_0 v_4}^4$, which is the number of pairs of paths v_0, v_1, v_2, v_3, v_4 and v_0, v_5, v_4. Not every such pair of paths can be combined to a 6-cycle. We have to subtract the number of cases, where v_5 is equal to one of the vertices v_1, v_2, or v_3.

Let's compute $\#H(v_2 = v_5)$. After 1 round (v_0, v_2) and (v_2, v_4) know the number of triangles in which they participate. These numbers are $p_{v_0 v_2}^1 p_{v_0 v_2}^2$ and $p_{v_2 v_4}^1 p_{v_2 v_4}^2$ respectively. After 2 rounds, (v_0, v_2) (for varying v_2) knows the multiset of these pairs of numbers. If v_0 is adjacent to v_4, then each triangle count is too high by 1, because (v_0, v_2) also counts the triangle v_0, v_2, v_4 and (v_2, v_4) also counts the triangle v_2, v_4, v_0. Both these triangles don't contribute to a collection of pairs of paths v_0, v_1, v_2, v_3, v_4 and v_0, v_5, v_4 intersecting only in the endpoints and in $v_2 = v_5$. Thus the number of bad subgraphs $H(v_2 = v_5)$ is

$$\#H(v_2 = v_5) = \sum_{v_2 \in V \setminus \{v_0, v_4\}} (p_{v_0 v_2}^1 p_{v_0 v_2}^2 - p_{v_0 v_4}^1)(p_{v_2 v_4}^1 p_{v_2 v_4}^2 - p_{v_0 v_4}^1).$$

This number is known to (v_0, v_4) after 2 rounds.

Now we compute $\#H(v_1 = v_5)$ of bad subgraphs with $v_1 = v_5$ for fixed v_0 and v_4, and varying v_1, v_2 and v_3. Let $\#H(v_1 = v_5, *)$ be the number of graphs obtained from a graph of type $H(v_1 = v_5)$ by possibly identifying v_2 or v_3 with v_0. Then we have

$$\#H(v_1 = v_5) = \#H(v_1 = v_5, *) - \#H(v_1 = v_5, v_0 = v_2) - \#H(v_1 = v_5, v_0 = v_3).$$

For $\ell \in \{1, 2, 3\}$ every pair (u, v) knows the number p_{uv}^ℓ of paths of length ℓ from u to v after $\ell - 1$ rounds. Thus in particular, after 2 rounds (v_0, v_1) knows $p_{v_0 v_1}^1$ and $p_{v_0 v_1}^2$, (v_1, v_4) knows $p_{v_1 v_4}^3$, and (v_0, v_4) knows $p_{v_0 v_4}^1$ and $p_{v_0 v_4}^2$. Thus

$$\#H(v_1 = v_5, *) = \sum_{v_1 \in V \setminus \{v_0, v_4\}} p_{v_0 v_1}^1 p_{v_1 v_4}^3 p_{v_1 v_4}^1.$$

It is easy to see that

$$\#H(v_1 = v_5, v_0 = v_2) = \binom{p_{v_0 v_4}^2}{2},$$

because v_0 and v_2 are opposite corners of a square, and

$$\#H(v_1 = v_5, v_0 = v_3) = p_{v_0 v_1}^1 p_{v_0 v_4}^1 p_{v_1 v_4}^1 (p_{v_0 v_1}^2 - 1).$$

After 2 rounds, the pair (v_0, v_4) knows $\#H(v_1 = v_5, v_0 = v_2)$ and $\#H(v_1 = v_5, v_0 = v_3)$.

The computation of $\#H(v_3 = v_5)$ is completely analog.

Now we determine the number $\#H(v_2 = v_5)$. Let $\#H(v_2 = v_5, *)$ be the number of graphs obtained from a graph of type $H(v_2 = v_5)$ by possibly identifying v_1 with v_3. Then we have

$$\#H(v_2 = v_5) = \#H(v_2 = v_5, *) - \#H(v_2 = v_5, v_1 = v_3).$$

After 1 round (v_0, v_2) and (v_2, v_4) know the number of triangles in which they participate. These numbers are $p_{v_0 v_2}^1 p_{v_0 v_2}^2$ and $p_{v_2 v_4}^1 p_{v_2 v_4}^2$ respectively. After 2 rounds, (v_0, v_2) (for varying v_2) knows the multiset of these pairs of numbers. If v_0 is adjacent to v_4, then each triangle count is too high by 1, because (v_0, v_2) also counts the triangle v_0, v_2, v_4 and (v_2, v_4) also counts the triangle v_2, v_4, v_0. Both these triangles don't contribute to a collection of pairs of paths v_0, v_1, v_2, v_3, v_4 and v_0, v_5, v_4 intersecting only in the endpoints and in $v_2 = v_5$. Thus the number of bad subgraphs for $v_2 = v_5$ is

$$\#H(v_2 = v_5, *) = \sum_{v_2 \in V \setminus \{v_0, v_4\}} (p_{v_0 v_2}^1 p_{v_0 v_2}^2 - p_{v_0 v_4}^1)(p_{v_2 v_4}^1 p_{v_2 v_4}^2 - p_{v_0 v_4}^1).$$

This number is known to (v_0, v_4) after 2 rounds.

Here, we hit a complication when we want to compute $\#H(v_2 = v_5, v_1 = v_3)$. The subgraph $H(v_2 = v_5, v_1 = v_3)$ is a square with a diagonal form v_1 to v_2. The other corners are v_0 and v_4. The pair (v_0, v_4) does not know $\#H(v_2 = v_5, v_1 = v_3)$. Therefore, (v_0, v_4) might not know the number of 6-cycles in which the distance from v_0 to v_4 on the cycle is 2. Luckily, we don't have to know this number, but only their sum over all (v_0, v_4). Thus instead of counting the number of $H(v_2 = v_5, v_1 = v_3)$ for fixed v_0 and v_4, we count this number for fixed v_1 and v_2. This number $n_{v_1 v_2}$ is 0, if v_1 and v_2 are not adjacent and $\binom{p}{2}$ for $p = p_{v_1 v_2}^2$ otherwise, and the pair (v_1, v_2) knows this number. Thus instead of computing the sum of $\#H(v_2 = v_5, v_1 = v_3)$ over all pairs (v_0, v_4), we compute the sum of $n_{v_1 v_2}$ over all (v_1, v_2) obtaining the same result.   □

3   WL[2] Does Not Count 4-Cliques

Proposition 1. *For $k \geq 2$ every k-tuple (and thus also every vertex) knows the multiset of colors of all k-tuples of vertices of the same graph.*

Note that the result does not hold for $k = 1$.
The definition of "knowing" immediately implies the following.

Corollary 1. *For $k \geq 2$ two graphs agree in the color of one k-tuple, iff they agree in the multiset of colors of all k-tuples.*

We consider the CFI construction with the global graph being the simplest regular degree 3 graph, the 4-clique K_4. Assume, the vertices $v_0, \ldots, v_3$ are arranged in consecutive corners of a square. For the vertices of G, we use double indices. The vertex v_{ij} is the jth vertex in the ith corner ($i, j \in \{0, 1, 2, 3\}$). Assume, we have assigned partition **a** to $(v_i, v_{i+1 \bmod 4})$, and partition **c** to $(v_{i+1 \bmod 4}, v_i)$. Thus partition **b** is assigned to the 4 diagonal directions.

For every global graph H, WL[2] produces edge colors with the same multiplicities in the nonisomophic graphs G and $\widetilde{G}$ produced by the CFI construction [8]. In fact this is the purpose of this construction. In our case with $H = K_4$, this is immediately clear, as the two graphs are strongly regular. Thus we have just 3 edge colors, one for original edges, one for non-adjacent disjoint pairs, and one for self-loops.

We say that two invariants are incomparable, if neither of them implies the other.

Theorem 3. *G and $\widetilde{G}$ differ in their number of occurrences of the subgraph K_4. WL[2] is incomparable with the invariant #4-cliques.*

Proof. Consider $\widetilde{G}$ having the flip along the $\{v_1, v_3\}$ diagonal edge. In both, G and $\widetilde{G}$ start with v_{00}. It is adjacent to v_{10} and v_{13}. The vertices v_{00} and v_{10} are adjacent to v_{20} in both G and $\widetilde{G}$. Likewise, the vertices v_{00} and v_{20} are adjacent to v_{30} in both G and $\widetilde{G}$. Finally, v_{30} is adjacent to v_{20} forming a clique in G, but v_{30} is not adjacent to v_{20} forming no clique in $\widetilde{G}$. Likewise, for every start in one of the vertices of v_0, there are two neighbors in v_1. Then there are twice unique common neighbors of two previously chosen vertices both in G and $\widetilde{G}$. Finally in G the chosen vertices in v_3 and v_1 are adjacent, but not in $\widetilde{G}$.

Considering some automorphisms, this can be verified by checking 2 cases instead of 8. The result is G has 8 K_4, while $\widetilde{G}$ has none, even though the edge colors have the same multiplicities in G and $\widetilde{G}$.

That counting K_4 is sometimes weaker than WL[2] is trivial, e.g., take a path of length 2 and a 3-cycle. □

4   Difficult Cycles

We show that WL[2] does not identify cycles of length 8. We give a clear argument about where to look for counter examples. But the actual example is created by computer.

Table 1. The number of cycles of length n in G and $\widetilde{G}$

n	Not twisted	Twisted	n	Not twisted	Twisted
1	0	0	9	34368	33920
2	48	48	10	91296	92256
3	32	32	11	211968	216192
4	60	60	12	417264	423216
5	288	288	13	670464	674304
6	1248	1248	14	822528	824448
7	4032	4032	15	678912	680960
8	11952	11688	16	284112	281232

It is difficult to find a counter example, because WL[2] is very powerful and identifies almost all graphs. Thus we use again our example from the previous section. The non-isomorphic graphs G and $\widetilde{G}$ are created with the CFI method from a tetrahedron. As the single flip in $\widetilde{G}$ can be pushed into any edge it is clear that the 2 graphs have the same number of occurrences of any subgraph that does not involve all the edges of the global graph. The global graph is a K_4. It has 6 edges, but the shortest closed walk through all edges has length 8. Thus the shortest cycles that might have a different count in G and $\widetilde{G}$ are necessarily of length at least 8. Indeed we succeed. For all even lengths k between the minimum 8 and the maximum 16 (Hamiltonian cycles), the counts in G and $\widetilde{G}$ are different.

As our graphs are small, we can use a pretty brute force algorithm to count the cycles starting at a fixed vertex. For each of the two graphs, the count C_k^v (the number of k cycles through any given start vertex v) does not depend on the chosen start vertex, because both graphs are vertex transitive. #k-cycles, the total number of cycles of length k is n times C_k^v divided by the length k of the cycles.

Theorem 4. *WL[2] does not identify the number of 8-cycles.*

Proof. By Computer.   ☐

Table 1 is the output of the Cycle Count program. It shows that for lengths up to 7, the number of cycles is equal in the two graphs. Starting at length 8, the number of cycles differ, i.e., WL[2], cannot count the number of 8-cycles. We knew that for this pair of graphs, the numbers have to be the same for lengths up to 6, because the graphs G and $\widetilde{G}$ are constructed such that WL[2] does not detect any difference between them. It is open whether WL[2] can always count the number of 7-cycles.

Interestingly enough, there is other evidence that the complexity changes after 7. For $k \leq 7$, Alon et al. [1] can count k-cycles in time $O(n^\omega)$, where $\omega < 2.373$ [19] is the exponent of matrix multiplication, while Flum and Grohe [10] show that with k as a parameter the problem of counting k-cycles is #W-complete.

References

1. Alon, N., Yuster, R., Zwick, U.: Finding and counting given length cycles. Algorithmica **17**(3), 209–223 (1997)
2. Alzaga, A., Iglesias, R., Pignol, R.: Spectra of symmetric powers of graphs and the Weisfeiler-Lehman refinements. J. Comb. Theory Ser. B **100**(6), 671–682 (2010)
3. Babai, L.: Graph isomorphism in quasipolynomial time. CoRR, abs/1512.03547 (2015)
4. Babai, L.: Graph isomorphism in quasipolynomial time (extended abstract). In: Proceedings of the 48th Annual ACM SIGACT Symposium on Theory of Computing (STOC), pp. 684–697. ACM (2016)
5. Babai, L., Erdős, P., Selkow, S.M.: Random graph isomorphism. SIAM J. Comput. **9**(3), 628–635 (1980)
6. Babai, L., Kučera, L.: Graph canonization in linear average time. In: Proceedings of the 20th Annual Symposium on Foundations of Computer Science (FOCS), pp. 39–46. IEEE Computer Society Press (1979)
7. Babai, L., Mathon, R.: Talk at the South-East Conference on Combinatorics and Graph Theory (1980)
8. Cai, J.-Y., Fürer, M., Immerman, N.: An optimal lower bound on the number of variables for graph identification. Combinatorica **12**(4), 389–410 (1992)
9. Cameron, P.J.: 6-transitive graphs. J. Comb. Theory Ser. B **28**(2), 168–179 (1980)
10. Flum, J., Grohe, M.: The parameterized complexity of counting problems. SIAM J. Comput. **33**(4), 892–922 (2004)
11. Fürer, M.: Graph isomorphism testing without numerics for graphs of bounded eigenvalue multiplicity. In: Proceedings of the 6th Annual ACM-SIAM Symposium on Discrete Algorithms (SODA), pp. 624–631 (1995)
12. Fürer, M.: On the power of combinatorial and spectral invariants. Linear Algebra Appl. **432**(9), 2373–2380 (2010)
13. Gol'fand, Y.Y., Klin, M.H.: On k-regular graphs. In: Algorithmic Research in Combinatorics, pp. 76–85. Nauka Publ., Moscow (1978)
14. Klin, M.C., Pöschel, R., Rosenbaum, K.: Angewandte Algebra. Vieweg & Sohn Publ., Braunschweig (1988)
15. Kučera, L.: Canonical labeling of regular graphs in linear average time. In: Proceedings of the 28th Annual Symposium on Foundations of Computer Science (FOCS), pp. 271–279. IEEE Computer Society Press (1987)
16. Luks, E.M.: Isomorphism of graphs of bounded valence can be tested in polynomial time. J. Comput. Syst. Sci. **25**, 42–65 (1982)
17. Read, R.C., Corneil, D.G.: The graph isomorphism disease. J. Graph Theory **1**(4), 339–363 (1977)
18. Weisfeiler, B. (ed.): On Construction and Identification of Graphs. Lecture Notes in Mathematics. Springer, Berlin (1976). With contributions by A. Lehman, G. M. Adelson-Velsky, V. Arlazarov, I. Faragev, A. Uskov, I. Zuev, M. Rosenfeld and B. Weisfeiler
19. Williams, V.V.: Multiplying matrices faster than Coppersmith-Winograd. In: Proceedings of the 44th Symposium on Theory of Computing Conference (STOC), pp. 887–898. ACM (2012)

Cost-Sharing in Generalised Selfish Routing

Martin Gairing[1], Konstantinos Kollias[2], and Grammateia Kotsialou[1(✉)]

[1] University of Liverpool, Liverpool, UK
{gairing,gkotsia}@liverpool.ac.uk
[2] Stanford University, Stanford, USA
kkollias@stanford.edu

Abstract. We study a generalisation of atomic selfish routing games where each player may control multiple flows which she routes seeking to minimise their aggregate cost. Such games emerge in various settings, such as traffic routing in road networks by competing ride-sharing applications or packet routing in communication networks by competing service providers who seek to optimise the quality of service of their customers. We study the existence of pure Nash equilibria in the induced games and we exhibit a separation from the single-commodity per player model by proving that the Shapley value is the only cost-sharing method that guarantees it. We also prove that the price of anarchy and price of stability is no larger than in the single-commodity model for general cost-sharing methods and general classes of convex cost functions. We close by giving results on the existence of pure Nash equilibria of a splittable variant of our model.

1 Introduction

Congestion games are a well-studied abstraction of a large collection of applications which includes several network routing games. Rosenthal proposed the model [26,27] and in the past 15 years, starting with [31], there has been a large body of work in the area (e.g., [2,4,6–8,10,11,13,17,18,28]). Network applications have been one of the main motivations behind the success of the model and *selfish routing* is the paradigmatic example in the study of existence and inefficiency of equilibrium solutions. A selfish routing game is played on a directed graph $G = (V, E)$. Each player i in the game is characterised by a start vertex s_i, a destination vertex t_i, and a flow size w_i. Player i must select an s_i–t_i path that minimises the sum of the edge costs along the path. The edge costs are increasing functions of the total flow on them and there is a predefined cost-sharing method that dictates how edge costs are distributed among each edge's users. The main assumption is that players reach a Nash equilibrium and the system performance is typically measured by comparing the worst or best Nash equilibrium to the optimal solution in terms of total cost. These metrics are termed the *price of anarchy* (POA) and *price of stability* (POS), respectively [3,23]. Existence of a pure Nash equilibrium (PNE) and POA/POS performance properties are very well understood for general cost-sharing methods in the selfish routing model [12,14,21,22,30].

D. Fotakis et al. (Eds.): CIAC 2017, LNCS 10236, pp. 272–284, 2017.
DOI: 10.1007/978-3-319-57586-5_23

In this paper, we study a generalisation of the selfish routing game, which we term *selfish routing with multi-commodity players*. In this generalisation, each player may control more than one flow in the network. Similar settings have been studied before in the context of scheduling games [1], in the context of integer splittable routing games [27,33], and for a special case of our model (where each commodity has the same flow size) in [10]. More specifically, player i is described by a set of commodities Q_i. Each commodity q has a starting vertex s_q, a destination vertex t_q and a flow size w_q. Each player i must pick how to route the flows in Q_i, each on a single path. Applications of our model include routing in road networks where ride-sharing platforms operate and routing in communication networks where connections are operated by service providers. Consider the example of ride-sharing platforms. The game is played on the road network and there is a continuous flow of rides using either platform between each pair of nodes in the graph. The route that each car follows is dictated centrally by the platform that seeks to optimise the aggregate travel time of its flows. In the packet routing application, network connections are routed by competing service providers. Each service provider wishes to optimise the quality of service of their clients and hence routes connections seeking to minimise their aggregate costs.

As a concrete example, consider a network with two nodes s, t, and two parallel edges e_1, e_2, from s to t. The joint cost of each edge is given as $C(x) = x^2$, with x the total flow on the edge. The game has two players. Player 1 wishes to route a flow of size 1 from s to t, while player 2 wishes to route two flows, each of size 1, from s to t. Suppose the cost-sharing method dictates that each commodity traveling through an edge pays an equal share of the joint cost. Player 2 has three options: route both commodities on the same edge that player 1 is using, route both commodities on the other edge, or route the two commodities on different edges. The corresponding costs for player 2 would be 6, 4, and 3, which establishes the latter option as the best response.

1.1 Our Results

In this work, we search for cost-sharing methods that guarantee the existence of pure Nash equilibria in multi-commodity selfish routing games. We also focus on the inefficiency of equilibria and we conduct a comprehensive study of the POA/POS, i.e., the ratio of the total cost in the worst/best Nash equilibrium over the optimal total cost. Our results hold for general cost functions and cost-sharing methods and they also extend to general congestion games.

Regarding the existence of pure Nash equilibria, we show that applying the *Shapley value* per edge, with the weight of a player on an edge being the sum of the commodity sizes she places on the edge, results in a potential game and, hence, guarantees the existence of a pure Nash equilibrium. On the contrary, we show that weighted Shapley values may induce games such that no pure Nash equilibrium exists, which exhibits a separation from the single-commodity case, where each player controls only one commodity. Given that the class of weighted Shapley values are the unique cost-sharing methods that guarantee pure Nash equilibria in the single-commodity player model [15], our results suggest that the Shapley value is essentially the unique anonymous cost-sharing method that guarantees pure Nash equilibria in the multi-commodity player model.

With respect to the inefficiency of equilibria, we prove upper bounds on the POA that match the ones from the single-commodity per player model. Our bounds work for general (convex) cost functions and for general cost sharing methods satisfying the following natural assumptions [12], which we briefly discuss afterwards and explain in more detail in Sect. 1.3:

1. Every cost function in the game is continuous, increasing and convex.
2. Cost-sharing is consistent when player sets generate costs in the same way.
3. For convex resource cost functions, the cost share of a player on a resource is a convex function of her flow on the resource.

Assumption 1 is standard in congestion-type settings. For example, linear cost functions have obvious applications in many network models, as do queueing delay functions, while higher degree polynomials (such as quartic) have been proposed as realistic models of road traffic [32]. With assumption 2, the cost sharing method only charges player s according to how they contribute in the total cost and there is no other way of discrimination between them. Assumption 3 asks that the curvature of the cost shares is consistent, i.e., given assumption 1, that the share of a player on a resource is a convex function of her weight (otherwise, we would get that the cost share of the player increases in a slower than convex way but the total cost of the constant weight of players increases in a convex way, which we view as unfair).

The POS is an interesting concept and it is very well motivated in cases where the players can be started in an initial configuration or where a trusted mediator can suggest a solution to the players. This suggests that the POS is especially interesting in cases where a pure Nash equilibrium exists. Therefore, on the POS side, we focus on the Shapley value, the only cost-sharing method that guarantees existence of a pure Nash equilibrium in our setting. We prove that the POS is equal to the POS of the single-commodity case for general classes of cost functions.

Finally, we study an extension to the *splittable* model, where players may split their commodities across different paths. In particular, we study the existence of pure Nash equilibria in that setting and mention interesting open problems.

1.2   Related Work

Previous works in [10,20,27,33] study settings that share similarities to multi-commodity routing games. In [27], Rosenthal studies weighted routing games where each player may split her integer flow size among different subflows of integer size. Focusing on the proportional cost-sharing method (that charges each player a cost proportional to her flow on an edge), he proves that there exist such games with no PNE. In [33], the authors identify special cases where PNE exist in Rosenthal's model. Our approach differs from the work in [27,33].

We study general multi-commodity players and not only players who control unit flows with the same start vertex. In [20], it is shown that there exist games where merging atomic players into a coalition (similarly into a multi-commodity player) may degrade the quality of the induced PNE when proportional sharing is used. In a small contrast, we focus on worst-case metrics and show that the POA and POS of multi-commodity player games is no worse than in the single-commodity case, for general cost-sharing methods. Finally, in [10], the authors focus on coalitions of atomic players in routing games (equivalent to multi-commodity players) and mostly on the objective of minimizing the maximum cost. For the sum of costs objective, which we consider in this paper, they prove that the game always admits a pure Nash equilibrium under proportional cost-sharing and quadratic edge cost functions. We provide more comprehensive results with respect to the existence of pure Nash equilibria for general methods and general classes of cost functions.

On the cost-sharing side, the authors in [15] characterise the class of (generalised) weighted Shapley values as the only methods to guarantee existence of a PNE when each player controls one commodity. We exhibit a separation from this result by showing that weighted Shapley values do not guarantee pure Nash equilibria existence in the multi-commodity extension. With respect to the POA and POS of cost-sharing in routing games, [12] provides general tight bounds, which, in this work, we generalise to the multi-commodity player model.

1.3 Preliminaries

In this section, we present the notation and preliminaries for our model in terms of a general *congestion game with multi-commodity players*. In such a game, there is a set Q of k commodities which are partitioned into $n \leq k$ non-empty and disjoint subsets $Q_1, Q_2, \ldots Q_n$. Each set of commodities Q_i, for $i = 1, 2, \ldots, n$, is controlled by an independent player. Denote $N = \{1, 2, \ldots, n\}$ the set of players. The players in N share access to a set of resources E. Each resource $e \in E$ has a flow-dependent cost function $C_e : \mathbb{R}_{\geq 0} \to \mathbb{R}_{\geq 0}$. As stated in assumption 1 (Sect. 1.1), we assume the cost functions of the game are drawn from a given set $\mathcal{C}$ of allowable cost functions, such that every $C \in \mathcal{C}$ must be continuous, increasing and convex. We also make the mild technical assumption that the set $\mathcal{C}$ is closed under (*i*) scaling and (*ii*) dilation, meaning that if $C(x) \in \mathcal{C}$, then (*i*) $C(a \cdot x) \in \mathcal{C}$ and also (ii) $a \cdot C(x) \in \mathcal{C}$, for every positive a.

Strategies. Each commodity $q \in Q$ has a set of possible strategies $\mathcal{P}^q \subseteq 2^E$. Associated with each commodity q is a weight w_q, which has to be allocated to a strategy in $\mathcal{P}^q$. For a player i, a strategy $P_i = (P_q)_{q \in Q_i}$ defines the strategy for each commodity q player i controls. An outcome $P = (P_1, P_2, \ldots, P_n)$ is a tuple of strategies of the n players.

Load. For an outcome P, the flow $f_e^i(P)$ of a player i on resource e equals the sum of the weights of all her commodities using e, i.e., $f_e^i(P) = \sum_{q \in Q_i, e \in P_q} w_q$. The total flow on a resource e is given as $f_e(P) = \sum_{i \in N} f_e^i(P)$. We use $x_e(P)$ for the set of players who assign positive flow on resource e on an outcome P.

Cost Shares. The cost sharing method of the game determines how the flow-dependent joint cost of a resource $C_e(f_e(P))$ is divided among its users. Given an outcome P, we write $\chi_{ie}(P)$ for the cost of player i on resource e, such that $\sum_{i \in N} \chi_{ie}(P) = C_e(f_e(P))$. The cost of a player i, $X_i(P)$, is the sum of her costs on each resource, $X_i(P) = \sum_{e \in E} \chi_{ie}(P)$. For any $T \subseteq N$, let $f_e^T(P)$ be the vector of the flows that each player in T assigns to e. Then the cost share of player i can also be defined as a function of the player's identity, the resource's cost function and the vector of flows assigned to e, i.e., $\chi_{ie}(P) = \xi(i, f_e^N(P), C_e)$.

In this paragraph we explain in more detail assumption 2 and 3 from Sect. 1.1, which are needed for our general POA results in Sect. 3. Assumption 2 states that the cost-sharing method only charges players based on how they contribute to the joint cost. More specifically, assume we scale the joint cost on a resource by a positive factor β, i.e., $C_e'(f_e(P)) = \beta \cdot C_e(f_e(P))$. Given that the same players use this resource, the new cost shares of the players would be a scaled by factor β version of their initial cost shares, i.e., $\xi'(i, f_e^N(P), C_e) = \beta \cdot \xi(i, f_e^N(P), C_e)$. This is given by scaling and replication arguments. Last, we make the fairness-related assumption 3 which states that the cost share of a player on a resource is a convex function of her flow.

We now define a specific class of cost-sharing methods, which is important in our analysis.

Weighted Shapley Values. The *weighted Shapley value* defines how the cost $C_e(\cdot)$ of resource e is distributed among the players using it. Given an ordering π of N, let $F_e^{<i,\pi}(P)$ be the sum of flows of the players preceding i in π. Then the marginal cost increase caused by player i is $C_e(F_e^{<i,\pi}(P) + f_e^i(P)) - C_e(F_e^{<i,\pi}(P))$. For a given distribution Π over orderings, the cost share of player i on resource e is $E_{\pi \sim \Pi}[C_e(F_e^{<i,\pi}(P) + f_e^i(P)) - C_e(F_e^{<i,\pi}(P))]$. For the weighted Shapley value, the distribution over orderings is given by a sampling parameter $\lambda_e^i(P)$ for each player i. The last player in the ordering is picked proportional to the sampling parameters $\lambda_e^i(P)$. Then this process is repeated iteratively for the remaining players.

As in [12], we study a parameterised class of weighted Shapley values defined by a parameter γ. For this class, $\lambda_e^i(P) = f_e^i(P)^\gamma$ for all players i and resources e. For $\gamma = 0$, this reduces to the (standard) *Shapley value*, where we have a uniform distribution over orderings.

Pure Nash Equilibrium. We now proceed with the definition of our solution concept. The *pure Nash equilibrium* (PNE) condition on an outcome P states that for every player i it must be the case that

$$X_i(P) \leq X_i(P_i', P_{-i}), \text{ for any other strategy } P_i'. \tag{1}$$

Social Cost. The social cost in the game is given by the sum of the player costs, i.e.,

$$SC(P) = \sum_{i \in N} X_i(P) = \sum_{i \in N} \sum_{e \in E} \xi(i, f_e^N(P), C_e) = \sum_{e \in E} C_e(f_e(P)). \tag{2}$$

Price of Anarchy and Price of Stability. Let $\mathcal{Z}$ be the set of outcomes and $\mathcal{Z}^N$ be the set of pure Nash equilibria outcomes of the game. Then the *price of anarchy* (POA) and the *price of stability* (POS) are defined as follows,

$$POA = \frac{\max_{P \in \mathcal{Z}^N} SC(P)}{\min_{P \in \mathcal{Z}} SC(P)} \quad \text{and} \quad POS = \frac{\min_{P \in \mathcal{Z}^N} SC(P)}{\min_{P \in \mathcal{Z}} SC(P)}. \tag{3}$$

The POA and POS for a class of games are defined as the largest such ratios among all games in the class.

2   Existence of Pure Nash Equilibria

Our first result proves that applying the (standard) Shapley value (with respect to the player flows $f_e^i(P)$) on each resource, induces a potential game. Recall that, for the Shapley value cost-sharing, we have a uniform distribution over orderings, i.e., we use the definition of weighted Shapley values in Sect. 1.3 with every sampling parameter equal to 1.

Theorem 1. *Congestion games with multi-commodity players under Shapley cost sharing are exact potential games.*

Proof. Consider any ordering π of the players in N and let $f_e^{\leq i,\pi}(P)$ denote the vector that we get after truncating $f_e^N(P)$ by removing all entries for players that succeed i in π. We prove that the following is a potential function of the game,

$$\Phi(P) = \sum_{e \in E} \sum_{i \in N} \xi(i, f_e^{\leq i,\pi}(P), C_e). \tag{4}$$

Hart and Mas-Colell [19] proved that (4) is independent of the ordering π in which players are considered. Let $P' = (P_i', P_{-i})$. It suffices to show that $\Phi(P) - \Phi(P')$ equals the change in the cost of player i. Focus on a single resource e and let π be one of the orderings that places the flow of player i, $f_e^i(P)$, in the last position. Then, the potential on resource e loses a term equal to

$$\xi(i, f_e^{\leq i,\pi}(P), C_e) = \xi(i, f_e^N(P), C_e)$$

and gains a term equal to

$$\xi(i, f_e^{\leq i,\pi}(P_i', P_{-i}), C_e) = \xi(i, f_e^N(P_i', P_{-i}), C_e),$$

which is precisely the change in the cost of player i on e. Summing over all edges gives the desirable $\Phi(P) - \Phi(P') = X_i(P) - X_i(P')$, which completes the proof. $\square$

One might expect that, similarly to standard congestion games, the same potential function argument would apply under weighted Shapley values as well. However, this is not the case.

Theorem 2. *There is a congestion game with multi-commodity players admitting no PNE for any weighted Shapley value defined by sampling weights of the form $f_e^i(P)^\gamma$ with $\gamma > 0$ or $\gamma < 0$.*

Proof. We prove this theorem by showing two examples admitting no PNE, for $\gamma > 0$ and $\gamma < 0$. Due to page limitations, we restrict to the description of the instances. For the $\gamma > 0$ case: Consider two players, 1 and 2, who compete for two parallel (meaning each commodity must pick exactly one of them) resources e, e' with identical cost functions $C_e(x) = C_{e'}(x) = x^{1+\delta}$ with $\delta > 0$ and $\frac{\gamma}{\delta}$ a large positive number (note that for $\delta = 0$, we have linear cost functions where in this case we have an equilibrium. As soon as we deviate from linearity, we use convexity to construct an example with no equilibrium). Player 1 controls a unit commodity $p \in Q_1$. Player 2 controls two commodities $q, q' \in Q_2$, with $w_{q'} = 1$ and $w_q = k$, for k a very large number. Recall, that the sampling weight of a player i on a resource e is given by $\lambda_e^i = (f_e^i)^\gamma$. This means that smaller weights are favoured when constructing the weighted Shapley ordering. In particular, for $k \to \infty$, if commodities p, q share the same resource, then the probability that q goes last in the Shapley ordering becomes 1 and the cost of commodity q would be $(k+1)^d - 1$.

We switch to the $\gamma < 0$ case: Consider players $i = 1, 2, \ldots, k$, who compete for two parallel resources e_1, e_2 with identical cost functions $C_{e_1}(x) = C_{e_2}(x) = x^3$. Player k controls two commodities $p, q \in Q_k$ with weights $w_p = k$ and $w_q = 1$. Each player $i < k$ controls only one commodity $r_i \in Q_i$ with $w_{r_i} = 1$. The sampling weight of a player i on a resource e is given by $\lambda_e^i = (f_e^i)^\gamma$, for $\gamma < 0$. □

2.1 Alternative Cost-Sharing Based on Commodity Weights

One might consider a different way of generalizing weighted Shapley values to multi-commodity congestion games: Apply a weighted Shapley value on the commodity weights by charging a player the sum of the weighted Shapley values of the commodities controlled by her. These cost-sharing methods coincide when all commodities have unit weights, which is equivalent to proportional cost-sharing, i.e., every player pays a cost-share that is proportional to her flow on any given resource. Below we use one such instance with unit commodities to prove that applying a weighted Shapley value method on commodity weights does not guarantee pure Nash equilibrium existence.

Our instance is based on an example in [10], where Fotakis et al. prove that network unweighted congestion games with linear resource cost functions and equal cardinality coalitions do not have the *finite improvement property*, therefore they admit no potential function. Their example translates to a restricted setting of our model where each player controls an equal number of unit commodities. We strengthen their result by proving non-existence of pure Nash equilibria for congestion games with multi-commodity players and cubic resource cost functions (we construct even a network congestion game with no pure Nash equilibrium).

A similar example has already been given by Rosenthal [27]. However, Rosenthal's example uses concave cost functions, which we disallow in our setting. In contrast, our proof uses only convex functions.

Theorem 3. *There is a congestion game with multi-commodity players and cubic cost functions admitting no pure Nash equilibrium under weighted Shapley sharing applied on commodity weights.*

3   The POA and POS of Multi-commodity Games

In this section we prove that the POA and POS of multi-commodity congestion games are no larger than those of their single-commodity counterparts, for any cost-sharing method and class of cost functions satisfying the natural assumptions in Sect. 1.3. Due to space limits, the POA proof is omitted. It follows along the lines of the proof for the single-commodity per player model [12].

Theorem 4. *The POA of multi-commodity congestion games under a cost-sharing method ξ and with costs drawn from a given class of increasing and convex cost functions C, such that ξ, C satisfy assumptions 1, 2, and 3, is equal to the POA of single-commodity congestion games induced by ξ and C.*

Theorem 5. *The POS of Shapley value based multi-commodity congestion games with costs drawn from a given class of increasing and convex cost functions C, is equal to the corresponding POS of the single-commodity case.*

Proof. We begin with the potential function of the game (4) and we prove the following lemma which we use to prove the upper bound on POS. Briefly, the lemma states the following. For any instance with N players and any strategy profile, we can construct a new instance with $N + 1$ players by splitting one player in half into two new players. Then this can only reduce the potential value of the game. More precisely, we do this by splitting in half the flow of each commodity controlled by a player i on a resource creating two new commodities, which we assign to the new players i' and i''.

Lemma 1. *Consider an outcome P of the game and assume that on a resource e, we substitute the total flow of a player i with the flows of two other players i', i'' such that $f_e^{i'}(\hat{P}) = f_e^{i''}(\hat{P}) = \frac{f_e^i(P)}{2}$. Then we claim that*

$$\Phi_e(P) \geq \Phi'_e(\hat{P}),$$

where $\Phi'_e(\hat{P})$ is the potential value of resource e after the substitution.

Proof. First, rename the flows such that the substituted one $f_e^i(P)$ to have the highest index. Assign indices i' and i'' to the new ones, with $i < i' < i''$ in ordering π. Then, for any resource e, the new potential value equals to

$$\Phi'_e(\hat{P}) = \sum_{j=1}^{i-1} \xi(j, f_e^{\leq j, \pi}(P), C_e) \; + \; \xi(i', (f_e^{<i,\pi}(P), f_e^{i'}(\hat{P})), C_e) \; +$$

$$+ \; \xi(i'', (f_e^{<i,\pi}(P), f_e^{i'}(\hat{P}), f_e^{i''}(\hat{P})), C_e).$$

Note that the contribution to the potential value of the players before player i is the same as before the substitution. Therefore it is enough to show that

$$\xi(i, f_e^{<i,\pi}(P), C_e) \geq \xi(i', (f_e^{<i,\pi}(P), f_e^{i'}(\hat{P})), C_e) +$$
$$+ \xi(i'', (f_e^{<i,\pi}(P), f_e^{i'}(\hat{P}), f_e^{i''}(\hat{P})), C_e). \qquad (5)$$

To simplify, in what follows call

$$\xi = \xi(i, f_e^N(P), C_e),$$
$$\xi' = \xi(i', (f_e^{<i,\pi}(P), f_e^{i'}(\hat{P})), C_e),$$
$$\xi'' = \xi(i'', (f_e^{<i,\pi}(P), f_e^{i'}(\hat{P}), f_e^{i''}(\hat{P})), C_e).$$

Define as $x_e^i(\pi)$ the set of players preceding player i in π. Then, for every ordering π and permutation τ^i of set $x_e^i(\pi) \cup \{i\}$, define as $F_e^{<i,\pi,\tau^i}(P)$ the sum of players' flows who precede i in both π and τ^i. Let now $|x_e(P)| = r$. By definition of Shapley values, we get

$$\xi = \frac{1}{r!} \sum_{\tau^i} \left(C_e \left(F_e^{<i,\pi,\tau^i}(P) + f_e^i(P) \right) - C_e \left(F_e^{<i,\pi,\tau^i}(P) \right) \right), \qquad (6)$$

$$\xi' = \frac{1}{r!} \sum_{\tau^i} \left(C_e \left(F_e^{<i,\pi,\tau^i}(P) + f_e^{i'}(\hat{P}) \right) - C_e \left(F_e^{<i,\pi,\tau^i}(P) \right) \right). \qquad (7)$$

For ξ'', since the position of $f_e^{i'}(\hat{P})$ in the ordering is unspecified, we give an upper bound for this value as follows. For any permutation τ, let $A(\tau)$ be the marginal cost increase caused by $f_e^{i''}(\hat{P})$ when she precedes $f_e^{i'}(\hat{P})$ in π, and $B(\tau)$ when she succeeds. That is

$$A(\tau) = C_e \left(F_e^{<i,\pi,\tau^i}(P) + f_e^{i''}(\hat{P}) \right) - C_e \left(F_e^{<i,\pi,\tau^i}(P) \right),$$
$$B(\tau) = C_e \left(F_e^{<i,\pi,\tau^i}(P) + f_e^{i'}(\hat{P}) + f_e^{i''}(\hat{P}) \right) - C_e \left(F_e^{<i,\pi,\tau^i}(P) + f_e^{i'}(\hat{P}) \right). \qquad (8)$$

Let now p equal the probability of $f_e^{i'}(\hat{P})$ preceding $f_e^{i''}(\hat{P})$. Then, the definition of the Shapley value gives

$$\xi'' = (1-p) \cdot \frac{1}{r!} \cdot \sum_{\tau^i} A(\tau) + p \cdot \frac{1}{r!} \cdot \sum_{\tau^i} B(\tau). \qquad (9)$$

Due to convexity, $A(\tau) \leq B(\tau)$. Therefore, by substituting $A(\tau)$ with $B(\tau)$ in definition (9), we get the following upper bound for ξ'',

$$\xi'' \leq \frac{1}{r!} \sum_{\tau^i} B(\tau). \qquad (10)$$

Towards proving inequality (5), we have

$$\xi' + \xi'' \overset{(7),(10)}{\le} \frac{1}{r!} \sum_{\tau^i} C_e \left(F_e^{<i,\pi,\tau^i}(P) + f_e^{i''}(\hat{P}) \right) - C_e \left(F_e^{<i,\pi,\tau^i}(P) \right)$$

$$+ C_e \left(F_e^{<i,\pi,\tau^i}(P) + f_e^{i'}(\hat{P}) + f_e^{i''}(\hat{P}) \right)$$

$$- C_e \left(F_e^{<i,\pi,\tau^i}(P) + f_e^{i'}(\hat{P}) \right).$$

Since $f_e^{i'}(\hat{P}) = f_e^{i''}(\hat{P}) = \frac{f_e^i(P)}{2}$, we get

$$\xi' + \xi'' \le \frac{1}{r!} \sum_{\tau^i} \left(C_e \left(F_e^{<i,\pi,\tau^i}(P) + f_e^i(P) \right) - C_e \left(F_e^{<i,\pi,\tau^i}(P) \right) \right) \overset{(6)}{=} \xi,$$

as desired. This completes Lemma's 1 proof. $\qquad\square$

We now continue to the proof for the POS upper bound. By repeatedly applying Lemma 1, we can break the total flow on each resource in flows of infinitesimal size without increasing the value of the potential. This implies that

$$\Phi_e(P) \ge \int_0^{f_e(P)} \frac{C_e(x)}{x} dx. \tag{11}$$

Now call P^* the optimal outcome and $P = \arg\min_{P'} \Phi(P')$ the minimiser of the potential function, which is, by definition, also a PNE. Then

$$SC(P^*) \overset{(4)}{\ge} \Phi(P^*) \overset{\text{Def.}P}{\ge} \Phi(P) \overset{(11)}{\ge} \sum_{e \in E} \int_0^{f_e(P)} \frac{C_e(x)}{x} dx$$

$$= \frac{\sum_{e \in E} \int_0^{f_e(P)} \frac{C_e(x)}{x} dx}{\sum_{e \in E} C_e(f_e(P))} \cdot SC(P) \ge \min_{e \in E} \frac{\int_0^{f_e(P)} \frac{C_e(x)}{x} dx}{C_e(f_e(P))} \cdot SC(P).$$

Rearranging yields the upper bound $POS \le \max_{C \in \mathcal{C}, x > 0} \frac{C(x)}{\int_0^x \frac{C(x')}{x'} dx'}$, which completes the proof of Theorem 5.

Corollary 1. *For polynomials with non-negative coefficients and degree at most d, the POS of the Shapley value is at most $d + 1$, which asymptotically matches the lower bound of [7] for single commodity per player.*

4   Splittable Games

We conclude the paper with a discussion of interesting open problems on cost-sharing in the *splittable* version [5,9,16,20,25,29] of congestion games with multi-commodity players and with some results. In the splittable version of such games, the weight w_q of a commodity $q \in Q$ can be split among its strategies in $\mathcal{P}^q$; i.e., a fractional strategy of commodity $q \in Q$ is a vector $P_q = (w_{q,P})_{P \in \mathcal{P}^q} \in \mathbb{R}_{\ge 0}^{|\mathcal{P}^q|}$

with $\sum_{P \in \mathcal{P}_q} w_{q,P} = w_q$. For the unsplittable version, vector P_q has only one non-zero and equal to w_q component, which is not necessarily the case for the splittable games. For the single-commodity per player model, it is known that the proportional sharing method, having players paying a cost share proportional to their flows on each resource, guarantees existence of a pure Nash equilibrium. Moreover, the POA of this simple cost-sharing method is well understood [29].

Understanding the POA of other cost-sharing methods both in the single- and multi-commodity models is an interesting open question. Similarly, it is interesting to study questions pertaining to the existence of pure Nash equilibria in such games, which we do next.

A result from Orda et al. [25] implies the existence of pure Nash equilibria in the multi-commodity splittable model, if the cost share of a player on a resource is a convex function of her flow on the resource. The result in [25] is based on the Kakutani Fixed Point theorem. This immediately gives us existence of pure Nash equilibria for the *standard* Shapley cost sharing. We strengthen this result by showing that such games are exact potential games [24] and thus best response dynamics converge to a pure Nash equilibrium. The proof of the following theorems can be found in Appendix.

Theorem 6. *Splittable congestion games with multi-commodity players under Shapley cost sharing are exact potential games.*

As soon as we deviate to the weighted Shapley value method, we prove that they do not guarantee PNE existence. Our proof uses the fact that the cost shares of the players are not necessarily convex anymore in this setting.

Theorem 7. *For parameterised weighted Shapley values with (finite) parameter γ, PNE are not guaranteed to exist for splittable congestion games with multi-commodity players.*

For $\gamma = \infty$, we can even construct a counter example that uses only single-commodity players.

Theorem 8. *For parameterised weighted Shapley values with parameter $\gamma = +\infty$, PNE are not guaranteed to exist even for single-commodity players.*

References

1. Abed, F., Correa, J.R., Huang, C.-C.: Optimal coordination mechanisms for multi-job scheduling games. In: Schulz, A.S., Wagner, D. (eds.) ESA 2014. LNCS, vol. 8737, pp. 13–24. Springer, Heidelberg (2014). doi:10.1007/978-3-662-44777-2_2
2. Aland, S., Dumrauf, D., Gairing, M., Monien, B., Schoppmann, F.: Exact price of anarchy for polynomial congestion games. SIAM J. Comput. **40**(5), 1211–1233 (2011)
3. Anshelevich, E., Dasgupta, A., Kleinberg, J., Tardos, E., Wexler, T., Roughgarden, T.: The price of stability for network design with fair cost allocation. SIAM J. Comput. **38**(4), 1602–1623 (2008)
4. Awerbuch, B., Azar, Y., Epstein, A.: The price of routing unsplittable flow. In: Proceedings of STOC, pp. 57–66. ACM (2005)

5. Bhaskar, U., Fleischer, L., Hoy, D., Huang, C.: Equilibria of atomic flow games are not unique. In: Proceedings of SODA, pp. 748–757 (2009)
6. Bhawalkar, K., Gairing, M., Roughgarden, T.: Weighted congestion games: price of anarchy, universal worst-case examples, and tightness. Proc. of ACM TEAC 2(4), 14 (2014)
7. Christodoulou, G., Gairing, M.: Price of stability in polynomial congestion games. In: Fomin, F.V., Freivalds, R., Kwiatkowska, M., Peleg, D. (eds.) ICALP 2013. LNCS, vol. 7966, pp. 496–507. Springer, Heidelberg (2013). doi:10.1007/978-3-642-39212-2_44
8. Christodoulou, G., Koutsoupias, E.: The price of anarchy of finite congestion games. In: Proceedings of STOC, pp. 67–73. ACM (2005)
9. Cominetti, R., Correa, J.R., Moses, N.E.S.: The impact of oligopolistic competition in networks. Oper. Res. 57(6), 1421–1437 (2009)
10. Fotakis, D., Kontogiannis, S., Spirakis, P.: Atomic congestion games among coalitions. ACM Trans. Algorithms (TALG) 4(4), 52 (2008)
11. Fotakis, D., Spirakis, P.G.: Cost-balancing tolls for atomic network congestion games. Internet Math. 5(4), 343–363 (2008)
12. Gairing, M., Kollias, K., Kotsialou, G.: Tight bounds for cost-sharing in weighted congestion games. In: Halldórsson, M.M., Iwama, K., Kobayashi, N., Speckmann, B. (eds.) ICALP 2015. LNCS, vol. 9135, pp. 626–637. Springer, Heidelberg (2015). doi:10.1007/978-3-662-47666-6_50
13. Gairing, M., Schoppmann, F.: Total latency in singleton congestion games. In: Deng, X., Graham, F.C. (eds.) WINE 2007. LNCS, vol. 4858, pp. 381–387. Springer, Heidelberg (2007). doi:10.1007/978-3-540-77105-0_42
14. Gkatzelis, V., Kollias, K., Roughgarden, T.: Optimal cost-sharing in weighted congestion games. In: Liu, T.-Y., Qi, Q., Ye, Y. (eds.) WINE 2014. LNCS, vol. 8877, pp. 72–88. Springer, Cham (2014). doi:10.1007/978-3-319-13129-0_6
15. Gopalakrishnan, R., Marden, J.R., Wierman, A.: Potential games are necessary to ensure pure Nash equilibria in cost sharing games. Math. Oper. Res. 39, 1252–1296 (2014)
16. Harks, T.: Stackelberg strategies and collusion in network games with splittable flow. Theory Comput. Syst. 48(4), 781–802 (2011)
17. Harks, T., Klimm, M.: On the existence of pure Nash equilibria in weighted congestion games. Math. Oper. Res. 37(3), 419–436 (2012)
18. Harks, T., Klimm, M., Möhring, R.H.: Characterizing the existence of potential functions in weighted congestion games. Theory Comput. Syst. 49(1), 46–70 (2011)
19. Hart, S., Mas-Colell, A.: Potential, value, and consistency. Econom.: J. Econom. Soc. 589–614 (1989)
20. Hayrapetyan, A., Tardos, É., Wexler, T.: The effect of collusion in congestion games. In: Proceedings of STOC, pp. 89–98 (2006)
21. Klimm, M., Schmand, D.: Sharing non-anonymous costs of multiple resources optimally. In: Paschos, V.T., Widmayer, P. (eds.) CIAC 2015. LNCS, vol. 9079, pp. 274–287. Springer, Cham (2015). doi:10.1007/978-3-319-18173-8_20
22. Kollias, K., Roughgarden, T.: Restoring pure equilibria to weighted congestion games. In: Aceto, L., Henzinger, M., Sgall, J. (eds.) ICALP 2011. LNCS, vol. 6756, pp. 539–551. Springer, Heidelberg (2011). doi:10.1007/978-3-642-22012-8_43
23. Koutsoupias, E., Papadimitriou, C.: Worst-case equilibria. Comput. Sci. Rev. 3(2), 65–69 (2009)
24. Monderer, D., Shapley, L.S.: Potential games. Games Econ. Behav. 14(1), 124–143 (1996)

25. Orda, A., Rom, R., Shimkin, N.: Competitive routing in multiuser communication networks. IEEE/ACM Trans. Netw. **1**(5), 510–521 (1993)
26. Rosenthal, R.W.: A class of games possessing pure-strategy Nash equilibria. Int. J. Game Theory **2**(1), 65–67 (1973)
27. Rosenthal, R.W.: The network equilibrium problem in integers. Networks **3**(1), 53–59 (1973)
28. Roughgarden, T.: Intrinsic robustness of the price of anarchy. In: Proceedings of STOC, pp. 513–522. ACM (2009)
29. Roughgarden, T., Schoppmann, F.: Local smoothness and the price of anarchy in splittable congestion games. J. Econ. Theory **156**, 317–342 (2015)
30. Roughgarden, T., Schrijvers, O.: Network cost-sharing without anonymity. In: Lavi, R. (ed.) SAGT 2014. LNCS, vol. 8768, pp. 134–145. Springer, Heidelberg (2014). doi:10.1007/978-3-662-44803-8_12
31. Roughgarden, T., Tardos, É.: How bad is selfish routing? J. ACM (JACM) **49**(2), 236–259 (2002)
32. Sheffi, Y.: Urban Transportation Networks: Equilibrium Analysis with Mathematical Programming Methods. Prentice-Hall, Upper Saddle River (1985)
33. Tran-Thanh, L., Polukarov, M., Chapman, A., Rogers, A., Jennings, N.R.: On the existence of pure strategy Nash equilibria in integer–splittable weighted congestion games. In: Persiano, G. (ed.) SAGT 2011. LNCS, vol. 6982, pp. 236–253. Springer, Heidelberg (2011). doi:10.1007/978-3-642-24829-0_22

Cache Oblivious Minimum Cut

Barbara Geissmann$^{(\boxtimes)}$ and Lukas Gianinazzi

Department of Computer Science, ETH Zurich, Zurich, Switzerland
`barbara.geissmann@inf.ethz.ch`

Abstract. We show how to compute the minimum cut of a graph cache-efficiently. Let B be the width of a cache line and M be the size of the cache. On a graph with V vertices and E edges, we give a cache oblivious algorithm that incurs $O(\lceil \frac{E}{B}(\log^4 E) \log_{M/B} E \rceil)$ cache misses and a simpler one that incurs $O(\lceil \frac{V^2}{B} \log^3 V \rceil)$ cache misses.

1 Introduction

Memory bandwidth has become a key limiting factor on the performance of computations. To improve upon this issue, modern microarchitectures feature a hierarchy of caches. In order to really achieve better performance, it is crucial that algorithms take these hierarchies into account: The major goal of such a *cache-efficient algorithm* is to keep the number of *cache misses* (data loadings into a cache) as small as possible. This can save a large amount of time, especially for non-trivial problems of large input size.

In a two-level hierarchy, one distinguishes between a *main memory* of unbounded size and a *cache* of size M with lines of width B. There are two ways of modeling algorithms in memory hierarchies: the *cache aware model* and the *cache oblivious model*. In the cache aware model, the cache parameters M and B are used to describe the algorithm, whereas in the cache oblivious model, these parameters must not be used in the algorithm description.

Graph problems can be especially challenging to solve cache-efficiently, since approaches that work well in RAM (like traversing a graph) have poor spatial locality. A common approach to design cache-efficient algorithms is to exploit a relation between parallel and cache-efficient algorithms and to run a so-called PRAM simulation [5,8]. The downside of PRAM simulation is that it only works well for problems for which work-efficient parallel algorithms are known. This is not the case for the minimum cut problem.

Improving on previous bounds, we present two cache oblivious minimum cut algorithms. On a graph with V vertices and E edges our algorithms incur $O(\lceil \frac{E}{B}(\log^4 E) \log_{M/B} E \rceil)$ and $O(\lceil \frac{V^2}{B} \log^3 V \rceil)$ cache misses, respectively.

Model. We consider two levels of memory: a *main memory* of unbounded size and a fully associative *cache* of size M with lines of width B. We assume an optimal replacement strategy: if a new line is loaded into the (full) cache, it evicts the line that is accessed farthest in the future. Such a data loading is called a *cache*

© Springer International Publishing AG 2017
D. Fotakis et al. (Eds.): CIAC 2017, LNCS 10236, pp. 285–296, 2017.
DOI: 10.1007/978-3-319-57586-5_24

miss or a *memory transfer*. This idealized cache has been described by Frigo et al. [9], who also introduced the two ways of modeling memory hierarchies, namely the *cache aware model* and the *cache oblivious model*.

Depending on the model, one may (cache aware) or may not (cache oblivious) use the cache parameters B and M in an algorithm description. The latter has the advantage that bounds proven for a two-level cache extend to a multilevel cache, which is not the case in the cache aware model. The cache aware model is equivalent to the *external memory* or *I/O model* [1], up to constant factors.

To assume an optimal replacement strategy allows us to consider any other strategy in order to prove upper bounds on the number of cache misses. Moreover, the number of cache misses changes only by a constant factor if the cache evicts the *least recently used* line instead of the one used farthest in the future [9].

Throughout this paper, we hold on the so-called *tall cache assumption*, that $M \in \Omega(B^2)$. A tall cache is required for cache-optimal matrix transposition [16] in the cache oblivious model. A weaker assumption, $M \in \Omega(B^{1+\varepsilon})$, is required for cache-optimal sorting [7]. With a tall cache, transposing an $N \times N$ matrix incurs $\Theta(\lceil \frac{N^2}{B} \rceil)$ and sorting N elements incurs $\Theta(\lceil \frac{N}{B} \lceil \log_{M/B} N \rceil \rceil)$ cache misses [9]. One way to sort cache-optimally is to use a cache oblivious priority queue [2], where each operation incurs $O(\frac{1}{B} \lceil \log_{M/B} N \rceil)$ amortized cache misses. Let:

$$\text{SORT}(N) := \lceil N/B \lceil \log_{M/B} N \rceil \rceil. \tag{1}$$

Graph Algorithms in Memory Hierarchies. Since many operations on graphs have poor locality, it is challenging to solve graph problems cache-efficiently. For example, although minimum spanning trees, connected components, and breadth first search are well understood in the RAM model, in the cache oblivious model tight bounds on the number of cache misses are still unknown for these problems [2,5,6]. In the cache aware model, the situation is similar [3,13].

Chiang et al. [8] showed how to simulate PRAM algorithms in external memory and Blelloch et al. [5] observed that this result generalizes to the cache oblivious model (as it relies only on scanning and sorting).

Lemma 1 (by Blelloch et al. *[5]* and Chiang et al. *[8]*). *A single time step of a PRAM algorithm using $O(N)$ contiguous space and $O(P)$ processors can be simulated in the cache oblivious model incurring $O(\lceil \frac{N}{B} + \text{SORT}(P) \rceil)$ cache misses.*

PRAM simulations give cache-efficient algorithms for depth-first search and other tree problems. In particular, depth-first search incurs $O(\text{SORT}(V))$ and answering N least common ancestor queries incurs $O(\text{SORT}(V+N))$ cache misses in a tree with V vertices [5,8].

Minimum Cut. In an undirected weighted graph $\mathcal{G} = (\mathcal{V}, \mathcal{E}, w)$ on $V = |\mathcal{V}|$ vertices, $E = |\mathcal{E}|$ edges, and edge weights w, a *cut* $\mathcal{C}$ is a non-empty proper subset of vertices ($\mathcal{C} \subset \mathcal{V}, \mathcal{C} \neq \emptyset$). Every cut $\mathcal{C}$ is associated with a *value* $\omega(\mathcal{C})$, which is the total weight of all edges in $\mathcal{G}$ that have one endpoint in $\mathcal{C}$ and another in $\mathcal{V} \setminus \mathcal{C}$ and we say that $\mathcal{C}$ *cuts* those edges. Similarly, we denote by $\omega(v)$ the total weights of all edges incident to a vertex v, which we call its *capacity*.

A *minimum cut* is a cut with minimum value. We will focus on computing the minimum cut value, rather than the cut itself, though all ideas can be extended to compute minimum cuts explicitly.

Minimum Cut Algorithms. The fastest known deterministic minimum cut algorithms take $O(VE + V^2 \log V)$ time [14,18] for graphs with weighted edges and $O(E \text{ polylog } V)$ time for graphs with unweighted edges [12].

Randomized algorithms achieve better bounds. In this paper, we consider two such algorithms, that find a minimum cut with high probability:

- KARGER [10]. This algorithm runs in $\Theta(E \log^3 V)$ time. A parallel (PRAM) variant of this algorithm takes $\Theta(\log^3 V)$ time on $\Theta(E+V^2/\log^2 V)$ processors using $O(V^2 \log V)$ space. Hence, a PRAM simulation of this algorithm incurs $O(\lceil \frac{V^2}{B} \log^4 V \rceil)$ cache misses. An external memory adaptation by Bushan and Sajith [4] incurs $O((c+\log V)\text{SORT}(E) \log V + \frac{V}{B}\text{SORT}(V) \log V)$ memory transfers on unweighted graphs, where c is the minimum cut value.
- KARGER AND STEIN [11]. This algorithm takes $\Theta(V^2 \log^3 V)$ time and incurs, naively implemented, $\Theta(V^2(\log^2 V) \log \lceil \frac{V}{M} \rceil)$ cache misses.

KARGER AND STEIN is notably good for dense graphs, while KARGER is faster for sparse graphs. For both algorithms, we present a cache oblivious variation.

Our Contribution. In Sect. 2, we adapt KARGER AND STEIN such that it incurs $O(\lceil \frac{V^2}{B}(\log^2 V) \log \lceil \frac{V}{M} \rceil \rceil)$ cache misses, thus improving on a naive implementation by a factor B and keeping the same running time of $O(V^2 \log^3 V)$.

In Sect. 3, we show how to transform RAM algorithms that access memory in a monotone order into cache-efficient algorithms, at a logarithmic time overhead. The technique allows a cache-efficient all-at-once execution of a *batch* of consecutive updates and queries on a datastructure, which appear inherently sequential (thus cannot easily be made cache-efficient using PRAM simulation). We solve a semistatic simplification of the dynamic tree problem [17], that we call *minimum path*: in a tree with V weighted vertices, we define a query that returns for a questioned vertex the smallest weight in its subtree, and we define an update that increases or decreases all weights in such a subtree. Such a query and update incur amortized $O(\frac{\log^3 V}{B \log M})$ and $O(\frac{\log^2 V}{B \log M})$ cache misses, respectively.

We use the minimum path structure in Sect. 4, where we adapt KARGER such that it incurs $O(\text{SORT}(E \log^4 E))$ cache misses and still takes $O(E \text{ polylog } V)$ time. On sparse graphs, this asymptotically improves upon the $\Omega(V^2)$ time and $\Omega(V^2/B^2)$ cache misses achieved by the external memory algorithm in [4].

2   A Simple Minimum Cut Algorithm for Dense Graphs

Here, we develop a cache oblivious variant of KARGER AND STEIN [11]. This algorithm repeatedly applies *random edge contractions* to the graph, where a random edge contraction consists of choosing an edge at random with probability proportional to its weight and then merging its two end vertices into a single vertex. Starting with two copies of the graph, the algorithm performs edge contractions until only $\lceil V/\sqrt{2} + 1 \rceil$ vertices are left. Then, each copy is duplicated again and the algorithm continues recursively until only a constant number of vertices is left for which the minimum cut is computed deterministically and the smallest such found value is returned. With high probability, $\Theta(\log^2 V)$ runs of this algorithm find the minimum cut value.

In the commonly known implementations, one run of this algorithm incurs $\Omega(V^2 \log \frac{V}{M})$ cache misses. We reduce this number by a factor B, by showing how to perform $O(V)$ random edge contractions using only $O(\lceil \frac{V^2}{B} \rceil)$ cache misses.

Let $\{1, \ldots, V\} = \mathcal{V}$ be the set of vertices in our graph. When we represent the graph as an adjacency matrix, a naive way to compute the contraction of an edge $\{i, j\} \in \mathcal{E}$ is to first add row j to row i and column j to column i, and second to put zeros into row and column j and mark them as deleted. This is not cache-efficient, since every contraction requires to traverse both rows and columns in the matrix. Thus, if the matrix is stored in row-major order (column-major order), traversing a column (row) leads to $\Omega(V)$ cache misses.

There are three tasks: (1) avoid the eagerly updating of columns and rows for every single edge contraction. (2) ensure that randomly selecting an edge does not incur too many cache misses. (3) shrink the adjacency matrix cache-efficiently to the new size of the graph before it is copied in the recursive process.

Lazy Edge Contraction. Let $\mathcal{G} = (\mathcal{V}, \mathcal{E}, w)$ be given as a $V \times V$ adjacency matrix A in row-major order, such that an entry $A_{i,j}$ stores the weight of the edge $\{i, j\}$. In addition, we use two arrays U and D of size V. In U we keep track of the merged vertices, akin to a union find structure, where $U_j = i$ means that j has been merged with i. We initialize $U_i := i$ for all $i \in \{1, \ldots, V\}$. D is used to store for each vertex its capacity. We call such a tuple (A, U, D) a *lazy graph representation*. The *lazy contraction* of an edge $\{i, j\} \in \mathcal{E}$ involves three steps:

1. In U, replace all occurrences of j by i.
2. In A, pairwise add row j to row i and set each $A_{i,k}$ to zero if $U_k = i$.
3. In D, set D_i to the sum of all entries in the i-th row of A, and set $D_j = 0$.

With the first step we keep track of the merged vertices, which is especially important since we do not add the columns. In the second step we merge new parallel edges together and remove loops. In the third step we update the capacities of the vertices, which is zero for j since it is now merged with i.

Observe that before and after an edge contraction, the current graph is implied by the lazy graph representation: if $\mathcal{G}' = (\mathcal{V}', \mathcal{E}', w)$ is the current graph and i' is the i-th vertex in $\mathcal{V}'$, then the adjacency matrix A' is given by

$$A'_{ij} = \sum_{k \in \{1, \ldots, V\},\ U_k = j'} A_{i'k} \qquad \text{for } i, j \in \{1, \ldots, |\mathcal{V}'|\} \ .$$

Since, in steps 1 to 3, we iterate a constant number of times through data sets of length V, each time causing $O(\lceil \frac{V}{B} \rceil)$ cache misses, we conclude:

Lemma 2. *In a graph with V vertices given in lazy graph representation, a lazy edge contraction incurs $O\left(\lceil \frac{V}{B} \rceil\right)$ cache misses.*

Selecting an Edge at Random. Equivalent to choose an edge at random with probability proportional to its weight is to select the first vertex i by its capacity, and among all adjacent vertices, select the second vertex j by the weight of $\{i, j\}$.

We adopt this procedure to our lazy graph representation as follows: First, we choose i with probability proportional to its capacity D_i. Second, we choose $j = U_k$ by selecting a column k in A with probability $\frac{A_{i,k}}{D_i}$, which works because the weight of $\{i, j\}$ is the sum of all entries $A_{i,k}$ with $U_k = j$.

To select an entry from an array proportional to its value, we compute the prefix sums of the array, generate a number uniformly at random between 1 and the sum of the array and perform a binary search for this value. This takes $O(\lceil \frac{n}{B} \rceil)$ cache misses (where n is the length of the array), hence:

Lemma 3. *In a graph with V vertices in lazy graph representation, the random selection of an edge by its weight incurs $O(\lceil \frac{V}{B} \rceil)$ cache misses.*

The above approach takes $O(n)$ time if the sum of all edge weights is in $O(2^n)$. We assume this is the case, but remark that if it is not, there is a technique [11], which implies that with high probability, a random edge can be selected incurring amortized $O(\lceil \frac{n}{B} \rceil)$ cache misses and taking $O(n)$ amortized time.

Multiple Contractions. The dimensions of a lazy graph representation are not changed by a lazy edge contraction, although the graph itself shrinks by one vertex. If we disregard this, a contraction in one of the numerous small graphs that appear during the recursive calls of the algorithm would incur the same cost as a contraction in the initial graph. This would lead to $\Omega(\frac{V^3}{B})$ cache misses. For this reason, we *compact* the graph before every recursive call, which means that we let the data structures shrink to the new size of the graph.

Before we can compact the graph representation, we have to combine the columns in A that belong to the same vertex. Since traversing columns of a matrix in row major order is expensive, we first transpose A, using $O(\lceil \frac{V^2}{B} \rceil)$ cache misses at once. Then, we combine rows instead of columns:

1. In A, remove every row i with $U_i \neq i$.
2. Transpose A.
3. Let i' be the i-th vertex in U with $U_{i'} = i'$. Create A' where row i is the sum of all rows j in A for which $U_j = i'$.
4. Let U and D shrink such that they only contain still existing vertices.

The first two steps incur $O(\lceil \frac{V^2}{B} \rceil)$, the third step $O(V \lceil \frac{V}{B} \rceil)$ and the last step $O(V)$ cache misses. Therefore:

Lemma 4. *To compact a graph in lazy graph representation from V to $V' < V$ vertices incurs $O(V \lceil \frac{V}{B} \rceil)$ cache misses.*

Finally, we conclude the total number of cache misses in a contracting phase:

Theorem 1. *In a graph with V vertices given in lazy graph representation, $O(V)$ random edges can be contracted with $O(\lceil \frac{V^2}{B} \rceil)$ cache misses.*

Proof. By Lemmas 2 to 4, to select and contract $O(V)$ random edges and shrink the graph to its new size costs $O(V\lceil\frac{V}{B}\rceil)$ cache misses. By the tall cache assumption, the matrix either fits into cache or $\frac{V^2}{B} \in \Omega(V)$, thus $V\lceil\frac{V}{B}\rceil \in O(\lceil\frac{V^2}{B}\rceil)$.
□

Let $Q(V)$ denote the number of cache misses in the entire algorithm (not only in one contracting phase), starting with a graph of V vertices. Assume that $\frac{M}{B} > d_0$, where d_0 is a constant such that the space used by the algorithm on V vertices is bounded by $d_0 V^2$. We get

$$Q(V) \leq \begin{cases} d_0 \cdot \left(\frac{V^2}{B}\right) + 1 & \text{if } V^2 \leq \frac{M}{d_0}, \\ 2 \cdot Q\left(\lceil\frac{V}{\sqrt{2}} + 1\rceil\right) + O\left(\frac{V^2}{B} + 1\right) & \text{otherwise,} \end{cases}$$

which resolves to $Q(V) \in O(\frac{V^2}{B}\log(\frac{V}{M} + 1) + 1)$. Note that if $V^2 \leq \frac{M}{d_0}$, the data used by the algorithm fits into cache, and no block needs to be read twice.

In order to boost the probability of returning a minimum cut, the algorithm is repeated $\Theta(\log^2(V))$ times, which altogether results in $O(\lceil\frac{V^2}{B}\log^2 V \log\lceil\frac{V}{M}\rceil\rceil)$ cache misses. The running time does not change compared to the original algorithm and is still $O(V^2 \log^3 V)$.

3   Simulating Random Access Datastructures

In this section, we first describe a general simulation technique to make certain RAM algorithms cache-efficient. Then, we apply this technique to a data structure problem which we call *minimum path* and which has not yet been solved cache-efficiently. Minimum path is a semistatic simplification of dynamic trees, which does not change the structure of the tree.

3.1   Monotone RAM simulation

The idea of our simulation technique is to ensure that operations access memory in a monotone order. This allows us to execute a batch of operations by going through the memory locations only once, since the state of a memory cell only depend on memory cells with a smaller address.

Definition 1. *An operation* op *is called a* monotone RAM operation *if it accesses the memory in some given total order* $\lhd$. *That is, if* op *accesses location* x *at some point and location* y *at a later point in time, then* $x \lhd y$.

We demand that $x \lhd y$ can be evaluated in constant time, $O(\frac{1}{B})$ amortized cache misses, and using $O(1)$ local variables. Since we *batch* operations together and execute them at once in order to save cache misses, the results of a query are available only after all operations have terminated. This technique is thus only suitable if the results are not required immediately. We will see that the orders in which queries and updates access the memory do not need to be equal.

In a batch of updates and queries, let $\lhd_u$ ($\lhd_q$) be the memory access order of the updates (queries). We define an explicit representation of the *intermediate state* of a single operation as a tuple (l, t, s), with a unique *timestamp* t as the position in the operation sequence of the batch, the next memory *location* l that it will access, and the *state* s of its $O(1)$ local variables.

The simulation consists of two phases: (1) run all updates and record every intermediate state, (2) sort the intermediate states by $\lhd_q$ and run all queries.

- *Phase 1.* A priority queue P contains the ongoing updates ordered by $\lhd_u$, with updates to the same location sorted by timestamps. An array D stores all changes. While P is not empty, we dequeue and run all updates to the next location l: (•) If an update reads location l, we return the current state of l. (•) If an update changes the current state to x at time t, we insert a new tuple (l, t, x) into D. (•) If an update needs to access a location l' next, we insert a new update tuple into P with the same timestamp but with location l'.
- *Phase 2.* A priority queue Q contains all queries, first ordered by $\lhd_q$ and second by timestamps. An array R stores all results. We first sort D by $\lhd_q$ and timestamps. Then, while Q is not empty, we dequeue all queries to the next location l and process these queries and D in tandem, such that we return to a query at time t the state of l at the closest preceding time appearing in a tuple of D. (•) If a query produces a result, we append it to R. (•) If a query accesses some location l' next, we insert the query into Q with the new location l'.

Theorem 2. *A batch of $O(N)$ monotone RAM updates and queries each taking $O(f)$ time can be performed incurring $O(\text{SORT}(Nf))$ cache misses.*

Proof. The number of operations in P and Q and the lengths of D and R are bounded by the total number of reads and writes in the original RAM structure, which is $O(Nf)$. Thus, we get $O(\text{SORT}(Nf))$ cache misses in total.   □

3.2   Minimum Path

We show how to apply the monotone RAM simulation technique to a data structure called *minimum path*. On a tree T on V weighted vertices $v_1, \ldots, v_V$ with weights $w_1, \ldots, w_V$ and with root ρ, the minimum path data structure supports the following two operations:

- MinPath(v_k) returns the smallest weight of a vertex on the path from v_k to ρ.
- AddPath(v_k, x) adds x to the weight of all vertices on the path from v_k to ρ.

Dynamic tree structures [17] achieve the two operations in $O(\log V)$ time each, but are not amenable to cache-efficient operations. We show how to apply a monotone RAM simulation to minimum path. In particular, we present a $O(\log^2 V)$ time monotone RAM structure that allows a cache-efficient execution of a batch of minimum path operations. By Theorem 2, this implies the following:

Theorem 3. *On a tree with V nodes, a batch of $O(N)$ minimum path operations is executed with $O(\text{SORT}(N \log^2 V + V))$ cache misses.*

We first consider the case where T is a path and then generalize for any tree.

Path. Let T be the path $(v_1, v_2, \ldots, v_V)$ with root $\rho = v_1$. For simplicity, we assume that V is a power of two. We define *pAddPath* and *pMinPath* analogously to *AddPath* and *MinPath*, with the difference that these new operations only work for paths (and not for trees in general).

We construct a complete binary tree B with $(v_1, v_2, \ldots, v_V)$ as its leaves and denote the root of B by ρ_B. For every node $b \in B$, we denote the set of its descendants by $b^{\downarrow}$ and the smallest weight among all descendant leaves by $\min(b)$. For efficient updates, we do not store this value explicitly. Instead, we only explicitly store $\min(\rho_B)$ and with every inner node b the difference $\Delta(b)$ between the smallest weight in its right and left subtree. For every inner node $b \in B$, let l_b and r_b denote its left and right child, respectively.

- *Initialization*: For each node $b \in B$, derive $\min(b)$ and $\Delta(b)$ bottom-up from the values of its children. This takes linear time in the number of vertices.
- *pMinPath*(v_k): For a node $b \in B$,

$$\delta(b) := (\min_{v_i \in b^{\downarrow}, i \leq k} w_i) - \min(b).$$

Hence, pMinPath$(v_k) = \delta(\rho_B) + \min(\rho_B)$. We compute $\delta(b)$ for each node b on the path from v_k to ρ_B. We start with $\delta(v_k) = 0$ and sequentially compute the other $\delta(b)$'s based on $\delta(l_b)$, $\delta(r_b)$ and $\Delta(b)$:

$$\delta(b) = \begin{cases} \delta(l_b) & \text{if } \Delta(b) > 0 , \\ \delta(r_b) & \text{if } \Delta(b) \leq 0 , v_k \in r_b^{\downarrow} , \text{ and } \delta(r_b) + \Delta(b) < 0 , \\ \delta(l_b) - \Delta(b) & \text{otherwise} . \end{cases}$$

Note that whenever we consider $\delta(l_b)$ or $\delta(r_b)$, we either already computed this value or it is zero. To show correctness, we make some observations.
(1) If $\Delta(b) > 0$, then $\min(b) = \min(l_b)$. Otherwise $\min(b) = \min(r_b)$.
(2) If $\Delta(b) \leq 0$ and $v_k \in r_b^{\downarrow}$, the node minimizing $\min_{v_i \in b^{\downarrow}, i \leq k} w_i$ is in b's right subtree if $\delta(r_b) + \Delta(b) < 0$ and in b's left subtree if $\delta(r_b) + \Delta(b) > 0$, since

$$\delta(r_b) + \Delta(b) = \min_{v_i \in r_b^{\downarrow}, i \leq k} w_i - \min(l_b).$$

(3) If $v_k \in l_b^{\downarrow}$, then $\{v_i | v_i \in b^{\downarrow}, i \leq k\} = \{v_i | v_i \in l_b^{\downarrow}, i \leq k\}$, because all leaves of b which are left of v_k are in $l_b^{\downarrow}$. The rest of the proof is case distinction and computation, which we omit here.
- *pAddPath*(v_k, x): Observe that $\Delta(b)$ of a node b only changes if v_k is a descendant of b. Therefore, it suffices to update only Δ values along the path from ρ_B to v_k. For every node b on this path, compute the difference $\varphi(b)$ between the new and the old minimum weight of its descendants: First, set $\varphi(v_k) = x$, then proceed upwards to ρ_B. If v_k is in b's left subtree, we already computed $\varphi(l_b)$ and the minimum in the right subtree did not change, thus $\varphi(r_b) = 0$. Otherwise, we already computed $\varphi(r_b)$ and we infer that $\varphi(l_b) = x$. Hence,

$$\Delta'(b) = \Delta(b) + \varphi(r_b) - \varphi(l_b) ,$$

$$\varphi(b) = \begin{cases} \varphi(l_b) & \text{if } \Delta(b) > 0 \text{ and } \Delta'(b) > 0 , \\ \varphi(l_b) - \Delta(b) & \text{if } \Delta(b) \le 0 \text{ and } \Delta'(b) > 0 , \\ \varphi(r_b) & \text{if } \Delta(b) \le 0 \text{ and } \Delta'(b) \le 0 , \\ \varphi(r_b) + \Delta(b) & \text{if } \Delta(b) > 0 \text{ and } \Delta'(b) \le 0 . \end{cases}$$

For $b \in B$, let $\min'(b)$ be the smallest weight among its descendants after the update ($\min(b)$ before the update). To explain correctness, we need to show that $\varphi(b) = \min'(b) - \min(b)$. This is simply done by a case distinction, always assuming that $\varphi(l_b) = \min'(l_b) - \min(l_b)$ and $\varphi(r_b) = \min'(r_b) - \min(r_b)$.

Lemma 5. *Given a path $(v_1, ..., v_V)$, pMinPath and pAddPath take $O(\log V)$ time in monotone RAM.*

Tree. We show now how to solve the minimum path problem for any tree T. This for, we partition the vertices of the tree into vertex disjoint paths, as follows:

Definition 2. *A path $(u_1, \ldots, u_k)$ in a rooted tree T is called a* bough *of T if*

1. *the first vertex u_1 is a leaf,*
2. *all other vertices $u_2, \ldots, u_k$ have a unique child in T, i.e. u_i has child u_{i-1},*
3. *the last vertex u_k does not have a parent with a unique child in T.*

Definition 3. *A path in a rooted tree T is a* principal branch *of T if it is a bough of T or if it is a principal branch of the tree given by contracting all edges incident to any vertex in any bough of T.*

Observe that any root to leaf path in T intersects at most $O(\log V)$ distinct principal branches, since the number of leaves is at least halved whenever all edges incident to a vertex in a bough are contracted.

For any vertex $v \in T$, both $MinPath(v)$ and AddPath(v, x) consist of a series of $O(\log V)$ pMinPath and pAddPath operations, respectively. Let P denote the set of principal branches that intersect with the path from v to ρ. For each principle branch $p \in P$, let c_p denote the vertex in p that is closest to the root, and except for the branch that contains ρ, let f_p denote the parent vertex of c_p.

- *Initialization:* Initialize a minimum path structure for each principal branch in T, which we identify by a depth-first traversal of T. This traversal also associates c_p and f_p with every vertex in a principal branch p.
- *MinPath(v):* Return the smallest value found by $pMinPath(v)$ in the principal branch of v and by $pMinPath(f_p)$ in each principal branch that contains an f_p with $p \in P$.
- *AddPath(v, x):* Run $pAddPath(v, x)$ on the principal branch that contains v and $pAddPath(f_p, x)$ each principal branch that contains an f_p with $p \in P$.

By Lemma 5 and since the size of P is in $O(\log V)$ we conclude:

Lemma 6. *In a tree T with V vertices, MinPath, AddPath, and initialization take $O(\log^2 V)$, $O(\log^2 V)$, and $O(V)$ time in monotone RAM, respectively.*

4   A Minimum Cut Algorithm for Sparse Graphs

We use the monotone RAM simulation from Sect. 3.2 to compute a minimum cut in sparse graphs cache-efficiently: We adapt KARGER [10] such that it incurs $O(\text{SORT}(E \log^4 V))$ cache misses. KARGER consists of two main steps:

1. Identify a set S of $O(\log V)$ spanning trees $T_1, T_2, \ldots T_{|S|}$ of $\mathcal{G}$.
2. For each $T_i \in S$, find the smallest cut in $\mathcal{G}$ that cuts at most two edges of T_i.

S is chosen such that, with high probability, the smallest cut found in the second step is indeed a minimum cut. KARGER finds a set of appropriate spanning trees in $O(\log^3 V)$ time using $O(E + V \log V)$ processors and $O(E \log V)$ space. PRAM simulation results in $O(\text{SORT}(E \log^3 V + V \log^4 V))$ cache misses for step 1. For step 2, we compute two cuts for every tree T_i in S: the smallest cut that cuts exactly one edge and the smallest cut that cuts exactly two edges of T_i. Whilst the former can be parallelized efficiently, no near-linear work polylogarithmic span solution is known to the latter. For the computations of the cuts, we follow the algorithms described in [10] and make them cache efficient.

Cutting One Edge of a Tree. We consider a fixed spanning tree T in S, which we root at some vertex ρ. We denote the including *set of descendants* of a vertex v in T by $v^{\downarrow}$, and we define for every edge $e = \{u, v\}$ in $\mathcal{G}$ its *least common ancestor* $a(e, T)$ in T: vertex w is the least common ancestor $a(e, T)$ of e if both u and v are in $w^{\downarrow}$ and there is no other descendant of w for which this also holds.

Theorem 4. *Given a spanning tree T of a graph $\mathcal{G}$ on V vertices and E edges, computing the smallest value of a cut in $\mathcal{G}$ that cuts exactly one edge of T incurs $O(\text{SORT}(E))$ cache misses.*

Proof. Assume the edge $\{i, j\} \in T$ is cut, where i is the parent of j, such that the cut is defined by the set $i^{\downarrow}$. Let $h(v)$ denote the sum of the weights of all edges in $\mathcal{G}$, which have v as their least common ancestor: $h(v) := \sum_{e \in E, a(e,T)=v} w(e)$. Then, the cut value is $\omega(i^{\downarrow}) = \sum_{v' \in i^{\downarrow}} \omega(v') - 2h(v')$.

We first sort all edges in $\mathcal{G}$ by their endvertices to enable an efficient precomputation of the ω terms in the sum. We compute for each edge its least common ancestor in T, using a PRAM simulation [15]. Then, we sort the edges by least common ancestors in order to compute the h values. This causes $O(\text{SORT}(E))$ cache misses. Finally, given all precomputed ω and h values, we compute all $\omega(v^{\downarrow})$ using treefix sums, which involves $O(\text{SORT}(V))$ cache misses. $\square$

Cutting Two Edges of a Tree. Assume the edges $\{i, j\}, \{i', j'\} \in T$, $j \neq j'$ are cut, where i (i') is the parent of j (j'). We distinguish two cases: Either j and j' lie on the same path from a leaf to the root or they don't. We focus on the second case, the first one is similar.

The algorithm repeatedly processes the boughs of T. The smallest cut of $\mathcal{G}$ which cuts exactly two edges of T is computed as follows.

Initialize a minimum path structure on the tree T where each vertex v has weight $\omega(v^{\downarrow})$. For every bough of T move on the path from the leaf that it included towards the root and do the following at every vertex v that we pass:

1. AddPath(v, ∞).
2. For each neighbor u of v, AddPath($u, -2w(\{u, v\})$).
3. For each neighbor u of v, MinPath(u) and store the minimum result as r_v.

After a bough has been processed, go through its vertices in reverse order and undo all modifications by changing the sign of the second argument in the AddPath operations. When all boughs have been processed, traverse each bough bottom-up and determine for every vertex v its value $C_v = \min_{v' \in v^{\downarrow}} r_{v'}$. Every such vertex gives an upper-bound $\omega(v^{\downarrow}) + C_v$ on the result. Then, contract all edges incident to the vertices in the boughs, and recursively repeat the procedure.

Theorem 5. *Given a spanning tree T of a graph $\mathcal{G}$ on V vertices and E edges, computing the smallest value of a cut in $\mathcal{G}$ that cuts exactly two edges of T incurs $O(\text{SORT}(E \log^3 V))$ cache misses.*

Proof. We follow the procedure described before. We shrink $\mathcal{G}$ and T at most $O(\log V)$ times, since there are at most $O(\log V)$ principal branches on a path from the root to a leaf. A shrinking phase costs $O(\text{SORT}(E))$ and a phase of processing boughs costs $O(\text{SORT}(E \log^2 V))$ cache misses. For the shrinking, we apply a PRAM simulation for E processors and $O(E)$ space that replaces every old edge by a new one using a mapping function. The bound follows by Lemma 1.

Since we need the results of the MinPath operations only after processing all boughs, we apply the batched approach from Sect. 3.2, Theorem 3. Every vertex of a bough is accessed exactly twice: once on the way from a leaf to the root and once in the reversed order to undo the AddPath operations. Since an edge is accessed only when one of its endpoints is accessed, we compute the order in which the vertices and edges are accessed by a depth-first traversal of the tree followed by sorting a constant number of duplicates of the edges and vertices. This allows to generate the batch of minimum path operations cache-efficiently.

The number of cache misses is dominated by the cost of $O(E)$ minimum path operations that, by Theorem 3, incur $O(\text{SORT}(E \log^2 V))$ cache misses.   □

Theorem 6. *Computing a minimum cut of a graph $\mathcal{G}$ on E edges with high probability incurs $O(\text{SORT}(E \log^4 V))$ cache misses.*

Proof. Using PRAM simulation, it costs $O(\text{SORT}(E \log^3 V + V \log^4 V))$ cache misses to find $O(\log V)$ appropriate spanning trees, and by Theorems 4 and 5, it costs $O(\text{SORT}(E \log^3 V))$ cache misses to find the cut with smallest value that cuts at most two edges of a spanning tree.   □

5 Conclusion

We compute a minimum cut in a number of cache misses equal to scanning the input a polylogarithmic number of times. It remains an open problem to find a minimum path structure which takes $O(\frac{1}{B} \lceil \log_{M/B} V \rceil)$ amortized cache misses per operation. Such a structure would improve the number of cache misses and the running time of our sparse minimum cut algorithm by a factor $\log^2 V$. The monotone RAM framework might be applicable to other semi-static data structure problems as they also arise in computational geometry.

References

1. Aggarwal, A., Vitter, J.S.: The input/output complexity of sorting and related problems. Commun. ACM **31**(9), 1116–1127 (1988)
2. Arge, L., Bender, M.A., Demaine, E.D., Holland-Minkley, B., Ian Munro, J.: An optimal cache-oblivious priority queue and its application to graph algorithms. SIAM J. Comput. **36**(6), 1672–1695 (2007)
3. Arge, L., Brodal, G.S., Toma, L.: On external-memory MST, SSSP, and multi-way planar graph separation. In: Halldórsson, M.M. (ed.) SWAT 2000. LNCS, vol. 1851, pp. 433–447. Springer, Heidelberg (2000). doi:10.1007/3-540-44985-X_37
4. Bhushan, A., Gopalan, S.: I/O efficient algorithms for the minimum cut problem on unweighted undirected graphs. Theor. Comput. Sci. **575**, 33–41 (2015)
5. Blelloch, G.E., Gibbons, P.B., Simhadri, H.V.: Low depth cache-oblivious algorithms. In: Proceedings of the Twenty-Second Annual ACM Symposium on Parallelism in Algorithms and Architectures, SPAA 2010, pp. 189–199. ACM, New York (2010)
6. Brodal, G.S., Fagerberg, R., Meyer, U., Zeh, N.: Cache-oblivious data structures and algorithms for undirected breadth-first search and shortest paths. In: Algorithm Theory - SWAT 2004, Proceedings of the 9th Scandinavian Workshop on Algorithm Theory, Humlebaek, Denmark, 8–10 July 2004, pp. 480–492 (2004). doi:10.1007/978-3-540-27810-8_41
7. Brodal, G.S., Fagerberg, R.: On the limits of cache-obliviousness. In: Proceedings of the Thirty-Fifth Annual ACM Symposium on Theory of Computing, STOC 2003, pp. 307–315. ACM, New York (2003)
8. Chiang, Y.-J., Goodrich, M.T., Grove, E.F., Tamassia, R., Vengroff, D.E., Vitter, J.S.: External-memory graph algorithms. In: Proceedings of the Sixth Annual ACM-SIAM Symposium on Discrete Algorithms, San Francisco, California, 22–24 January 1995, pp. 139–149 (1995)
9. Frigo, M., Leiserson, C.E., Prokop, H., Ramachandran, S.: Cache-oblivious algorithms. ACM Trans. Algorithms **8**(1), 4:1–4:22 (2012)
10. Karger, D.R.: Minimum cuts in near-linear time. J. ACM **47**(1), 46–76 (2000)
11. Karger, D.R., Stein, C.: A new approach to the minimum cut problem. J. ACM **43**(4), 601–640 (1996)
12. Kawarabayashi, K., Thorup, M.: Deterministic global minimum cut of a simple graph in near-linear time. In: Proceedings of the Forty-Seventh Annual ACM on Symposium on Theory of Computing, STOC 2015, Portland, OR, USA, 14–17 June 2015, pp. 665–674 (2015)
13. Munagala, K., Ranade, A.: I/O-complexity of graph algorithms. In: Proceedings of the Tenth Annual ACM-SIAM Symposium on Discrete Algorithms, SODA 1999, Philadelphia, PA, USA, 1999, pp. 687–694. Society for Industrial and Applied Mathematics (1999)
14. Nagamochi, H., Ibaraki, T.: Computing edge-connectivity in multigraphs and capacitated graphs. SIAM J. Discret. Math. **5**(1), 54–66 (1992)
15. Schieber, B., Vishkin, U.: On finding lowest common ancestors: simplification and parallelization. SIAM J. Comput. **17**(6), 1253–1262 (1988)
16. Silvestri, F.: On the limits of cache-oblivious matrix transposition. In: Montanari, U., Sannella, D., Bruni, R. (eds.) TGC 2006. LNCS, vol. 4661, pp. 233–243. Springer, Heidelberg (2007). doi:10.1007/978-3-540-75336-0_15
17. Sleator, D.D., Tarjan R.E.: A data structure for dynamic trees. In: Proceedings of the 13th Annual ACM Symposium on Theory of Computing, Milwaukee, Wisconsin, USA, 11–13 May 1981, pp. 114–122 (1981)
18. Stoer, M., Wagner, F.: A simple min-cut algorithm. J. ACM **44**(4), 585–591 (1997)

Enumeration of Maximal Irredundant Sets for Claw-Free Graphs

Petr A. Golovach[1]([✉]), Dieter Kratsch[2], and Mohamed Yosri Sayadi[2]

[1] Department of Informatics, University of Bergen, Bergen, Norway
petr.golovach@ii.uib.no,
[2] Université de Lorraine, LITA, Metz, France
{dieter.kratsch,yosri.sayadi}@univ-lorraine.fr

Abstract. Domination is one of the classical subjects in structural graph theory and in graph algorithms. The MINIMUM DOMINATING SET problem and many of its variants are NP-complete and have been studied from various algorithmic perspectives. One of those variants called irredundance is highly related to domination. For example, every minimal dominating set of a graph G is also a maximal irredundant set of G. In this paper we study the enumeration of the maximal irredundant sets of a claw-free graph. We show that an n-vertex claw-free graph has $O(1.9341^n)$ maximal irredundant sets and these sets can be enumerated in the same time. We complement the aforementioned upper bound with a lower bound by providing a family of graphs having 1.5848^n maximal irredundant sets.

1 Introduction

Many problems of great theoretical importance and/or arising from real-world applications turn out to be intractable in the general case. These problems are typically in one of the following forms. An optimization problem asks for a best solution (defined under some criteria) from a set of feasible solutions. A counting problem asks for the number of feasible solutions. An enumeration problem asks to list all feasible solutions. While optimization is ubiquitous in algorithms, in some applications finding one or even all optimal solutions is not always satisfactory. Optimization might simply not be the goal. To give an example, experts of the domain of origin of the problem to solve it in an algorithmic way, for example biologists, may prefer to have a large set of feasible solutions that under their hard-to-formalize-objectives is used to find the "good solution(s)" in a non-algorithmic way. Another motivation for the study of enumeration algorithms is the fact that for many problems, even well-studied ones, we cannot exclude that the essentially best algorithm is one solving a corresponding enumeration problem.

This work is supported by the Research Council of Norway (the project CLASSIS) and the French National Research Agency (ANR project GraphEn/ANR-15-CE40-0009).

D. Fotakis et al. (Eds.): CIAC 2017, LNCS 10236, pp. 297–309, 2017.
DOI: 10.1007/978-3-319-57586-5_25

The classical algorithmic approach to enumeration (sometimes called output-sensitive) has been studied since more than 50 years with an ever growing interest in the last years. The holy grail of output-sensitive algorithms are those of polynomial (or even linear) delay. Recently classical worst-case running time analysis has been applied successfully to enumeration. Those algorithms typically have exponential running times due to the fact that, contrary to optimization, the output-size is exponential in the input-size. Often these (input-sensitive) enumeration algorithms are branching ones and a sophisticated time analysis like Measure & Conquer is needed. The running time of an enumeration algorithm implies in a trivial way a combinatorial upper bound on the output-size, i.e. the (maximum) number of objects to be enumerated in an input of size n, called an *upper bound*. The ultimate aim of input-sensitive enumeration are algorithms of best possible (worst case) running times, i.e. there are inputs that indeed show optimality of the established running time. This motivates the search for lower bounds as a measure for optimality which are typically achieved via the explicit construction of families of inputs having a certain number of objects which is called a *lower bound*. Thus one aims at so-called matching upper and lower bounds.

The number of papers on domination in graphs is in the thousands, and several well known surveys and books are dedicated to the topic (see, e.g., [14,15]). Concerning enumeration, Fomin et al. gave an $O(1.7159^n)$ time algorithm to enumerate all (inclusion) minimal dominating sets in an n-vertex graph, thereby showing that the maximum number of minimal dominating sets in such a graph is at most 1.7159^n; they also provided the until today best known lower bound of 1.5704^n [9]. This cornerstone paper of input-sensitive enumeration initiated a sequence of papers on enumerating all minimal dominating sets in graphs of various graph classes achieving matching bounds for all those classes except chordal graphs [1,6,7,22]. Motivated by a long-standing open question concerning the output-polynomial enumeration of the minimal transversals in a hypergraph, this enumeration problem has also been studied extensively from output-sensitive perspective [12,13,18–21].

Irredundance in graphs is strongly related to domination. A set of vertices D of a graph is *irredundant* if each vertex $u \in D$ has a *private vertex* v in the closed neighborhood of u that is not dominated by any other vertex of D except u (we refer to Sect. 2 for definitions). It is folklore that every minimal dominating set is an (inclusion) maximal irredundant set. While there is a large number of graph-theoretic papers on irredundance (see e.g. [2,14,15]), little is know about irredundance from the algorithmic viewpoint. For optimization, the *lower* and *upper irredundance numbers* of a graph denoted by $ir(G)$ and $IR(G)$ were considered. These parameters are defined as minimum and maximum size of a maximal irredundant set respectively. Unsurprisingly, both corresponding decision problems are NP-complete [11,16] and computing $ir(G)$ is known to be NP-hard on bipartite and split graphs [16,23]. On bipartite and chordal graphs, it holds that $IR(G)$ coincides with the independence number, and thus computing $IR(G)$ can be done in polynomial time [5,17] for these graph classes. Binkele et al. [2] provided a variety of results on the exact and parameterized complexity of irredundance, among others, algorithms computing $\text{ir}(G)$ and $\text{IR}(G)$ in times $O(1.9956^n)$ and $O(1.8475^n)$ respectively. The results of Binkele et al. [2] indicate that the irredundance problems are more complicated than the closely related domination problems. Clearly, a set of vertices D is a dominating set if and only if for each vertex u of the graph, D contains a vertex in the closed neighborhood of u, and, respectively, the minimal dominating sets are exactly the (inclusion) minimal sets with this property. The all aforementioned enumeration algorithms for minimal dominating sets [1, 6, 7, 9, 22] heavily depend on this *local* property. It is not the case for irredundant sets. In particular, a graph can have a vertex that is not dominated by any vertex of a maximal irredundant set and this non-locality creates difficulties for the application of standard techniques.

We consider the problem of enumeration of maximal irredundant sets for claw-free graphs. Claw-free graphs, which form a superclass of line graph, are well-known for their structural and algorithmic properties; to mention only the series of papers on the structure of claw-free graphs by Chudnovsky et al., see e.g. [4], and the polynomial time algorithms for the (weighted) independent set problem by Minty and Sbihi [24, 25]. While it is not known whether the maximum number of maximal irredundant sets of a graph on n vertices can be upper bounded by $(2 - \epsilon)^n$ for some $\epsilon > 0$, we show in this paper that an n-vertex claw-free graph has $O(1.9341^n)$ maximal irredundant sets and that these sets can be enumerated in time $O(1.9341^n)$. We complement the aforementioned upper bound with a lower bound by providing a family of graphs having 1.5848^n maximal irredundant sets. Due to space limitations we only sketch the proofs of our results in this extended abstract.

2 Preliminaries

We consider finite undirected graphs without loops or multiple edges. Throughout the paper we denote by $n = |V(G)|$ and $m = |E(G)|$ the numbers of vertices and edges of the input graph G respectively. For a graph G and a subset $U \subseteq V(G)$ of vertices, we write $G[U]$ to denote the subgraph of G induced by U. We write $G - U$ to denote the subgraph of G induced by $V(G) \setminus U$, and we write $G - u$ instead of $G - \{u\}$ for a single element set. For a vertex v, we denote by $N_G(v)$ the *(open) neighborhood* of v, i.e., the set of vertices that are adjacent to v in G. The *closed neighborhood* $N_G[v] = N_G(v) \cup \{v\}$. For a set of vertices $U \subseteq V(G)$, $N_G[U] = \cup_{v \in U} N_G[v]$ and $N_G(U) = N_G[U] \setminus U$. The *degree* of a vertex v is $d_G(v) = |N_G(v)|$. A graph is *claw-free* if it does not contain the *claw*, i.e. $K_{1,3}$, as an induced subgraph.

A vertex v of a graph G *dominates* a vertex u if $u \in N_G[v]$; similarly v dominates a set of vertices U if $U \subseteq N_G[v]$. For two sets $D, U \subseteq V(G)$, the set D dominates U if $U \subseteq N_G[D]$. Let $D \subseteq V(G)$. A vertex v is a *private vertex* (or, simply, a *private*) for $u \in D$, if v is dominated by u but v is not dominated by any other vertex of D; notice that a vertex can be a private for itself. A set of vertices D is an *irredundant set* of G if every vertex $v \in D$ has a private. An irredundant set D is *(inclusion) maximal* if D is irredundant but any proper superset of D has not this property.

We obtain upper bounds for the number of maximal irredundant sets via constructing recursive branching enumeration algorithms. For the analysis of the running time and the number of sets that are produced by such an algorithm, we use a technique based on solving recurrences for branching steps and branching rules respectively. We refer to the book [10] for a detailed introduction.

The following proposition provides a lower bound for the number of maximal irredundant sets of a claw-free graph.

Proposition 1. *For every $k \geq 1$, there is a claw-free graph with $n = 5k$ vertices with at least $10^{n/5}$ maximal irredundant sets.*

Since $10^{1/5} \geq 1.5848$, we obtain that there are claw-free graphs with at least 1.5848^n maximal irredundant sets.

To obtain an upper bound for the number of maximal irredundant sets of a claw-free graph, we need the following lemma.

Lemma 1. *If D is an irredundant set in a claw-free graph G, then $G[D]$ is a disjoint union of complete graphs.*

In our algorithm, we have to find a set W of minimum size in a graph G such that $G - W$ is a disjoint union of complete graphs. This problem is well-known under the name CLUSTER VERTEX DELETION. Combining the recent result about parameterized complexity of the problem by Boral et al. [3] and the approach for constructing exact algorithm proposed by Fomin et al. [8], we obtain the following lemma.

Lemma 2 [8]. *The* Cluster Vertex Deletion *problem can be solved in time $O(1.4765^n)$.*

3   Enumeration of Maximal Irredundant Sets for Claw-Free Graphs

Now we are ready to state our main result and give a sketch of its proof.

Theorem 1. *A claw-free graph has $O(1.9341^n)$ maximal irredundant sets, and these can be enumerated in time $O(1.9341^n)$.*

Proof. To prove the theorem, we construct a recursive branching algorithm that enumerates all maximal irredundant sets of the input graph.

Consider the unique positive real root $\lambda \approx 1.8393$ of the polynomial $x^3 - x^2 - x - 1$ and let $\varepsilon \approx 0.4006$ be the unique root of the equation

$$\lambda^x \cdot 2^{1-x} = 3^x \cdot 3^{(1-x)/3}.$$

We will explain the choice of λ and ε later; now we just observe that ε is used to balance running times for two base cases and proceed with the algorithm.

Let G be a claw-free graph. We use Lemma 2 to find a set $W \subseteq V(G)$ of minimum size such that $G - W$ is a disjoint union of complete graphs. We have two cases depending on whether $|W| \leq \varepsilon n$ or $|W| > \varepsilon n$. For each of these cases, we construct enumeration algorithms that give the bound for the number of maximal irredundant sets.

Let $|W| > \varepsilon n$. In this case, we use the fact that by Lemma 1, each maximal irredundant set D induces a disjoint union of complete graphs. It follows that if P is an induced path on 3 vertices in G with $V(P) = \{v_1 v_2 v_3\}$, then either $v_1 \notin D$, or $v_1 \in D$ and $v_2 \notin D$, or $v_1, v_2 \in D$ and $v_3 \notin D$. Following this idea, we construct a branching recursive algorithm that finds induced paths on 3 vertices in the considered graph and branches as indicated above. If at some moment we cannot find induced paths to branch any more, then we just consider all possible variants for the remaining undecided vertices by brute-force.

This leads to the following branching recursive algorithm ENUMCF-A(S, F, X), where $S, F, X \subseteq V(G)$ compose a partition of $V(G)$; note that some sets could be empty. The algorithm enumerates the maximal irredundant sets D of G such that $S \subseteq D$ and $D \cap F = \emptyset$. We say that the vertices of F are *forbidden* as they cannot be included in an irredundant set and the vertices of X are *free*. The measure of an instance (S, F, X) is the number of free vertices $|X|$. As it is standard, each step of the algorithm is executed only if the previous steps do not apply.

ENUMCF-A(S, F, X).

1. If $X = \emptyset$, then check whether S is a maximal irredundant set of G and output it if it holds; then stop.
2. If there is an induced path $P = v_1 v_2 v_3$ such that $v_1, v_2, v_3 \in S$, then stop.
3. If there is an induced path $P = v_1 v_2 v_3$ such that $v_1, v_2, v_3 \in S \cup X$, then let $\{x_1, \ldots, x_h\} = \{v_1, v_2, v_3\} \cap X$ and branch:
   (i) call ENUMCF-A$(S, F \cup \{x_1\}, X \setminus \{x_1\})$;
   (ii) if $h \geq 2$, then call ENUMCF-A$(S \cup \{x_1\}, F \cup \{x_2\}, X \setminus \{x_1, x_2\})$;
   (iii) if $h = 3$, then call ENUMCF-A$(S \cup \{x_1, x_2\}, F \cup \{x_3\}, X \setminus \{x_1, x_2, x_3\})$.
4. For each $Y \subseteq X$, call ENUMCF-A$(S \cup Y, F \cup (X \setminus Y), \emptyset)$.

To enumerate the maximal irredundant sets of a graph G, ENUMCF-A $(\emptyset, \emptyset, V(G))$ is called. Clearly, all the sets generated by the algorithm are maximal irredundant sets, because at Step 1, where we output sets, we verify whether a generated set is a maximal irredundant set. To show that every maximal irredundant set D of G is generated by the algorithm, we apply inductive arguments. To evaluate the running time, observe that we branch on Steps 3 and 4. The branching vector of Step 3 is either $(2, 1)$ or $(3, 2, 1)$ and the maximum branching number is λ. The algorithm uses Steps 1–3 while $G[S \cup X]$ has induced paths on three vertices. Suppose that we come to Step 4. Then $G[S \cup X] = G - F$ has no induced paths on three vertices and, therefore, $G[S \cup X]$ is a disjoint union of complete graphs. Since $W \subseteq V(G)$ is a set of minimum size such that $G - W$ is a disjoint union of complete graphs, $|F| \geq |W| > \varepsilon n$. Hence, $|X| \leq |S \cup X| = n - |F| \leq (1 - \varepsilon)n$ and the measure s of the instance for a node of the search tree where Step 4 is called is at most $(1 - \varepsilon)n$. It can be seen that for each $s \leq n$, the search tree has $O^*(\lambda^{n-s})$ nodes where Step 4 is called. On Step 4, we check $2^{|X|}$ subsets of X. Since $\lambda < 2$, we conclude that the number of leaves of the search tree is $O^*(\lambda^{\varepsilon n} \cdot 2^{(1-\varepsilon)n})$. The direct computations shows that $\lambda^{\varepsilon} \cdot 2^{(1-\varepsilon)} < 1.9341$. Then the number of the leaves is $O(1.9341^n)$. It follows that G has $O(1.9341^n)$ maximal irredundant sets and these set can be enumerated in time $O(1.9341^n)$.

From now we assume that $|W| \leq \varepsilon n$. Recall that $G - W$ is a disjoint union of complete graphs. Denote by $H_1, \ldots, H_k$ the components of $G - W$. We consider $3^{|W|}$ partitions (A, B, C) (some sets could be empty) of W, and for each partition, we enumerate all maximal irredundant sets D of G such that (i) $D \cap W = A$, (ii) the vertices of B are private vertices of some vertices of D, and (iii) each vertex of D has at least one private in $V(G) \setminus C$, i.e., the vertices of C are irrelevant in the sense that they are not included in D and not needed to ensure that each vertex of D has a private. Note that $B \cap D = \emptyset$ and a vertex of B could be a private for a vertex of A. Since $D \cap C = \emptyset$ and each vertex of D has at least one private in $V(G) \setminus C$ for the considered irredundant sets, it is sufficient to enumerate all maximal irredundant sets of $G - C$ satisfying (i) and (ii), because every maximal irredundant set of G satisfying (i)–(iii) is a maximal irredundant set of $G - C$ satisfying (i) and (ii).

Similarly to the case $|W| > \varepsilon n$, we construct the following branching recursive algorithm $\text{ENUMCF-B}(S, p, F, X)$, where $S, F, X \subseteq V(G)$ compose a partition of $V(G)$ (note that some sets could be empty) and $p \colon S \to 2^{V(G) \setminus C}$. The algorithm enumerates the maximal irredundant sets D of $G - C$ such that (a) $A \subseteq S \subseteq D$, (b) $B \cup C \subseteq F$ and $D \cap F = \emptyset$, and (c) for each vertex $v \in D$, there is a private for v in $p(v)$ with respect to D. We say that the vertices of F are *forbidden* as they cannot be included in an irredundant set and the vertices of X are *free*. The measure of an instance (S, p, F, X) is the number of free vertices $|X|$.

First, we perform the initialization as follows:

- set $G = G - C$ and let $W = A \cup B$,
- set $S = A$,
- set $F = B$,
- set $X = V(G) \setminus W$, and
- for each vertex $v \in S$, $p(v) = N_G[v] \setminus N_G[S \setminus \{v\}]$.

It is straightforward to observe that if D is an irredundant set of $G - C$ satisfying (i) and (ii), then D should be enumerated by $\text{ENUMCF-B}(S, p, F, X)$ for the above initial assignment. Notice that initially we have $p(v) \subseteq N_G[v] \setminus N_G[S \setminus \{v\}]$. Our algorithm maintains this property, i.e., the vertices of $p(v)$ are dominated only by v on each step. We say that a vertex $u \in S$ has *fixed privates* if $N_G[p(u)] \setminus \{u\} \subseteq F$. Notice that if u has fixed privates, then for any extension of S, u has a private in $p(u)$. The algorithm works in three stages. On Stage 1, we pick neighbors of vertices of B that have vertices of B as their privates, and on Stages 2 and 3, we extend the obtained sets.

On the first stage, we pick neighbors of vertices of B that have vertices of B as their privates and check the consistency of the choice. Recall that $B \cap D = \emptyset$

for a maximal irredundant set D that is constructed by the algorithm and each vertex of B should be a private for some vertex of D. Therefore, each vertex v of B should have a unique neighbor $u \in D$. As $(N_G(v) \setminus \{u\}) \cap D = \emptyset$, the vertices of $N_G(v) \setminus \{u\}$ should be forbidden if u is chosen to be in D. Clearly, if v is adjacent to two vertices in S, then v cannot be a private for any vertex of D and our choice is inconsistent. If v is adjacent to a unique vertex of S, then v is a private for this vertex. Otherwise, we branch on each free vertex u in $N_G(v)$ by including it in a (possible) maximal irredundant set and declaring the vertices of $N_G(v) \setminus \{u\}$ to be forbidden. Notice that because the vertices of $N_G(v) \setminus \{u\}$ become forbidden, we obtain that v is a private for u for any extension of the current set. Hence, we can safely set $p(u) = \{v\}$ and say that u has fixed privates. Now we are ready to describe this stage of the algorithm formally. When we select a neighbor of a vertex $v \in B$ such that this neighbor is included in an irredundant set, we say that v gets *assigned*; otherwise, the vertex $v \in B$ is *unassigned*.

Stage 1. Assignment.

1. If there is $v \in B$ such that $|N_G(v) \cap S| \geq 2$, then stop.
2. If there is an unassigned vertex $v \in B$ such that $|N_G(v) \cap S| = 1$, then set $p'(x) = p(x)$ for $x \in S \setminus \{u\}$ and $p'(u) = \{v\}$ for $\{u\} = N_G(v) \cap S$, set v to be assigned, set $F' = F \cup (N_G(v) \setminus \{u\})$, $X' = X \setminus N_G(v)$. Then if $F' \neq F$, call ENUMCF-B(S, p', F', X'); otherwise, set $p = p'$.
3. If there is an unassigned vertex $v \in B$ such that $N_G(v) \cap X = \emptyset$, then stop.
4. If there is an unassigned vertex $v \in B$, then set v assigned and for each $u \in N_G(v) \cap X$, branch: set $S' = S \cup \{u\}$, set $p'(x) = p(x) \setminus N_G[u]$ for each $x \in S$ and set $p'(u) = \{v\}$, set $F' = F \cup (N_G(v) \setminus \{u\})$ and $X' = X \setminus N_G(v)$, and then call ENUMCF-B(S', p', F', X').

Before we proceed with the next stage, recall that $H_1, \ldots, H_k$ are the components of $G - W$ and each H_i is a complete graph. After the first stage, the current set S contains the vertices of A and some vertices of $H_1, \ldots, H_k$. Notice that if some H_i contains two vertices of a maximal irredundant set D, then these vertices have their privates in W, because H_i is a complete graphs. Then these privates are in B, but all the vertices having their privates in B are selected on Stage 1. Hence, the set S can be extended only by selecting at most one free vertex in each H_i that does not contain vertices of S. Notice that if we select a vertex v of H_i to be included in a maximal irredundant set, we have to ensure that it has a private in H_i and the selection of v does not kill the privates of some already selected vertex. It is easy to observe that we can find a private for v if and only if H_i contains a vertex not dominated by the vertices of A and this condition is easy to verify. Hence, the main issue is to guarantee that by selecting v, we do not destroy privates of other vertices. It can be seen that v can only affect privates of the vertices of A. Notice that if $u \in A$ has fixed privates, then they cannot be destroyed. Therefore, we should take care only about the vertices of A that have non-fixed privates.

On Stage 2, we mainly apply a number of reduction steps whose aim is to fix privates for some vertices of S and forbid some vertices of $G - W$. For the first task, we use the property that if a vertex $u \in S$ has a private v such that $N_G[v] \setminus \{u\} \subseteq F$, then it is safe to fix v to be a private for u. We are forbidding a vertex of $G - W$ if its inclusion in an irredundant set impossible, because it destroys the privates for some vertex of S. We call H_i *finalized* if $V(H_i) \cap X = \emptyset$. Clearly, only vertices of non-finalized graphs H_i can be added to S.

Stage 2. Reduction.

1. If $X = \emptyset$, then check whether S is a maximal irredundant set of G and output it if it holds; then stop.
2. If there is $v \in S$ such that $p(v) = \emptyset$, then stop.
3. If for a vertex $u \in S$, $u \in p(u) \neq \{u\}$ and $N_G(u) \subseteq F$, then set $p(u) = \{u\}$.
4. If for a vertex $u \in S$, there is a finalized component H_i such that $p(u) \cap V(H_i) \neq \emptyset$, then set $p(u) = p(u) \cap V(H_i)$.
5. If there is a non-finalized component H_i such that $V(H_i) \subseteq N_G[S]$, then set $F' = F \cup (X \cap V(H_i))$, $X' = X \setminus V(H_i)$ and call ENUMCF-B(S, p, F', X').
6. If for a vertex $u \in S$, $p(u) \subseteq V(H_i)$ for a non-finalized component H_i, then set $F' = F \cup (X \cap V(H_i))$, $X' = X \setminus V(H_i)$ and call ENUMCF-B(S, p, F', X').
7. If for a vertex $u \in S$, there is a unique non-finalized component H_i with $N_G(u) \cap V(H_i) \neq \emptyset$ and if it holds that $N_G(u) \cap V(H_i) \cap X \neq \emptyset$ and $u \in p(u)$, then set $F' = F \cup (N_G(u) \cap V(H_i))$, set $p'(x) = p(x)$ for $x \in S \setminus \{u\}$ and $p'(u) = \{u\}$, set $X' = X \setminus (N_G(u) \cap V(H_i))$ and call ENUMCF-B(S, p', F', X').
8. If there is a vertex $u \in S$ such that $u \in p(u)$ and u has neighbors in two non-finalized components H_i and H_j but $V(H_j) \cap N_G(u) \subseteq N_G[S \setminus \{u\}]$ and $N_G(u) \cap V(H_i) \cap X \neq \emptyset$, then set $F' = F \cup (N_G(u) \cap V(H_i))$, set $X' = X \setminus (N_G(u) \cap V(H_i))$ and call ENUMCF-B(S, p, F', X').

Recall that in the selection of vertices in $H_1, \dots, H_k$ for the inclusion in an irredundant set, we should ensure that we do not destroy the privates for a vertex of A that has no fixed privates. Notice that because G is claw-free, every vertex of A has neighbors in at most two graphs from $H_1, \dots, H_k$. It can be shown that after Stage 2, we have the following property: if a vertex $u \in A$ has no fixed privates, then u has neighbors in exactly two distinct non-finalized H_i and H_j for $i, j \in \{1, \dots, k\}$. Moreover, u is an isolated vertex of $G[W]$, that is, it has neighbors only in H_i and H_j. We call such a vertex $u \in A$ an *important* vertex and we say that H_i and H_j are *adjacent* to u. Notice that at least one neighbor of an important vertex u should be free, because otherwise we would fix u to be a private for itself. If $v \in N_G(u) \cap V(H_i) \cap X$ for an important vertex $u \in A$ for some H_i, we say that v is a *pivot* of u in H_i. Clearly, each important vertex u has at least one pivot. We use pivots for branching on the vertices of graphs H_i exploiting the following two observations.

Suppose that $v \in V(H_i) \cap X$ is a pivot of an important vertex u adjacent to H_i and H_j. If v is included in a maximal irredundant set, then we have that v is the unique vertex of the set in H_i and, respectively, other vertices of H_i should be forbidden and, moreover, the vertices of H_j should be forbidden as well, because u should have its privates in H_j.

Suppose that $v \in V(H_i) \cap X$ is not a pivot of any important vertex u adjacent to H_i and H_j. Then there is an important vertex u adjacent to H_i and H_j that has a pivot w in H_j. If v is included in a maximal irredundant set, then the other vertices of H_j should be forbidden and the vertex w should be forbidden as well, because the inclusion of w in the irredundant set would kill all the privates of u.

For branching, we consider the auxiliary graph $\mathcal{H}$ whose nodes are non-finalized graphs $H_1, \ldots, H_k$ (we call them *nodes* to distinguish them from the vertices of G), and two graphs H_i and H_j are adjacent in $\mathcal{H}$ if and only if there is an important vertex $u \in A$ that is adjacent to H_i and H_j. First, we branch on non-isolated nodes of $\mathcal{H}$ using the aforementioned observations about pivots. If $\mathcal{H}$ has only isolated nodes, then we have no important vertices and, therefore, all the vertices of A have fixed privates. In this case, every non-finalized H_i should contain exactly one vertex of a maximal irredundant set and we branch respectively.

Stage 3. Branching.

1. If there are non-finalized H_i and H_j that are adjacent in $\mathcal{H}$ such that $|(V(H_i) \cup V(H_j)) \cap X| \geq 4$, then select a pivot v of an important vertex adjacent to H_i and H_j and branch:
   (i) set $S' = S \cup \{v\}$, $F' = F \cup ((V(H_i) \cup V(H_j)) \cap (X \setminus \{v\}))$, $X' = X \setminus (V(H_i) \cup V(H_j))$, $p'(x) = p(x) \setminus N_G[v]$ for $x \in S$ and $p'(v) = N_G[v] \setminus N_G[S]$, and call ENUMCF-B$(S', p', F', X')$,
   (ii) set $F' = F \cup \{v\}$, $X' = X \setminus \{v\}$, and call ENUMCF-B(S, p, F', X').
   Notice that from now we can assume that for any two H_i and H_j adjacent in $\mathcal{H}$, $|V(H_i) \cap X| + |V(H_j) \cap X| \leq 3$.

2. If there is a node H_i of $\mathcal{H}$ of degree at least three, then do the following. For $v \in V(H_i) \cap X$, let $I_v \subseteq \{1, \ldots, k\}$ be the set of all indices such that for every $j \in I_v$, there is an important vertex u adjacent to H_i and H_j having v as its pivot. Denote by R the sets of pivots of the important vertices adjacent to H_i. Then select $v \in V(H_i) \cap X$ such that $|I_v|$ is maximum and branch:
   (i) set $S' = S \cup \{v\}$, $F' = F \cup (V(H_i) \setminus \{v\}) \cup R \cup \bigcup_{j \in I_v} V(H_j)$, $X' = X \setminus (V(H_i) \cup R \cup \bigcup_{j \in I_v} V(H_j))$, $p'(x) = p(x) \setminus N_G[v]$ for $x \in S$ and $p'(v) = N_G[v] \setminus N_G[S]$, and call ENUMCF-B$(S', p', F', X')$,
   (ii) set $F' = F \cup \{v\}$, $X' = X \setminus \{v\}$, and call ENUMCF-B(S, p, F', X').

3. If there is a node H_i of $\mathcal{H}$ of degree two such that $|V(H_i) \cap X| = 2$, then let H_h and H_j be its neighbors in $\mathcal{H}$, and let $\{w_1\} = V(H_h) \cap X$, $\{w_2\} = V(H_j) \cap X$.

Then we branch as follows.

If $V(H_i)$ does not contain pivots, then we select an arbitrary vertex $v \in V(H_i) \cap X$, if there is a $v \in X \cap V(H_i)$ such that one of the following holds:
   – v is a pivot of two important vertices adjacent to H_h and H_j respectively,
   – v is a pivot of an important vertex adjacent to H_h and there is an important vertex adjacent to H_i and H_j that has its pivot in H_j,
   – v is a pivot of an important vertex adjacent to H_j and there is an important vertex adjacent to H_i and H_h that has its pivot in H_h,
   then select such v. Then branch:

(i) set $S' = S \cup \{v\}$, $F' = F \cup (V(H_i) \setminus \{v\}) \cup \{w_1, w_2\}$, $X' = X \setminus (V(H_i) \cup \{w_1, w_2\})$, $p'(x) = p(x) \setminus N_G[v]$ for $x \in S$ and $p'(v) = N_G[v] \setminus N_G[S]$, and call ENUMCF-B$(S', p', F', X')$,

(ii) set $F' = F \cup \{v\}$, $X' = X \setminus \{v\}$, and call ENUMCF-B(S, p, F', X').

If $V(H_i) \cap X = \{v_1, v_2\}$, where v_1 and v_2 are pivots of two important vertices u_1 and u_2 adjacent to H_h and H_j respectively such that $u_1 v_1, u_2 v_2 \in E(G)$, then branch:

(i) set $S' = S \cup \{v_1\}$, $F' = F \cup \{v_2, w_1\}$, $X' = X \setminus \{v_1, v_2, w_1\}$, $p'(x) = p(x) \setminus N_G[v_1]$ for $x \in S$ and $p'(v_1) = N_G[v_1] \setminus N_G[S]$, and call ENUMCF-B$(S', p', F', X')$,

(ii) set $S' = S \cup \{v_2\}$, $F' = F \cup \{v_1, w_2\}$, $X' = X \setminus \{v_1, v_2, w_2\}$, $p'(x) = p(x) \setminus N_G[v_2]$ for $x \in S$ and $p'(v_2) = N_G[v_2] \setminus N_G[S]$, and call ENUMCF-B$(S', p', F', X')$,

(iii) set $S' = S \cup \{w_1, w_2\}$, $F' = F \cup \{v_1, v_2\}$, $X' = X \setminus \{v_1, v_2, w_1, w_2\}$, $p'(x) = p(x) \setminus (N_G[w_1] \cup N_G[w_2])$ for $x \in S$ and $p'(w_1) = N_G[w_1] \setminus N_G[S]$, $p'(w_2) = N_G[w_2] \setminus N_G[S]$, and call ENUMCF-B$(S', p', F', X')$.

Notice that from now we can assume that for any H_i of degree two in $\mathcal{H}$, $|V(H_i) \cap X| = 1$.

4. If there is a component $\mathcal{C}$ of $\mathcal{H}$ that is a cycle, then select distinct nodes $H_{h_1}, H_{h_2}, H_{h_3}$ of $\mathcal{C}$ such that $H_{h_1} H_{h_2}, H_{h_2} H_{h_3} \in E(\mathcal{C})$ and do the following. Let H_{h_0} be the neighbor of H_{h_1} in $\mathcal{C}$ distinct from H_{h_2} and let H_{h_4} be the neighbor of H_{h_3} distinct from H_{h_2} (note that it can happen that $H_{h_0} = H_{h_3}$ or $H_{h_0} = H_{h_4}$ or $H_{h_1} = H_{h_4}$). For $i \in \{1, 2, 3\}$, we select $v_i \in V(H_{h_i}) \cap X$ and branch as follows for $i = 1, 2, 3$: set $S' = S \cup \{v_i\}$, $F' = F \cup \{v_{i-1}, v_{i+1}\}$, $X' = X \setminus \{v_{i-1}, v_i, v_{i+1}\}$, $p'(x) = p(x) \setminus N_G[v_i]$ for $x \in S$ and $p'(v_i) = N_G[v_i] \setminus N_G[S]$, and call ENUMCF-B$(S', p', F', X')$.

5. If there is a component of $\mathcal{H}$ that is a path $\mathcal{P} = H_{h_1} \ldots H_{h_r}$ of length at least one with $|V(H_{h_1}) \cap X| = |V(H_{h_2}) \cap X| = 1$, then let $v_1 \in V(H_{h_1}) \cap X$, $v_2 \in V(H_{h_2}) \cap X$ and branch for $i = 1, 2$: set $S' = S \cup \{v_i\}$, $F' = F \cup (\{v_1, v_2\} \setminus \{v_i\})$, $X' = X \setminus \{v_1, v_2\}$, $p'(x) = p(x) \setminus N_G[v_i]$ for $x \in S$ and $p'(v_i) = N_G[v_i] \setminus N_G[S]$, and call ENUMCF-B$(S', p', F', X')$.

6. If there is a component of $\mathcal{H}$ that is a path $\mathcal{P} = H_{h_1} \ldots H_{h_r}$ of length at least one with $|V(H_{h_1}) \cap X| = 2$ and $|V(H_{h_2}) \cap X| = 1$, then let $V(H_{h_1}) \cap X = \{v_1, v_2\}$ and $V(H_{h_2}) \cap X = \{v_3\}$ and branch as follows depending on the properties of v_1, v_2, v_3.

If v_1 and v_2 are pivots of some important vertices adjacent to H_{h_1} and H_{h_2} or v_3 is a pivot of an important vertex adjacent to H_{h_1} and H_{h_2}, then branch for $i = 1, 2, 3$: set $S' = S \cup \{v_i\}$, $F' = F \cup (\{v_1, v_2, v_3\} \setminus \{v_i\})$, $X' = X \setminus \{v_1, v_2, v_3\}$, $p'(x) = p(x) \setminus N_G[v_i]$ for $x \in S$ and $p'(v_i) = N_G[v_i] \setminus N_G[S]$, and call ENUMCF-B$(S', p', F', X')$.

Otherwise, if v_3 and exactly one of the vertices v_1, v_2 are not pivots of any important vertex adjacent to H_{h_1} and H_{h_2}, branch for $i = 1, 2$: set $S' = S \cup \{v_i\}$, $F' = F \cup (\{v_1, v_2\} \setminus \{v_i\})$, $X' = X \setminus \{v_1, v_2\}$, $p'(x) = p(x) \setminus N_G[v_i]$ for $x \in S$ and $p'(v_i) = N_G[v_i] \setminus N_G[S]$, and call ENUMCF-B$(S', p', F', X')$.

7. If every component of $\mathcal{H}$ is an isolated vertex, then branch: for every non-finalized H_i and for every $v \in V(H_i) \cap X$, set $S' = S \cup \{v\}$, $F' = F \cup (V(H_i) \setminus \{v\})$, $X' = X \setminus V(H_i)$, $p'(x) = p(x) \setminus N_G[v]$ for $x \in S$ and $p'(v) = N_G[v] \setminus N_G[S]$, and call ENUMCF-B$(S', p', F', X')$.

It is straightforward to see that all sets generated by the algorithm are maximal irredundant sets of $G - C$, because at Step 1 of Stage 2, where we output sets, we verify whether a generated set is a maximal irredundant set. To show that every maximal irredundant set D of $G - C$ such that (i) $W \cap D = A$ and (ii) the vertices of B are privates of some vertices of D, we use inductive arguments.

Now we evaluate the running time of ENUMCF-B. To do it, we use the standard approach (see [10]) and compute branching vectors for all branching steps of the algorithm. Recall that the measure of an instance is the number of free vertices $s = |X|$. By the analysis of the branching steps we obtain that the branching numbers are at most $3^{1/3}$. Hence, for each partition (A, B, C) of W, the algorithm ENUMCF-B produces $O^*(3^{(n-|W|)/3})$ maximal irredundant sets in time $O^*(3^{(n-|W|)/3})$. Since there are at most $3^{|W|}$ partitions (A, B, C), we conclude that G has $O^*(3^{|W|} \cdot 3^{(n-|W|)/3})$ maximal irredundant sets that can be enumerated in time $O^*(3^{|W|} \cdot 3^{(n-|W|)/3})$. Recall that $|W| \leq \varepsilon n$. As $3^{\varepsilon} \cdot 3^{(1-\varepsilon)/3} < 1.9341$, G has $O(1.9341^n)$ maximal irredundant sets and these sets can be enumerated in time $O(1.9341^n)$.

4   Conclusions

We have shown that the maximum number of maximal irredundant sets of a claw-free graph is upper bounded by $O(1.9341^n)$ and lower bounded by 1.5848^n. The upper bound has been established by a branching algorithm to enumerate all maximal irredundant sets of a claw-free graph having running time $O(1.9341^n)$.

The following tasks related to our research are challenging: improving upon the 2^n trivial upper bound for the number of maximal irredundant sets in general n-vertex graphs, and upon the upper bound $O(1.9341^n)$ for line graphs which are all claw-free. It would also be interesting to know whether it is possible to establish an output-polynomial algorithm to enumerate the maximal irredundant sets of claw-free graphs. Note that all cobipartite graphs are claw-free and that the existence of an output-polynomial algorithm for cobipartite graphs would imply the existence of an output-polynomial time enumeration algorithm for general graphs which can be shown using the construction of the corresponding result for enumeration of the minimal dominating sets [19].

References

1. Abu-Khzam, F.N., Heggernes, P.: Enumerating minimal dominating sets in chordal graphs. Inf. Process. Lett. **116**(12), 739–743 (2016)
2. Binkele-Raible, D., Brankovic, L., Cygan, M., Fernau, H., Kneis, J., Kratsch, D., Langer, A., Liedloff, M., Pilipczuk, M., Rossmanith, P., Wojtaszczyk, J.O.: Breaking the 2^n-barrier for Irredundance: two lines of attack. J. Discret. Algorithms **9**(3), 214–230 (2011)

3. Boral, A., Cygan, M., Kociumaka, T., Pilipczuk, M.: A fast branching algorithm for cluster vertex deletion. Theory Comput. Syst. **58**(2), 357–376 (2016)
4. Chudnovsky, M., Seymour, P.: Claw-free graphs. V. Global structure. J. Comb. Theory Ser. B **98**, 1373–1410 (2008)
5. Cockayne, E., Favaron, O., Payan, C., Thomason, A.: Contributions to the theory of domination, independence and irredundance in graphs. Discret. Math. **33**, 249–258 (1981)
6. Couturier, J.F., Heggernes, P., van't Hof, P., Kratsch, D.: Minimal dominating sets in graph classes: combinatorial bounds and enumeration. Theoret. Comput. Sci. **487**, 82–94 (2013)
7. Couturier, J.-F., Letourneur, R., Liedloff, M.: On the number of minimal dominating sets on some graph classes. Theoret. Comput. Sci. **562**, 634–642 (2015)
8. Fomin, F.V., Gaspers, S., Lokshtanov, D., Saurabh, S.: Exact algorithms via monotone local search. In: STOC 2016, pp. 764–775 (2016)
9. Fomin, F.V., Grandoni, F., Pyatkin, A.V., Stepanov, A.A.: Combinatorial bounds via measure and conquer: bounding minimal dominating sets and applications. ACM Trans. Algorithms **5**(1), 9:1–9:17 (2008)
10. Fomin, F.V., Kratsch, D.: Exact exponential algorithms. Texts in Theoretical Computer Science. An EATCS Series. Springer, Heidelberg (2010)
11. Fellows, M.R., Fricke, G., Hedetniemi, S.T., Jacobs, D.P.: The private neighbor cube. SIAM J. Discret. Math. **7**, 41–47 (1994)
12. Golovach, P.A., Heggernes, P., Kanté, M.M., Kratsch, D., Villanger, Y.: Enumerating minimal dominating sets in chordal bipartite graphs. Discret. Appl. Math. **199**, 30–36 (2016)
13. Golovach, P.A., Heggernes, P., Kratsch, D., Villanger, Y.: An incremental polynomial time algorithm to enumerate all minimal edge dominating sets. Algorithmica **72**, 836–859 (2015)
14. Haynes, T.W., Hedetniemi, S.T., Slater, P.: Domination in Graphs: Advanced Topics. Marcel Dekker, New York (1997)
15. Haynes, T.W., Hedetniemi, S.T., Slater, P. (eds.): Fundamentals of Domination in Graphs. CRC Press, New York (1998)
16. Hedetniemi, S.T., Laskar, R., Pfaff, J.: Irredundance in graphs: a survey. Congr. Numerantium **48**, 183–193 (1985)
17. Jacobson, M.S., Peters, K.: Chordal graphs and upper irredundance, upper domination and independence. Discret. Math. **86**, 59–69 (1990)
18. Kanté, M.M., Limouzy, V., Mary, A., Nourine, L.: On the neighbourhood helly of some graph classes and applications to the enumeration of minimal dominating sets. In: Chao, K.-M., Hsu, T., Lee, D.-T. (eds.) ISAAC 2012. LNCS, vol. 7676, pp. 289–298. Springer, Heidelberg (2012). doi:10.1007/978-3-642-35261-4_32
19. Kanté, M.M., Limouzy, V., Mary, A., Nourine, L.: On the enumeration of minimal dominating sets and related notions. SIAM J. Discret. Math. **28**(4), 1916–1929 (2014)
20. Kanté, M.M., Limouzy, V., Mary, A., Nourine, L., Uno, T.: On the enumeration and counting of minimal dominating sets in interval and permutation graphs. In: Cai, L., Cheng, S.-W., Lam, T.-W. (eds.) ISAAC 2013. LNCS, vol. 8283, pp. 339–349. Springer, Heidelberg (2013). doi:10.1007/978-3-642-45030-3_32
21. Kanté, M.M., Limouzy, V., Mary, A., Nourine, L., Uno, T.: Polynomial delay algorithm for listing minimal edge dominating sets in graphs. In: Dehne, F., Sack, J.-R., Stege, U. (eds.) WADS 2015. LNCS, vol. 9214, pp. 446–457. Springer, Cham (2015). doi:10.1007/978-3-319-21840-3_37

22. Krzywkowski, M.: Trees having many minimal dominating sets. Inform. Proc. Lett. **113**, 276–279 (2013)
23. Laskar, R., Pfaff, J.: Domination and irredundance in graphs, Technical report 434, Clemson University, Department of Mathematical (1983)
24. Minty, G.J.: On maximal independent sets of vertices in claw-free graphs. J. Comb. Theory Ser. B **28**, 284–304 (1980)
25. Sbihi, N.: Algorithme de recherche d'un stable de cardinalité maximum dans un graphe sans étoile. Discret. Math. **29**, 53–76 (1980)

Approximate Maximin Share Allocations in Matroids

Laurent Gourvès[⊠] and Jérôme Monnot

Université Paris-Dauphine, PSL Research University, CNRS, UMR 7243, LAMSADE, 75016 Paris, France
{laurent.gourves,jerome.monnot}@dauphine.fr

Abstract. The maximin share guarantee is, in the context of allocating indivisible goods to a set of agents, a recent fairness criterion. A solution achieving a constant approximation of this guarantee always exists and can be computed in polynomial time. We extend the problem to the case where the goods collectively received by the agents satisfy a matroidal constraint. Polynomial approximation algorithms for this generalization are provided: a 1/2-approximation for any number of agents, a $(1 - \varepsilon)$-approximation for two agents, and a $(8/9 - \varepsilon)$-approximation for three agents. Apart from the extension to matroids, the $(8/9 - \varepsilon)$-approximation for three agents improves on a $(7/8 - \varepsilon)$-approximation by Amanatidis *et al.* (ICALP 2015).

Keywords: Approximation algorithms · Fair division · Matroids

1 Introduction

This article deals with the allocation of a set $\mathcal{X}$ of indivisible goods to a set N of n agents. The agents typically have different valuations for the elements of $\mathcal{X}$ and the goal is to find a *fair* allocation. As opposed to *cake-cutting* (when *divisible* resources have to be shared), not every instance with indivisible goods admits a solution that is *envy-free* (everyone finds her share at least as good as the share of another agent) or *proportional* (everyone values her share at least her valuation for $\mathcal{X}$ divided by n), even for two agents [1]. Is there any fairness criterion that can be satisfied for indivisible goods? In the hierarchy of fairness provided by Bouveret and Lemaître [2], the *maximin share guarantee* is known to be less demanding than envy-freeness and proportionality. Suppose an agent is given the opportunity to partition $\mathcal{X}$ in n parts but she is adversarially allocated her least preferred subset in this partition. The maximin share of an agent is, within the set of all partitions, her maximum value for the least preferred part. The maximin share criterion is satisfied when there exists an allocation in which every agent gets at least her maximin share. Every 2-agent instance satisfies this fairness criterion popularized by Budish [3], but for $n \geq 3$, Procaccia and Wang [4] (see also [5]) have provided a family of intricate counterexamples.

Supported by the project ANR CoCoRICo-CoDec.

D. Fotakis et al. (Eds.): CIAC 2017, LNCS 10236, pp. 310–321, 2017.
DOI: 10.1007/978-3-319-57586-5_26

The impossibility to guarantee the existence of a fair (i.e. satisfying envy-freeness, proportionality or maximin share) allocation of indivisible goods is quite embarrassing. Fortunately, approximation can circumvent the obstacle. There exist approximation results in this respect, e.g. [4,6–9], but envy-freeness and proportionality are not amenable to multiplicative approximation within a constant factor. Nevertheless, the recent works on maximin share have provided algorithms with constant approximation factors, i.e. polynomial algorithms that return an allocation in which every agent values her share at least ρ times her maximin share. One can mention the 2/3-approximation algorithm of Procaccia and Wang [4] which is polynomial for n constant. Recently, Amanatidis *et al.* [9] have proposed two (polynomial in n) algorithms with approximation factors $1/2$ and $2/3 - \varepsilon$, respectively. Better ratios can be reached if n is small. The *cut-and-choose* protocol gives a $(1 - \varepsilon)$-approximation [2,10] for two agents and every $\varepsilon > 0$. For three agents, a $(3/4 - \varepsilon)$-approximation has been proposed [4], followed by an improved $(7/8 - \varepsilon)$-approximation [9].

The need for fair solutions appears in various contexts, see e.g. [11], and these contexts impose some constraints on which objects are collectively allocated to the agents. The classical framework of fair division rarely integrates sophisticated constraints. However, putting constraints on a set of discrete objects (the indivisible goods) can give rise to a rich combinatorial structure that can be exploited. The specificity of the present work is that the objects collectively received by the agents must satisfy a given feasibility constraint modeled with a set system, i.e. a collection $\mathcal{F}$ of subsets of $\mathcal{X}$, such that $F \subseteq \mathcal{X}$ is feasible if, and only if, $F \in \mathcal{F}$. Let us illustrate and motivate the model with simple examples.

- *Budget:* There can be an upper bound on the number of objects that the agents collectively receive.
- *Mutual exclusion:* For given pairs of objects, at most one element in each pair can be allocated.
- *Storage:* Suppose the allocated objects must be stored in some places (each place can accommodate at most one object within a given subset); a collective set of objects is feasible if, and only if, all its elements can be stored.

The set system model clearly generalizes the allocation of indivisible goods. However, it is too large to allow the existence of a general approximation algorithm. Indeed, one can easily exhibit a special case in which computing a feasible allocation is, without any consideration of fairness, an intractable problem. Therefore, a reasonable task is to identify a significant subclass in which approximation is possible.

This article deals with the set system model when $(\mathcal{X}, \mathcal{F})$ is a matroid (defined in Sect. 2) which extends the allocation of indivisible goods (indeed, the *free matroid* is such that $\mathcal{F} = 2^{\mathcal{X}}$). The above examples (budget, mutual exclusion, and storage) are matroidal constraints. We resort to approximation to provide allocations that satisfy the feasibility constraint and the maximin share guarantee up to a constant multiplicative factor. A polynomial 1/2-approximation algorithm that extends the one of [9] is presented in Sect. 3. It works for any number of agents, but if we restrict ourselves to a small number of agents, better ratios can be reached: For any ε, polynomial approximation algorithms with factors $1 - \varepsilon$ and $\frac{8}{9} - \varepsilon$ can be achieved for two and three agents, respectively. These results are given in Sects. 4 and 5, respectively. The $(\frac{8}{9} - \varepsilon)$-approximation improves on the known $(\frac{7}{8} - \varepsilon)$-approximation. We conclude with possible improvements for the 3-agent case, and possible extensions of the model to greedoids, independence systems, and the intersection of two matroids. Due to space limitations, some proofs are omitted.

2   Set Systems and Maximin Share Fairness

A *set system* $(\mathcal{X}, \mathcal{F})$ consists of a finite set of elements $\mathcal{X} = \{x_1, \ldots, x_m\}$ and a collection $\mathcal{F} \subseteq 2^{\mathcal{X}}$ which defines the feasible solutions. Let $N = \{1, \ldots, n\} = [n]$ be a set of n agents. We suppose that every agent i has an additive valuation function $\nu_i : \mathcal{X} \to \mathbb{R}_{\geq 0}$. For every $P \subseteq \mathcal{X}$, $\nu_i(P)$ is defined as $\sum_{x \in P} \nu_i(x)$, and $\mathcal{F}_{|P}$ denotes $2^P \cap \mathcal{F}$. Thus, $(P, \mathcal{F}_{|P})$ is the restriction of $(\mathcal{X}, \mathcal{F})$ to P.

Definition 1. *For a set system* $(\mathcal{X}, \mathcal{F})$ *and* $P \subseteq \mathcal{X}$, *a feasible allocation* $T = (T_1, \ldots, T_{|N|})$ *of* P *for a set of agents* N *satisfies*

- $T_i \cap T_{i'} = \emptyset$ *for every pair of agents* $(i, i') \in N^2$;
- $(\cup_{i \in N} T_i) \subseteq P$;
- $(\cup_{i \in N} T_i) \in \mathcal{F}_{|P}$.

T_i is what agent i receives (her share). The set of all feasible allocations of P over n agents is denoted by $\Pi_n(P, \mathcal{F}_{|P})$. The fact that we define a feasible allocation for a subset of $\mathcal{X}$, and not only for $\mathcal{X}$, will become clear in the next sections.

For a set system $(\mathcal{X}, \mathcal{F})$, an agent i and $P \subseteq \mathcal{X}$, $\beta^i(P, \mathcal{F}_{|P})$ denotes a member of $\mathcal{F}_{|P}$ for which agent i has maximum valuation. Since $\nu_i(x_j) \geq 0$ for every i, j, we will assume w.l.o.g. that $\beta^i(P)$ is maximal for inclusion.

Definition 2. *Given* $d \in [n]$, *a set system* $(\mathcal{X}, \mathcal{F})$, *and* $P \subseteq \mathcal{X}$, *the* d-*maximin share of an agent* i *with respect to* P, *is denoted by* $\mu_i(d, P, \mathcal{F}_{|P})$ *and defined as* $\max_{T \in \Pi_d(P, \mathcal{F}_{|P})} \min_{T_j \in T} \nu_i(T_j)$. *When* $d = n$ *and* $P = \mathcal{X}$, $\mu_i(n, \mathcal{X}, \mathcal{F})$ *is simply called the* maximin share *of agent* i.

Definition 3. *For* $\rho \in (0, 1]$, *a group* N *of* n *agents, and a set system* $(\mathcal{X}, \mathcal{F})$, *a* ρ-*approximate maximin share allocation is an allocation* $T \in \Pi_n(\mathcal{X}, \mathcal{F})$ *such that* $\nu_i(T_i) \geq \rho \cdot \mu_i(n, \mathcal{X}, \mathcal{F})$, *for every agent* $i \in N$. *A* 1-*approximate maximin share allocation is simply called a* maximin share allocation.

We say that an allocation T *achieves* the maximin share of a given agent i when $\min_{T_j \in T} \nu_i(T_i) = \mu_i(n, \mathcal{X}, \mathcal{F})$.

The contribution of this article are approximation algorithms for the significant subclass of matroids. Some elements of matroid theory are provided for the sake of presentation.

2.1 Basic Notions of Matroids

A matroid is a set system $(\mathcal{X}, \mathcal{F})$ satisfying the next three properties:

(M1) $\emptyset \in \mathcal{F}$;
(M2) if $F_2 \subseteq F_1$ and $F_1 \in \mathcal{F}$, then $F_2 \in \mathcal{F}$;
(M3) if $F_1, F_2 \in \mathcal{F}$ such that $|F_1| < |F_2|$, then there exists $x \in F_2 \backslash F_1$ such that $F_1 \cup \{x\} \in \mathcal{F}$.

The elements of $\mathcal{F}$ are called *independent sets*. Inclusionwise maximal independent sets are called *bases*. The *rank* of a subset $P \subseteq \mathcal{X}$ is defined as $\mathrm{rank}(P) = \max\{|F| : F \subseteq P, F \in \mathcal{F}\}$. All the bases of a matroid have the same cardinality $\mathrm{rank}(\mathcal{X})$, also called the rank of the matroid.

Given a subset $\mathcal{X}' \subset \mathcal{X}$, the *restriction* of $(\mathcal{X}, \mathcal{F})$ to $\mathcal{X}'$ is a matroid $(\mathcal{X}', \mathcal{F}')$ where $\mathcal{F}' = \{F \in \mathcal{F} : F \subseteq \mathcal{X}'\}$. If $G \in \mathcal{F}$, then the *contraction* of $(\mathcal{X}, \mathcal{F})$ by G, denoted by $(\mathcal{X}, \mathcal{F})/G$, is a matroid $(\mathcal{X} \setminus G, \mathcal{F}')$ where $\mathcal{F}' = \{F \subseteq \mathcal{X} \setminus G : F \cup G \in \mathcal{F}\}$.

Matroids satisfy the *bases exchange property* [12]: let B_1, B_2 be two distinct bases. Then for every $e_1 \in B_1 \backslash B_2$, there exists $e_2 \in B_2 \backslash B_1$ such that $B_1 - e_1 + e_2$ and $B_2 - e_2 + e_1$ are two bases. Matroids also satisfy the *multiple bases exchange properties* [13–15]. Let A, B be two distinct bases.

- Then for every partition (A_1, A_2) of A, there exists a partition (B_1, B_2) of B such that $A_1 \cup B_2$ and $A_2 \cup B_1$ are two bases.
- Then for every n-partition $(A_1, \ldots, A_n)$ of A, there exists a partition $(B_1, \ldots, B_n)$ of B such that $A \setminus A_i \cup B_i$ is a base, $\forall i \in [n]$. The construction of $(B_1, \ldots, B_n)$, for $(A_1, \ldots, A_n)$ given, can be done in polynomial time [16].

Typical examples of matroids are *uniform matroids* (corresponding to the previous example called *Budget*), *free matroids* (corresponding to the "classical" allocation of indivisible goods), *partition matroids* (generalizing the previous example called *Mutual exclusion*), *transversal matroids* (corresponding to the previous example called *Storage*), see [17] for the definitions.

When every element $x \in \mathcal{X}$ has a weight $w(x) \in \mathbb{R}_{\geq 0}$, a typical optimization problem consists in computing a base B that maximizes $\sum_{x \in B} w(x)$. This problem is solved by the GREEDY algorithm given in Algorithm 1.

Algorithm 1. GREEDY

   Data: $(\mathcal{X}, \mathcal{F})$, $w : \mathcal{X} \to \mathbb{R}_{\geq 0}$
1 Let $\mathcal{X} = \{x_1, \ldots, x_m\}$ such that $w(x_i) \geq w(x_{i+1})$, $\forall i \in \{1, \ldots, m-1\}$
2 $B \leftarrow \emptyset$
3 **for** $i = 1$ *to* m **do**
4   **if** $B + x_i \in \mathcal{F}$ **then**
5        $B \leftarrow B + x_i$

6 **return** B

The time complexity of matroid algorithms depends on the time for testing if a set is independent. Here, we assume that the test runs in polynomial time with respect to the input data. For the ease of presentation, we often write that a matroid is part of the input of an algorithm but concretely, we only require $\mathcal{X}$ and the test.

2.2 The Matroidal Set System Model

In this article, we suppose that the set system $(\mathcal{X}, \mathcal{F})$ is a matroid. The maximin share of any agent ℓ can be estimated as follows. Use GREEDY to compute a base $\beta^\ell(\mathcal{X}, \mathcal{F})$ that has maximum valuation for agent ℓ. The next step relies on the following lemma.

Lemma 1 [18]. *Let $(\mathcal{X}, \mathcal{F})$ be a matroid and $w : \mathcal{X} \to \mathbb{R}_{\geq 0}$ an additive weight function. Given a base A of maximum weight and a partition of another base B into n parts $B_1, \ldots, B_n$, there exists a partition $A_1, \ldots, A_n$ of A satisfying $\min_{i \in [n]} w(A_i) \geq \min_{i \in [n]} w(B_i)$.*

Therefore, $\beta^\ell(\mathcal{X}, \mathcal{F})$ can be partitioned in $(\gamma_1, \ldots, \gamma_n)$ in such a way that $\min_{i \in [n]} \nu_\ell(\gamma_i) = \min_{i \in [n]} \nu_\ell(B_i)$ where $(B_1, \ldots, B_n)$ achieves the maximin share allocation for agent ℓ, i.e. $\min_{i \in [n]} \nu_\ell(B_i) = \mu_\ell(n, \mathcal{X}, \mathcal{F})$. With $\varepsilon \in [0, 1)$, use the PTAS of Woeginger [10] to partition $\beta^\ell(\mathcal{X}, \mathcal{F})$ in n bundles $(\beta_1^\ell, \ldots, \beta_n^\ell)$ so as to maximize the value of the lightest bundle. We obtain a $(1 - \varepsilon)$-approximation of $\min_{i \in [n]} \nu_\ell(\gamma_i) = \mu_\ell(n, \mathcal{X}, \mathcal{F})$, together with a feasible allocation $(\beta_1^\ell, \ldots, \beta_n^\ell)$ achieving this value.

Proposition 1. *For every $\ell \in N$ and $\varepsilon \in [0, 1)$, $\beta^\ell(\mathcal{X}, \mathcal{F})$ can be partitioned in n bundles $(\beta_1^\ell, \ldots, \beta_n^\ell)$ such that $\min_{i \in [n]} \nu_\ell(\beta_i^\ell) \geq (1 - \varepsilon)\mu_\ell(n, \mathcal{X}, \mathcal{F})$. Moreover, $(\beta_1^\ell, \ldots, \beta_n^\ell)$ can be built in polynomial time.*

3   A $\frac{1}{2}$-Approximation for Any Number of Agents

We propose an adaptation of the $\frac{1}{2}$-approximation algorithm of Amanatidis *et al.* [9] to the matroidal set system model. The algorithm is given in Algorithm 4 but beforehand, preliminary properties and algorithms are provided.

Claim 1. *Given a matroid $(\mathcal{X}, \mathcal{F})$, for every $i \in N$ and every $P \subseteq \mathcal{X}$,*

$$\mu_i(|N|, P, \mathcal{F}_{|P}) \leq \frac{\nu_i(\beta^i(P, \mathcal{F}_{|P}))}{|N|} \leq \frac{\nu_i(\beta^i(\mathcal{X}, \mathcal{F}))}{|N|}.$$

The input of the problem is a matroid $(\mathcal{X}, \mathcal{F})$ but for the moment we consider a contraction $(\mathcal{X}', \mathcal{F}')$ of $(\mathcal{X}, \mathcal{F})$, which is also matroid. Let $S = (S_1, \ldots, S_{|N|})$ be the feasible allocation returned by Greedy Round-Robin (see Algorithm 2) with input $(\mathcal{X}', \mathcal{F}')$.

Algorithm 2. Greedy Round-Robin

Data: a matroid $(\mathcal{X}', \mathcal{F}')$, a set of agents N, and their valuations $(\nu_i)_{i \in N}$

1 $S_i = \emptyset$ for every $i \in N$
2 Consider the agents by ascending index, proceeding in a round-robin fashion
3 **while** $rank(\cup_i S_i) < rank(\mathcal{X}')$ **do**
4      $S_i \leftarrow S_i \cup \{x_j\}$, where i is the next agent to be examined in the current round and x_j is agent i's most desired element among the currently unallocated elements that can be added to $\cup_i S_i$ without violating the independence property (i.e. $\{x_j\} \cup (\cup_i S_i) \in \mathcal{F}'$)
5 **return** $S = (S_1, \ldots, S_{|N|})$

Take any agent $k \in N$. Algorithm 3 does not directly contribute to the construction of the final $\frac{1}{2}$-approximate solution but it is used in the analysis. Algorithm 3 finds a base $\beta^k(\mathcal{X}', \mathcal{F}')$ for which agent k has maximum valuation, and returns a partition $(\beta_1^k, \ldots, \beta_{|N|}^k)$ of it with the help of the output $(S_1, \ldots, S_{|N|})$ of Greedy Round-Robin. We denote by $\{x_1, \ldots, x_t\}$ the elements of $\cup_i S_i$ and we suppose w.l.o.g. that x_i is inserted right before x_{i+1} in Greedy Round-Robin, for $i = 1 \ldots t - 1$. Algorithm 3 gradually constructs $(\beta_1^k, \ldots, \beta_{|N|}^k)$ with the elements of $\beta^k(\mathcal{X}', \mathcal{F}')$. Meanwhile B, which is initially equal to $\beta^k(\mathcal{X}', \mathcal{F}')$, is gradually modified to eventually become equal to $\{x_1, \ldots, x_t\}$. In the end of Algorithm 3, each x_i is mapped with an element $f(x_i) \in \beta^k(\mathcal{X}', \mathcal{F}')$; f is a bijection.

In fact, Algorithm 3 represents the main difference between the original approximation algorithm and its extension to the matroidal set system. It allows to translate an allocation produced by an agent i into another allocation for which another agent has a valuation.

Let ν_{max} be $\max_{i,j} \nu_i(x_j)$ where $\{x_j\} \in \mathcal{F}'$. It is the maximum valuation of an agent for an element belonging to a feasible allocation.

Theorem 1. *For any agent $k \in N$, and $(S_1, \ldots, S_{|N|})$ and $(\beta_1^k, \ldots, \beta_{|N|}^k)$ the allocations returned by Algorithms 2 and 3, respectively, it holds that*

$$\nu_k(S_k) \geq \frac{\nu_k(\beta^k(\mathcal{X}', \mathcal{F}'))}{|N|} - \nu_{max} \geq \mu_k(|N|, \mathcal{X}', \mathcal{F}') - \nu_{max}.$$

Proof. Fix an agent k. Algorithm 3 manipulates a base $\cup_i S_i = \{x_1, \ldots, x_t\}$ and a base B initially equal to $\beta^k(\mathcal{X}', \mathcal{F}')$. Algorithm 3 gradually transforms B into $\{x_1, \ldots, x_t\}$. For every $i \in [t]$, B contains $\{x_1, \ldots, x_{i-1}, f(x_i)\}$ and $\{x_1, \ldots, x_{i-1}, x_i\}$, just before, and right after round i of the "for" loop, respectively (step 11 of Algorithm 3). Thus, for every $i \in [t]$, there is a moment where B is a superset of $\{x_1, \ldots, x_{i-1}, f(x_i)\}$. In other words, adding $f(x_i)$ to $\{x_1, \ldots, x_{i-1}\}$ gives an independent set.

Take a positive integer r such that $rk < t$. During round rk of Greedy Round-Robin, agent k gets her rth element, which is agent k's most valued element that can be added to $\{x_1, \ldots, x_{rk-1}\}$. Thus, every $e \in \beta^k(\mathcal{X}', \mathcal{F}') \setminus \{x_1, \ldots, x_{rk-1}\}$

Algorithm 3. Partition of a most valued base for agent k

Data: a matroid $(\mathcal{X}', \mathcal{F}')$, a feasible allocation $(S_1, \ldots, S_{|N|})$, and ν_k

1   $\beta^k(\mathcal{X}', \mathcal{F}') \leftarrow$ GREEDY$((\mathcal{X}', \mathcal{F}'), \nu_k)$

2   $B \leftarrow \beta^k(\mathcal{X}', \mathcal{F}')$

3   $\beta_1^k = \beta_2^k = \ldots = \beta_{|N|}^k = \emptyset$

4   Let $\{x_1, \ldots, x_t\}$ be the elements of $\cup_i S_i$ such that x_i is inserted before x_{i+1} in Greedy Round-Robin

5   **for** $i = 1$ *to* t **do**

6   Let j be such that $x_i \in S_j$

7   **if** $x_i \in B$ **then**

8          $f(x_i) \leftarrow x_i$

9   **else**

10   Find $f(x_i) \in B \setminus \{x_1, \ldots, x_t\}$ such that $B - f(x_i) + x_i$ and $\{x_1, \ldots, x_t\} - x_i + f(x_i)$ are bases (note that $f(x_i)$ exists by the bases exchange property)

11         $B \leftarrow B - f(x_i) + x_i$

12       $\beta_j^k \leftarrow \beta_j^k + f(x_i)$

13   **return** $(\beta_1^k, \ldots, \beta_{|N|}^k)$

---

such that $\{x_1, \ldots, x_{rk-1}\} + e \in \mathcal{F}'$ satisfies $\nu_k(x_{rk}) \geq \nu_k(e)$, and this is the case for $f(x_{rk}), f(x_{rk+1}), \ldots, f(x_t)$.

$$\forall j > rk, \; \nu_k(x_{rk}) \geq \nu_k(f(x_j)) \tag{1}$$

Take a second agent ℓ. If $\ell \geq k$ then agent ℓ does not appear before agent k in the ordering of Greedy Round-Robin and $|S_k| = |\beta_k^k| \geq |\beta_\ell^k|$. The elements of β_ℓ^k can be paired with some elements of S_k as follows: the jth element inserted in β_ℓ^k with the jth element inserted in S_k. By Inequality (1), agent k prefers the latter to the former. Therefore, $\nu_k(S_k) \geq \nu_k(\beta_\ell^k)$ by the additivity of ν_k. Now suppose $\ell < k$; agent ℓ appears before agent k in the ordering of Greedy Round-Robin and $|S_k| + 1 = |\beta_k^k| + 1 \geq |\beta_\ell^k|$. Again, some elements of β_ℓ^k can be paired with the elements of S_k as follows: the second element inserted in β_ℓ^k with the first element inserted in S_k, the third element inserted in β_ℓ^k with the second element inserted in S_k, and so on. If e denotes the first element inserted in β_ℓ^k, then Inequality (1) gives $\nu_k(S_k) \geq \nu_k(\beta_\ell^k) - \nu_k(e)$. Since $\nu_k(e) \leq \nu_{max}$, we get that $\nu_k(S_k) \geq \nu_k(\beta_\ell^k) - \nu_{max}$.

In all, $\nu_k(S_k) \geq \nu_k(\beta_\ell^k) - \nu_{max}$ for all $\ell \neq k$. Use this inequality for all $\ell \in N$ to get that $|N| \cdot \nu_k(S_k) \geq \left(\sum_{\ell \in N} \nu_k(\beta_\ell^k)\right) - |N| \cdot \nu_{max} = \nu_k(\beta^k(\mathcal{X}', \mathcal{F}')) - |N| \cdot \nu_{max}$ where the last equality is due to the additivity of ν_k. Finally, use Claim 1 to obtain the expected result. $\square$

The next result (proof omitted) applies to a matroid $(\mathcal{X}, \mathcal{F})$ and a set N of $n \geq 2$ agents.

Lemma 2 (Monotonicity property). *For any agent $k \in N$ and any element $x \in \mathcal{X}$, it holds that $\mu_k(n - 1, \mathcal{X} \setminus \{x\}, \mathcal{F}) \geq \mu_k(n, \mathcal{X}, \mathcal{F})$.*

Algorithm 4. $\frac{1}{2}$-Approximation Maximin Share

Data: A matroid $(\mathcal{X}, \mathcal{F})$, a set of agents N and their valuations $(\nu_i)_{i \in N}$

1   $(\mathcal{X}', \mathcal{F}') \leftarrow (\mathcal{X}, \mathcal{F})$

2   **for** $i = 1$ *to* $|N|$ **do**

3      $\alpha_i \leftarrow \frac{\nu_i(\beta^i(\mathcal{X}', \mathcal{F}'))}{|N|}$

4   **while** $\exists i \in N$ *and* $\{x_j\} \in \mathcal{F}'$ *such that* $\nu_i(\{x_j\}) \geq \alpha_i/2$ **do**

5   Define $\{x_j\}$ as the share of agent i, i.e. $S_i = \{x_j\}$

6      $N \leftarrow N \setminus \{i\}$

7      $(\mathcal{X}', \mathcal{F}') \leftarrow$ contraction of $(\mathcal{X}', \mathcal{F}')$ by $\{x_j\}$

8   Recompute the α_is

9   **if** $N = \emptyset$ **then**

10   **return** $S = (S_1, \ldots, S_{|N|})$

11   **else**

12   Run Greedy Round-Robin (Algorithm 2) on instance $\langle (\mathcal{X}', \mathcal{F}'), N, (\nu_i)_{i \in N} \rangle$

Theorem 2. *Algorithm 4 produces an allocation* $S = (S_1, \ldots, S_n)$ *such that*

$$\nu_i(S_i) \geq \frac{1}{2}\mu_i(n, \mathcal{X}, \mathcal{F}), \ \forall i \in N.$$

Proof. Fix an agent i and suppose she was allocated a single element during the first phase of Algorithm 4 (out of Greedy Round-Robin). Suppose at the time when i was allocated her element, there were n_1 active agents, $n_1 \leq n$, and that $(\mathcal{X}', \mathcal{F}')$ was the current (contracted) matroid. This means that agent i values her element at least $\alpha_i/2 = \frac{\nu_i(\beta^i(\mathcal{X}', \mathcal{F}'))}{2n_1} \geq \frac{\mu_i(n_1, \mathcal{X}', \mathcal{F}')}{2}$ where the inequality is by Claim 1. If we apply the monotonicity property (Lemma 2) repeatedly, we get that $\mu_i(n_1, \mathcal{X}', \mathcal{F}') \geq \mu_i(n, \mathcal{X}, \mathcal{F})$ and we are done.

Suppose now that the share of agent i was constructed during the second phase (Line 12 of Algorithm 4), i.e. Greedy Round-Robin. Let n_2 be the number of active agents at that point, and $(\mathcal{X}', \mathcal{F}')$ the input matroid of Greedy Round-Robin. We know that ν_{max} at that point is less than half the current value of α_i for agent i. Hence, by the additive guarantee of Greedy Round-Robin, we have that agent i values her share at least

$$\frac{\nu_i(\beta^i(\mathcal{X}', \mathcal{F}'))}{n_2} - \nu_{max} > \frac{\nu_i(\beta^i(\mathcal{X}', \mathcal{F}'))}{n_2} - \frac{\alpha_i}{2} = \frac{\nu_i(\beta^i(\mathcal{X}', \mathcal{F}'))}{2n_2} \geq \frac{1}{2}\mu_i(n_2, \mathcal{X}', \mathcal{F}').$$

Again, after applying the monotonicity property, we get that $\mu_i(n_2, \mathcal{X}', \mathcal{F}') \geq \mu_i(n, \mathcal{X}, \mathcal{F})$, which completes the proof. $\qquad\square$

4   A $(1 - \varepsilon)$-Approximation for Two Agents

As pointed out in [2], one can adapt *cut-and-choose* to construct a $(1 - \varepsilon)$-approximate maximin share allocation in case of allocating indivisible goods.

A similar approach can be used for the matroidal model (proof omitted). In Algorithm 5, $(\mathcal{X}, \mathcal{F})/S$ denotes the contraction of $(\mathcal{X}, \mathcal{F})$ to S.

Proposition 2. *Algorithm 5, which runs in polynomial time, outputs a $(1 - \varepsilon)$-approximate maximin share allocation for 2 agents.*

Algorithm 5.

Data: $(\mathcal{X}, \mathcal{F})$, $N = \{1, 2\}$, $\varepsilon \in [0, 1)$

1 Compute an allocation (S_1, S_2) that achieves the $(1 - \varepsilon)$-maximin share of agent 1 (see Proposition 1)

2 $B_1 \leftarrow$ GREEDY $((\mathcal{X}, \mathcal{F})/S_2, \nu_2)$ and $B_2 \leftarrow$ GREEDY $((\mathcal{X}, \mathcal{F})/S_1, \nu_2)$

3 **if** $\nu_2(B_1) \geq \nu_2(B_2)$, **then return** (S_2, B_1), **else return** (S_1, B_2)

---

5   A $(8/9 - \varepsilon)$-Approximation for Three Agents

We propose an algorithm (see Algorithm 6) inspired of the *divide-ask-and-choose* protocol of [18]. For every $\varepsilon \in [0, 1)$, Algorithm 6 achieves a guarantee of $8/9 - \varepsilon$ which improves on the best known guarantee of $7/8 - \varepsilon$ for three agents [9].

Algorithm 6 starts by computing an allocation (A_1, A_2, A_3) that achieves the $(1 - \varepsilon)$-approximate maximin share of agent 1 in the presence of 3 agents (see Proposition 1). We get that $\nu_1(A_i) \geq \mu_1^*$, $\forall i \in \{1, 2, 3\}$ where $\mu_1^* = \min_{i \in \{1,2,3\}} \{\nu_1(A_i)\}$ and $\mu_1^* \geq (1 - \varepsilon)\mu_1(3, \mathcal{X}, \mathcal{F})$. Similarly, μ_2^* and μ_3^* will denote the $(1 - \varepsilon)$-approximate maximin shares of agent 2 and 3, respectively.

According to Proposition 1, one can partition the most valued base $\beta^\ell(\mathcal{X}, \mathcal{F})$ of every agent ℓ to get an allocation that achieves her maximin share. For agent 2, such a base is $\beta^2(\mathcal{X}, \mathcal{F})$, and by the multiple bases exchange property, there exists a partition $(\beta_1^2, \beta_{-1}^2)$ of $\beta^2(\mathcal{X}, \mathcal{F})$ such that $\beta_1^2 \cup A_2 \cup A_3$ and $A_1 \cup \beta_{-1}^2$ are two bases of $(\mathcal{X}, \mathcal{F})$. As previously mentioned, $(\beta_1^2, \beta_{-1}^2)$ can be obtained in polynomial time [16]. Consider the matroid $(\mathcal{X}, \mathcal{F})/\beta_1^2$ for which β_{-1}^2 and $A_2 \cup A_3$ are two bases. Again, by the multiple bases exchange property, there exists (and one can construct it in polynomial time) a partition (β_2^2, β_3^2) of β_{-1}^2 such that $\beta_2^2 \cup A_3$ and $A_2 \cup \beta_3^2$ are two bases of $(\mathcal{X}, \mathcal{F})/\beta_1^2$. We eventually get four bases of $(\mathcal{X}, \mathcal{F})$: $(\beta_1^2 \cup \beta_2^2 \cup A_3)$, $(A_1 \cup \beta_2^2 \cup \beta_3^2)$, $(\beta_1^2 \cup A_2 \cup \beta_3^2)$, and $(\beta_1^2 \cup \beta_2^2 \cup \beta_3^2)$.

Next result (proof omitted) relies on a puzzle: given a 3×3 array of non-negative reals such that each row sums to 1, select two columns of the array and output a bipartition of these two columns in order to maximize the lightest part.

Lemma 3. *There exists $i \in \{1, 2, 3\}$ such that $(\mathcal{X}, \mathcal{F})/A_i$ admits a base B which can be partitioned in (B_j, B_k) and $\min\{\nu_2(B_j), \nu_2(B_k)\} \geq \frac{8}{9}\mu_2^*$.*

For an agent ℓ, if every element x of a matroid $(\mathcal{X}, \mathcal{F})$ is given the weight $\nu_\ell(x)$, then $OPT_\ell(\mathcal{X}, \mathcal{F})$ denotes the maximum value of a base. The following lemma will be used to prove the guarantee of Algorithm 6.

Algorithm 6.

Data: $(\mathcal{X}, \mathcal{F})$, $N = \{1, 2, 3\}$, $\varepsilon \in [0, 1)$

1   $\rho \leftarrow 8/9$

2   Compute an allocation (A_1, A_2, A_3) which achieves the $(1 - \varepsilon)$-maximin share for agent 1 (see Proposition 1)

3   Find $i \in \{1, 2, 3\}$ such that $(\mathcal{X}, \mathcal{F})/A_i$ admits a base B which can be partitioned in (B_j, B_k) and $\min\{\nu_2(B_j), \nu_2(B_k)\} \geq \rho\mu_2^*$ (see Lemma 3)

4   Let $\{j, k\} = \{1, 2, 3\} \setminus \{i\}$

5   Use GREEDY to compute $OPT_3((\mathcal{X}, \mathcal{F})/A_i)$

6   **if** $OPT_3((\mathcal{X}, \mathcal{F})/A_i) \geq 2\rho\mu_3^*$ **then**

7 | $C_k \leftarrow$ GREEDY $((\mathcal{X}, \mathcal{F})/(A_i \cup B_j), \nu_3)$

8 | $C_j \leftarrow$ GREEDY $((\mathcal{X}, \mathcal{F})/(A_i \cup B_k), \nu_3)$

9 | **if** $\nu_3(C_k) \geq \nu_3(C_j)$, **then return** (A_i, B_j, C_k), **else return** (A_i, B_k, C_j)

10   **else**

11 | $B'_k \leftarrow$ GREEDY $((\mathcal{X}, \mathcal{F})/(A_i \cup A_j), \nu_2)$

12 | $B'_j \leftarrow$ GREEDY $((\mathcal{X}, \mathcal{F})/(A_i \cup A_k), \nu_2)$

13 | **if** $\nu_2(B'_k) \geq \nu_2(B'_j)$ **then**

14 | $C_i \leftarrow$ GREEDY $((\mathcal{X}, \mathcal{F})/(A_j \cup B'_k), \nu_3)$

15 | **return** (A_j, B'_k, C_i)

16 | **else**

17 | $C_i \leftarrow$ GREEDY $((\mathcal{X}, \mathcal{F})/(A_k \cup B'_j), \nu_3)$

18 | **return** (A_k, B'_j, C_i)

Lemma 4. [18] *Let S be an independent set of a matroid $\mathcal{M}$ such that $OPT(\mathcal{M}) \geq \rho_0$ and $OPT(\mathcal{M}/S) < \rho_1 \leq \rho_0$. Then for every base T of $\mathcal{M}/S$, $OPT(\mathcal{M}/T) \geq \rho_0 - \rho_1$.*

Theorem 3. *Algorithm 6, which runs in polynomial time, outputs a feasible $(\frac{8}{9} - \varepsilon)$-approximate maximin share allocation.*

Proof. We provide a bound of $(1 - \varepsilon)8/9$, which is equivalent to $8/9 - \varepsilon$ because $\varepsilon > 0$ is a constant as small as possible.

Suppose the final allocation is the one of step 9. Agent 1 receives A_i that she values at least μ_1^* (see line 2). Agent 2 receives B_j or B_k that she values at least $\rho\mu_2^*$ (see line 3). By the multiple bases exchange property, there exists a bipartition (T_j, T_k) of $\beta^3((\mathcal{X}, \mathcal{F})/A_i)$ such that both $B_j \cup T_k$ and $B_k \cup T_j$ are bases of $(\mathcal{X}, \mathcal{F})/A_i$. Both B_j and T_j (resp., B_k and T_k) are bases of $(\mathcal{X}, \mathcal{F})/(A_i \cup B_k)$ (resp., $(\mathcal{X}, \mathcal{F})/(A_i \cup B_j)$). Since C_k and C_j are bases of maximum valuation for agent 3, we get that $\nu_3(C_k) \geq \nu_3(T_k)$ and $\nu_3(C_j) \geq \nu_3(T_j)$. Thus, $\max\{\nu_3(C_k), \nu_3(C_j)\} \geq \max\{\nu_3(T_k), \nu_3(T_j)\} \geq \frac{1}{2}(\nu_3(T_k) + \nu_3(T_j)) = \frac{1}{2}\nu_3(\beta^3((\mathcal{X}, \mathcal{F})/A_i))$. Since $\nu_3(\beta^3((\mathcal{X}, \mathcal{F})/A_i))$ is equal to $OPT_3((\mathcal{X}, \mathcal{F})/A_i)$ and $OPT_3((\mathcal{X}, \mathcal{F})/A_i) \geq 2\rho\mu_3^*$ (see line 6), we get that $\max\{\nu_3(C_k), \nu_3(C_j)\} \geq \rho\mu_3^*$.

Now suppose the final allocation is the one of step 15 or 18. Agent 1 receives A_j or A_k that she values at least μ_1^* (see line 2). Agent 2 receives B'_j or B'_k and we are going to see that she values each of them at least $\rho\mu_2^*$. Since $\min\{\nu_2(B_j),$

$\nu_2(B_k)\} \geq \mu_2^*$ (see line 3), we know that $OPT_2((\mathcal{X}, \mathcal{F})/A_i) \geq 2\mu_2^*$. By the multiple bases exchange property, there exists a bipartition (D_j, D_k) of $\beta^2((\mathcal{X}, \mathcal{F})/A_i)$ such that $D_j \cup A_k$ and $D_k \cup A_j$ are two bases of $(\mathcal{X}, \mathcal{F})/A_i$. Both B_j' and D_j (resp., B_k' and D_k) are bases of $(\mathcal{X}, \mathcal{F})/(A_i \cup A_k)$ (resp., $(\mathcal{X}, \mathcal{F})/(A_i \cup A_j)$). Since B_k' and B_j' are base of maximum valuation for agent 2, we get that $\nu_2(B_k') \geq \nu_2(D_k)$ and $\nu_2(B_j') \geq \nu_2(D_j)$. Thus, $\max\{\nu_2(B_k'), \nu_2(B_j')\} \geq \max\{\nu_2(D_k), \nu_2(D_j)\} \geq \frac{1}{2}\nu_2(D_k \cup D_j) = \frac{1}{2}\nu_2(\beta^2((\mathcal{X}, \mathcal{F})/A_i))$. Since $\min\{\nu_2(B_j), \nu_2(B_k)\} \geq \rho\mu_2^*$ (see line 3), we get that $\nu_2(\beta^2((\mathcal{X}, \mathcal{F})/A_i)) = OPT_2((\mathcal{X}, \mathcal{F})/A_i) \geq 2\rho\mu_2^*$ and then $\max\{\nu_2(B_k'), \nu_2(B_j')\} \geq \rho\mu_2^*$.

It remains to bound the valuation of agent 3 for C_i. We know that $OPT_3(\mathcal{X}, \mathcal{F}) \geq 3\mu_3^*$ and $OPT_3((\mathcal{X}, \mathcal{F})/A_i) < 2\rho\mu_3^*$ (see line 6). By Lemma 4, we get that $OPT_3((\mathcal{X}, \mathcal{F})/T) \geq (3 - 2\rho)\mu_3^* \geq \mu_3^*$ for every base T of $(\mathcal{X}, \mathcal{F})/A_i$. Since $A_j \cup B_k'$ and $A_k \cup B_j'$ are bases of $(\mathcal{X}, \mathcal{F})/A_i$, we get that $OPT_3((\mathcal{X}, \mathcal{F})/(A_j \cup B_k')) \geq \mu_3^*$ and $OPT_3((\mathcal{X}, \mathcal{F})/(A_k \cup B_j')) \geq \mu_3^*$. By construction, C_i is an optimal base of $(\mathcal{X}, \mathcal{F})/(A_j \cup B_k')$ or $(\mathcal{X}, \mathcal{F})/(A_k \cup B_j')$ for agent 3, so $\nu_3(C_i) \geq \mu_3^*$.

To conclude, each agent i gets a share that she values at least $\rho\mu_i^*$ with $\rho = 8/9$, and $\mu_i^* \geq (1 - \varepsilon)\mu_i$. $\square$

6   Conclusion

This article deals with a matroidal extension of maximin share allocations of indivisible goods. We are able to extend the approximation algorithm with guarantee $1/2$ for all n, but not the one with guarantee $2/3 - \varepsilon$ [9]. The difficulty resides in finding disjoint parts that approximate the agents' maximin share, and appending these parts must form an independent set. Each task taken separately is manageable, but their combination is problematic for more than two agents.

For the moment, we can improve the $1/2$-approximation when $n \in \{2, 3\}$. Since matroids generalize the allocation of indivisible goods, our $(8/9 - \varepsilon)$-approximation (Theorem 3) improves on the $(7/8 - \varepsilon)$-approximation of [9]. Note that any improvement on the $8/9$ guarantee of Lemma 3 (i.e. the described puzzle) implies an improvement on Theorem 3.

We have shown that within the set system model, matroids are particularly amenable to (multiplicative) approximate maximin share allocations. Is it the same for other, more general, set systems? In this respect, we can list greedoids, independence systems and the intersection of two matroids [17, 19]. We were able to build 2-agent instances (omitted due to space limitations) showing that the existence of a ρ-approximate maximin share for any $\rho \in (0, 1]$ is not always guaranteed with greedoids, independence systems, and the intersection of two matroids. Thus, further extending the approximation results to set systems that generalize matroids seems unlikely.

Finally, some special cases are known to admit a maximin share allocation for indivisible goods, see e.g. [2, 5, 9], and it would be interesting to study them under the matroidal set system model.

References

1. Brams, S.J., Taylor, A.D.: Fair Division: From Cake Cutting to Dispute Resolution. Cambridge University Press, Cambridge (1996)
2. Bouveret, S., Lemaître, M.: Characterizing conflicts in fair division of indivisible goods using a scale of criteria. Auton. Agents Multi-agent Syst. **30**, 259–290 (2016)
3. Budish, E.: The combinatorial assignment problem: approximate competitive equilibrium from equal incomes. J. Polit. Econ. **119**, 1061–1103 (2011)
4. Procaccia, A.D., Wang, J.: Fair enough: guaranteeing approximate maximin shares. In: Babaioff, M., Conitzer, V., Easley, D. (eds.) ACM Conference on Economics and Computation, EC 2014, Stanford, CA, USA, 8–12 June 2014, pp. 675–692. ACM (2014)
5. Kurokawa, D., Procaccia, A.D., Wang, J.: When can the maximin share guarantee be guaranteed? In: Proceedings of the Thirtieth AAAI Conference on Artificial Intelligence, 12–17 February 2016, Phoenix, Arizona, USA, pp. 523–529. AAAI Press (2016)
6. Lipton, R., Markakis, E., Mossel, E., Saberi, A.: On approximately fair allocations of indivisible goods. In: Breese, J., Feigenbaum, J., Seltzer, M. (eds.) ACM Conference on Electronic Commerce, pp. 125–131. ACM (2004)
7. Asadpour, A., Saberi, A.: An approximation algorithm for max-min fair allocation of indivisible goods. SIAM J. Comput. **39**, 2970–2989 (2010)
8. Markakis, E., Psomas, C.-A.: On worst-case allocations in the presence of indivisible goods. In: Chen, N., Elkind, E., Koutsoupias, E. (eds.) WINE 2011. LNCS, vol. 7090, pp. 278–289. Springer, Heidelberg (2011). doi:10.1007/978-3-642-25510-6_24
9. Amanatidis, G., Markakis, E., Nikzad, A., Saberi, A.: Approximation algorithms for computing maximin share allocations. In: Halldórsson, M.M., Iwama, K., Kobayashi, N., Speckmann, B. (eds.) ICALP 2015. LNCS, vol. 9134, pp. 39–51. Springer, Heidelberg (2015). doi:10.1007/978-3-662-47672-7_4
10. Woeginger, G.J.: A polynomial-time approximation scheme for maximizing the minimum machine completion time. Oper. Res. Lett. **20**, 149–154 (1997)
11. Spliddit: provably fair solutions (2017). http://www.spliddit.org/
12. Brualdi, R.: Comments on bases in different structures. Bull. Austral. Math. Soc. **1**, 161–167 (1969)
13. Greene, C.: A multiple exchange property for bases. Proc. Am. Math. Soc. **39**, 45–50 (1973)
14. Woodall, D.: An exchange theorem for bases of matroids. J. Comb. Theory (B) **16**, 227–228 (1974)
15. Greene, C., Magnanti, T.L.: Some abstract pivot algorithms. SIAM J. Appl. Math. **29**, 530–539 (1975)
16. Bixby, R.E., Cunningham, W.H.: Matroid optimization and algorithms. In: Handbook of Combinatorics, North Holland (1995)
17. Korte, B., Vygen, J.: Combinatorial Optimization: Theory and Algorithms, 4th edn. Springer Publishing Company, Incorporated, Heidelberg (2007)
18. Gourvès, L., Monnot, J., Tlilane, L.: A protocol for cutting matroids like cakes. In: Chen, Y., Immorlica, N. (eds.) WINE 2013. LNCS, vol. 8289, pp. 216–229. Springer, Heidelberg (2013). doi:10.1007/978-3-642-45046-4_18
19. Korte, B., Lovász, L., Schrader, R.: Greedoids. Springer, Heidelberg (1991)

Space-Efficient Euler Partition
and Bipartite Edge Coloring

Torben Hagerup$^{(\boxtimes)}$, Frank Kammer, and Moritz Laudahn

Institut für Informatik, Universität Augsburg, 86135 Augsburg, Germany
{hagerup,kammer,moritz.laudahn}@informatik.uni-augsburg.de

Abstract. We describe space-efficient algorithms for two problems on undirected multigraphs: Euler partition (partitioning the edges into a minimum number of trails); and bipartite edge coloring (coloring the edges of a bipartite multigraph with the minimum number of colors). Let n, m and $\Delta \geq 1$ be the numbers of vertices and of edges and the maximum degree, respectively, of the input multigraph. For Euler partition we reduce the amount of working memory needed by a logarithmic factor, to $O(n+m)$ bits, while preserving a running time of $O(n+m)$. For bipartite edge coloring, still using $O(n+m)$ bits of working memory, we achieve a running time of $O(n + m \min\{\Delta, \log \Delta(\log^* \Delta + (\log m \log \Delta)/\Delta)\})$. This is $O(m \log \Delta \log^* \Delta)$ if $m = \Omega(n \log n \log \log n/\log^* n)$, to be compared with $O(m \log \Delta)$ for the fastest known algorithm.

1 Introduction

Continuing an investigation of space-efficient yet reasonably time-efficient graph algorithms begun in [6], we illustrate new techniques for the design of such algorithms. Our model of computation is a word RAM with read-only access to its input, write-only access to an output medium and a read-write working memory. We say that an algorithm works with s bits if it can operate correctly with a working memory of that size. If s is small (compared to the space requirements of competing algorithms), the algorithm is *space-efficient*. By a *classic* algorithm we mean one of the widely known standard algorithms in whose design economy of space was not a primary concern.

Many reasons for the study of space-efficient algorithms have been advanced. Maybe the input is available in the Internet for queries, but is so huge that it is impossible or impractical to copy it to the memory of a local computer. Maybe an algorithm is supposed to run on a handheld or embedded device that has only a tiny amount of general-purpose memory. Or maybe most of the available memory is of a kind for which writing is much slower than reading or can be performed only a limited number of times. Of course, there is also the purely intellectual challenge of discovering whether the time bounds of classic algorithms can be met, up to a constant factor, while at the same time the space requirements are reduced by more than a constant factor.

All algorithms considered here allow graphs to have several edges with the same two endpoints. To emphasize this, we sometimes use the term "multigraph". In the following discussion we denote by n and m the numbers of vertices

© Springer International Publishing AG 2017
D. Fotakis et al. (Eds.): CIAC 2017, LNCS 10236, pp. 322–333, 2017.
DOI: 10.1007/978-3-319-57586-5_27

and of edges, respectively, of an input graph under consideration. Many classic graph algorithms work in linear time and space. "Linear time" usually means $O(n+m)$ time on a random-access machine capable of operating on (e.g., adding) integers of $\Theta(\log(n+m))$ bits in constant time. In modern parlance, the model used is a *word RAM* with a word length w of $\Theta(\log(n+m))$. Correspondingly, "linear space" means $O(n)$ or sometimes $O(n+m)$ words of $\Theta(\log(n+m))$ bits each. We use the same model and assume that w is large enough that the input and working memories can be addressed with w bits.

The problems considered in [6] include depth-first search and its applications as well as the computation of minimum spanning forests and single-source shortest-paths trees. The space bounds achieved there range from $O(n)$ bits to $O(n\log(n+m))$ bits. Here we study "edge-centered" graph problems for which efficient solutions appear to require at least $n+m$ bits of working memory and classic algorithms use $\Omega((n+m)\log(n+m))$ bits.

1.1   Related Work and New Results and Techniques

Let $G = (V,E)$ be an undirected graph with $E \neq \emptyset$. A *trail* in G is a walk in G whose edges are pairwise distinct, and an *Euler partition* of G is a collection of minimal cardinality of (open or closed) trails in G, the edge sets of which form a partition of E. An algorithm for computing Euler partitions was described more than a hundred years ago by Hierholzer [12]. The algorithm is easily implemented to run in $O(n+m)$ time, but it needs $\Theta((n+m)\log(n+m))$ bits of working space. The most straightforward implementation of an algorithm of Fleury [7] for the special case of connected graphs with at most two vertices of odd degree runs in $O(m^2+1)$ time with $\Theta(m+n\log(n+m))$ bits. With more effort, one can achieve a space bound of $O(m)$ bits or, using a dynamic-connectivity algorithm of Thorup [17], an expected-time bound of $O(m\log n(\log\log n)^3)$. In Sect. 3 we describe a different implementation of Hierholzer's algorithm that still works in $O(n+m)$ time, but needs only $O(n+m)$ bits of working space. A central component of our solution is a *trail structure* that can keep track of all the trails that pass through a vertex with just a constant number of bits per edge.

In Sect. 4 we apply our Euler-partition results to bipartite edge coloring. Take an *edge coloring* of an undirected graph $G = (V,E)$ with $E \neq \emptyset$ to be a partition of E into sets $E_1, \ldots, E_k$, each of which induces a graph of maximum degree 1. The edge coloring is *optimal* if its cardinality k is as small as possible. Equivalently, the edges in E_i can be viewed as colored with the color i, for $i = 1, \ldots, k$, no two edges with a common endpoint may be assigned the same color, and the goal is to minimize the number of colors used. Whereas computing an optimal edge coloring of a general graph is NP-hard [13], the edges of a bipartite graph with maximum degree $\Delta \geq 1$ can be colored in polynomial time with Δ colors, which is obviously optimal. The fastest algorithm known for this task, due to Cole et al. [4], works in $O(n + m\log\Delta)$ time using $\Theta((n+m)\log(n+m))$ bits. We reduce the amount of working memory to $O(n+m)$ bits while achieving a running time of $O(n + m\min\{\Delta, \log\Delta(\log^*\Delta + (\log m\log\Delta)/\Delta)\})$.

If $m = \Omega(n \log n \log \log n / \log^* n)$, the space reduction comes at a price in the running time of only a factor of $O(\log^* \Delta)$.

Besides the new Euler-partition algorithm, ingredients of our algorithm for bipartite edge coloring include space-efficient data structures and tricks of the trade invented for earlier algorithms for the problem [1,3,8,9,16], the closest ancestors of our algorithm being those of Schrijver [16] and Alon [1]. All of these algorithms use the divide-and-conquer paradigm, and to support this paradigm in a space-efficient setting we introduce a general machinery with additional applications outside of this paper for working with recursive calls on subgraphs without storing these in full.

2   The Representation of Input Graphs

Let $G = (V, E)$ be an undirected input graph with n vertices. We assume, as is standard in graph algorithms, that $V = \{1, \ldots, n\}$. Take $\mathbb{N} = \{1, 2, \ldots\}$ and $\mathbb{N}_0 = \mathbb{N} \cup \{0\}$. For $u \in V$, let d_u be the degree of u and, for all $v \in V$, denote by $m_{u,v}$ the number of edges in G with endpoints u and v. Take $L = \{(u, k) \in V \times \mathbb{N} \mid 1 \leq k \leq d_u\}$. We assume that the representation of G makes it possible to determine n in constant time and supports constant-time evaluation of functions $deg : V \to \mathbb{N}_0$, $head : L \to V$ and $mate : L \to L$ with the following properties:

- For all $u \in V$, $deg(u) = d_u$;
- For all $u, v \in V$, $|\{(u, k) \in L : head(u, k) = v\}| = m_{u,v}$;
- For all $(u, k) \in L$, $mate(u, k) = (head(u, k), \ell)$ for some ℓ and $mate(mate(u, k)) = (u, k)$.

The undirected graph G is represented essentially as its *directed version*, i.e., as the directed graph on the vertex set V that, for each undirected edge in G with endpoints u and v, say, has a directed *arc* from u to v and one from v to u. Informally, every vertex u has an *incidence array* with an entry for each arc out of u, the entry for an arc a being the head of a. The operation *head* allows us to index into an incidence array, and *deg* returns its size. The representation of an undirected graph additionally has *cross links*, i.e., for all vertices u and v, every arc from u to v is matched to an arc from v to u, called its *mate*, and cross links (realized through the function *mate*) allow us to find the mate of a given arc.

3   Computing Euler Partitions

It will be convenient to view the elements of an Euler partition as directed trails that we call *Euler trails*. When including an (undirected) edge e with endpoints u and v in a trail, we *mark* the corresponding arc from u to v if we view e as traversed in the direction from u to v and the corresponding arc from v to u otherwise. We consider an (undirected) edge to be marked exactly if one of its corresponding arcs is marked. Call a vertex *white* if it is not isolated and all of its incident edges are unmarked, *gray* if it has both marked and unmarked incident edges, and *black* if all of its incident edges are marked. Also call the vertex *odd* or *even* according as the number of its incident unmarked edges is odd or even.

3.1 Hierholzer's Algorithm

Let $G = (V, E)$ be an undirected graph with $E \neq \emptyset$. A simple algorithm for computing an Euler partition of G that essentially goes back to Hierholzer [12] can be formulated as follows:

As long as not all vertices are black, repeatedly execute an *iteration* that comprises the following steps: First select a vertex u that is odd if possible, otherwise gray if possible, and white if there are neither odd nor gray vertices. We call u the *start vertex* of the iteration. If u is gray and even, remember a marked arc a_{old} directed out of u—there always is one. Choose an arbitrary unmarked edge incident with u, mark its corresponding arc a out of u and initialize a new Euler trail T to consist only of a. With v equal to the head of a, execute *extend*(v), which does the following: If v has no incident unmarked edges, stop; v is the end vertex of T. Otherwise choose an arbitrary unmarked edge incident with v, mark its corresponding arc a' out of v, add a' to T and call *extend*(v') recursively, where v' is the head of a'. This process extends T greedily as far as possible to an (open or closed) trail that begins at u and is composed exclusively of arcs corresponding to formerly unmarked edges.

If an arc a_{old} was remembered, replace the Euler trail T_{old} that contains a_{old} by a combination of T and T_{old}. This is possible because T is necessarily closed and consists in the insertion of (the arcs of) T before a_{old} and, if a_{old} has a predecessor a_{pred} on T_{old}, after a_{pred}. This ends the iteration.

Hierholzer's algorithm can easily be seen to be correct. Take $n = |V|$ and $m = |E|$. Storing each relevant set in a doubly-linked list, we can maintain the sets of white, of gray and of odd vertices so that a suitable start vertex u of an iteration can be found in constant time. A suitable arc a_{old} out of a given gray and even start vertex u can be located in constant time in the same manner, now with a doubly-linked list of incident arcs for each vertex, and the same is true of the first arc a of a new Euler trail and the arc a' needed by *extend*. Since every call of *extend* marks a previously unmarked edge, the time spent in calls of *extend* sums to $O(m)$ over the whole execution. Finally, the combination of two Euler trails into a single Euler trail takes constant time. In conclusion, Hierholzer's algorithm can be executed in $O(n + m)$ time.

The analysis above paid no attention to space issues. Even if only $O(n + m)$ bits of working space are available, most of the arguments still go through. In particular, the various sets that need to be maintained can be realized with choice dictionaries [11, Theorem 5.4]. Only the representation of the Euler trails themselves is troublesome. A straightforward representation of each trail as a list of integers that represent vertices or arcs requires $\Omega(m \log(n + m))$ bits.

3.2 The Trail Structures

In order to reduce the space requirements to $O(n + m)$ bits, we equip each vertex v with a data structure D_v, called its *trail structure*. If v is of degree $d \geq 1$, D_v maintains a partition of $\{1, \ldots, d\}$ into three sets, I, O and U, as well as a matching, each edge of which has one endpoint in I and one in O.

Initially, I and O are both empty. The connection to the construction of Euler trails is immediate: Each integer k in $\{1, \ldots, d\}$ belongs to O if the kth arc a in the incidence array of v is marked, to I if the mate of a is marked, and to U if neither a nor its mate is marked, and two integers in $\{1, \ldots, d\}$ are matched if and only if they correspond to edges that are consecutive on some Euler trail and have v as a common endpoint.

D_v operates in two *phases*, with different sets of operations supported in the two phases. In both phases D_v supports *simple queries* that ask whether U is empty or whether a given element of $\{1, \ldots, d\}$ is matched or belongs to a named set among I, O and U. In the first phase, D_v supports two additional operations: First, *leave*, which selects an (arbitrary) element of U, moves it from U to O, leaves it unmatched and returns it to the caller; *leave* may be called only when $U \neq \emptyset$. Second, *enter*, which takes an argument $i \in U$ and moves i from U to I. If subsequently $U \neq \emptyset$, the operation proceeds to select an (arbitrary) element $o \in U$, move o from U to O, match i and o, and return o; otherwise it returns nothing. When U becomes empty, D_v enters its second phase.

In its second phase, in addition to the simple queries, D_v supports an operation that returns the element matched to a given matched element in $I \cup O$ and an operation *marry* that takes as its arguments elements $i \in I$ and $o \in O$ and matches i and o while making any other elements previously matched to i or o unmatched. Over the life of D_v, the operation *marry* may be called at most twice; thus the matching is nearly invariant during the second phase. We demonstrate in Subsect. 3.4 how to implement D_v in $O(d)$ bits so that all operations take constant time, except that $\Theta(d)$ time is needed between the two phases.

Recall that a *rank-select structure* for a bit sequence $B = (b_1, \ldots, b_N)$ supports two types of queries: $rank_B(j)$ ($j \in \{1, \ldots, N\}$), which returns $\sum_{i=1}^{j} b_i$; and $select_B(k)$ ($k \in \{1, \ldots, \sum_{i=1}^{N} b_i\}$), which returns the smallest $j \in \{1, \ldots, N\}$ with $rank_B(j) = k$. Storing trail structures of sizes $p_1, \ldots, p_n$ compactly in an array, we can find the start address of the kth trail structure, for $k = 1, \ldots, n$, as $select_B(k)$, where B is a bit vector of size $N = \sum_{j=1}^{n} p_j$ with 1s precisely in the positions $1 + \sum_{i=1}^{j-1} p_i$, for $j = 1, \ldots, n$. Rank-select structures for sequences of N bits that execute every operation in constant time and occupy $O(N)$ bits were first described by Clark [2].

3.3 Hierholzer's Algorithm with Trail Structures

Using the trail structures postulated in the previous subsection, we can realize the manipulation of Euler trails needed in Hierholzer's algorithm as follows:

Consider an iteration with start vertex u. Instead of remembering the arc a_{old}, it will be convenient to remember the position k_{old} of a_{old} in the incidence array of u. To initialize T to consist of a single arc out of u, execute the operation *leave* on D_u and remember the integer returned as k_{first}. Informally, k_{first} represents the first arc of T. To carry out *extend*(v) when the arc preceding v on T is a, execute *enter*(k) on D_v, where k is the position of the mate of a in the incidence array of v (found via a cross link). If T cannot be extended beyond v, this fact is signaled by the absence of a return value from the call $enter(k)$. In that case proceed, if a value k_{old} was remembered, to remember the last argument of $enter$ as k_{last}—informally, the last arc of T—and to combine the trails T_{old} and T. Since D_v has now entered its second phase, this can be done by executing on D_v the operations $marry(k_{last}, k_{old})$ and, if k_{old} was matched to some element k_{pred} before the first call of $marry$, additionally $marry(k_{pred}, k_{first})$.

Note that even a closed Euler trail has a unique first arc. To output the Euler partition, step through the arcs of the input graph, test for each whether it is the first arc of an Euler trail and, if so, output the arcs of that trail in order. If the position of an arc a in the incidence array of a vertex u is k, a is the first arc of an Euler trail exactly if k is unmatched in D_u. Similarly, if the position of the mate of an arc a in the incidence array of a vertex v is k, a is the last arc of an Euler trail exactly if k is unmatched in D_v. If a is followed by a' on its Euler trail, the position of a' in the incidence array of v can be obtained as the element matched to k in D_v. Therefore the Euler partition can be output in constant time per vertex and arc, for a total time of $O(n + m)$.

3.4 The Realization of the Trail Structures

The key to a space-efficient implementation of D_v is to make good use of the freedom that the operation $enter$ leaves concerning the choice of o. We follow a simple rule: Viewing $\{1, \ldots, d\}$ as cyclic (i.e., 1 is the successor of d), in each call of $enter$ that starts with $|U| \geq 2$ we choose o as the cyclically first element of U that follows i. To support this, we maintain the elements of U in a doubly-linked cyclic list by storing for each element of U the smallest cyclic distances in the forward and backward directions to another element of U. Since the distances sum to d for each of the two directions, they can be stored in binary in a total of $O(d)$ bits. Because the distances have binary representations of varying lengths, they must in fact be stored as self-delimiting numeric values, which essentially means that each binary representation is extended by a unary representation of its length that can be decoded with table lookup. The number of bits needed remains $O(d)$. In order to support the simple queries, we also maintain the sets I, O and U and the set of matched elements in choice dictionaries [11, Theorem 5.4]. The implementation of $leave$ and $enter$ is now straightforward.

Define a run of D_v as a maximal linear contiguous subsequence y of the cyclic sequence $\{1, \ldots, d\}$ that consists only of elements of $I \cup O$ and associate with each run y the word over the alphabet $\{(,)\}$ obtained from y by replacing each matched element in I by an opening parenthesis, (, each matched element of O by a closing parenthesis,), and each unmatched element by the empty word. As long as $U \neq \emptyset$, the word associated with an arbitrary run is a $Dyck\ word$, i.e., a balanced sequence of parentheses, and each pair of matching parentheses in the Dyck word corresponds to elements of $\{1, \ldots, d\}$ that are also matched in D_v. To see this, note that each pair of new parentheses, by induction, encloses a Dyck word. The corresponding run may merge with the run that precedes it and/or the run that follows it. Since the set of Dyck words is closed under concatenation, in either case the new run is again associated with a Dyck word.

When D_v enters its second phase, let y^* be the Dyck word of the (d-element) run that ends with the last element to leave U. We initialize two static $O(d)$-bit data structures, namely a rank-select structure that allows us to translate in constant time between a position in y^* and the element of $\{1, \ldots, d\}$ from which it originated and a structure that allows us, given the position of a parenthesis in y^*, to find the position of its matching parenthesis in constant time. Both data structures can be constructed in $O(d)$ time [2, 10, 15]. Used in conjunction, they enable us, given a matched element of $I \cup O$, to locate the element of $O \cup I$ matched to it in constant time until the first call of *marry*.

Because *marry* may be called at most twice, we can support this final operation in a completely naive way: We store the arguments of all calls of *marry* in a table of $O(\log d)$ bits and direct all operations in the second phase to begin by searching this table of exceptions in constant time and take appropriate action if one or more of their arguments are found there.

Theorem 1. *An Euler partition of an undirected multigraph with n vertices and m edges can be computed in $O(n + m)$ time with $O(n + m)$ bits.*

4   Bipartite Edge Coloring

In this section we describe a space-efficient algorithm for the following problem, called *bipartite edge coloring*: Given a bipartite undirected graph G with maximum degree $\Delta \geq 1$, output the Δ sets in an optimal edge coloring of G, one by one, while following the elements of each set by a set-end indicator.

As observed by Gabow and Kariv [9], if we allow the number of colors used to increase from Δ to $2^{\lceil \log_2 \Delta \rceil}$, the problem reduces very simply to Euler partition: If $\Delta = 1$, output the edge set E of the given graph G as its single color class. Otherwise partition E into two sets E_1 and E_2 by assigning the edges on each trail in an Euler partition of G alternately to E_1 and to E_2. The two subgraphs of G induced by E_1 and E_2 have maximum degrees at most $\lceil \Delta/2 \rceil$, and processing them recursively leads to the desired result. The maximum depth of recursion is $O(\log \Delta)$, and every edge in E appears in at most one recursive instance at every level of recursion. With $\Theta(m \log(n + m))$ bits available, mainly for storing subgraphs on a recursion stack, the algorithm can therefore be executed in $O(n + m \log \Delta)$ time on graphs with n vertices and m edges. Wanting to get by with $O(n + m)$ bits, we have to work harder and to pay a small price in the running time. This is described in the next subsection.

4.1   Recursion on Subgraphs

Consider an algorithm $\mathcal{A}$ that inputs a graph $G = (V, E)$ and calls itself recursively a number of times on subgraphs of G. For simplicity, assume that these subgraphs as well as G itself have no isolated vertices. Take $n = |V|$ and $m = |E|$. We show, for frequently occurring cases, how to manage the recursion stack of $\mathcal{A}$ using only $O(n + m)$ bits, i.e., less space than what would be needed to store even a single incidence-array representation of a (large) subgraph of G. Because our method has applications beyond the present paper, we describe it in slightly greater generality than what is needed here.

Recall that, by convention, $V = \{1, \ldots, n\}$. It will be convenient also to number the arcs of G consecutively starting at 1: For $j = 1, \ldots, n$, let d_j be the degree of the vertex j. Then, for $j = 1, \ldots, n$ and $k = 1, \ldots, d_j$, we assign the number $g(j, k) = k + \sum_{i=1}^{j-1} d_i$ to the kth arc in j's incidence array. Let us call g the *arc-numbering function* of G. In order to be able to evaluate g and g^{-1} in constant time, we store a rank-select structure for a bit sequence P of length $2m$ with 1s precisely in the positions $1 + \sum_{i=1}^{j-1} d_i$, for $j = 1, \ldots, n$—as is easy to see, $g(j, k) = select_P(j) + k - 1$ and, for $r = 1, \ldots, 2m$, $g^{-1}(r) = (j, r - select_P(j) + 1)$, where $j = rank_P(r)$. By assumption, cross links allow us to map the number of an arc to the number of its mate in constant time.

The input graph $H = (V_H, E_H)$ of every call C of $\mathcal{A}$ other than the top-level call is stored incrementally with respect to the input graph $\overline{H} = (V_{\overline{H}}, E_{\overline{H}})$ of some proper ancestor call $\overline{C}$ of C (thus H is a subgraph of $\overline{H}$). In concrete terms, C pushes on the recursion stack a stack frame that contains a rank-select structure for a bit sequence B_V of length $|V_{\overline{H}}|$ whose ith bit is 1, for $i = 1, \ldots, |V_{\overline{H}}|$, if and only if the ith smallest element of $V_{\overline{H}}$ (according to the original vertex order $1, \ldots, n$) is still present in V_H. If C numbers a vertex k, $\overline{C}$'s number for the vertex is $select_{B_V}(k)$. Conversely, if $\overline{C}$ numbers a vertex j, the vertex is present in H exactly if the jth bit in B_V is 1, and then C's number for the vertex is $rank_{B_V}(j)$.

We call $\overline{H}$ the *reference graph* of H. The reference relation induces a *reference tree* on the calls of $\mathcal{A}$ that is similar to $\mathcal{A}$'s recursion tree, but has "shortcuts" (as after path compression). Given the number used for a vertex v by a call C of $\mathcal{A}$ in a depth of t in the reference tree, the original number of v (i.e., v itself) can be found in $O(t+1)$ time by repeated translation along the path in the reference tree from C to the root, with constant time spent in each tree node along the way, and a translation in the opposite direction can also happen in $O(t + 1)$ time. A completely analogous rank-select structure for a bit sequence of length $2|E_{\overline{H}}|$ allows the corresponding translations among arc numbers within the same time bounds. In addition, we equip each recursive call with a data structure for evaluating the arc-numbering function g of its input graph as well as g^{-1}.

Now standard algorithms can operate on the input graph of a call at a depth of t in the reference tree (*locally*, say), albeit with a slowdown of $O(t + 1)$. A central operation is, given (the local number n_u of) a vertex u and a positive integer k bounded by the (local) degree of u, to determine (the local number n_v of) the head v of the kth arc in u's (fictitious) local incidence array. To carry out the operation, the pair (n_u, k) is translated via the local arc-numbering function to a local arc number, that local arc number is translated in $O(t+1)$ time to an original arc number, the latter is translated using the inverse g^{-1} of the original arc-numbering function to a pair (u, k'), the one and only true input graph is consulted to determine v, and finally v is translated in $O(t+1)$ time to obtain n_v. The mate of a given edge and the (local) degree of a vertex can also be found in $O(t + 1)$ time (the latter even in constant time).

Suppose that the algorithm $\mathcal{A}$ under consideration has the common property that the size of the input graph decreases geometrically with the recursive depth, i.e., there are constants $c \in \mathbb{N}$ and $\epsilon > 0$ such that if $H = (V_H, E_H)$ is the input graph of a call of $\mathcal{A}$ whose recursive depth is larger by c than that of an ancestor call with input graph $\overline{H} = (V_{\overline{H}}, E_{\overline{H}})$, then $|V_H| + |E_H| \leq (1 - \epsilon)(|V_{\overline{H}}| + |E_{\overline{H}}|)$. If reference graphs are consistently chosen at small depths in the recursion tree, the reference tree will become shallow, i.e., the slowdown will be small. On the other hand, because reference graphs at a smaller recursive depth are larger, the space requirements will be high. Choosing the reference graphs judiciously and guided in part by the same principles as in [5, Sect. 5], we can guarantee a slowdown of $O(\log^* t)$ for accesses to graphs at a recursive depth of t, while using only $O(n + m)$ bits and $O(S + (1 + N/\log(n + m))(n + m))$ time for the input graphs on the recursion stack, where S is the sum of $|V_H| + |E_H|$ over all subgraphs (V_H, E_H) on which $\mathcal{A}$ is called recursively and N is the number of such calls with $(|V_H| + |E_H|) \log(n + m) \geq n + m$. We obtain the following result.

Theorem 2. *An edge coloring of an undirected bipartite multigraph G with n vertices, m edges and maximum degree $\Delta \geq 1$ that uses at most $2^{\lceil \log_2 \Delta \rceil} \leq 2\Delta - 1$ colors can be computed in $O(n + m \log \Delta \log^* \Delta)$ time with $O(n + m)$ bits.*

Proof. Execute the algorithm described in the beginning of Sect. 4 with the space-efficient recursion stack developed above, but taking care to assign the first edge of each Euler trail to the currently smallest set among E_1 and E_2. This latter specialization of the algorithm ensures that a child graph of a graph with m' edges has at most $\lceil m'/2 \rceil$ edges, i.e., the size of an input graph decreases geometrically with the recursive depth, as required.

If the total effort expended by $\mathcal{A}$ in a given recursive depth also decreases geometrically with the depth, the slowdown of $O(\log^* t)$ can easily be "swallowed" by the geometric decrease, so that $\mathcal{A}$ can be executed entirely without a time penalty and with the graphs on the recursion stack occupying $O(n + m)$ bits.

4.2   Reduction to Top Matching

The algorithm of Theorem 2 reaches our goal of coloring a bipartite undirected graph $G = (V, E)$ of maximum degree Δ with Δ colors only if Δ is a power of 2. Otherwise we proceed as first suggested by Gabow [8] and Gabow and Kariv [9]: If Δ is odd, compute a matching M in G that matches at least the vertices of maximum degree—let us call such a matching a *top matching*—and remove the edges in M after outputting them as a first color class. Now, whether or not a top matching was removed, the maximum degree is even, and we use Euler partition as in the proof of Theorem 2 to partition the remaining edges into two set E_1 and E_2 such that the graphs G_1 and G_2 induced by E_1 and E_2 are both of maximum degree $\lfloor \Delta/2 \rfloor$. We color G_1 recursively, but output only $2\lfloor \Delta/2 \rfloor - 2^{\lfloor \log_2 \Delta \rfloor}$ of its color classes and transfer the edges of the remaining color classes from E_1 to E_2. This "fills up" G_2 to maximum degree $2^{\lfloor \log_2 \Delta \rfloor}$, so that it can be colored optimally with the algorithm of Theorem 2. In summary, to color a graph G induced by m edges and with maximum degree Δ, we must compute a top matching in G if Δ is odd, recursively color a subgraph of G induced by at most $\lceil m/2 \rceil$ edges and with maximum degree $\lfloor \Delta/2 \rfloor$, and spend $O(m \log \Delta \log^* \Delta)$ additional time. Over all recursive calls, this amounts to computing top matchings in subgraphs of G induced by at most $\lceil m/2^i \rceil$ edges and with maximum degree $\lfloor \Delta/2^i \rfloor$, for $i = 0, \ldots, \lfloor \log_2 \Delta \rfloor$, and spending $O(m \log \Delta \log^* \Delta)$ additional time. Because the recursion tree is unary, the condition set out in the last paragraph of Subsect. 4.1 is satisfied for our computation of top matchings, so we can ignore the slowdown caused by using the space-efficient recursion stack.

The following two subsections describe a solution to the remaining problem of finding a top matching in bipartite graphs. First, following Schrijver [16], the problem is reduced to perfect matching in regular bipartite graphs. Subsequently, using a technique introduced by Cole and Hopcroft [3] and called *edge-sparsification* by Makino et al. [14], the given Δ-regular graph G with n vertices is replaced by a subgraph G' of G on the same vertex set and with $O(n \log(\Delta + 1))$ edges, each with a nonnegative integer *weight*, in which regularity is replaced by *weight-regularity*: For each vertex v, the weights of the edges incident on v sum to Δ. It suffices to find a perfect matching in G', which is also a perfect matching in G. In the full version of the paper we show how to implement the sparsification so that it runs in $O(m \log(\Delta + 1))$ time using $O(m)$ bits and describe a realization of Schrijver's algorithm [16] for perfect matching that runs in $O(m\Delta)$ time using $O(m)$ bits. Here we describe a space-efficient version of Alon's perfect-matching algorithm [1]. Together the algorithms of Schrijver and Alon work in $O(m \min\{\Delta, (\log m \log \Delta)/\Delta\})$ time.

Theorem 3. *An optimal edge coloring of an undirected bipartite multigraph with n vertices, m edges and maximum degree $\Delta \geq 1$ can be computed in $O(n + m \min\{\Delta, \log \Delta(\log^* \Delta + (\log m \log \Delta)/\Delta)\})$ time with $O(n + m)$ bits.*

4.3   Reduction to Perfect Matching in Regular Bipartite Graphs

In this subsection we show how to merge vertices and add edges in a bipartite input graph $G = (V, E)$ with n vertices, m edges and maximum degree $\Delta \geq 1$ to turn G into a Δ-regular bipartite graph $\overline{G} = (\overline{V}, \overline{E})$ with $O(m)$ edges, m of which are *old* edges that correspond bijectively to those of G. At a cost of $O(n)$ time and $O(n)$ bits, we assume that G has no isolated vertices.

Suppose that V is composed of *left* vertices and *right* vertices such that every edge in E has one left and one right endpoint. Let n_L be the number of left vertices and, for $i = 1, \ldots, n_L$, let d_i be the degree of the ith smallest left vertex (in the usual vertex order $1, \ldots, n$). Let P_L and Q_L be bit vectors of $N = \lceil 2m/\Delta \rceil$ *sectors* of Δ bits each, initialized to contain only 0s and only 1s, respectively. $\overline{V}$ will consist of N left and N right *supernodes*. For $i = 1, \ldots, N$, the ith left supernode u corresponds to the ith sector in P_L, and every 1 in that sector will represent one of the left vertices in V merged to obtain u. Q_L will be a bit-vector representation of the set of new arcs that go from left to right, i.e., every 1 in Q_L corresponds to a new arc and every 0 corresponds to an old arc.

P_L and Q_L are computed by taking $s = 0$ and then, for $i = 1, \ldots, n_L$, doing the following: First, if $\lfloor (s + d_i)/\Delta \rfloor > \lfloor s/\Delta \rfloor$ (informally, $s + d_i$ is in the next sector, i.e., the ith left vertex does not fit in the current supernode), add $\Delta - (s \bmod \Delta)$ to s (informally, fill up the current supernode with new arcs and step to the next supernode). Whether or not s was increased, now set the $(s+1)$st bit in P_L to 1, set the bits numbered $s + 1, \ldots, s + d_i$ in Q_L to 0, and add d_i to s. It is easy to see that the total increase in s is bounded by $\sum_{i=1}^{n} 2d_i = 2m$, so that P_L and Q_L are large enough for their intended use. Similar vectors P_R and Q_R of $N\Delta$ bits each are computed for the right supernodes.

We can number the supernodes and the arcs in $\overline{G}$ consecutively, starting at 1, and pair the arcs with mates in a natural way that corresponds to the following conventions: For $i = 1, \ldots, N$, the supernode numbered i is a left supernode. For $k = 1, \ldots, \Delta$, its kth arc a is numbered $r = (i - 1)\Delta + k$. If bit number r in Q_L is 1, a is a new arc, and its mate is numbered $N\Delta + select_{Q_R}(rank_{Q_L}(r))$. Otherwise a is an old arc, and its corresponding arc in G is numbered $g(j, r - select_{P_L}(j) + 1)$, where g is the arc-numbering function of G and $j = rank_{P_L}(r)$. Storing rank-select structures for P_L, Q_L, P_R and Q_R, we can evaluate all of these expressions and the corresponding expressions for right supernodes in constant time and therefore navigate in $\overline{G}$ as though it were given in an incidence-array representation. The reduction altogether works in $O(n + m)$ time.

4.4   Alon's Perfect Matching in Weight-Regular Bipartite Graphs

Assume that a weight-regular bipartite input graph $\widetilde{G} = (V, \widetilde{E})$ has n vertices and total edge weight $m = n\Delta/2$, for some $\Delta \geq 1$. Take $t = \lceil \log_2 m \rceil$. Calling the edges in $\widetilde{E}$ *good*, Alon's algorithm first multiplies the weight of every good edge by $\alpha = \lfloor 2^t/\Delta \rfloor$ and introduces $n/2$ new *bad edges* that induce a matching on V and respect the bipartiteness of $\widetilde{G}$. Every bad edge is given a weight of $\beta = 2^t \bmod \Delta$. Let $G = (V, E)$ be the resulting bipartite weighted graph. Define the total bad weight of an edge-weighted graph with good and bad edges to be the sum of the weights of its bad edges. If every vertex has weight q and the total bad weight is less than q, for some $q \in \mathbb{N}$, we say that the graph is q-*good*. Since $\Delta\alpha + \beta = 2^t$, G is 2^t-good.

The algorithm next computes copies $G_t, G_{t-1}, \ldots, G_0$ of G with new weights such that G_i is 2^i-good, for $i = t, \ldots, 0$. It finishes by returning G_0, stripped of its zero-weight edges, which is a 1-regular subgraph of G without bad edges, i.e., a perfect matching in the original graph $\widetilde{G}$. G_t is simply G. For $i = t - 1, \ldots, 0$, G_i is computed from G_{i+1} as one of two graphs G_i' and G_i'' that are copies of G_{i+1}, but have different weights. If the weight of an edge e in G_{i+1} is k, e is assigned an initial weight of $\lfloor k/2 \rfloor$ in each of G_i' and G_i''. The edges of odd weight in G_{i+1} induce a graph H in which every vertex has even degree, and the edges of each (closed) trail in an Euler partition of H—computed with the algorithm of Sect. 3—subsequently increase their weights by 1 alternately in G_i' and in G_i''. Finally G_i is chosen as one of G_i' and G_i'' of total bad weight less than 2^i.

The running time is linear in the total number of edges in $G_t, \ldots, G_0$. If the input graph $\widetilde{G}$ resulted from the sparsification mentioned in Subsect. 4.2, this number is $O(1 + tn \log \Delta) = O(1 + m(\log m \log \Delta)/\Delta)$. In order to store the edge weights in $O(m)$ bits, we observe that for a given initial weight, only a limited number of weights is possible at an arbitrary later time. For $i = t-1, \ldots, 0$, if the weight of some edge in G_{i+1} is k, then in G_i it is either $\underline{r}(k) = \lfloor k/2 \rfloor$ or $\overline{r}(k) = \lceil k/2 \rceil$. For $j \in \mathbb{N}$, $\underline{r}^{(j)}(k) = \lfloor k/2^j \rfloor$ and $\overline{r}^{(j)}(k) = \lceil k/2^j \rceil \le \underline{r}^{(j)}(k) + 1$, where the superscript $^{(j)}$ denotes j-fold repeated function application. As a consequence, it suffices to store for each edge in E, in addition to its initial weight, a single bit that indicates whether the weight of the edge is "low" or "high".

References

1. Alon, N.: A simple algorithm for edge-coloring bipartite multigraphs. Inform. Process. Lett. **85**(6), 301–302 (2003)
2. Clark, D.: Compact Pat trees. Ph.D. thesis, University of Waterloo (1996)
3. Cole, R., Hopcroft, J.: On edge coloring bipartite graphs. SIAM J. Comput. **11**(3), 540–546 (1982)
4. Cole, R., Ost, K., Schirra, S.: Edge-coloring bipartite multigraphs in $O(E \log D)$ time. Combinatorica **21**(1), 5–12 (2001)
5. Cole, R., Vishkin, U.: Deterministic coin tossing with applications to optimal parallel list ranking. Inform. Control **70**(1), 32–53 (1986)
6. Elmasry, A., Hagerup, T., Kammer, F.: Space-efficient basic graph algorithms. In: Mayr, E.W., Ollinger, N. (eds.) Proceedings of the 32nd International Symposium on Theoretical Aspects of Computer Science (STACS 2015). LIPIcs, vol. 30, pp. 288–301. Schloss Dagstuhl – Leibniz-Zentrum für Informatik (2015)
7. Fleury, M.: Deux problèmes de géométrie de situation. J. Math. Élém. 2nd Ser. **2**, 257–261 (1883)
8. Gabow, H.N.: Using Euler partitions to edge color bipartite multigraphs. Int. J. Comput. Inform. Sci. **5**(4), 345–355 (1976)
9. Gabow, H.N., Kariv, O.: Algorithms for edge coloring bipartite graphs and multigraphs. SIAM J. Comput. **11**(1), 117–129 (1982)
10. Geary, R.F., Rahman, N., Raman, R., Raman, V.: A simple optimal representation for balanced parentheses. Theor. Comput. Sci. **368**(3), 231–246 (2006)
11. Hagerup, T., Kammer, F.: Succinct choice dictionaries. Computing Research Repository (CoRR) arXiv:1604.06058 [cs.DS] (2016)
12. Hierholzer, C.: Ueber die Möglichkeit, einen Linienzug ohne Wiederholung und ohne Unterbrechung zu umfahren. Math. Ann. **6**(1), 30–32 (1873). (communicated by Wiener, C.)
13. Holyer, I.: The NP-completeness of edge-coloring. SIAM J. Comput. **10**(4), 718–720 (1981)
14. Makino, K., Takabatake, T., Fujishige, S.: A simple matching algorithm for regular bipartite graphs. Inform. Process. Lett. **84**(4), 189–193 (2002)
15. Munro, J.I., Raman, V.: Succinct representation of balanced parentheses and static trees. SIAM J. Comput. **31**(3), 762–776 (2001)
16. Schrijver, A.: Bipartite edge coloring in $O(\Delta m)$ time. SIAM J. Comput. **28**(3), 841–846 (1998)
17. Thorup, M.: Near-optimal fully-dynamic graph connectivity. In: Yao, F., Luks, E. (eds.) Proceedings of the 32nd Annual ACM Symposium on Theory of Computing (STOC 2000), pp. 343–350. ACM (2000)

Minimum Point-Overlap Labeling

Yuya Higashikawa[1], Keiko Imai[1], Yusuke Matsumoto[2], Noriyoshi Sukegawa[1], and Yusuke Yokosuka[3(✉)]

[1] Department of Information and System Engineering, Chuo University,
Tokyo, Japan
`{higashikawa,imai,sukegawa}@ise.chuo-u.ac.jp`
[2] Tokyo Software & Systems Development Laboratory, IBM, Tokyo, Japan
`pinecone@jp.ibm.com`
[3] Information and System Engineering Course,
Graduate School of Science and Engineering, Chuo University, Tokyo, Japan
`yusuke.yokosuka@gmail.com`

Abstract. In the air-traffic control, the information related to each air-plane needs to be always displayed as the label. Motivated by this application, de Berg and Gerrits (Comput. Geom. 2012) presented *free-label maximization* problem, where the goal is to maximize the number of intersection-free labels. In this paper, we introduce an alternative labeling problem for the air-traffic control, called *point-overlap minimization*. In this problem, we focus on the number of overlapping labels at a point in the plane, and minimize the maximum among such numbers. Instead of maximizing the number of readable labels as in the free-label maximization, we here minimize the cost required for making unreadable labels readable. We provide a 4-approximation algorithm using LP rounding for arbitrary rectangular labels and a faster combinatorial 8-approximation algorithm for unit-square labels.

Keywords: Map labeling · Air-traffic control · Approximation algorithm

1 Introduction

Map labeling is the problem of placing text or symbol labels corresponding to graphical features on input maps. This problem is important in several areas, such as geographic information system (GIS), cartography, and graph drawing. On maps, labels of regions, rivers, stations, etc., are placed in appropriate positions so that the corresponding features in the map can be understood. In map labeling, points, polylines, and polygons are considered as graphical features. In this paper, we consider map labeling for points only.

Usually, map labeling considers to place labels so that the labels are pairwise disjoint. In the air-traffic control, however, all labels have to be displayed because they contain important information (e.g. altitude, velocity) other than the air-plane position. Furthermore, the label size is fixed. In order to read the unreadable labels, air-traffic controllers move labels by hand wasting time.

© Springer International Publishing AG 2017
D. Fotakis et al. (Eds.): CIAC 2017, LNCS 10236, pp. 334–344, 2017.
DOI: 10.1007/978-3-319-57586-5_28

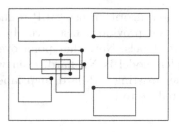
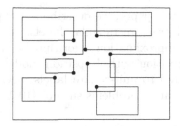

(a) Free-label maximization   (b) Point-overlap minimization

Fig. 1. The difference between optimal solutions for (a) free-label maximization and (b) point-overlap minimization

Given the above background, de Berg and Gerrits [4] considered *free-label maximization* problem. The goal of this problem is to maximize the number of *free* labels which have no intersection with any other labels. By this problem setting, the number of readable labels without moving other labels is increased. For unreadable labels, however, the cost of moving other labels may be increased. Figure 1(a) shows an optimal solution for free-label maximization. Here, the number of free labels is five. However, in the sense of real application for air-traffic control, this solution may not be good since there are overlapping labels which make each other illegible, and so an air-traffic controller has to move at least one among such labels by hand wasting time.

As an alternative approach to decrease the number of moving labels by hand, we introduce the following new problem:

Definition 1 (point-overlap minimization). *Given an instance I which consists of a set of n points $\{p_1, \ldots, p_n\}$ (which is called as* label points*) in the plane and a set of n axis-parallel rectangular labels $\{\ell_1, \ldots, \ell_n\}$. Here, each label point p_i is associated to ℓ_i. For any point p in the plane, let a function $\lambda(p)$ be the number of labels overlapping with p. The* point-overlap minimization *for I is to find a placement of all labels which minimizes the maximum $\lambda(p)$ over all the points p in the plane, and satisfies that each label contains the label point on its boundaries.*

Figure 1(b) shows an optimal solution for point-overlap minimization. Here, the optimal value is two and the number of free labels is two. Compared with free-label maximization, however, all labels can be read without moving any label.

In map labeling research, two models have been considered with respect to the number of label candidates for each label point: The *fixed-position model* [10] and the *slider model* [15]. In both models, each label is placed so that the corresponding label point is on the boundary of the label. The fixed-position model has a finite number of label candidates (e.g., the 2-position and 4-position models). The label candidates of the slider model are the specified sides of the labels (e.g., in the 2-slider (4-slider) model, two (four) sides of the label serve as a set of label candidates). We only consider 4-position model.

In this paper, we describe a 4-approximation algorithm using LP rounding for arbitrary rectangular labels. In addition, we provide a faster combinatorial 8-approximation algorithm for unit square labels. Note that point-overlap minimization with 4-position model or 4-slider model is NP-hard, even when restricted to unit-square labels, because other existing NP-hard map labeling optimization problems such as [10,15] easily reduce to it.

1.1   Related Work

Various types of map labeling have been studied so far (see e.g. [18]). There are two typical optimization problems in map labeling. One is the *label number maximization problem* of finding the placement of a maximum cardinality subset of labels with fixed size. The other is the *label size maximization problem* of placing all labels such that the sizes of the labels are maximized under a global scale factor. Free-label maximization and point-overlap minimization are different from them.

It is known that the label size maximization problems, except for the 1-position and 2-position model, are APX-hard, even for unit square labels [10]. Furthermore, the label number maximization problems are known to be NP-hard (e.g., [10,15]). Therefore, many approximation algorithms have already been provided for both problems (e.g., [2,10,14,15]). The most work only considered unit-square or unit-height rectangular labels. On the other hand, *maximum independent set of rectangles* (MISR) has been studied so far [1,7–9], and this problem can be applied to label number maximization for arbitrary rectangular labels.

In the past decade, map labeling for "dynamic" maps (*dynamic map labeling*) was considered because the importance of dynamic maps has increased due to several applications (e.g. personal mapping systems). There are several dynamic cases, for example, zooming maps [3,12,16,20] and rotating maps [13,19]. Although the above problems are basically considered for 1-position model, Buchin and Gerrits [6] showed dynamic map labeling for 4-position, 2-slider, and 4-slider models is strongly PSPACE-complete.

For the trajectory of a point which is one of the dynamic case, Gemsa et al. [11] treated the problem of maximizing sum of active ranges, where the active range of a label is a contiguous range the label is displayed. This problem is NP-hard and W[1]-hard, and they provided approximation algorithms. For the case that all points are moving, de Berg and Gerrits [4] introduced free-label maximization in "static" map as the first step. Furthermore, de Berg and Gerrits [5] considered about a trade-off between label speed and label overlap for moving points. As in [4], point-overlap minimization is static.

2   Problem Formulation and a 4-Approximation Algorithm

In this section, we formulate point-overlap minimization with 4-position model by IP, and describe a 4-approximation algorithm using LP rounding. This algorithm can treat a set of n arbitrary rectangular labels.

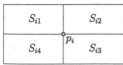

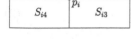

S_{i1}	S_{i2}
S_{i4} p_i	S_{i3}

Fig. 2. Label candidates

cells of i-th slab

Fig. 3. Slabs and cells

We first fix some notations. Let $\mathcal{I}$ be the set of instances of point-overlap minimization with 4-position model. For a given instance $I \in \mathcal{I}$, we denote by $\mathsf{ALG}(I)$ the objective value of the solution of Algorithm ALG. Similarly, we denote by $\mathsf{OPT}(I)$ the optimal value of this problem.

Let us give an IP formulation for point-overlap minimization with 4-position model. For a label point p_i, let S_{ij} be the j-th candidate region in which the label ℓ_i is placed. Here, $j = 1, 2, 3$, and 4 correspond to "upper left", "upper right", "lower right", and "lower left" label candidates, respectively (Fig. 2). The IP have an indicator variable x_{ij} denoting whether a label ℓ_i is placed at S_{ij}. We define $\mathcal{C}$ to be a set of cells where a *cell* is defined as follows. We draw horizontal lines through top and bottom sides of each label candidate. This partitions the plane into horizontal *slabs* (Fig. 3), and the number of slabs is $O(n)$. Then, for each slab, we draw vertical lines through left and right sides of each label candidate intersecting the slab. We define a cell as a region between consecutive vertical lines. Since the number of cells in a slab is $O(n)$, the number of all cells in the plane is $O(n^2)$. The objective of the IP is to minimize τ which indicates the maximum number of labels overlap with a point in the plane. The IP is the following:

$$
\begin{aligned}
&\text{mininmize} && \tau \\
&\text{subject to} && \sum_{S_{ij} \cap C \neq \emptyset} x_{ij} \leq \tau && C \in \mathcal{C} \\
& && \sum_{j=1}^{4} x_{ij} = 1 && i = 1, \ldots, n \\
& && x_{ij} \in \{0, 1\}\ i = 1, \ldots, n, && j = 1, 2, 3, 4
\end{aligned}
\tag{1}
$$

In this IP, the first set of constraints ensures that the number of labels which overlap with each cell is at most τ. The constraints consider only cells instead of all points in the plane because all possible cases that labels intersect at a point is expressed by the cells. Similar constraints are used to design approximation algorithms for MISR [7,8]. The second set of constraints ensures that each label has to be placed at one of the label candidates. Therefore, the number of constraints is $O(n^2)$.

The LP relaxation of IP (1) is:

$$
\begin{aligned}
\text{mininmize} \quad & \tau \\
\text{subject to} \quad & \sum_{S_{ij} \cap C \neq \emptyset} x_{ij} \leq \tau \; C \in \mathcal{C} \\
& \sum_{j=1}^{4} x_{ij} = 1 \; i = 1, \dots, n \\
& x_{ij} \geq 0 \; i = 1, \dots, n, \quad j = 1, 2, 3, 4
\end{aligned}
\tag{2}
$$

Using this LP, we obtain a simple 4-approximation algorithm for point-overlap minimization. The algorithm is as follows:

Step 1. Find an optimal solution to the LP (2).
Step 2. For each label point p_i, pick one variable x_{ij} such that $x_{ij} \geq 1/4$, and set it to 1. Other variables are set to 0.

This algorithm is called LP-rounding (LPR).

Theorem 1. *For any instance $I \in \mathcal{I}$, $\mathsf{LPR}(I) \leq 4\mathsf{OPT}(I)$.*

Proof. Algorithm LPR places just one label for each label point. Thus, the placement of labels by Algorithm LPR is valid.

We consider the objective value $\mathsf{LPR}(I)$ of a solution found by Algorithm LPR. We denote the optimal value of LP (2) by $\mathsf{OPT_{LP}}(I)$. Since x_{ij} is at least $1/4$ for each label point p_i, $\mathsf{LPR}(I)$ is at most $4\mathsf{OPT_{LP}}(I)$. By $\mathsf{OPT_{LP}}(I) \leq \mathsf{OPT}(I)$, we have $\mathsf{LPR}(I) \leq 4\mathsf{OPT}(I)$, and this completes the proof. □

The computation time of Algorithm LPR is as follows: In Step 1, we need to construct $\mathcal{C}$. It takes $O(n^2)$ time. Moreover, solving LP (2) takes $O(N^3 L)$ [17] where N is the number of variables in LP (2), and L is the bit length of LP (2). Since $N = O(n)$ and $L = O(n^3)$ in LP (2), Step 1 can be done in $O(n^6)$ time in total. Step 2 takes $O(n)$ time. Summarizing the above argument, Algorithm LPR runs in $O(n^6)$ time.

3 A Faster Combinatorial 8-Approximation Algorithm for Unit Square Labels

In this section, we discuss the problem with restriction on the form of labels to the unit square, i.e., each label is a square of size 1. For ease of explanation, we assume that each coordinate of each label point is not an integer but positive. Such label points can be obtained only by translation. Once an instance is given, we set a square grid such that each square has size 1 and any grid point has integral coordinates. In the following, we use the notation $(\cdot, \cdot)$ to denote coordinates.

For positive integers s and t, let $B(s, t)$ denote a square box of the grid whose corners are $(s-1, t-1), (s-1, t), (s, t)$ and $(s, t-1)$ (Fig. 4). For convenience,

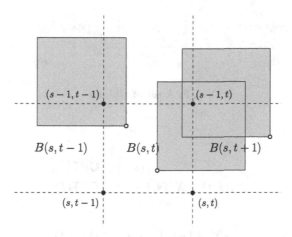

Fig. 4. $B(s, t)$ and its corners

we abuse $B(s,t)$ to denote a set of points inside box $B(s,t)$. A box is said to be *empty* if there is no label point in the box; otherwise, it is said to be *non-empty*. We also define *row s* (resp. *column t*) as a set of boxes $B(s,h)$ (resp. $B(h,t)$) for any h. For positive integers s, t and t' with $t \leq t'$, let $m_s(t,t')$ denote the number of label points which belong to $\bigcup_{t \leq h \leq t'} B(s,h)$. Let $s_{\max}$ (resp. $t_{\max}$) be the maximum integer such that there is at least one label point in row $s_{\max}$ (resp. column $t_{\max}$).

Lemma 1. *For any instance $I \in \mathcal{I}$,*

$$D \equiv \max \left\{ \left\lceil \frac{m_s(t,t')}{2(t'-t+2)} \right\rceil \;\middle|\; 1 \leq s \leq s_{\max}, 1 \leq t \leq t' \leq t_{\max} \right\} \leq \mathsf{OPT}(I).$$

Proof. Let x^* be an optimal solution of the IP formulation. Recall that $x^*_{ij} = 1$ iff the label point i receives the j-th label, where $j = 1, 2, 3$, and 4 correspond to "upper left", "upper right", "lower right", and "lower left", respectively. For a given box $B(s,t)$, let $X^*_j(s,t)$ denote the number of label points belonging to $B(s,t)$ and receive the j-th label, i.e., $\sum_{i:p_i \in B(s,t)} x^*_{ij}$.

Now, arbitrarily choose a row s and two columns t and t' with $t \leq t'$. Observe that in the optimal solution x^*, the point $(s-1, t-1)$ is overlapped by at least $X^*_1(s,t)$ labels. This means that $X^*_1(s,t) \leq \mathsf{OPT}(I)$. Similarly, focusing on the point $(s-1, t)$, we have $X^*_2(s,t) + X^*_1(s,t+1) \leq \mathsf{OPT}(I)$ (see Fig. 4). Applying the similar argument to all the $2(t'-t+2)$ corners of the boxes $B(s,t), \ldots, B(s,t')$,

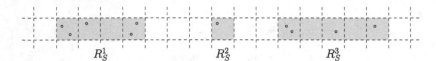

Fig. 5. A set of ribbons

we obtain

$$X_1^*(s,t) \leq \mathsf{OPT}(I),$$
$$X_2^*(s,t) + X_1^*(s,t+1) \leq \mathsf{OPT}(I),$$
$$\vdots$$
$$X_3^*(s,t'-1) + X_4^*(s,t') \leq \mathsf{OPT}(I),$$
$$X_3^*(s,t') \leq \mathsf{OPT}(I).$$

In the above inequalities, each X^* variable appears in exactly one inequality, therefore, the sum of the left-hand sides is $m_s(t,t')$. On the other hand, the sum of the right-hand sides is $2(t'-t+2)\mathsf{OPT}(I)$. To conclude, $m_s(t,t')/(2(t'-t+2)) \leq \mathsf{OPT}(I)$. Then, by the integrality of $\mathsf{OPT}(I)$, $\lceil m_s(t,t')/(2(t'-t+2)) \rceil \leq \mathsf{OPT}(I)$, which completes the proof. □

3.1 Algorithm

In the rest of this section, we propose a combinatorial algorithm with the approximation ratio of 8, which we call *Place-Upperside-of-Ribbon* (PUR). As a preprocessing, it first computes *ribbons*, based on which we can efficiently compute D.

Preprocessing. We define *ribbon* as a maximal sequence of consecutive boxes in a row in which the leftmost and rightmost boxes are non-empty and one of any consecutive two boxes is non-empty. See Fig. 5 for an image of the ribbons. Note that by the definition, every non-empty box is covered by some ribbon and the union of all the ribbons consists of $O(n)$ boxes. The following preprocessing finds out all ribbons and simultaneously computes D.

P-Step 1: For each label point, compute the row-coordinate s and the column-coordinate t, that is, s and t such that $B(s,t)$ includes the point.

P-Step 2: Sort all the label points for their row-coordinates.

P-Step 3: For each row s with at least one label point,

    3-1: sort the label points belonging to row s for their column-coordinates,

    3-2: make a set of ribbons $\{R_s^1, R_s^2, \ldots, R_s^{q(s)}\}$ whose union covers all the label points belonging to row s, and

    3-3: compute $\tilde{D}_s$, the maximum of $\lceil m_s(t,t')/(2(t'-t+2)) \rceil$ over all (t,t') such that $t \leq t'$ and both $B(s,t)$ and $B(s,t')$ belong to some ribbons. Update $\tilde{D} = \max\{\tilde{D}, \tilde{D}_s\}$.

Lemma 2. $D = \tilde{D}$.

Proof. Since $\tilde{D}$ is the value of D when restricted to the boxes in the ribbons, $D \geq \tilde{D}$. Suppose that D is attained at (u, v, v'), i.e., $D = \lceil m_u(v, v')/(2(v' - v + 2)) \rceil$. Then, $B(u, v)$ and $B(u, v')$ are non-empty. If otherwise, by reducing an empty box, we obtain a strictly better solution, which contradicts to the optimality of D. Since the ribbons include all the non-empty boxes, (u, v, v') must be considered in P-Step 3-3, which implies that $D = \tilde{D}$. □

Labeling Strategy. For label points in each isolated ribbon, PUR places every label at its upper left or upper right position. In the following, we use the notation $m_s(t)$ to denote $m_s(t, t)$, and for a given box $B(s, t)$, let $l_s(t)$ (resp. $r_s(t)$) denote the number of label points in $B(s, t)$ whose labels are placed at their upper left (resp. upper right) positions. Note that $m_s(t) = l_s(t) + r_s(t)$. We below show the strategy of PUR for a particular ribbon consisting of boxes $B(s, t), \ldots, B(s, t')$.

L-Step 0: For label points in $B(s, t)$, place labels so that $l_s(t) = \min\{2D, m_s(t)\}$ and $r_s(t) = m_s(t) - l_s(t)$.

L-Step h: For $h = \{1, 2, \ldots, t' - t\}$, for label points in $B(s, t + h)$, place labels so that $l_s(t + h) = \min\{2D - r_s(t + h - 1), m_s(t + h)\}$ and $r_s(t + h) = m_s(t + h) - l_s(t + h)$.

At any L-Step h for $h \geq 1$, $2D - r_s(t + h - 1)$ should be non-negative; otherwise the algorithm returns a negative as $l_s(t + h)$. The following guarantees that.

Lemma 3. *For label points in ribbon consisting of boxes $B(s, t), \ldots, B(s, t')$, PUR places labels so that $r_s(t + h) \leq 2D$ for any integer h with $0 \leq h \leq t' - t$.*

Proof. We prove the claim by induction on h, that is, we prove (i) $r_s(t) \leq 2D$, and (ii) $r_s(t + k + 1) \leq 2D$ assuming $r_s(t + h) \leq 2D$ for $0 \leq h \leq k$.

(i) By the maximality of D, we have $m_s(t)/4 \leq \lceil m_s(t)/4 \rceil \leq D$, that is,

$$m_s(t) \leq 4D. \tag{3}$$

If $m_s(t) \leq 2D$, $r_s(t) = 0 \leq 2D$. Otherwise, $r_s(t) = m_s(t) - 2D \leq 2D$ by (3).

(ii) Observe that by the definition of l_s, $r_s(t + k) + l_s(t + k + 1) \leq 2D$. If $r_s(t + k) + l_s(t + k + 1) < 2D$, then $r_s(t + k + 1) = 0 \leq 2D$. Now, suppose that $r_s(t + k) + l_s(t + k + 1) = 2D$. In this case, there is an integer i with $-1 \leq i \leq k$ satisfying $r_s(t + i - 1) + l_s(t + i) < 2D$. Choosing the maximum of such integers as i,

$$r_s(t + i) = 0. \tag{4}$$

By the maximality of i, we have the following $(k - i + 1)$ equations:

$$r_s(t + i) + l_s(t + i + 1) = 2D,$$

$$\vdots$$

$$r_s(t + k) + l_s(t + k + 1) = 2D.$$

Taking the sum of these, we have

$$r_s(t+i) + m_s(t+i+1, t+k) + l_s(t+k+1) = 2(k-i+1)D$$
$$\Longleftrightarrow \quad m_s(t+i+1, t+k) + l_s(t+k+1) = 2(k-i+1)D \quad \text{(by (4))}$$
$$\Longleftrightarrow \quad m_s(t+i+1, t+k+1) - r_s(t+k+1) = 2(k-i+1)D$$
$$\Longleftrightarrow \quad r_s(t+k+1) = m_s(t+i+1, t+k+1) - 2(k-i+1)D. \tag{5}$$

On the other hand, by the maximality of D, we have

$$\frac{m_s(t+i+1, t+k+1)}{2(k-i+2)} \leq \left\lceil \frac{m_s(t+i+1, t+k+1)}{2(k-i+2)} \right\rceil \leq D$$
$$\Longleftrightarrow \quad m_s(t+i+1, t+k+1) \leq 2(k-i+2)D. \tag{6}$$

By (5) and (6), we obtain $r_s(t+k+1) \leq 2D$. □

Theorem 2. *For any instance* $I \in \mathcal{I}, \mathsf{PUR}(I) \leq 8\mathsf{OPT}(I)$.

Proof. We consider a point $p \in B(s,t)$ with $1 \leq s \leq s_{\max}$ and $1 \leq t \leq t_{\max}$. By algorithm PUR, p can be overlapped only by labels which cover corners of $B(s,t)$, i.e., $(s,t), (s-1,t), (s,t-1)$ and $(s-1,t-1)$. Note that by PUR, a grid point (u,v) is overlapped by exactly $r_{u+1}(v) + l_{u+1}(v+1)$ labels, which is at most $2D$ labels. Therefore p is overlapped by at most $8D$ labels, which implies $\mathsf{PUR}(I) \leq 8D$. This and Lemma 1 conclude the proof. □

For the computation time, we observe the following.

Theorem 3. *Algorithm* PUR *runs in* $O(\max\{k^2, n \log n\})$ *time, where k is the number of non-empty boxes.*

Proof. P-Step 1 is easy and can be done in $O(n)$ time. Also, P-Step 2 can be done in $O(n \log n)$ time. If we denote by n_s the number of the label points belonging to row s, then P-Step 3-1 can be done in $O(n_s \log n_s)$ time, and by scanning in the order, P-Step 3-2 can be done in $O(n_s)$ time. Since $n = \sum_{s=1}^{s_{\max}} n_s$, in total, P-Step 3-1 and P-Step 3-2 require $O(n \log n)$ time in total. For P-Step 3-3, we first prepare an array $A_s = \{m_s(t_s^1, t) \mid B(s,t) \in \bigcup_{1 \leq q \leq q(s)} R_s^q\}$ for every row s with at least one label point, where t_s^1 is the smallest (leftmost) column such that $B(s, t_s^1)$ is non-empty. It clearly takes $O(n)$ time to construct all such arrays since the label points in each row have been sorted for their column-coordinates in P-Step 3-1. Using A_s, $m_s(t, t')$ can be computed in constant time, thus it is easy to observe that an exhaustive search requires $O(k^2)$ time in total. Once D is computed as above, every L-Step h for a ribbon in row s runs in constant time (again using A_s), which means that all labels can be placed in $O(n)$ time. Summarizing the above argument, we complete the proof. □

Since k is at most n, the overall computation time is $O(n^2)$. However, for non-trivial instances, especially for real-world instance, k must be small and hence the computation time will be like $O(n \log n)$ in practice.

4   Conclusion

In this paper, we introduced the point-overlap minimization problem as a new direction for map labeling. Point-overlap minimization seems to possess a promising and somewhat better concept for the air-traffic control compared to free-label maximization considered in de Berg and Gerrits [4], as illustrated in Fig. 1. For point-overlap minimization with 4-position model, we proposed a 4-approximation algorithm based on LP rounding, which runs in strongly polynomial time but its complexity is high. Alternatively, with restriction on the form of labels to the unit square, we also proposed a fully combinatorial 8-approximation algorithm, which runs much faster than the above one.

Acknowledgments. The work of the second author was supported in part by Grant-in-Aid for Scientific Research of Japan Society for the Promotion of Science.

References

1. Adamaszek, A., Wiese, A.: Approximation schemes for maximum weight independent set of rectangles. In: 54th Annual IEEE Symposium on Foundations of Computer Science, FOCS 2013, Berkeley, CA, USA, 26–29 October 2013, pp. 400–409 (2013)
2. Agarwal, P.K., van Kreveld, M.J., Suri, S.: Label placement by maximum independent set in rectangles. Comput. Geom. **11**(3–4), 209–218 (1998)
3. Been, K., Nöllenburg, M., Poon, S.H., Wolff, A.: Optimizing active ranges for consistent dynamic map labeling. Comput. Geom. **43**(3), 312–328 (2010)
4. de Berg, M., Gerrits, D.H.P.: Approximation algorithms for free-label maximization. Comput. Geom. **45**(4), 153–168 (2012)
5. de Berg, M., Gerrits, D.H.P.: Labeling moving points with a trade-off between label speed and label overlap. In: Bodlaender, H.L., Italiano, G.F. (eds.) ESA 2013. LNCS, vol. 8125, pp. 373–384. Springer, Heidelberg (2013). doi:10.1007/978-3-642-40450-4_32
6. Buchin, K., Gerrits, D.H.P.: Dynamic point labeling is strongly PSPACE-complete. Int. J. Comput. Geom. Appl. **24**(4), 373–395 (2014)
7. Chalermsook, P., Chuzhoy, J.: Maximum independent set of rectangles. In: Proceedings of the Twentieth Annual ACM-SIAM Symposium on Discrete Algorithms, SODA 2009, New York, NY, USA, 4–6 January 2009, pp. 892–901 (2009)
8. Chan, T.M., Har-Peled, S.: Approximation algorithms for maximum independent set of pseudo-disks. Discret. Comput. Geom. **48**(2), 373–392 (2012)
9. Chuzhoy, J., Ene, A.: On approximating maximum independent set of rectangles. In: IEEE 57th Annual Symposium on Foundations of Computer Science, FOCS 2016, Hyatt Regency, New Brunswick, New Jersey, USA, 9–11 October 2016, pp. 820–829 (2016)
10. Formann, M., Wagner, F.: A packing problem with applications to lettering of maps. In: Proceedings of the Seventh Annual Symposium on Computational Geometry, North Conway, NH, USA, 10–12 June 1991, pp. 281–288 (1991)
11. Gemsa, A., Niedermann, B., Nöllenburg, M.: Trajectory-based dynamic map labeling. In: Cai, L., Cheng, S.-W., Lam, T.-W. (eds.) ISAAC 2013. LNCS, vol. 8283, pp. 413–423. Springer, Heidelberg (2013). doi:10.1007/978-3-642-45030-3_39

12. Gemsa, A., Nöllenburg, M., Rutter, I.: Sliding labels for dynamic point labeling. In: Proceedings of the 23rd Annual Canadian Conference on Computational Geometry, Toronto, Ontario, Canada, 10–12 August 2011 (2011)
13. Gemsa, A., Nöllenburg, M., Rutter, I.: Consistent labeling of rotating maps. JoCG **7**(1), 308–331 (2016)
14. Jung, J.W., Chwa, K.Y.: Labeling points with given rectangles. Inf. Process. Lett. **89**(3), 115–121 (2004)
15. van Kreveld, M.J., Strijk, T., Wolff, A.: Point labeling with sliding labels. Comput. Geom. **13**(1), 21–47 (1999)
16. Liao, C., Liang, C., Poon, S.: Approximation algorithms on consistent dynamic map labeling. Theor. Comput. Sci. **640**, 84–93 (2016)
17. Renegar, J.: A polynomial-time algorithm, based on Newton's method, for linear programming. Math. Program. **40**(1–3), 59–93 (1988)
18. Wolff, A., Strijk, T.: The map-labeling bibliography (2009). http://i11www.iti.uni-karlsruhe.de/map-labeling/bibliography/
19. Yokosuka, Y., Imai, K.: Polynomial time algorithms for label size maximization on rotating maps. In: Proceedings of the 25th Canadian Conference on Computational Geometry, CCCG 2013, Waterloo, Ontario, Canada, 8–10 August 2013, pp. 187–192 (2013)
20. Zhang, X., Poon, S., Li, M., Lee, V.C.S.: On maxmin active range problem for weighted consistent dynamic map labeling. In: Proceedings of the Seventh International Conference on Advanced Geographic Information Systems, Applications, and Services, GEOProcessing 2015, Lisbon, Portugal, 22–27 February 2015, pp. 32–37 (2015)

Fine-Grained Parameterized Complexity Analysis of Graph Coloring Problems

Lars Jaffke[1]([✉]) and Bart M.P. Jansen[2]

[1] University of Bergen, Postboks 7803, 5020 Bergen, Norway
l.jaffke@uib.no
[2] Eindhoven University of Technology, P.O. Box 513,
5600 MB Eindhoven, The Netherlands
b.m.p.jansen@tue.nl

Abstract. The q-COLORING problem asks whether the vertices of a graph can be properly colored with q colors. Lokshtanov et al. [SODA 2011] showed that q-COLORING on graphs with a feedback vertex set of size k cannot be solved in time $\mathcal{O}^*((q - \varepsilon)^k)$, for any $\varepsilon > 0$, unless the Strong Exponential-Time Hypothesis (SETH) fails. In this paper we perform a fine-grained analysis of the complexity of q-COLORING with respect to a hierarchy of parameters. We show that unless ETH fails, there is no universal constant θ such that q-COLORING parameterized by vertex cover can be solved in time $\mathcal{O}^*(\theta^k)$ for all fixed q. We prove that there are $\mathcal{O}^*((q - \varepsilon)^k)$ time algorithms where k is the vertex deletion distance to several graph classes $\mathcal{F}$ for which q-COLORING is known to be solvable in polynomial time, including all graph classes whose $(q + 1)$-colorable members have bounded treedepth. In contrast, we prove that if $\mathcal{F}$ is the class of paths – some of the simplest graphs of unbounded treedepth – then no such algorithm can exist unless SETH fails.

1 Introduction

In an influential paper from 2011, Lokshtanov et al. showed that for several problems, straightforward dynamic programming algorithms for graphs of bounded treewidth are essentially optimal unless the Strong Exponential Time Hypothesis (SETH) fails [12]. (Section 2 gives the definitions of the two Exponential Time Hypotheses, see [3, Chap. 14] or the survey [14] for further details.) Some of the lower bounds, as the one for q-COLORING, even hold for parameters such as the feedback vertex number, which form an upper bound on the treewidth but may be arbitrarily much larger. For other problems such as DOMINATING SET, the tight lower bound of $\Omega((3 - \varepsilon)^k)$ holds for the parameterization pathwidth, but is not known for the parameterization feedback vertex set. In general, moving

---

This research was partially funded by the Networks programme via the Dutch Ministry of Education, Culture and Science through the Netherlands Organisation for Scientific Research. The research was done while the first author was at CWI, Amsterdam. The second author was supported by NWO Veni grant "Frontiers in Parameterized Preprocessing".

© Springer International Publishing AG 2017
D. Fotakis et al. (Eds.): CIAC 2017, LNCS 10236, pp. 345–356, 2017.
DOI: 10.1007/978-3-319-57586-5_29

to a parameterization that takes larger values might enable running times with
a smaller base of the exponent. In this paper, we therefore investigate the parameterized complexity of the q-COLORING and q-LIST-COLORING problems from
a more fine-grained perspective.

In particular, we consider a hierarchy of graph parameters — ordered by their
expressive strength — which is a common method in parameterized complexity,
see e.g. [6] for an introduction. One of the strongest parameters for a graph
problem is the number of vertices in a graph, in the following denoted by n.
Björklund et al. showed that the chromatic number $\chi(G)$ (the smallest number of
colors q such that G is q-colorable) of a graph G can be computed in time $\mathcal{O}^*(2^n)$
[1], so the base of the exponent in the runtime of the algorithm is independent
of the value of $\chi(G)$. We show that if you consider a slightly weaker parameter,
the size k of a vertex cover of G, it is very unlikely that there is a constant θ,
such that q-COLORING can be solved in time $\mathcal{O}^*(\theta^k)$ for all fixed $q \in \mathcal{O}(1)$: It
would imply that ETH is false.

However, we show that there is a simple algorithm that solves q-COLORING
parameterized by vertex cover, and for which the base of the exponential in its
runtime is strictly smaller than the base q that is potentially optimal for the
treewidth parameterization.

Proposition 1 (★). *There is an algorithm which decides whether a graph G is
q-colorable and runs in time $\mathcal{O}^*((q-1.11)^k)$, where k denotes the size of a given
vertex cover of G.*

On the other hand, the above algorithm does not obviously generalize to other
parameterizations. To derive more general results about obtaining non-trivial
runtime bounds for parameterized q-COLORING, we study graph classes with
small *vertex modulators* to several graph classes $\mathcal{F}$: Given a graph $G = (V, E)$, a
vertex modulator $X \subseteq V$ to $\mathcal{F}$ is a subset of its vertices such that if we remove
X from G the resulting graph is a member of $\mathcal{F}$, i.e. $G - X \in \mathcal{F}$. If $|X| \leq k$, we
say that $G \in \mathcal{F} + kv$. (For example, graphs that have a vertex cover of size at
most k are INDEPENDENT $+ kv$ graphs.) Hence, we study the following problems
which were first investigated in this parameterized setting by Cai [2].

q-(LIST-)COLORING ON $\mathcal{F} + kv$ GRAPHS
Input: An undirected graph G and a modulator $X \subseteq V(G)$ such that
$G - X \in \mathcal{F}$ (and lists $\Lambda \colon V \to 2^{[q]}$).
Parameter: $|X| = k$, the size of the modulator.
Question: Can we assign each vertex v a color from $[q]$ (on its list $\Lambda(v)$)
such that adjacent vertices have different colors?

Given a NO-instance (G, Λ) of q-LIST-COLORING we call (G', Λ') a NO-
subinstance of (G, Λ) if its answer is also NO and G' is an induced subgraph of G
with $\Lambda(v) = \Lambda'(v)$ for all $v \in V(G')$. We show that if a graph class $\mathcal{F}$ has small
NO-certificates for q-LIST-COLORING (Definition 4) then q-(LIST-)COLORING on
$\mathcal{F} + kv$ graphs can be solved in time $\mathcal{O}^*((q-\varepsilon)^k)$, for some $\varepsilon > 0$. This notion was

introduced by Jansen and Kratsch to prove the existence of polynomial kernels
for said parameterizations [11].

In addition to that, we give some further structural insight into hereditary
graph classes $\mathcal{F}$, for which $\mathcal{F} + kv$ graphs have non-trivial algorithms: We show
that if the $(q+1)$-colorable members of $\mathcal{F}$ have bounded treedepth, then $\mathcal{F} + kv$
has $\mathcal{O}^*((q-\varepsilon)^k)$ time algorithms for q-COLORING when parameterized by the size
k of a given modulator, for some $\varepsilon > 0$. We prove that this *treedepth-boundary* is
in some sense tight: Arguably the most simple graphs of unbounded treedepth
are paths. We show that q-COLORING cannot be solved in time $\mathcal{O}^*((q-\varepsilon)^k)$
for any $\varepsilon > 0$ on PATH $+ kv$ graphs, unless SETH fails — strengthening the
lower bound for FOREST $+ kv$ graphs [12] via a somewhat simpler construction.
Using this strengthened lower bound, we prove that if a hereditary graph class
$\mathcal{F}$ excludes a complete bipartite graph $K_{t,t}$ for some constant t, then $\mathcal{F} + kv$
has $\mathcal{O}^*((q - \varepsilon)^k)$ time algorithms for q-(LIST-)COLORING *if and only if* the
$(q+1)$-colorable members of $\mathcal{F}$ have bounded treedepth.

Throughout the paper, proofs of statements marked with (★) are deferred
to the full version [10] due to space restrictions.

2   Preliminaries

We assume the reader to be familiar with the basic notions in graph theory and
parameterized complexity and refer to [3–5,7] for an introduction. We now give
the most important definitions which are used throughout the paper.

We use the following notation: For $a, b \in \mathbb{N}$ with $a < b$, $[a] = \{1, \ldots, a\}$ and
$[a..b] = \{a, a+1, \ldots, b\}$. The $\mathcal{O}^*$-notation suppresses polynomial factors in the
input size n, i.e. $\mathcal{O}^*(f(n, \cdot)) = \mathcal{O}(f(n, \cdot) \cdot n^{\mathcal{O}(1)})$. For a function $f\colon X \to Y$, we
denote by $f_{|X'}$ the restriction of f to $X' \subseteq X$.

Graphs and Parameters. Throughout the paper a graph G with vertex set
$V(G)$ and edge set $E(G)$ is finite and simple. We sometimes shorthand $'V(G)'$
$('E(G)')$ to $'V'$ $('E')$ if it is clear from the context. For graphs G, G' we denote
by $G' \subseteq G$ that G' is a subgraph of G, i.e. $V(G') \subseteq V(G)$ and $E(G') \subseteq E(G)$. We
often use the notation $n = |V|$ and $m = |E|$. For a vertex $v \in V(G)$, we denote
by $N_G(v)$ (or simply $N(v)$, if G is clear from the context) the set of neighbors
of v in G, i.e. $N_G(v) = \{w \mid \{v, w\} \in E(G)\}$.

For a vertex set $V' \subseteq V(G)$, we denote by $G[V']$ the subgraph *induced* by
V', i.e. $G[V'] = (V', E(G) \cap V' \times V')$. A graph class $\mathcal{F}$ is called *hereditary*, if it is
closed under taking induced subgraphs. We now list a number of graph classes
which will be important for the rest of the paper. A graph G is *independent*, if
$E(G) = \emptyset$. A *cycle* is a connected graph all of whose vertices have degree two.
A graph is a *forest*, if it does not contain a cycle as an induced subgraph and
a *linear forest* if additionally its maximum degree is at most two. A connected
forest is a *tree* and a tree of maximum degree at most two is a *path*. A graph G
is a *split graph*, if its vertex set $V(G)$ can be partitioned into sets $W, Z \subseteq V(G)$
such that $G[W]$ is a clique and $G[Z]$ is independent. We define the class $\bigcup$SPLIT
containing all graphs that are disjoint unions of split graphs. A graph G is a

cograph if it does not contain P_4, a path on four vertices, as an induced subgraph. A graph is *chordal*, if it does not have a cycle of length at least four as an induced subgraph. A cochordal graph is the edge complement of a chordal graph and the class ⋃COCHORDAL contains all disjoint unions of cochordal graphs.

Let Σ be an alphabet. A *parameterized problem* is a set $\Pi \subseteq \Sigma^* \times \mathbb{N}$, the second component being the *parameter* which usually expresses a structural measure of the input. A parameterized problem is (strongly uniform) *fixed-parameter tractable* (fpt) if there exists an algorithm to decide whether $\langle x, k \rangle \in \Pi$ in time $f(k) \cdot |x|^{\mathcal{O}(1)}$ where f is a computable function.

The main focus of our research is how the function $f(k)$ behaves for q-COLORING w.r.t. different structural graph parameters, such as the size of a vertex cover.

In this paper we study a *hierarchy of parameters*, a term which we will now discuss. For a detailed introduction we refer to [6, Sect. 3]. For notational convenience, we denote by Π_p a parameterized problem with parameterization p. Suppose we have a graph problem and two parameterizations $p(G)$ and $p'(G)$ regarding some structural graph measure. We call parameterization $p'(G)$ *larger* than $p(G)$ if there is a function f, such that $f(p'(G)) \geq p(G)$ for all graphs G. Modulo some technicalities, we can then observe that if a problem Π_p is fpt, then $\Pi_{p'}$ is also fpt. This induces a partial ordering on all parameterizations based on which a hierarchy can be defined.

Exponential-Time Hypotheses. In 2001, Impagliazzo et al. made two conjectures about the complexity of q-SAT — the problem of finding a satisfying assignment for a Boolean formula in conjunctive normal form with clauses of size at most q [8,9]. These conjectures are known as the Exponential-Time Hypothesis (ETH) and Strong Exponential-Time Hypothesis (SETH), formally defined below. For a survey of conditional lower bounds based on such conjectures, see [14].

Conjecture 2 (ETH [8]). There is an $\varepsilon > 0$, such that 3-SAT on n variables cannot be solved in time $\mathcal{O}^*(2^{\varepsilon n})$.

Conjecture 3 (SETH [8,9]). For every $\varepsilon > 0$, there is a $q \in \mathcal{O}(1)$ such that q-SAT on n variables cannot be solved in time $\mathcal{O}^*((2 - \varepsilon)^n)$.

3 Upper Bounds

In this section we present upper bounds for parameterized q-COLORING. In particular, in Sect. 3.1 we show that if a graph class $\mathcal{F}$ has NO-certificates of constant size, then there exist $\mathcal{O}^*((q - \varepsilon)^k)$ time algorithms for q-COLORING on $\mathcal{F} + kv$ graphs for some $\varepsilon > 0$ depending on $\mathcal{F}$. In Sect. 3.2 we show that if the $(q + 1)$-colorable members of a hereditary graph class $\mathcal{F}$ have bounded treedepth, then $\mathcal{F}$ has NO-certificates of small size.

3.1 Small No-Certificates

In earlier work [11], Jansen and Kratsch studied the kernelizability of q-COLORING and established a generic method to prove the existence of polynomial kernels for several parameterizations of q-COLORING. We now show that we can use their method to prove the existence of $\mathcal{O}^*((q-\varepsilon)^k)$ time algorithms, for some $\varepsilon > 0$, for several graph classes $\mathcal{F} + kv$ as well.

We first introduce the necessary terminology. Let (G, Λ) be an instance of q-LIST-COLORING. We call (G', Λ') a *subinstance* of (G, Λ), if G' is an induced subgraph of G and $\Lambda(v) = \Lambda'(v)$ for all $v \in V(G')$.

Definition 4 ($g(q)$-size No-certificates). *Let $g : \mathbb{N} \to \mathbb{N}$ be a function. A graph class $\mathcal{F}$ is said to have $g(q)$-size No-certificates for q-LIST-COLORING if for all No-instances (G, Λ) of q-LIST-COLORING with $G \in \mathcal{F}$ there is a No-subinstance (G', Λ') on at most $g(q)$ vertices.*

Theorem 5. *Let $\mathcal{F}$ be a graph class with $g(q)$-size No-certificates for q-LIST-COLORING. Then, there is an $\varepsilon > 0$, such that q-LIST-COLORING (and hence, q-COLORING) on $\mathcal{F} + kv$ graphs can be solved in time $\mathcal{O}^*((q - \varepsilon)^k)$ given a modulator to $\mathcal{F}$ of size at most k. In particular, the algorithm runs in time $\mathcal{O}^*\left(\sqrt[g(q) \cdot q]{q^{g(q) \cdot q} - 1}^{k} \right)$, where the degree of the hidden polynomial depends on $g(q)$.*

Proof. Let $G \in \mathcal{F} + kv$ with vertex modulator X, such that $\mathcal{F}$ has $g(q)$-size No-certificates for q-LIST-COLORING. The idea of the algorithm is to enumerate partial colorings of X, except some colorings for which it is clear that they cannot be extended to a proper coloring of the entire instance. The latter can occur as follows: After choosing a coloring for some vertices of X and removing the chosen colors from the lists of their neighbors, a No-subinstance appears in the graph $G - X$. If the minimal No-subinstances have constant size, then for any given instance, either *all* proper colorings on X can be extended onto $G - X$, or there is a way to find a constant-size set $X' \subseteq X$ of vertices for which at least one of the $q^{|X'|}$ colorings would trigger a No-subinstance and can therefore be discarded. Branching on the remaining relevant colorings for X' then gives a nontrivial running time. An outline is given in Algorithm 1.

The main condition (line 2) checks whether the input graph G contains the graph of a minimal No-instance as an induced subgraph. If so, we look for a neighborhood of $V(G')$ in X (the sets $X_1, \ldots, X_q$), which can block the colors that are on the lists Λ but not on the lists of the minimal No-instance. If these conditions are satisfied, then we know that we can exclude the coloring on $X_1, \ldots, X_q$ which assigns each vertex $v \in X_c$ the color c (for all $c \in [q]$): This coloring induces a No-subinstance on (G, Λ). It suffices to use sets X_c of at most $g(q)$ vertices each. To induce the No-instance, in the worst case we need a different vertex in X_c for each of the $g(q)$ vertices in H that do not have c on their list. Hence, as described from line 4 on, we enumerate all colorings $\gamma : \mathcal{X} \to [q]$ (where $\mathcal{X} = \bigcup_i X_i$) except the one we just identified as not being extendible to $G - X$. For each such γ, we make a copy of the current instance and 'assign'

Algorithm 1. q-LIST-COLORING for $\mathcal{F} + kv$ graphs where $\mathcal{F}$ has $g(q)$-size No-certificates.

Input : A graph $G \in \mathcal{F} + kv$ with vertex modulator X and $\Lambda \colon V \to 2^{[q]}$.
Output: YES, if G is q-list-colorable, No otherwise.

1 Let ζ be the set of No-instances of q-LIST-COLORING for $\mathcal{F}$ of size at most $g(q)$, which is computed once by complete enumeration;

2 **if** *there exist* $(H, \Lambda_H) \in \zeta$, $G' \subseteq G - X$ *and* $X_1, \ldots, X_q \subseteq X$ *of size at most* $g(q)$ *each such that:*

    1. $\exists$ *isomorphism* $\varphi \colon V(G') \to V(H)$

    2. *For all* $c \in [q]$ *and* $v \in X_c$ *we have* $c \in \Lambda(v)$

    3. $(\forall v \in V(G'))(\forall c \in \Lambda(v) \setminus \Lambda_H(\varphi(v)))\, \exists w \in X_c$ *with* $\{v, w\} \in E(G)$

3 **then**

4   **foreach** *proper coloring* $\gamma \colon \mathcal{X} \to [q]$ *where* $\mathcal{X} = \bigcup_i X_i$ *and* $\forall v \in \mathcal{X}$: $\gamma(v) \in \Lambda(v)$ **do**

5   **if** $(\forall c \in [q])(\forall v \in X_c)$: $\gamma(v) = c$ **then**

6   Skip this coloring, it is not extendible to $G - X$;

7   **else**

8   Create a copy (G'', Λ'') of (G, Λ) and denote by $\mathcal{X}''$ the vertex set in G'' corresponding to $\mathcal{X}$ in G;

9   For each vertex $v \in \mathcal{X}''$ and each neighbor w of v: Remove $\gamma(v)$ from $\Lambda''(w)$;

10   Recurse on $(G'' - \mathcal{X}'', \Lambda'')$;

11   **if** *the recursive call returns* YES **then**

12   **Return** YES and terminate the algorithm;

13   **Return** No;

14 **else**

15   Decide whether $(G[X], \Lambda_{|X})$ is q-list-colorable and if so, return YES;

---

each vertex v corresponding to a vertex in $\mathcal{X}$ the color $\gamma(v)$: We remove $\gamma(v)$ from the lists of its neighbors and then remove v from the copy instance. In the worst case we therefore recurse on $q^{q \cdot g(q)} - 1$ instances with the size of the vertex modulator decreased by $q \cdot g(q)$. If during a branch in the computation, the condition in line 2 is not satisfied, then we know that there is no coloring on the modulator that cannot be extended to the vertices outside the modulator and hence it is sufficient to compute whether $G[X]$ is q-list-colorable using the standard $\mathcal{O}^*(2^n)$ algorithm for computing the chromatic number [1]. As soon as one branch returns YES, we can terminate the algorithm, since we found a valid list coloring.

Claim 6 ($\bigstar$). *If the condition of line 2 does not hold, then G is q-list-colorable if and only if $G[X]$ is q-list-colorable.*

Claim 7 ($\bigstar$). *If the condition of line 5 holds, then the coloring γ cannot be extended to a proper q-list-coloring of G.*

The correctness of the procedure and the runtime bound are established in the full version [10]. $\square$

The following corollary follows by combining Theorem 5 with known bounds on the sizes of NO-certificates given in Corollaries 1–2 and Lemmas 2–4 of [11].

Corollary 8. *There is an $\varepsilon > 0$, such that the q-COLORING and q-LIST-COLORING problems on $\mathcal{F} + kv$ graphs can be solved in time $\mathcal{O}^*((q - \varepsilon)^k)$ given a modulator to $\mathcal{F}$ of size k, where $\mathcal{F}$ is one of the following classes: INDEPENDENT, $\bigcup$SPLIT, $\bigcup$COCHORDAL and COGRAPH.*

3.2 Bounded Treedepth

We now show that if the $(q + 1)$-colorable members of a hereditary graph class $\mathcal{F}$ have treedepth at most t, then $\mathcal{F}$ has q^t-size NO-certificates. For a detailed introduction to the parameter treedepth, we refer to [13, Chap. 6].

Definition 9 (Treedepth). *Let G be a connected graph. A treedepth decomposition $\mathcal{T} = (V(G), F)$ is a rooted tree on the vertex set of G such that the following holds. For $v \in V(G)$, let $\mathcal{A}_v$ denote the set of ancestors of v in $\mathcal{T}$. Then, for each edge $\{v, w\} \in E(G)$, either $v \in \mathcal{A}_w$ or $w \in \mathcal{A}_v$.*

The depth of $\mathcal{T}$ is the number of vertices on a longest path from the root to a leaf. The treedepth of a connected graph is the minimum depth of all its treedepth decompositions. The treedepth of a disconnected graph is the maximum treedepth of its connected components.

Lemma 10 (★). *Let $\mathcal{F}$ be a hereditary graph class whose $(q+1)$-colorable members have treedepth at most t. Then, $\mathcal{F}$ has q^t-size NO-certificates for q-LIST-COLORING.*

To see the versatility of Lemma 10, observe that the vertices of a $(q + 1)$-colorable split graph can be partitioned into a clique of size at most $(q + 1)$ and an independent set, which makes it easy to see that they have treedepth at most $q + 2$. Since the treedepth of a disconnected graph equals the maximum of the treedepth of its connected components, we then get a finite (q^{q+2}) bound on the size of minimal NO-instances for q-LIST-COLORING on $\bigcup$SPLIT graphs. An ad-hoc argument was needed for this in earlier work [11, Lemma 2], albeit resulting in a better bound $(q + 4^q)$.

4 Lower Bounds

In this section we prove lower bounds for q-COLORING in the parameter hierarchy. Since in the following, the '$\mathcal{F} + kv$'-notation is more convenient for the presentation of our results, we will mostly refer to graphs which have a vertex cover of size k as INDEPENDENT $+ kv$ graphs and graphs that have a feedback vertex set of size k as FOREST $+ kv$ graphs.

In Sect. 4.1 we show that there is no universal constant θ, such that q-COLORING on INDEPENDENT $+ kv$ graphs can be solved in time $\mathcal{O}^*(\theta^k)$ for all fixed $q \in \mathcal{O}(1)$, unless ETH fails. We generalize the lower bound modulo SETH for FOREST $+ kv$ graphs [12] to LINEAR FOREST $+ kv$ (and PATH $+ kv$) graphs in Sect. 4.2. Note that by the constructions we give in their proofs, the lower bounds also hold in case a modulator of size k is given in the input.

4.1   No Universal Constant for Independent + kv Graphs

The following theorem shows that, unless ETH fails, the runtime of any algorithm for q-COLORING parameterized by vertex cover (equivalently, on INDEPENDENT + kv graphs), always has a term depending on q in the base of the exponent.

Theorem 11. *There is no (universal) constant θ, such that for all fixed $q \in \mathcal{O}(1)$, q-COLORING on INDEPENDENT + kv graphs can be solved in time $\mathcal{O}^*(\theta^k)$, unless ETH fails.*

Proof. Assume we can solve q-COLORING on INDEPENDENT + kv graphs in time $\mathcal{O}^*(\theta^k)$. We will use this hypothetical algorithm to solve 3-SAT in $\mathcal{O}^*(2^{\varepsilon n})$ time for arbitrarily small $\varepsilon > 0$, contradicting ETH. We present a way to reduce an instance φ of 3-SAT to an instance of $3q$-LIST-COLORING for q an arbitrary power of 2. The larger q is, the smaller the vertex cover of the constructed graph will be. It will be useful to think of a color $c \in [q]$ ($q = 2^t$ for some $t \in \mathbb{N}$) as a bitstring of length t, which naturally encodes a truth assignment to t variables. The entire color range $[3q]$ partitions into three consecutive blocks of q colors, so that the same truth assignment to t variables can be encoded by three distinct colors $c, c+q$, and $c+2q$ for some $c \in [q]$. The reason for the threefold redundancy is that clauses in φ have size three and will become clear later.

Given an instance φ of 3-SAT, we create a graph G_{3q} and lists $\Lambda \colon V(G_{3q}) \to [3q]$ as follows. First, we add $\lceil n/\log q \rceil$ vertices $v_{1,i}$ (where $i \in [\lceil n/\log q \rceil]$) to $V(G_{3q})$, whose colorings will correspond to the truth assignments of the variables $x_1, \ldots, x_n$ in φ. We let $\Lambda(v_{1,i}) = [q]$ for all these vertices. In particular, the variable x_i will be encoded by vertex $v_{1,\lceil i/\log q \rceil}$. We add two more layers of vertices $v_{2,i}, v_{3,i}$ (where $i \in [\lceil n/\log q \rceil]$) to G_{3q} whose lists will be $\Lambda(v_{2,i}) = [(q+1)..2q]$ and $\Lambda(v_{3,i}) = [(2q+1)..3q]$, respectively (for all i). Throughout the proof, we denote the set of all these *variable vertices* by $\mathcal{V} = \bigcup_{i,j} v_{i,j}$, where $i \in [3]$ and $j \in [\lceil n/\log q \rceil]$.

For each $i \in [2]$ and $j \in [\lceil n/\log q \rceil]$ we do the following. For each pair of colors $c \in [((i-1)q+1)..(i \cdot q)]$ and $c' \in [(i \cdot q+1)..((i+1)q)]$ such that $c+q \neq c'$, we add a vertex $u_{c,c'}^{i,j}$ with list $\Lambda(u_{c,c'}^{i,j}) = \{c, c'\}$ and make it adjacent to both $v_{i,j}$ and $v_{i+1,j}$. Note that this way, we add $\mathcal{O}(q^2)$ and hence a constant number of vertices for each such i and j. We denote the set of all vertices $u_{\cdot,\cdot}^{i,j}$ for all i and j by $\mathcal{U}$.

Claim 12 ($\bigstar$). *Let $i \in [2]$ and $j \in [\lceil n/\log q \rceil]$. In any proper list-coloring of G_{3q}, the color $c \in [((i-1)q+1)..(i \cdot q)]$ appears on $v_{i,j}$ if and only if the color $c+q$ appears on $v_{i+1,j}$. If color $c \in [((i-1)q+1)..(i \cdot q)]$ appears on $v_{i,j}$ and $c' = c+q$ appears on $v_{i+1,j}$, then all vertices $u_{\cdot,\cdot}^{i,j}$ can be assigned a color from their list that does not appear on a neighbor.*

Claim 12 shows that in any proper list-coloring of $\mathcal{V}$, there is a threefold redundancy: If color c appears on $v_{1,i}$, then color $c+q$ appears on $v_{2,i}$ and $c+2q$ appears on $v_{3,i}$. We associate a proper list-coloring of $\mathcal{V}$ with the truth assignment whose

TRUE/FALSE assignment to the i-th block of $\log q$ consecutive variables follows the $1/0$-bit pattern in the least significant $\log q$ bits of the binary expansion of the color of vertex $v_{1,i}$. Conversely, given a truth assignment to $x_1, \ldots, x_n$ we associate it to the coloring of $\mathcal{V}$ where the color of vertex $v_{1,i}$ is given by the number whose least significant $\log q$ bits match the truth assignment to the i-th block of $\log q$ variables, and any remaining bits are set to 0. The colors of $v_{2,i}$ and $v_{3,i}$ are q and $2q$ higher than the color of $v_{1,i}$.

For each clause $C_j \in \varphi$ we will now add a number of *clause vertices* to ensure that if C_j is not satisfied by a given truth assignment of its variables, then the corresponding coloring of the vertices $\mathcal{V}$ cannot be extended to (at least) one of these clause vertices.

Let $C_j \in \varphi$ be a clause with variables x_{j_1}, x_{j_2}, and x_{j_3}. Then, $v_{1,\lceil j_1/\log q \rceil}$, $v_{1,\lceil j_2/\log q \rceil}$, and $v_{1,\lceil j_3/\log q \rceil}$ denote the vertices whose colorings encode the truth assignments of the respective variables. In the following, let $j_i' = \lceil j_i/\log q \rceil$ for $i \in [3]$. Note that there is precisely one truth assignment of the variables x_{j_1}, x_{j_2}, and x_{j_3} that does not satisfy C_j. Choose $\ell_1, \ell_2, \ell_3 \in \{0,1\}$ such that $\ell_i = 0$ if and only if the i-th variable in C_j appears negated. For $i \in [3]$ let $F_i \subseteq [q]$ be those colors whose binary expansion differs from ℓ_i at the $(j_i \bmod (\log q))$-th least significant bit, and define $F_i^{+q} := \{q + c \mid c \in F_i\}$ and $F_i^{+2q} := \{2q + c \mid c \in F_i\}$. This implies that the truth assignment encoded by a proper list-coloring of $\mathcal{V}$ falsifies the i-th literal of C_j if and only if it uses a color from F_i on vertex $v_{1,j_i'}$. By Claim 12, this happens if and only if it uses a color from F_i^{+q} on vertex $v_{2,j_i'}$, which happens if and only if it uses a color of F_i^{+2q} on vertex $v_{3,j_i'}$. Hence the assignment encoded by a proper list-coloring satisfies clause C_j if and only if the colors appearing on $(v_{1,j_1'}, v_{2,j_2'}, v_{3,j_3'})$ do not belong to the set $F_1 \times F_2^{+q} \times F_3^{+2q}$. To encode the requirement that C_j be satisfied into the graph G_{3q}, for each $(\gamma_1, \gamma_2, \gamma_3) \in F_1 \times F_2^{+q} \times F_3^{+2q}$ we add a vertex $w_{\gamma_1,\gamma_2,\gamma_3}$ to G_{3q} that is adjacent to $v_{1,j_1'}, v_{2,j_2'}$, and $v_{3,j_3'}$ and whose list is $\{\gamma_1, \gamma_2, \gamma_3\}$. The threefold redundancy we incorporated ensures that the three colors in each forbidden triple are all distinct. Therefore, if one of the three neighbors of $w_{\gamma_1,\gamma_2,\gamma_3}$ does not receive its forbidden color, then $w_{\gamma_1,\gamma_2,\gamma_3}$ can properly receive that color. This would not hold if there could be duplicates among the forbidden colors. The reduction is finished by adding these vertices for each clause $C_j \in \varphi$. We denote the set of clause vertices by $\mathcal{W}$.

Claim 13. *The formula φ has a satisfying assignment, if and only if the graph G_{3q} obtained via the above reduction is $3q$-list-colorable.*

Proof. Suppose φ has a satisfying assignment $\psi \colon [n] \to \{0,1\}$. Let γ_ψ be the corresponding proper coloring of $\mathcal{V}$, as described above. We argue that γ_ψ can be extended to the vertices $\mathcal{W}$ as well. Let $C_j \in \varphi$ be a clause on variables x_{j_1}, x_{j_2}, and x_{j_3} and let $w_{\gamma_1,\gamma_2,\gamma_3} \in \mathcal{W}$ be a vertex we introduced in the construction above for C_j. For $i \in [3]$, let $\gamma_\psi^i = \gamma_\psi(v_{i,\lceil j_i/\log q \rceil})$.

Since γ_ψ encodes a satisfying assignment, we know that there exists an $i^* \in [3]$, such that $\gamma_\psi^{i^*} \neq \gamma_{i^*}$ (since otherwise, ψ is not a satisfying assignment to φ). Hence, the color γ_{i^*} is not blocked from the list of vertex $w_{\gamma_1,\gamma_2,\gamma_3}$ which can then be properly colored. By Claim 12 we know that the remaining vertices $\mathcal{U}$ can be properly list-colored as well.

Conversely, suppose that G_{3q} is properly list-colored. We show that each proper coloring must correspond to a truth assignment that satisfies φ. For the sake of a contradiction, suppose that there is a proper list-coloring $\gamma_\psi \colon V(G) \to [3q]$ which encodes a truth assignment ψ that does not satisfy φ. Let $C_j \in \varphi$ denote a clause which is not satisfied by ψ on variables x_{j_1}, x_{j_2}, and x_{j_3}. For $i \in [3]$, we denote by $\gamma_\psi^i = \gamma_\psi(v_{i, \lceil j_i / \log q \rceil})$ the colors of the variable vertices encoding the truth assignment of the variables in C_j. Since ψ does not satisfy C_j we know that we added a vertex $w_{\gamma_\psi^1, \gamma_\psi^2, \gamma_\psi^3}$ to $\mathcal{W}$, which is adjacent to $v_{1, \lceil j_1 / \log q \rceil}$, $v_{2, \lceil j_2 / \log q \rceil}$, and $v_{3, \lceil j_3 / \log q \rceil}$. This means that the colors $\gamma_\psi^1, \gamma_\psi^2$, and γ_ψ^3 appear on a vertex which is adjacent to $w_{\gamma_\psi^1, \gamma_\psi^2, \gamma_\psi^3}$ and hence the coloring γ_ψ is improper, a contradiction.   ⌟

We have shown how to reduce an instance of 3-SAT to an instance of $3q$-LIST-COLORING. We modify the graph G_{3q} to obtain an instance of q-COLORING which preserves the correctness of the reduction. We add a clique K_{3q} of $3q$ vertices to G_{3q}, each of whose vertices represents one color. We make each vertex in $v \in \mathcal{V} \cup \mathcal{W} \cup \mathcal{U}$ adjacent to each vertex in K_{3q} that represents a color which does not appear on v's list in the list-coloring instance. (The same trick was used in the proof of Theorem 6.1 in [12].) It follows that the graph without K_{3q} has a proper list-coloring if and only if the new graph has a proper $3q$-coloring.

We now compute the size of G_{3q} in terms of n and q and give a bound on the size of a vertex cover of G_{3q}. We observe that $|\mathcal{V}| = 3\lceil n / \log q \rceil$, $|\mathcal{U}| = \mathcal{O}(q^2 \cdot \lceil n / \log q \rceil)$, and clearly, $|V(K_{3q})| = 3q$. To bound the size of $\mathcal{W}$, we observe that for each clause C_j, we added $(2^{\log q - 1})^3$ vertices (since we considered all triples of bitstrings of length $\log q$ where one character is fixed in each string) and hence $|\mathcal{W}| = \mathcal{O}(q^3 \cdot m)$ with m the number of clauses in ϕ. It is easy to see that $\mathcal{V} \cup V(K_{3q})$ is a vertex cover of G_{3q} and hence G_{3q} has a vertex cover of size $3\lceil n / \log q \rceil + 3q$.

Assuming there is an algorithm that solves q-COLORING on INDEPENDENT $+$ kv graphs in time $\mathcal{O}^*(\theta^k)$ together with an application of the above reduction (whose correctness follows from Claim 13) would yield an algorithm for 3-SAT that runs in time

$$\theta^{3\lceil n / \log q \rceil + 3q} \cdot ((q^2 + 3)\lceil n / \log q \rceil + 3q + q^3 \cdot m)^{\mathcal{O}(1)} = \mathcal{O}^* \left(2^{\frac{3 \log \theta}{\log q} n} \right).$$

Hence, for any $\varepsilon > 0$ we can choose a constant q large enough such that $(3 \log \theta)/(\log q) < \varepsilon$ and Theorem 11 follows.   □

4.2   No Nontrivial Runtime Bound for Path $+$ kv Graphs

We now strengthen the lower bound for FOREST $+$ kv graphs due to [12] to the more restrictive class of LINEAR FOREST $+$ kv graphs. The key idea in our reduction is that we treat the clause size in a satisfiability instance as a constant, which allows for constructing a graph of polynomial size. The following lemma describes the clause gadget that will be used in the reduction.

Lemma 14 (★). *For each $q \geq 3$ there is a polynomial-time algorithm that, given $(c_1, \ldots, c_m) \in [q]^m$, outputs a q-list-coloring instance (P, Λ) where P is a path of size $\mathcal{O}(m)$ containing distinguished vertices $(\pi_1, \ldots, \pi_m)$, such that the following holds. For each $(d_1, \ldots, d_m) \in [q]^m$ there is a proper list-coloring γ of P in which $\gamma(\pi_i) \neq d_i$ for all i, if and only if $(c_1, \ldots, c_m) \neq (d_1, \ldots, d_m)$.*

Theorem 15 (★). *For any $\varepsilon > 0$ and constant $q \geq 3$, q-COLORING on LINEAR FOREST $+\, kv$ graphs cannot be solved in time $\mathcal{O}^*((q - \varepsilon)^k)$, unless SETH fails.*

The reduction given in the proof of Theorem 15 can easily be modified to give a lower bound for PATH $+\, kv$ graphs as well, for the details the reader is referred to the full version [10].

Corollary 16. *For any $\varepsilon > 0$ and constant $q \geq 3$, q-COLORING on PATH $+\, kv$ graphs cannot be solved in time $\mathcal{O}^*((q - \varepsilon)^k)$, unless SETH fails.*

5   A Tighter Treedepth Boundary

In Lemma 10 we showed that if the $(q + 1)$-colorable members of a hereditary graph class $\mathcal{F}$ have bounded treedepth, then $\mathcal{F}$ has constant-size No-certificates for q-LIST-COLORING and hence $\mathcal{F} + kv$ has nontrivial algorithms for q-(LIST-)COLORING parameterized by the size of a given modulator to $\mathcal{F}$. One might wonder whether a graph class $\mathcal{F} + kv$ has nontrivial algorithms for q-COLORING parameterized by a given modulator to $\mathcal{F}$ *if and only if* all $(q+1)$-colorable members in $\mathcal{F}$ have bounded treedepth. However, this is not the case. In [11, Lemma 4] the authors showed that q-COLORING parameterized by the size of a modulator to the class COGRAPH has nontrivial algorithms. Clearly, complete bipartite graphs are cographs and it is easy to see that (the 2-colorable balanced biclique) $K_{n,n}$ has treedepth $n + 1$. In this section we show that, unless SETH fails, bicliques are in some sense the only obstruction to this treedepth boundary.

Theorem 17 (★). *Let $\mathcal{F}$ be a hereditary class of graphs for which there exists a $t \in \mathbb{N}$ such that $K_{t,t}$ is not contained in $\mathcal{F}$, let $q \geq 3$, and suppose SETH is true. Then, q-COLORING parameterized by a given vertex modulator to $\mathcal{F}$ of size k has $\mathcal{O}^*((q - \varepsilon)^k)$ time algorithms for some $\varepsilon > 0$, if and only if all $(q+1)$-colorable graphs in $\mathcal{F}$ have bounded treedepth.*

6   Conclusion

In this paper we have presented a fine-grained parameterized complexity analysis of the q-COLORING and q-LIST-COLORING problems. We showed that if a graph class $\mathcal{F}$ has NO-certificates for q-LIST-COLORING of bounded size or if the $(q+1)$-colorable members of $\mathcal{F}$ (where $\mathcal{F}$ is hereditary) have bounded treedepth, then there is an algorithm that solves q-COLORING on graphs in $\mathcal{F} + kv$ (graphs with vertex modulators of size k to $\mathcal{F}$) in time $\mathcal{O}^*((q-\varepsilon)^k)$ for some $\varepsilon > 0$ (depending on $\mathcal{F}$). The parameter treedepth revealed itself as a boundary in some sense: We showed that PATH $+ kv$ graphs do not have $\mathcal{O}^*((q-\varepsilon)^k)$ time algorithms for any $\varepsilon > 0$ unless SETH is false — and paths are arguably the simplest graphs of unbounded treedepth. Furthermore we proved that if a graph class $\mathcal{F}$ does not have large bicliques, then $\mathcal{F} + kv$ graphs have $\mathcal{O}^*((q-\varepsilon)^k)$ time algorithms, for some $\varepsilon > 0$, if and only if $\mathcal{F}$ has bounded treedepth.

Treedepth is an interesting graph parameter which in many cases also allows for polynomial space algorithms where e.g. for treewidth this is typically exponential. It would be interesting to see how the problems studied by Lokshtanov et al. [12] behave when parameterized by treedepth. Naturally, a fine-grained parameterized complexity analysis as we did might be interesting for other problems as well.

References

1. Björklund, A., Husfeldt, T., Koivisto, M.: Set partitioning via inclusion-exclusion. SIAM J. Comput. **39**(2), 546–563 (2009). Based on two extended abstracts appearing in FOCS 2006
2. Cai, L.: Parameterized complexity of vertex colouring. Discret. Appl. Math. **127**(3), 415–429 (2003)
3. Cygan, M., Fomin, F.V., Kowalik, L., Lokshtanov, D., Marx, D., Pilipczuk, M., Pilipczuk, M., Saurabh, S.: Parameterized Algorithms, 1st edn. Springer, Cham (2015)
4. Diestel, R.: Graph Theory, Graduate Texts in Mathematics, vol. 173, 4th edn. Springer, Heidelberg (2010). Corrected reprint 2012
5. Downey, R.G., Fellows, M.R.: Fundamentals of Parameterized Complexity. Texts in Computer Science. Springer, London (2013)
6. Fellows, M.R., Jansen, B.M.P., Rosamond, F.: Towards fully multivariate algorithmics: parameter ecology and the deconstruction of computational complexity. Eur. J. Comb. **34**(3), 541–566 (2013). Previously app. in IWOCA 2009
7. Flum, J., Grohe, M.: Parameterized Complexity Theory. Texts in Theoretical Computer Science. Springer, Heidelberg (2006)
8. Impagliazzo, R., Paturi, R.: On the complexity of k-sat. J. Comput. Syst. Sci. **62**(2), 367–375 (2001)
9. Impagliazzo, R., Paturi, R., Zane, F.: Which problems have strongly exponential complexity? J. Comput. Syst. Sci. **63**(4), 512–530 (2001)
10. Jaffke, L., Jansen, B.M.P.: Fine-grained parameterized complexity analysis of graph coloring problems. ArXiv e-prints arXiv:1701.06985 (2017)
11. Jansen, B.M.P., Kratsch, S.: Data reduction for graph coloring problems. Inf. Comput. **231**, 70–88 (2013). Previously app. in FCT 2011
12. Lokshtanov, D., Marx, D., Saurabh, S.: Known algorithms for graphs of bounded treewidth are probably optimal. In: SODA, pp. 777–789. SIAM (2011)
13. Nešetřil, J., Ossona de Mendez, P.: Sparsity. Graphs, Structures and Algorithms, Algorithms and Combinatorics, vol. 28. Springer, Heidelberg (2012)
14. Williams, V.V.: Hardness of easy problems: basing hardness on popular conjectures such as the strong exponential time hypothesis. In: IPEC. LIPIcs, vol. 43, pp. 16–28 (2015)

Structural Parameters for Scheduling with Assignment Restrictions

Klaus Jansen[1], Marten Maack[1(✉)], and Roberto Solis-Oba[2]

[1] Christian-Albrechts-Universität zu Kiel, 24118 Kiel, Germany
{kj,mmaa}@informatik.uni-kiel.de
[2] Western University, London, ON N6A 3K7, Canada
solis@csd.uwo.ca

Abstract. We consider scheduling on identical and unrelated parallel machines with job assignment restrictions. These problems are NP-hard and they do not admit polynomial time approximation algorithms with approximation ratios smaller than 1.5 unless $P = NP$. However, if we impose limitations on the set of machines that can process a job, the problem sometimes becomes easier in the sense that algorithms with approximation ratios better than 1.5 exist. We introduce three graphs, based on the assignment restrictions and study the computational complexity of the scheduling problem with respect to structural properties of these graphs, in particular their tree- and rankwidth. We identify cases that admit polynomial time approximation schemes or FPT algorithms, generalizing and extending previous results in this area.

1 Introduction

We consider the problem of makespan minimization for scheduling on unrelated parallel machines. In this problem a set $\mathcal{J}$ of n jobs has to be assigned to a set $\mathcal{M}$ of m machines via a schedule $\sigma : \mathcal{J} \to \mathcal{M}$. A job j has a processing time p_{ij} for every machine i and the goal is to minimize the makespan $C_{\max}(\sigma) = \max_i \sum_{j \in \sigma^{-1}(i)} p_{ij}$. In the three-field notation this problem is denoted by $R||C_{\max}$. On some machines a job might have a very high, or even infinite processing time, so it should never be processed on these machines. This amounts to assignment restrictions in which for every job j there is a subset $M(j)$ of machines on which it may be processed. An important special case of $R||C_{\max}$ is given if the machines are identical in the sense that each job j has the same processing time p_j on all the machines on which it may be processed, i.e., $p_{ij} \in \{p_j, \infty\}$. This problem is sometimes called restricted assignment and is denoted as $P|M(j)|C_{\max}$ in the three-field notation.

We study versions of $R||C_{\max}$ and $P|M(j)|C_{\max}$ where the restrictions are in some sense well structured. In particular we consider three different graphs that are defined based on the job assignment restrictions and study how structural

This work was partially supported by the DAAD (Deutscher Akademischer Austauschdienst) and by the German Research Foundation (DFG) project JA 612/15-1.

© Springer International Publishing AG 2017
D. Fotakis et al. (Eds.): CIAC 2017, LNCS 10236, pp. 357–368, 2017.
DOI: 10.1007/978-3-319-57586-5_30

properties of these graphs affect the computational complexity of the corresponding scheduling problems. We briefly describe the graphs. In the *primal graph* the vertices are the jobs and two vertices are connected by an edge, iff there is a machine on which both of the jobs can be processed. In the *dual graph*, on the other hand, the machines are vertices and two of them are adjacent, iff there is a job that can be processed by both machines. Lastly we consider the *incidence graph*. This is a bipartite graph and both the jobs and machines are vertices. A job j is adjacent to a machine i, if $i \in M(j)$. In Fig. 1 an example of each graph is given. These graphs have also been studied in the context of constraint satisfaction (see e.g. [1] or [2]) and we adapted them for machine scheduling.

We consider the above scheduling problems in the contexts of parameterized and approximation algorithms. For $\alpha > 1$ an α-*approximation* for a minimization problem computes a solution of value $A(I) \leq \alpha\mathrm{OPT}(I)$, where $\mathrm{OPT}(I)$ is the optimal value for a given instance I. A family of algorithms consisting of $(1+\varepsilon)$-approximations for each $\varepsilon > 0$ with running times polynomial in the input length (and $1/\varepsilon$) is called a *(fully) polynomial time approximation scheme* (F)PTAS. Let π be some parameter defined for a given problem, and let $\pi(I)$ be its value for instance I. The problem is said to be *fixed-parameter tractable* (FPT) for π, if there is an algorithm that given I and $\pi(I) = k$ solves I in time $\mathcal{O}(f(k)|I|^c)$, where c is a constant, f any computable function and $|I|$ the input length. This definition can easily be extended to multiple parameters.

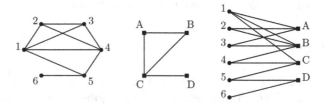

Fig. 1. Primal, dual and incidence graph for an instance with 6 jobs and 4 machines.

Related Work. In 1990 Lenstra, Shmoys and Tardos [3] showed, in a seminal work, that there is a 2-approximation for $R||C_{\max}$ and that the problem cannot be approximated with a ratio better than 1.5 unless $\mathsf{P} = \mathsf{NP}$. Both bounds also hold for $P|M(j)|C_{\max}$ and have not been substantially improved since that time. The case where the number of machines is constant is weakly NP-hard and there is an FPTAS for this case [4]. A special case of the restricted assignment problem called graph balancing was studied by Ebenlendr et al. [5]. In this variant each job can be processed by at most 2 machines and therefore an instance can be seen as a (multi-)graph where the machines are vertices and the jobs edges. They presented a 1.75 approximation for this problem and also showed that the 1.5 inapproximability result remains true. Lee et al. [6] studied the version of graph balancing where (in our notation) the dual graph is a tree and showed that there is an FPTAS for it. Moreover, the special case of graph balancing where the graph is simple has been considered. For this problem Asahiro et al. [7] presented—among other things—a pseudo-polynomial time algorithm for the case of graphs with bounded treewidth. For certain cases of $P|M(j)|C_{\max}$ with job assignment restrictions that are in some sense well-structured, PTAS results are known. In particular for the *path- and tree-hierarchical* cases [8,9] in which the machines can be arranged in a path or tree and the jobs can only be processed on subpaths starting at the leftmost machine or at the root machine respectively, and the *nested* case [10], where $M(j) \subseteq M(j')$, $M(j') \subseteq M(j)$ or $M(j) \cap M(j') = \emptyset$ holds for each pair of jobs (j, j').

The study of $R||C_{\max}$ from the FPT perspective has started only recently. Mnich and Wiese [11] showed that $R||C_{\max}$ is FPT for the pair of parameters m and the number of distinct processing times. The problem is also FPT for the parameter pair $\max p_{ij}$ and the number of machine types [12]. Two machines have the same type, if each job has the same processing time on them. Furthermore Szeider [13] showed that graph balancing on simple graphs with unary encoding of the processing times is not FPT for the parameter treewidth under usual complexity assumptions.

Results. Tree and branch decompositions are associated with certain structural *width parameters*. We consider two of them: treewidth and rankwidth. In the following we denote the treewidth of the primal, dual and incidence graph with tw_p, tw_d and tw_i, respectively. For the definitions of these concepts we refer to Sect. 2.

Let $J(i)$ be the set of jobs the machine i can process. In the context of parameterized algorithms we show the following.

Theorem 1. $R||C_{\max}$ *is FPT for the parameter* tw_p.

Theorem 2. $R||C_{\max}$ *is FPT for the parameters* k_1, k_2 *with* $k_1 \in \{tw_d, tw_i\}$ *and* $k_2 \in \{OPT, \max_i |J(i)|\}$.

Note that $R||C_{\max}$ with constant k_2 remains NP-hard [5]. In the context of approximation we get:

Theorem 3. $R||C_{\max}$ *is weakly NP-hard, if* tw_d *or* tw_i *is constant and there is an FPTAS for both of these cases.*

The hardness is due to the hardness of scheduling on two identical parallel machines $P2||C_{\max}$. The result for the dual graph is a generalization of the result in [6] and resolves cases that were marked as open in that paper. All results mentioned so far are discussed in Sect. 3. In the following section we consider the rankwidth:

Theorem 4. *There is a PTAS for instances of* $P|M(j)|C_{\max}$ *where the rankwidth of the incidence graph is bounded by a constant.*

It can be shown that instances of $P|M(j)|C_{\max}$ with path- or tree-hierarchical, or nested restrictions are special cases of the case when the incidence graph is a bicograph. Bicographs are known [14] to have a rankwidth of at most 4 and a suitable branch decomposition can be found very easily. Therefore we generalize and unify the known PTAS results for $P|M(j)|C_{\max}$ with structured job assignment restrictions.

Due to space limitations, we will leave out a lot of the details which can be found in a longer version of the paper [15].

2   Preliminaries

In the following I will always denote an instance of $R||C_{\max}$ or $P|M(j)|C_{\max}$ and most of the time we will assume that it is feasible. We call an instance feasible if $M(j) \neq \emptyset$ for every job $j \in \mathcal{J}$. A schedule is feasible if $\sigma(j) \in M(j)$. For a subset $J \subseteq \mathcal{J}$ of jobs and a subset $M \subseteq \mathcal{M}$ of machines we denote the subinstance of I induced by J and M with $I[J, M]$. Furthermore, for a set S of schedules for I we let $\mathrm{OPT}(S) = \min_{\sigma \in S} C_{\max}(\sigma)$, and $\mathrm{OPT}(I) = \mathrm{OPT}(S)$ if S is the set of all schedules for I. We will sometimes use $\mathrm{OPT}(\emptyset) = \infty$. Note that there are no schedules for instances without machines. On the other hand, if I is an instance without jobs, we consider the empty function a feasible schedule (with makespan 0), and have therefore $\mathrm{OPT}(I) = 0$ in that case.

Dynamic Programs for $R||C_{\max}$. We sketch two basic dynamic programs that will be needed as subprocedures in the following. The first one is based on iterating through the machines. Let $\mathrm{OPT}(i, J) = \mathrm{OPT}(I[\mathcal{J} \setminus J, [i]])$ for $J \subseteq \mathcal{J}$ and $i \in [m] := \{1, \ldots, m\}$, assuming $\mathcal{M} = [m]$. Then it is easy to see that $\mathrm{OPT}(i, J) = \min_{J \subseteq J' \subseteq \mathcal{J}} \max\{\mathrm{OPT}(i-1, J'), \sum_{j \in J' \setminus J} p_{ij}\}$. Using this recurrence relation a simple dynamic program can be formulated that computes the values $\mathrm{OPT}(i, J)$. It holds that $\mathrm{OPT}(I) = \mathrm{OPT}(m, \emptyset)$ and as usual for dynamic programs an optimal schedule can be recovered via backtracking. The running time of such a program can be bounded by $2^{\mathcal{O}(n)} \times \mathcal{O}(m)$, yielding the following trivial result:

Remark 1. $R||C_{\max}$ is FPT for the parameter n.

The second dynamic program is based on iterating through the jobs. Let $\lambda \in \mathbb{Z}_{\geq 0}^{\mathcal{M}}$. We call λ a *load vector* and say that a schedule σ fulfils λ, if $\lambda_i = \sum_{j \in \sigma^{-1}(i)} p_{ij}$. For $j \in [n]$ let $\Lambda(j)$ be the set of load vectors that are fulfilled by some schedule for the subinstance $I[[j], \mathcal{M}]$, assuming $\mathcal{J} = [n]$. Then $\Lambda(j)$ can also be defined recursively as the set of vectors λ with $\lambda_{i^*} = \lambda'_{i^*} + p_{i^* j}$ and $\lambda_i = \lambda'_i$ for $i \neq i^*$, where $i^* \in M(j)$ and $\lambda' \in \Lambda(j-1)$. Using this, a simple dynamic program can be formulated that computes $\Lambda(j)$ for all $j \in [n]$. $\mathrm{OPT}(I)$ can be recovered from $\Lambda(n)$ and a corresponding schedule can be found via backtracking. Let there be a bound L for the number of distinct loads that can occur on each machine, i.e. $|\{\sum_{j \in \sigma^{-1}(i)} p_{ij} | \sigma \text{ schedule for } I\}| \leq L$ for each $i \in \mathcal{M}$. Then the running time can be bounded by $L^{\mathcal{O}(m)} \times \mathcal{O}(n)$, yielding:

Remark 2. $R||C_{\max}$ is FPT for the pair of parameters m and k with $k \in \{\text{OPT}, \max_i |J(i)|\}$.

For this note that both OPT and $2^{\max_i |J(i)|}$ are bounds for the number of distinct loads that can occur on any machine. This dynamic program can also be used to get a simple FPTAS for $R||C_{\max}$ for the case when the number of machines m is constant. For this let B be an upper bound of OPT(I) with $B \le 2\text{OPT}$. Such a bound can be found with the 2-approximation by Lenstra et al. [3]. Moreover let $\varepsilon > 0$ and $\delta = \varepsilon/2$. By rounding the processing time of every job up to the next integer multiple of $\delta B/n$ we get an instance I' whose optimum makespan is at most $\varepsilon\text{OPT}(I)$ bigger than OPT(I). The dynamic program can easily be modified to only consider load vectors for I', where all loads are bounded by $(1 + \delta/n)B$. Therefore there can be at most $n/\delta + 2$ distinct load values for any machine and an optimal schedule for I' can be found in time $(n/\varepsilon)^{\mathcal{O}(m)} \times \mathcal{O}(n)$. The schedule can trivially be transformed into a schedule for the original instance without an increase in the makespan.

Tree Decompostion and Treewidth. A *tree decomposition* of a graph G is a pair $(T, \{X_t | t \in V(T)\})$, where T is a tree, $X_t \subseteq V(G)$ for each $t \in V(t)$ is a set of vertices of G, called a *bag*, and the following three conditions hold:

1. $\bigcup_{t \in V(T)} X_t = V(G)$
2. $\forall \{u, v\} \in E(G) \exists t \in V(T) : u, v \in X_t$
3. For every $u \in V(G)$ the set $T_u := \{t \in V(T) | u \in X_t\}$ induces a connected subtree of T.

The *width* of the decomposition is $\max_{t \in V(T)}(|X_t| - 1)$, and the *treewidth* tw$(G)$ of G is the minimum width of all tree decompositions of G. It is well known that forests are exactly the graphs with treewidth one, and that the treewidth of G is at least as big as the size of the biggest clique in G minus 1. More precisely, for each set of vertices $V' \subseteq V(G)$ inducing a clique in G, there is a node $t \in V(T)$ with $V' \subseteq X_t$ (see e.g. [16]). For a given graph and a value k it can be decided in FPT time (and linear in $|V(G)|$) whether the treewidth of G is at most k and in the affirmative case a corresponding tree decomposition with $\mathcal{O}(k|V(G)|)$ nodes can be computed [17]. However, deciding whether a graph has a treewidth of at most k, is NP-hard [18].

Branch Decomposition and Rankwidth. The treewidth is a concept appropriate for sparse graphs, while for dense graphs G other parameters like the clique- and rankwidth rw(G) typically are considered. We focus on the latter.

A *cut* of G is a partition of $V(G)$ into two subsets. For $X, Y \subseteq V(G)$ let $A_G[X, Y] = (a_{xy})$ be the $|X| \times |Y|$ adjacency submatrix induced by X and Y, i.e., $a_{xy} = 1$ if $\{x, y\} \in E(G)$ and $a_{xy} = 0$ otherwise for $x \in X$ and $y \in Y$. The *cut rank* of (X, Y) is the rank of $A_G[X, Y]$ over the field with two elements GF(2) and denoted by cutrk$_G(X, Y)$. A *branch decomposition* of $V(G)$ is a pair (T, η), where T is a tree with $|V(G)|$ leaves whose internal nodes have all degree 3, and η is a bijection from $V(G)$ to the leafs of T. For each $e = \{s, t\} \in E(T)$ there is an induced cut $\{X_s, X_t\}$ of G: For $x \in \{s, t\}$ the set X_x contains exactly the nodes $\eta^{-1}(\ell)$, where $\ell \in V(T)$ is a leaf that is in the same connected component of T as x, if e is removed. Now the *width* of e (with respect to cutrk$_G$) is cutrk$_G(X_s, X_t)$ and the *rankwidth* of the decomposition (T, η) is the maximum width over all edges of T. The *rankwidth* of G is the minimum rankwidth of all branch decompositions of G. It is well known that the rankwidth of a complete graph is equal to 1, $\mathrm{rw}(G) \leq \mathrm{tw}(G)$, and the treewidth can not be bounded by a function of rankwidth. For a given graph and fixed k there is an algorithm that finds a branch decomposition of width k in FPT-time (cubic in $|V(G)|$), or reports correctly that none exists [19].

3 Treewidth Results

We start with some basic relationships between different restriction parameters for $R||C_{\max}$, especially the treewidths of the different graphs for a given instance. Similar relationships have been determined for the three graphs in the context of constraint satisfaction.

Remark 3. $\mathrm{tw_p} \geq \max_i |J(i)| - 1$ and $\mathrm{tw_d} \geq \max_j |M(j)| - 1$.

To see this, note that the sets $J(i)$ and $M(j)$ are cliques in the primal and dual graphs, respectively.

Remark 4. $\mathrm{tw_i} \leq \mathrm{tw_p} + 1$ and $\mathrm{tw_i} \leq \mathrm{tw_d} + 1$. On the other hand $\mathrm{tw_p} \leq (\mathrm{tw_i} + 1) \max_i |J(i)| - 1$ and $\mathrm{tw_d} \leq (\mathrm{tw_i} + 1) \max_j |M(j)| - 1$.

These properties were pointed out by Kalaitis and Vardi [20] in a different context. Note that this Remark together with Theorem 1 implies the results of Theorem 2 concerning the parameter $\max_i |J(i)|$. Furthermore, in the case of $P||C_{\max}$ with only 1 job and m machines, or n jobs and only 1 machine, the primal graph has treewidth 0 or $n - 1$, and the dual $m - 1$ or 0, respectively, while the incidence graph in both cases has treewidth 1.

We show how a tree decomposition $(T, \{X_t | t \in V(T)\})$ of width k for any one of the three graphs can be used to design a dynamic program for the corresponding instance I of $R||C_{\max}$. Selecting a node as the root of the decompostion, the dynamic program works in a bottom-up manner from the leaves to the root. We assume that the decomposition has the following simple form: For each leaf node $t \in V(T)$ the bag X_t is empty and we fix one of these nodes as the root a of T. Furthermore each internal node t has exactly two children $\ell(t)$ and $r(t)$ (left and right), and each node $t \neq a$ has one parent $p(t)$. We denote the descendants of t with $\mathrm{desc}(t)$. A decomposition of this form can be generated from any other one without increasing the width and growing only linearly in size through the introduction of dummy nodes. The bag of a dummy node is either empty or identical to the one of its parent.

For each of the graphs and each node $t \in V(T)$ we define sets $\check{J}_t \subseteq \mathcal{J}$ and $\check{M}_t \subseteq \mathcal{M}$ of *inactive* jobs and machines along with sets J_t and M_t of *active* jobs and machines. The active jobs and machines in each case are defined based on the respective bag X_t, and the inactive ones have the property that they were active for a descendant $t \in \mathrm{desc}(t)$ of t but are not at t. In addition there are *nearly inactive* jobs $\tilde{J}_t$ and machines $\tilde{M}_t$, which are the jobs and machines that are deactivated when going from t to its parent $p(t)$ (for $t = a$ we assume them to be empty). The sets are defined so that certain conditions hold. The first two are that the (nearly) inactive jobs may only be processed on active or inactive machines, and the (nearly) inactive machines can only process active or inactive jobs:

$$M(\check{J}_t \cup \tilde{J}_t) \subseteq M_t \cup \check{M}_t \tag{1}$$

$$J(\check{M}_t \cup \tilde{M}_t) \subseteq J_t \cup \check{J}_t \tag{2}$$

Where $M(J^*) = \bigcup_{j \in J^*} M(j)$ and $J(M^*) = \bigcup_{i \in M^*} J(i)$ for any sets $J^* \subseteq \mathcal{J}$ and $M^* \subseteq \mathcal{M}$. Furthermore the (nearly) inactive jobs and machines of the children of an internal t form a disjoint union of the inactive jobs and machines of t, respectively:

$$\check{J}_t = \check{J}_{\ell(t)} \,\dot{\cup}\, \tilde{J}_{\ell(t)} \,\dot{\cup}\, \check{J}_{r(t)} \,\dot{\cup}\, \tilde{J}_{r(t)} \tag{3}$$

$$\check{M}_t = \check{M}_{\ell(t)} \,\dot{\cup}\, \tilde{M}_{\ell(t)} \,\dot{\cup}\, \check{M}_{r(t)} \,\dot{\cup}\, \tilde{M}_{r(t)} \tag{4}$$

Where $A \,\dot{\cup}\, B$ for any two sets A, B emphasizes that the union $A \cup B$ is disjoint, i.e., $A \cap B = \emptyset$. We now consider each of the three graphs.

The Primal Graph. In the primal graph all the vertices are jobs, and we define the active jobs of a tree node t to be exactly the jobs that are included in the respective bag, i.e., $J_t = X_t$. The inactive jobs are those that are not included in X_t but are in a bag of some descendant of t and the nearly inactive one are those that are active at t but inactive at $p(t)$, i.e., $\check{J}_t = \{j \in \mathcal{J} | j \notin X_t \wedge \exists t' \in \mathrm{desc}(t) : j \in X_{t'}\}$ and $\tilde{J}_t = J_t \setminus J_{p(t)}$. Moreover the inactive machines are the ones on which some inactive job may be processed, and the (nearly in-)active machines are those that can process (nearly in-)active jobs and are not inactive, i.e., $\check{M}_t = M(\check{J}_t)$, $M_t = M(J_t) \setminus \check{M}_t$ and $\tilde{M}_t = M(\tilde{J}_t) \setminus \check{M}_t$. For these definitions the conditions (1)–(4) hold, as well as:

$$J(\tilde{M}_t) \subseteq J_t \tag{5}$$

$$M(\check{J}_t \cup \tilde{J}_t) = \tilde{M}_t \cup \check{M}_t \tag{6}$$

For $J \subseteq \mathcal{J}$ and $M \subseteq \mathcal{M}$ let $\Gamma(J, M) = \{J' \subseteq J | \forall j \in J' : M(j) \cap M \neq \emptyset\}$. Let $t \in V(T)$, $J \in \Gamma(\check{J}_t, \check{M}_t)$ and $J' \in \Gamma(J_t \setminus \tilde{J}_t, \check{M}_t \cup \tilde{M}_t)$. We set $S(t, J)$ and $\tilde{S}(t, J')$ to be the sets of feasible schedules for the instances $I[\check{J}_t \cup J, \check{M}_t]$ and $I[\check{J}_t \cup \tilde{J}_t \cup J', \check{M}_t \cup \tilde{M}_t]$ respectively. We will consider $\mathrm{OPT}(S(t, J))$ and $\mathrm{OPT}(\tilde{S}(t, J'))$.

First note that $\mathrm{OPT}(I) = \mathrm{OPT}(S(a, \emptyset))$, where a is the root of T. Moreover, for a leaf node t there are neither jobs nor machines and $\mathrm{OPT}(S(t, \emptyset)) =$

$\text{OPT}(\tilde{S}(t,\emptyset)) = \text{OPT}(\{\emptyset\}) = 0$ holds. Hence let t be a non-leaf node. We first consider how $\text{OPT}(S(t,J))$ can be computed from the children of t. Due to Property 3 of the tree decomposition and (1) the jobs from J are already active on at least one of the direct descendants of t. Because of this and (4), J may be split in two parts $J_\ell \dot{\cup} J_r = J$, where $J_s \in \Gamma(J_{s(t)} \setminus \tilde{J}_{s(t)}, \check{M}_{s(t)} \cup \tilde{M}_{s(t)})$ for $s \in \{\ell, r\}$. Let $\Phi(J)$ be the set of such pairs (J_ℓ, J_r). From (3), (4) and (6) we get:

Lemma 1. $\text{OPT}(S(t,J)) = \min_{(J_\ell,J_r)\in\Phi(J)} \max_{s\in\{\ell,r\}} \text{OPT}(\tilde{S}(s(t), J_s))$.

Consider the computation of $\text{OPT}(\tilde{S}(t,J'))$. We may split J' and $\tilde{J}_t$ into a set going to the nearly inactive and a set going to the inactive machines. We set $\Psi(J')$ to be the set of pairs (A, X) with $J' \cup \tilde{J}_t = A \dot{\cup} X$, $A \in \Gamma(\tilde{J}_t \cup J', \check{M}_t)$ and $X \in \Gamma(\tilde{J}_t \cup J', \check{M}_t)$. Because of (3)–(5) we have:

Lemma 2.

$$\text{OPT}(\tilde{S}(t,J')) = \min_{(A,X)\in\Psi(J')} \max\{\text{OPT}(S(t,X)), \text{OPT}(I[A,\check{M}_t])\}.$$

Note that the values $\text{OPT}(I[A,\check{M}_t])$ can be computed using the first dynamic program from Sect. 2 in time $2^{\mathcal{O}(k)} \times \mathcal{O}(m)$.

The Dual Graph. For the dual graph the (in-)active jobs and machines are defined dually: The active machines for a tree node t are the ones in the respective bag, the inactive machines are those that were active for some descendant but are not active for t, and the nearly inactive machines are those that are active at t but inactive at its parent, i.e., $M_t = X_t$, $\check{M}_t = \{i \in \mathcal{M} | i \notin M_t \wedge \exists t' \in \text{desc}(t) : i \in X_{t'}\}$ and $\tilde{M}_t = M_t \setminus \check{M}_{p(t)}$. Furthermore the inactive jobs are those that may be processed on some inactive machine and the (nearly in-)active ones are those that can be processed on some (nearly in-)active machine and are not inactive, i.e., $\check{J}_t = J(\check{M}_t)$, $J_t = J(M_t) \setminus \check{J}_t$ and $\tilde{J}_t = J(\tilde{M}_t) \setminus \check{J}_t$. With these definitions the conditions (1)–(4) hold, as well as:

$$M(\tilde{J}_t) \subseteq M_t \tag{7}$$

$$J(\check{M}_t \cup \tilde{M}_t) = \tilde{J}_t \cup \check{J}_t \tag{8}$$

We will need some extra notation. Like we did in Sect. 2 we will consider load vectors $\lambda \in \mathbb{Z}_{\geq 0}^M$, where $M \subseteq \mathcal{M}$ is a set of machines. We say that a schedule σ fulfils λ, if $\lambda_i = \sum_{j\in\sigma^{-1}(i)} p_{ij}$ for each $i \in M$. For any set S of schedules for I we denote the set of load vectors for M that are fulfilled by at least one schedule from S with $\Lambda(S, M)$. Furthermore we denote the set of all schedules for I with $S(I)$, and for a subset of jobs $J \subseteq \mathcal{J}$, we write $\Lambda(J, M)$ as a shortcut for $\Lambda(S(I[J,M]), M)$. Let $t \in V(T)$. We set $S(t) = S(I[\check{J}_t, \check{M}_t \cup M_t])$ and $\tilde{S}(t) = S(I[\tilde{J}_t \cup \check{J}_t, \check{M}_t \cup M_t])$. Moreover, for $\lambda \in \Lambda(S(t), M_t)$ and $\lambda' \in \Lambda(\tilde{S}(t), M_t)$ we set $S(t,\lambda) \subseteq S(t)$ and $\tilde{S}(t,\lambda') \subseteq \tilde{S}(t)$ to be those schedules that fulfil λ and λ' respectively. We now consider $\text{OPT}(S(t,\lambda))$ and $\text{OPT}(\tilde{S}(t,\lambda'))$.

First note $\mathrm{OPT}(I) = \mathrm{OPT}(S(a, \emptyset))$. Moreover, for a leaf node t we have neither jobs nor machines and $\Lambda(S(t), M_t) = \Lambda(\tilde{S}(t), M_t) = \{\emptyset\}$. Therefore $\mathrm{OPT}(S(t, \emptyset)) = \mathrm{OPT}(\tilde{S}(t, \emptyset)) = \mathrm{OPT}(\{\emptyset\}) = 0$. Hence, let t be a non-leaf node. Again, we first consider how $\mathrm{OPT}(S(t, \lambda))$ can be computed from the children of t. Because of (3), λ may be split into a left and a right part. For two machine sets M, M' let $\tau_{M,M'} : \mathbb{Z}_{\geq 0}^M \to \mathbb{Z}_{\geq 0}^{M'}$ be a trasformation function for load vectors, where the i-th entry of $\tau_{M,M'}(\lambda)$ equals λ_i for $i \in M \cap M'$ and 0 otherwise. We set $\Xi(\lambda)$ to be the set of pairs $(\lambda_\ell, \lambda_r)$ with $\lambda = \tau_{M_{\ell(t)}, M_t}(\lambda_\ell) + \tau_{M_{r(t)}, M_t}(\lambda_r)$, and $\lambda_s \in \Lambda(\tilde{S}(s(t)), M_{s(t)})$ for $s \in \{\ell, r\}$. Because of (1), (3) and (4), we have:

Lemma 3. $\mathrm{OPT}(S(t, \lambda)) = \min_{(\lambda_\ell, \lambda_r) \in \Xi(\lambda)} \max_{s \in \{\ell, r\}} \mathrm{OPT}(\tilde{S}(s(t), \lambda_s))$.

Now we consider $\mathrm{OPT}(\tilde{S}(t, \lambda'))$. We may split λ' into the load due to inactive and that due to nearly inactive jobs. Note that the nearly inactive jobs can only be processed by active machines (7). We set $\Upsilon(\lambda')$ to be the set of pairs (α, ξ) with $\lambda' = \alpha + \xi$, $\alpha \in \Lambda(\tilde{J}_t, M_t)$ and $\xi \in \Lambda(S(t), M_t)$. Now (3), (4) and (7) yield:

Lemma 4. $\mathrm{OPT}(\tilde{S}(t, \lambda')) = \min_{(\alpha, \xi) \in \Upsilon(\lambda')} \max(\{\mathrm{OPT}(S(t, \xi))\} \cup \{\lambda'_i | i \in M_t\})$.

The set $\Lambda(\tilde{J}_t, M_t)$ can be computed using the second dynamic program described in Sect. 2 in time $L^{\mathcal{O}(k)} \times \mathcal{O}(n)$ if L is again a bound on the number of distinct loads that can occur on each machine.

The Incidence Graph. For the incidence graph the ideas used for the two other graphs can be combined, although the situation is slightly more complicated, because we have to handle the jobs and machines simultaneously. If the job sets are defined like in the primal, and the machine sets like in the dual graph case, the conditions (1)–(4) hold and recurrence relations similar to the first two cases can be formulated. The details are omitted in this version of the paper.

Results. Using above arguments, we can design dynamic programs with running time $2^{\mathcal{O}(k)} \times \mathcal{O}(nm)$ in the primal case and $L^{\mathcal{O}(k)} \times \mathcal{O}(nm)$ in the dual and incidence graph cases. Optimal schedules can be found via backtracking proving the Theorems 1 and 2. Theorem 3 follows by the combination of the dynamic programs and a rounding scheme similar to that in Sect. 2.

4   Rankwidth Results

We study the case when the rankwidth of the incidence graph is bounded by a constant k. First, we give some intuition on why a bounded rankwidth is useful and then discuss the main ideas for the design of a dynamic program utilizing a corresponding branch decomposition of the incidence graph.

For any Graph (V, E) and $X \subseteq V$, we say that $u, v \in V$ have the same *connection type with respect to* X, if $N(u) \cap X = N(v) \cap X$. If X is clear from the context, we say that u and v have the same connection type. Now, let $e =$

$\{a, b\} \in E(T)$ be some edge of the branch decomposition and $\{X_{e,a}, X_{e,b}\}$ the respective cut of T, i.e., $X_{e,x}$ for $x \in \{a, b\}$ is the set of vertices of T that are in the same connected component as x when the edge e is removed. Then $\{X_{e,a}, X_{e,b}\}$ induces a partition of both the jobs and machines by $J_{e,x} := \{j \in \mathcal{J} | \eta(j) \in X_{e,x}\}$ and $M_{e,x} := \{i \in \mathcal{M} | \eta(i) \in X_{e,x}\}$ for $x \in \{a, b\}$.

Remark 5. Let $x, y \in \{a, b\}$ with $x \neq y$. The number of distinct connection types of $J_{e,x}$ with respect to $M_{e,y}$ is bounded by 2^k.

Each edge of a branch decomposition corresponds to a partition of the job and machine sets and an optimal solution may be found by trying all possible ways of moving jobs between partitions. At the machine-leafs all arriving jobs have to be processed, with no jobs going out, and at the job-leafs all jobs have to be send away, with no jobs coming in. From this the procedure can work up to some root edge. This idea can be used to design a dynamic program with exponential running time. However, it can be argued that it is sufficient to consider only certain locally defined classes of job sets. The crucial part here is that the number of these classes can be polynomially bounded, if the number d of distinct sizes and the number of connection types of jobs are constant. We have already argued that the number of connection types of jobs is bounded by a function of k and can assume d to be constant using the following result. Let $\mathcal{I}$ be some class of instances of $P|M(j)|C_{\max}$, which is invariant with respect to changing the processing times of jobs and the introduction of copies of jobs.

Lemma 5 (Rounding Lemma). *If there is a PTAS for instances from $\mathcal{I}$, for which the number of distinct processing times is bounded by a constant, then there is also a PTAS for any instance from $\mathcal{I}$.*

It can be easily seen that the class of instances of $P|M(j)|C_{\max}$, for which the rankwidth of the incidence graph is bounded by a constant k, is such a class $\mathcal{I}$.

Dynamic Program. We now develop the recurrence relations needed for the dynamic program. Due to space limitations we leave out a lot of the details. In particular we leave out the considerations concerning the leafs of the branch decomposition—which are rather simple—and concerning the splitting of job classes at inner nodes—which are comparatively complicated.

Let $e = \{a, b\} \in E(T)$ again be some edge of the tree T and $\{X_{e,a}, X_{e,b}\}$ the corresponding cut of T. For $x, y \in e$ with $x \neq y$ let $\vec{J}_{x,y} \subseteq J_{e,x}$ and $I_{e,x}(\vec{J}_{x,y}, \vec{J}_{y,x}) = I[(J_{e,x} \setminus \vec{J}_{x,y}) \cup \vec{J}_{y,x}, M_{e,x}]$. The intuition here is that $\vec{J}_{x,y}$ is the set of jobs that is scheduled on—or *send to*—machines from $M_{e,y}$ by some schedule σ. There are some cases in which different sets $\vec{J}_{y,x}, \vec{J}'_{y,x} \subseteq J_{e,x}$ are in some sense similar and it holds that $\mathrm{OPT}(I_{e,x}(\vec{J}_{x,y}, \vec{J}_{y,x})) = \mathrm{OPT}(I_{e,x}(\vec{J}_{x,y}, \vec{J}'_{y,x}))$. This is the case if there is a bijection $\alpha : \vec{J}_{y,x} \rightarrow \vec{J}'_{y,x}$ such that j and $\alpha(j)$ have the same connection type with respect to $M_{e,x}$ and $p_j = p_{\alpha(j)}$ for each $j \in \vec{J}_{y,x}$. By this, equivalence relations $\sim_{e,y}$ can be defined and we get:

Lemma 6.

$$\text{OPT}(I) = \min_{([\vec{J}_{a,b}],[\vec{J}_{b,a}])} \max\{\min_{\vec{J}'_{a,b}} \text{OPT}(I_{e,a}(\vec{J}'_{a,b}, \vec{J}_{b,a})), \min_{\vec{J}'_{b,a}} \text{OPT}(I_{e,b}(\vec{J}'_{b,a}, \vec{J}_{a,b}))\}$$

Note that in this equation equivalence classes $[\vec{J}_{y,x}]$ are considered belonging to the relation $\sim_{e,y}$ and $\vec{J}_{y,x}$ is an arbitrary representative of this class. We now develop a sensible representation for the equivalence classes.

We assume some ordering of the different processing times, with $p(i)$ denoting the i-th processing time for $i \in [d]$. Any set of jobs J' induces a vector $\lambda \in \mathbb{Z}_{\geq 0}^d$ where λ_i is the number of jobs in J' that have the i-th processing time, i.e., $\lambda_i = |\{j \in J'|p_j = p(i)\}|$. Let $\kappa(e,b)$ be the number of connection types of jobs from $J_{e,b}$ with respect to $M_{e,a}$. Note that due to Remark 5 we get $\kappa(e,b) \leq 2^k$. Again assuming some ordering, for $i \in [\kappa(e,b)]$ let $\varphi_{e,b}(i)$ be the size vector induced by the i-th connection type of $J_{e,b}$ with respect to $M_{e,a}$. Moreover, let $\varphi_{e,b} = (\varphi_{e,b}(1),\ldots,\varphi_{e,b}(\kappa(e,b)))$. Now the equivalence classes of $\sim_{e,b}$ can naturally be represented and characterized by vectors $\iota \leq \varphi_{e,b}$.

Remark 6. For each $e \in E(T)$ and $b \in e$ there are at most $n^{\kappa(e,b)d}$ different vectors $\iota \leq \varphi_{e,b}$.

We now use an intuition of up and down with $e = \{a,b\} \in E(T)$, a above and b below. Let $\hat{\iota} \leq \varphi_{e,b}$ and $\check{\iota} \leq \varphi_{e,a}$ be candidate job classes to be send up and down e. Considering Lemma 6, we set $\text{OPT}(e,\hat{\iota},\check{\iota})$ to be the minimum value $\text{OPT}(I_{e,b}(\hat{J},\check{J}))$ where $\hat{J}$ is represented by $\hat{\iota}$ and $\check{J}$ by $\check{\iota}$. Furthermore, we assume that b is an inner node with two further (lower) neighbors ℓ and r, as well as connecting edges $e_\ell = \{\ell,b\}$ and $e_r = \{r,b\}$. We argue that $\text{OPT}(e,\hat{\iota},\check{\iota})$ can be computed from values $\text{OPT}(e_\ell,\hat{\lambda},\check{\lambda})$ and $\text{OPT}(e_r,\hat{\rho},\check{\rho})$ using a recurrence relation of the following form:

Lemma 7. $\text{OPT}(e,\hat{\iota},\check{\iota}) = \min_{(\hat{\lambda},\check{\lambda},\hat{\rho},\check{\rho})} \max\{\text{OPT}(e_\ell,\hat{\lambda},\check{\lambda}), \text{OPT}(e_r,\hat{\rho},\check{\rho})\}$.

It is not immediately clear which tuples $(\hat{\lambda},\check{\lambda},\hat{\rho},\check{\rho})$ should be considered. This detail is comparatively complicated and will not be covered here. Difficulties arise e.g. from the fact that classes of jobs defined for one edge translate into other classes for neighboring edges. Furthermore, any set of jobs that is sent through an edge and arrives at an internal node, will be split into two parts: one going forth through the second and one through the third edge. And looking at it the other way around: Any set that is sent by a schedule through an edge coming from an inner node, is put together from two parts, one coming from the second and one coming from the third edge. This has to be taken into account as well.

Results. With the above considerations a dynamic program for $P|M(j)|C_{\max}$ can be defined. This can be done in a way such that its running time is in $\mathcal{O}(m^2 n^{\mathcal{O}(d2^k)})$, proving Theorem 4 together with the Rounding Lemma and the considerations of Sect. 2.

Acknowledgements. The Rounding Lemma in the presented form was formulated by Lars Rohwedder and Kevin Prohn as part of a student project.

References

1. Szeider, S.: On fixed-parameter tractable parameterizations of SAT. In: Giunchiglia, E., Tacchella, A. (eds.) SAT 2003. LNCS, vol. 2919, pp. 188–202. Springer, Heidelberg (2004). doi:10.1007/978-3-540-24605-3_15
2. Samer, M., Szeider, S.: Constraint satisfaction with bounded treewidth revisited. J. Comput. Syst. Sci. **76**(2), 103–114 (2010)
3. Lenstra, J.K., Shmoys, D.B., Tardos, É.: Approximation algorithms for scheduling unrelated parallel machines. Math. Program. **46**(1–3), 259–271 (1990)
4. Horowitz, E., Sahni, S.: Exact and approximate algorithms for scheduling nonidentical processors. J. ACM (JACM) **23**(2), 317–327 (1976)
5. Ebenlendr, T., Krčál, M., Sgall, J.: Graph balancing: a special case of scheduling unrelated parallel machines. Algorithmica **68**(1), 62–80 (2014)
6. Lee, K., Leung, J.Y.T., Pinedo, M.L.: A note on graph balancing problems with restrictions. Inf. Process. Lett. **110**(1), 24–29 (2009)
7. Asahiro, Y., Miyano, E., Ono, H.: Graph classes and the complexity of the graph orientation minimizing the maximum weighted outdegree. Discret. Appl. Math. **159**(7), 498–508 (2011)
8. Ou, J., Leung, J.Y.T., Li, C.L.: Scheduling parallel machines with inclusive processing set restrictions. Naval Res. Logist. (NRL) **55**(4), 328–338 (2008)
9. Epstein, L., Levin, A.: Scheduling with processing set restrictions: PTAS results for several variants. Int. J. Prod. Econ. **133**(2), 586–595 (2011)
10. Muratore, G., Schwarz, U.M., Woeginger, G.J.: Parallel machine scheduling with nested job assignment restrictions. Oper. Res. Lett. **38**(1), 47–50 (2010)
11. Mnich, M., Wiese, A.: Scheduling and fixed-parameter tractability. Math. Program. **154**(1–2), 533–562 (2015)
12. Knop, D., Koutecký, M.: Scheduling meets n-fold integer programming. arXiv preprint arXiv:1603.02611 [cs.DS] (2016)
13. Szeider, S.: Not so easy problems for tree decomposable graphs. In: International Conference on Discrete Mathematics (2008)
14. Giakoumakis, V., Vanherpe, J.M.: Linear time recognition of weak bisplit graphs. Electron. Notes Discret. Math. **5**, 138–141 (2000)
15. Jansen, K., Maack, M., Solis-Oba, R.: Structural parameters for scheduling with assignment restrictions. arXiv preprint arXiv:1701.07242 [cs.DS] (2017)
16. Bodlaender, H.L.: A partial k-arboretum of graphs with bounded treewidth. Theoret. Comput. Sci. **209**(1), 1–45 (1998)
17. Bodlaender, H.L.: A linear-time algorithm for finding tree-decompositions of small treewidth. SIAM J. Comput. **25**(6), 1305–1317 (1996)
18. Arnborg, S., Corneil, D.G., Proskurowski, A.: Complexity of finding embeddings in a k-tree. SIAM J. Algebraic Discret. Methods **8**(2), 277–284 (1987)
19. Hlineny, P., Oum, S.I.: Finding branch-decompositions and rank-decompositions. SIAM J. Comput. **38**(3), 1012–1032 (2008)
20. Kolaitis, P.G., Vardi, M.Y.: Conjunctive-query containment and constraint satisfaction. J. Comput. Syst. Sci. **61**, 302–332 (1998)

On the Exact Complexity of Hamiltonian Cycle and q-Colouring in Disk Graphs

Sándor Kisfaludi-Bak[1]([✉]) and Tom C. van der Zanden[2]

[1] Department of Mathematics and Computer Science, TU Eindhoven,
Eindhoven, The Netherlands
s.kisfaludi.bak@tue.nl
[2] Department of Computer Science, Utrecht University, Utrecht, The Netherlands
T.C.vanderZanden@uu.nl

Abstract. We study the exact complexity of the Hamiltonian Cycle and the q-Colouring problem in disk graphs. We show that the Hamiltonian Cycle problem can be solved in $2^{O(\sqrt{n})}$ on n-vertex disk graphs where the ratio of the largest and smallest disk radius is $O(1)$. We also show that this is optimal: assuming the Exponential Time Hypothesis, there is no $2^{o(\sqrt{n})}$-time algorithm for Hamiltonian Cycle, even on unit disk graphs. We give analogous results for graph colouring: under the Exponential Time Hypothesis, for any fixed q, q-Colouring does not admit a $2^{o(\sqrt{n})}$-time algorithm, even when restricted to unit disk graphs, and it is solvable in $2^{O(\sqrt{n})}$-time on disk graphs.

1 Introduction

Exact algorithms for NP-hard problems have received considerable attention in recent years. The goal of research in this area is to develop 'moderately exponential' algorithms and to prove matching lower bounds under complexity-theoretic assumptions. Most work in this direction concerns fundamental graph problems.

The *square-root phenomenon* is a well-documented occurrence among algorithms on planar graphs [13]. The term illustrates that many problems that have $2^{O(n)}$ algorithms on general graphs can be solved in $2^{O(\sqrt{n})}$ in planar graphs. Moreover, matching lower bounds can be found based on the Exponential Time Hypothesis, i.e., for most of these problems, there are no algorithms with running time $2^{o(n)}$ resp. $2^{o(\sqrt{n})}$, unless the Exponential Time Hypothesis fails.

An important question about the square-root phenomenon is whether we can generalize the results on planar graphs to larger graph classes. One possible direction is to extend to *disk graphs*: the vertices are disks in $\mathbb{R}^2$, and two vertices are adjacent if their disks intersect. Note that disk graphs where the interiors of the disks are disjoint are exactly the planar graphs [12]. *Unit disk graphs* are disk graphs where all radii are one; *bounded-ratio disk graphs* are disk graphs where the ratio of the largest and smallest radius is bounded by some constant.

The research of the first author was supported by the Netherlands Organisation for Scientific Research (NWO) under project no. 024.002.003.

D. Fotakis et al. (Eds.): CIAC 2017, LNCS 10236, pp. 369–380, 2017.
DOI: 10.1007/978-3-319-57586-5_31

In this paper, we demonstrate the square-root phenomenon for the Hamiltonian Cycle problem in bounded-ratio disk graphs that are given by their geometric representations. Note that in planar graphs, the problem has a $2^{O(\sqrt{n})}$-time algorithm [13], and a matching $2^{\Omega(\sqrt{n})}$ lower bound conditional on the Exponential Time Hypothesis. The main obstacles for Hamiltonian Cycle in bounded-ratio disk graphs are the following.

- On the algorithmic side, the $2^{O(\sqrt{n})}$ running time often follows from the fact that planar graphs have treewidth $O(\sqrt{n})$ (see e.g. [4,6,15]). In our setting, bounded-ratio disk graphs are dense and may have unbounded treewidth.
- The lower bounds are based on reductions that planarize a graph by replacing each crossing of edges with a crossover gadget. Since there may be quadratically many crossings in a general graph, these reductions blow up an n-vertex graph to an n^2-vertex one, which results in the $2^{\Omega(\sqrt{n})}$ lower bound. In our setting, the NP-hardness of Hamiltonian Cycle was previously only known through its NP-hardness on grid graphs [8]. However, this reduction has a cubic blowup, giving only a $2^{\Omega(\sqrt[3]{n})}$ lower bound and – to our knowledge – it is an open problem whether this lower bound can be improved to match the best known ($2^{O(\sqrt{n})}$) algorithm [6].

The cubic time blowup of the reduction in [8] showing the hardness of Hamiltonian Cycle in grid graphs follows from two factors: the need to deal with crossings (introducing one factor n) and the need to replace long edges with some suitable 'path structure' (introducing another factor n). Compared to grid graphs, creating a reduction for disk graphs we have one major advantage: even though disk graphs have a structure somewhat similar to planar graphs, they can be (locally) non-planar and the Hamiltonian cycle in a solution can cross itself. Even so, our reduction still uses crossover gadgets and has to replace edges with path structures.

A key technique of our reduction is that in replacing long edges with some other structure, we need to ensure that all the vertices of this structure can be visited even if the edge is not used in the Hamiltonian cycle. This can be achieved with a $2 \times n$ grid (a *snake*), which can either be traversed in a zigzag manner (corresponding to using the edge in the cycle) or traversed going back and forth (corresponding to not using the edge). Our snakes are almost identical to the ones proposed by Itai et al. [8]. Unfortunately, it does not appear to be possible to create a crossover gadget for two snakes. To overcome this, we modify the reduction to ensure that some edges will certainly be included in the solution (which we can thus replace with a simple path rather than a snake) and build our reductions such that we only have crossings between simple paths and between simple paths and snakes (for which we *can* build crossover gadgets).

To complement our results for Hamiltonian Cycle, we also show that the same upper and lower bounds hold for q-colouring on disk graphs in the case where q is a constant. The algorithm follows from the observation that q-colorable graphs do not have large cliques, and a separator theorem due to Miller et al. [14]; the lower bound uses an adaptation of a reduction due to Gräf et al. [7].

Some proofs are omitted from this extended abstract.

2   Algorithm for Hamiltonian Cycle in Bounded-Ratio Disk Graphs

In this section, we show that the Hamiltonian Cycle problem can be solved in $2^{O(\sqrt{n})}$ time on bounded-ratio disk graphs. Our algorithm uses techniques due to Ito and Kadoshita [9], who show that Hamiltonian Cycle can be solved in $2^{O(\alpha)}n^{O(1)}$ time on unit disk graphs, where α is the area of a bounding square of the set of disks.

Theorem 1. *There exists a $2^{O(\sqrt{n})}$-time algorithm for Hamiltonian Cycle on bounded-ratio disk graphs (where the graphs are given by their geometric representation).*

Lemma 1. *Given a disk graph of ratio $\beta = O(1)$ with its representation, there are values $\gamma = \gamma(\beta)$ and $\Delta = \Delta(\beta)$ such that if we tessellate the plane using squares of diameter γ, the vertices in each tile induce a clique and the vertices in any given square have neighbours in at most Δ distinct other squares.*

Proof. Ito and Kadoshita [9] prove this lemma for unit disk graphs, where $\gamma = 1$ and $\Delta = 18$. The proof generalizes to bounded-ratio disk graphs.   □

Given a bounded-ratio disk graph G, the lemma gives a *clique partition* $Q_1, \ldots, Q_r$ of G, that is, a partition of the vertices of G into cliques, such that the vertices of each clique have neighbours in at most $\Delta = O(1)$ other cliques.

Given a graph G and sets $A, B \subseteq V(G)$, we let $E(A, B)$ denote the set of edges between a vertex in A and a vertex in B. Using the notion of *canonical Hamiltonian cycle* (which we do not need to consider), Ito and Kadoshita [9] then prove the following lemma:

Lemma 2 (Ito and Kadoshita [9]). *Let G have clique partition $Q_1, \ldots, Q_r$ defined by a tessellation as in Lemma 1. Then for each $i \neq j$, we can remove all but $O(\Delta^2)$ edges of $E(Q_i, Q_j)$ to obtain G', such that G' has a Hamiltonian cycle if and only if G has a Hamiltonian cycle.*

If G' is connected, removing the vertices from each clique of the clique partition in G' that do not have an edge to a vertex of some other clique in the partition preserves the Hamiltonicity of G'. We thus obtain the reduced graph G'', which contains at most $O(\Delta^3)$ vertices per tile.

Lemma 3. *The reduced graph has treewidth $O(\sqrt{n})$. A tree decomposition of treewidth $O(\sqrt{n})$ can be found in polynomial time.*

Proof. Alber and Fiala [1] show that unit disk graphs have balanced separators where the disks of the vertices in the separator cover an area of at most $O(\sqrt{n})$. This also holds for bounded-ratio disk graphs, as we can consider the supergraph obtained by making the radius of each disk equal to the largest radius. Since in the reduced graph, each tile contains at most a constant number of points, this gives a balanced separator of size $O(\sqrt{n})$ (in terms of vertices). These separators in turn imply that the reduced graph has treewidth $O(\sqrt{n})$. Using these separators we can also build a tree decomposition of width $O(\sqrt{n})$ in polynomial time [16]. Note that the hidden constant depends on β.   □

Theorem 2 (Bodlaender et al. [2], Cygan et al. [5]). *Given a graph with a tree decomposition of width w, there exists an algorithm solving Hamiltonian Cycle in $2^{O(w)}n^{O(1)}$ time.*

Applying this algorithm to the reduced graph of Lemma 3 finishes the proof of Theorem 1.   □

The techniques described in this section can also be used to solve Longest Path and Exact Path (which respectively are the problems of finding simple path of maximum length and finding a path between two specified vertices (u, v) of given length k):

Theorem 3. *There exists a $2^{O(\sqrt{n})}$-time algorithm for Longest Path and Exact Path on bounded-ratio disk graphs.*

Proof. Lemma 2 also holds for Longest Path and Exact Path. However, the subsequent step of removing vertices from the cliques no longer works: the information of how many vertices can be visited within each clique is essential. Instead, from each clique of the partition, we remove every vertex that does not have an edge to a vertex in some other clique of the partition. The removed vertices are then replaced by a path of the same number of vertices, and every vertex of the path is made adjacent to every (remaining) vertex of the clique. This preserves the longest (or exact) path, while only increasing the treewidth by a constant (compared to the reduced graph from Lemma 3). Bodlaender et al. [2] and Cygan et al. [5] also give algorithms for Longest Path and Exact Path parameterized by treewidth, similar to those of Theorem 2. Thus we obtain $2^{O(\sqrt{n})}$-time algorithms for Longest Path and Exact Path, by modifying the graph such that it has treewidth $O(\sqrt{n})$ and then applying one of these algorithms.   □

3   Lower Bound for Hamiltonian Cycle in Unit Disk Graphs

In this section, we give a tight lower bound for the running time of a Hamiltonian cycle algorithm in UDGs, assuming the Exponential Time Hypothesis. We use a reduction from 3-SAT.

We begin with a well-known reduction from 3-SAT to directed Hamiltonian cycle [11], and modify it significantly. We introduce the construction briefly; see Fig. 1 for an example of the construction with the formula $(x_1 \vee \bar{x}_2 \vee x_3) \wedge (x_2 \vee \bar{x}_3 \vee \bar{x}_4)$. Let n be the number of variables, and m be the number of clauses. For each variable x_i we introduce the vertices $v_i^1, \ldots v_i^b$, where $b = 3m + 3$. On these vertices we add a double chain (a directed path through the vertices and back); this double chain is the *row* of this variable. The Hamiltonian cycle can traverse this row left to right or right to left, which will indicate the truth setting of this variable. We add edges from the beginning and end of a variable's row to the beginning and end of the next variable's row, and we add a starting and ending point v_{start} and v_{end}, the arc (v_{end}, v_{start}), and arcs (v_{start}, v_1^1), (v_{start}, v_1^b), (v_n^1, v_{end}), (v_n^b, v_{end}). In order to check the clauses, we add a vertex c_j for each clause $j = 1, \ldots, m$. We connect the vertices v_i^{3j} and v_i^{3j+1} to clause j if a literal of x_i is present in the j-th clause. By orienting this arc pair correctly (depending on the sign of the literal), we make it possible for the Hamiltonian cycle to make a detour to c_j while traversing the variable's row in the direction (left or right) corresponding to the sign of the literal. For more details about this construction we refer the reader to the write-up in [11].

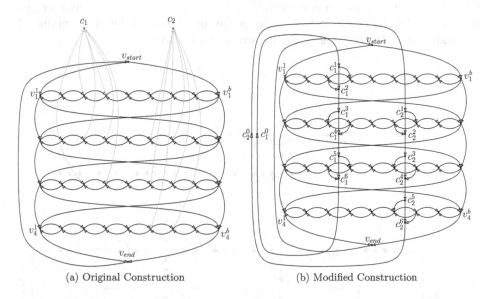

(a) Original Construction   (b) Modified Construction

Fig. 1. (a) The construction for $(x_1 \vee \bar{x}_2 \vee x_3) \wedge (x_2 \vee \bar{x}_3 \vee \bar{x}_4)$. (b) Modified construction with a directed cycle for each clause.

Our Construction. We replace the clause vertices by a different gadget: for each clause c_i, we introduce a directed cycle containing seven vertices, $c_i^0, \ldots, c_i^6$ (see Fig. 1b). If the first literal is the j-th variable, then we add the arcs $v_j^{3i} c_i^2$ and $c_i^1 v_j^{3i+1}$; if the literal is negated, we add $v_j^{3i+1} c_i^2$ and $c_i^1 v_j^{3i}$. Similarly, we add entry and exit arcs for the second and third literals at c_i^3, c_i^4 and c_i^5, c_i^6. This modified graph has a directed Hamiltonian cycle if and only if the original formula is satisfiable.

Next, we reduce to undirected Hamiltonian cycle. (From this point onward, we use the abbreviation HC for Hamiltonian cycle.) To do this, we start by replacing each vertex u of the construction with three vertices on a path: u^-, u^0 and u^+. An arc previously going from u to w is represented in the new graph by the edge $u^+ w^-$. This reduction from directed to undirected HC is already present in Karp's famous 1972 paper [10]. It is again routine to prove that the graph resulting from this construction has a HC if and only if the original formula is satisfiable. We denote the undirected graph that was obtained in this way by G.

We consider a specific drawing of G depicted on Fig. 2 of the resulting graph; we plan to emulate its properties in a unit disk graph. Intuitively, we would like to replace the edges of this graph by paths — this can be done by using unit disks that induce a path, making sure that the new graph has a HC if and only if the old graph has a HC. We call the set of disks used to represent an edge in such a way a *thread*. The difficulty stems from edges that are not used in the HC: substituting such edges with threads is not allowed. Essentially, we can only use threads if it is guaranteed by the construction that every HC has to pass through. If this cannot be guaranteed, we use *snakes*, which are constructions that allow the HC to either 'use' the edge uv, or to make a detour from one of the endpoints into the gadget, visiting every vertex inside.

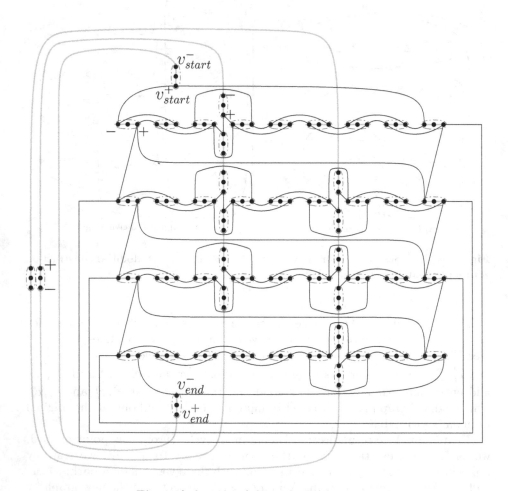

Fig. 2. A drawing of the undirected version.

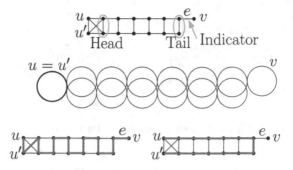

Fig. 3. A snake and a corresponding unit disk realization. Below: simulating a HC that passes or avoids edge uv.

Lemma 4. *The maximum vertex degree in G is four, and vertices of degree four induce a subgraph in which the maximum degree is two.*

Proof. The upper bound on the vertex degree follows from the fact that the original directed graph has maximum indegree and maximum outdegree three.

For the second statement, notice that vertices of degree four are either in- or out-vertices inside the row of a variable x_i, i.e., they are of the form $(v_i^j)^+$ or $(v_i^j)^-$. Moreover, notice that every vertex of degree four has a neighbour of degree two – the middle vertex $(v_i^j)^0$. Thus it is sufficient to show that for any degree four vertex v there is an additional neighbour of degree at most three.

The proof is for the case $(v_i^j)^+$. If $j = 1$ or $j = b = 3m + 3$, then $(v_i^j)^+$ has $(v_i^2)^-$ or $(v_i^{b-1})^-$ as a neighbour, and these are vertices of degree three: in the directed construction, the corresponding vertices had in- and outdegree two, because they are vertices of the form v_i^j, $j \equiv 2 \pmod 3$. If $1 < j < n$, then the vertex has a neighbour in the clause loop, where the maximum degree is three. $\qquad\square$

Representing Edges with Snakes. The snake is simply a $2 \times k$ grid graph for some $k \in \mathbb{N}$, with an extra disk at the *head* of the snake. In Fig. 3 we illustrate how a snake replacing an edge uv works. We need to add a disk u' which has the same neighbourhood as u - (this can be done by taking an identical or slightly perturbed copy of the disk of u). At the other end of the snake (at the *tail*) no such operation on v is required. Through the snake a HC can simulate passing the edge uv and it can also make a detour from u that covers all inner vertices if uv is not in the original HC. We define the indicator edge of the snake to be the edge connecting the snake to v, indicated by e in Fig. 3. It is easy to verify that if e is not used then we must detour (corresponding to avoiding uv in the HC), otherwise we must zigzag (corresponding to using uv).

Crossing Gadgets. Notice that in all the crossings in Fig. 2, exactly one of the crossing edges is a thick golden edge. These edges share the property that at least one of their endpoints has degree two, thus any Hamiltonian cycle of the

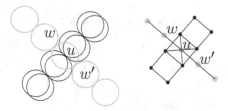

Fig. 4. Crossing a snake and a thread.

construction must pass through the golden edges. Therefore, we can replace the golden edges by threads; all other edges can be replaced by snakes. We have a crossing gadget for thread–snake crossings that we describe below.

The crossing gadget is depicted in Fig. 4. A Hamiltonian cycle passing through the snake in any way cannot enter the edges spanned by the thread: it can only enter at vertex u, and continue on one of the outgoing thread edges; that would render one of the points w and w' unreachable to the HC.

We note that snakes and threads can be used to represent bending edges, and a bend introduces only constant overhead; furthermore, a vertex can be the starting or ending point of up to five internally disjoint snakes or threads; since the maximum degree of G is four by Lemma 4, this threshold is not reached. We place snakes so that vertices of degree four have at most one connecting snake head – this can be done since the vertices of degree four in G span a collection of vertex disjoint paths and cycles by Lemma 4.

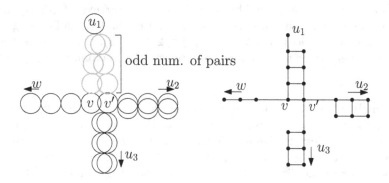

Fig. 5. Construction for degree four vertices.

Modifying the Neighbourhood of Vertices of Degree Four with a Snake Head. All degree four vertices have a neighbour of degree two (the in- and outvertices u^+ and u^- are connected to the degree two vertex u^0), thus the connecting edge is always used by all HCs in G. Let v be a degree four vertex with neighbours w, u_1, u_2 and u_3; let w be the neighbour of degree 2, and let u_1 be the neighbour whose snake head is at v. If the vertex has a connecting snake head, then we modify the neighbourhood of v so that v' is not a duplication of the disk representing v. Connect v to the tail of the w_1v snake, and connect it to the vw_2 snake together with v; the two remaining snake tails are only connected to v' (see Fig. 5 for the case when wv and u_1v are consecutive edges of v in the drawing). We can also ensure that the length of the vw_2 snake is odd, i.e., the snake is a $2 \times 2k + 1$ grid for some $k \in \mathbb{N}$, not including v and v'. This is required to make sure that the HC must pass both v and v' when the snake is used.

Finally, if wv and u_1v are non-consecutive edges around v in the drawing of G (Fig. 2), then we can change the drawing by introducing a new crossing between vw and vu_2 to make wv and u_1v consecutive around v. This requires a new snake-thread crossing, for which we can use our crossing gadget.

Fig. 6. Adding an extra thread-snake crossing around degree four vertices might be necessary. Here we added an extra crossing around the brown vertex w^- by changing a thread incident to w^- (in violet). (Color figure online)

The Final Construction. We begin by recreating a drawing of G in the plane with integer coordinates for all vertices, similar to the one seen in Fig. 2. This fits in a rectangle of size $O(n + m) \times O(n + m)$: there are $O(n)$ variable rows of length $O(m)$, and they require $O(1)$ vertical space each; together with the n long edges, we can fit these in $O(n + m)$ horizontal and $O(n)$ vertical space. The loop edges require $O(m)$ more vertical and horizontal space.

We apply a large constant scaling to make enough room for gadgets. Next, we define an orientation on the snakes so that degree four vertices have at most one snake head. Such an orientation exists due to Lemma 4. We also introduce extra crossings around degree four vertices when needed to ensure that the thread edge and the snake head are neighbours (see the change around w^- in Fig. 6). Finally, we exchange the golden edges with threads and the snake edges with snakes (Fig. 7); if our initial constant scaling was large enough, we have enough space to bend threads and snakes, without introducing intersections between independent snakes and threads. For disks representing vertices, for every incoming snake we introduce slightly perturbed disks (according to the original definition of snakes). These extra disks are indicated by a green number in Fig. 7.

Lemma 5. *Given an initial undirected graph G corresponding to a 3-CNF formula of n variables and m clauses, the unit disk graph G' constructed above is computable in time polynomial in $n + m$, has $O((m + n)^2)$ vertices and it is equivalent to G in the sense that G' has a HC if and only if G has a HC.*

Proof. First, we show that if G' has a HC then G has a HC. (The other implication is trivial.) Let H' be a HC in G'. In each thread we designate an arbitrary

Fig. 7. The part of the final construction corresponding to Fig. 6. Snake heads are represented by two red disks, plus multiplicity of the end vertex when needed. Golden disks correspond to threads. (Color figure online)

inner edge as indicator. Mark an edge in G if the indicator edge of the corresponding thread or snake is contained in H'. We claim that the set of marked edges (denoted by H) is a HC in G. Observe that a cut $C \subseteq E(G)$ corresponds to a cut of the same size in G': the indicator edges corresponding to the threads or snakes of the edges in C define a cut of G'. Consequently, for any cut C the number of H-edges contained in it is an even, positive number, since the corresponding cut C' is crossed by H' an even, positive number of times. It follows that H is a spanning connected Eulerian subgraph.

It remains to show that the maximum degree in H is two; since the maximum degree of G is four, it is sufficient to show that any vertex v of degree four has degree two in H. If v has no snake heads, then this follows from the fact that the disk corresponding to v has four independent neighbours in G'.

Let v be a vertex of degree four with a snake head. We denote by $S(x, y)$ the snake from x to y, with the head at x. We use the notation w, u_1, u_2, u_3 for the neighbours of v, where $\deg(w) = 2$ and $S(v, u_1)$ is the snake whose head is at v (see Fig. 5). Since there is a thread between w and v, the edge (w, v) is marked. If H' uses $S(v, u_1)$, then by the odd length of the snake and our construction, vv' is an edge of H' — this can be verified by stepping back through $S(v, u_1)$ from the indicator edge at u_1. So in this case, H' must detour on both $S(u_2, v')$ and $S(u_3, v')$. Otherwise (if $S(v, u_1)$ is only a detour in H'), then one of the neighbours of v' in H' is inside $S(v, u_1)$, so H' can use only one of $S(u_2, v')$ and $S(u_3, v')$. Thus, the degree of v in H is two in both cases.

The construction can be created in polynomial time from an initial graph G. It is placed in a rectangle of size $O(n + m) \times O(n + m)$; since every point in this rectangle is covered by at most four disks (which can occur if there are two snake heads at a degree three vertex), it follows that the number of disks used is $O((n + m)^2)$. □

Theorem 4. *There is no $2^{o(\sqrt{n})}$ algorithm for Hamiltonian cycle in unit disk graphs, unless the ETH fails.*

Proof. Suppose that the initial formula has $\hat{m}$ clauses and $\hat{n}$ variables. Without loss of generality, suppose that $\hat{m} = \Theta(\hat{n})$ (see the Sparsification Lemma in [3]). The graph G can be created in polynomial time starting from our formula, and by Lemma 5 we can create G' in polynomial time from G. The resulting unit disk graph G' has a HC if and only if the original formula is satisfiable. Since the resulting UDG has $O(\hat{n}^2)$ vertices, a $2^{o(\sqrt{n})} = 2^{o(\sqrt{\hat{n}^2})}$ algorithm would mean that we could decide the satisfiability of the formula in $2^{o(\hat{n})}$ time, which contradicts the Exponential Time Hypothesis. □

4   Colouring Disk Graphs

To complement our results on Hamiltonian Cycle, we show that the square root phenomenon also holds for q-colouring on disk graphs when q is a constant.

Theorem 5. *(a) For any constant q, there is an algorithm running in time $O(2^{O(\sqrt{n})})$ that solves the q-colouring problem on disk graphs. (b) There is no $2^{o(\sqrt{n})}$ algorithm for q-colouring in unit disk graphs for any constant $q \geq 3$, unless the ETH fails.*

5   Conclusions

We have shown that the HC problem and q-colouring both have $2^{O(\sqrt{n})}$ algorithms in bounded-ratio disk graphs, and matching lower bounds $2^{\Omega(\sqrt{n})}$ if ETH holds. We have also seen that in case of the colouring problem, the same result applies in general disk graphs.

Some preliminary work shows that it should be possible to get a $2^{\Omega(\sqrt{n})}$ lower bound for HC in the more restricted case of grid graphs, although the proof and the gadgets used will be more complicated.

A major remaining open problem is to find a $2^{O(\sqrt{n})}$ algorithm for HC in disk graphs. Finally, we remind the reader that reducing the coefficient of $\sqrt{n}$ in the exponents of these running times is also a worthwhile effort; in the case of Hamiltonian Cycle on general graphs, a steady wave of improvements yielded impressive results; can the community achieve something similar for these square-root type algorithms?

Acknowledgements. This work was initiated at the Lorentz Center workshop 'Fixed-Parameter Computational Geometry'. We are grateful to Hans L. Bodlaender and Mark de Berg for discussions and their help with improving this paper.

References

1. Alber, J., Fiala, J.: Geometric separation and exact solutions for the parameterized independent set problem on disk graphs. J. Algorithms **52**(2), 134–151 (2004). http://dx.doi.org/10.1016/j.jalgor.2003.10.001
2. Bodlaender, H.L., Cygan, M., Kratsch, S., Nederlof, J.: Deterministic single exponential time algorithms for connectivity problems parameterized by treewidth. Inf. Comput. **243**, 86–111 (2015)
3. Cygan, M., Fomin, F.V., Kowalik, L., Lokshtanov, D., Marx, D., Pilipczuk, M., Pilipczuk, M., Saurabh, S.: Parameterized Algorithms. Springer, Heidelberg (2015). doi:10.1007/978-3-319-21275-3
4. Cygan, M., Kratsch, S., Nederlof, J.: Fast hamiltonicity checking via bases of perfect matchings. In: Proceedings of the Forty-Fifth Annual ACM Symposium on Theory of Computing, STOC 2013, pp. 301–310. ACM, New York (2013). http://doi.acm.org/10.1145/2488608.2488646
5. Cygan, M., Nederlof, J., Pilipczuk, M., Pilipczuk, M., van Rooij, J.M.M., Wojtaszczyk, J.O.: Solving connectivity problems parameterized by treewidth in single exponential time. In: Proceedings of the 2011 IEEE 52nd Annual Symposium on Foundations of Computer Science, FOCS 2011, Washington, DC, USA, pp. 150–159 (2011). http://dx.doi.org/10.1109/FOCS.2011.23
6. Deineko, V.G., Klinz, B., Woeginger, G.J.: Exact algorithms for the hamiltonian cycle problem in planar graphs. Oper. Res. Lett. **34**(3), 269–274 (2006). http://dx.doi.org/10.1016/j.orl.2005.04.013
7. Gräf, A., Stumpf, M., Weißenfels, G.: On coloring unit disk graphs. Algorithmica **20**(3), 277–293 (1998). http://dx.doi.org/10.1007/PL00009196
8. Itai, A., Papadimitriou, C.H., Szwarcfiter, J.L.: Hamilton paths in grid graphs. SIAM J. Comput. **11**(4), 676–686 (1982). http://dx.doi.org/10.1137/0211056
9. Ito, H., Kadoshita, M.: Tractability and intractability of problems on unit disk graphs parameterized by domain area. In: Proceedings of the 9th International Symposium on Operations Research and Its Applications (ISORA 2010), pp. 120–127. Citeseer (2010)
10. Karp, R.M.: Reducibility among combinatorial problems. In: Miller, R.E., Thatcher, J.W., Bohlinger, J.D. (eds.) Proceedings of a Symposium on the Complexity of Computer Computations. The IBM Research Symposia Series, pp. 85–103. Springer, Heidelberg (1972). doi:10.1007/978-1-4684-2001-2_9
11. Kleinberg, J.M., Tardos, É.: Algorithm Design. Addison-Wesley, Boston (2006)
12. Koebe, P.: Kontaktprobleme der konformen Abbildung. Hirzel (1936)
13. Marx, D.: The square root phenomenon in planar graphs. In: Automata, Languages, and Programming - 40th International Colloquium, p. 28 (2013)
14. Miller, G.L., Teng, S., Thurston, W.P., Vavasis, S.A.: Separators for sphere-packings and nearest neighbor graphs. J. ACM **44**(1), 1–29 (1997). http://doi.acm.org/10.1145/256292.256294
15. Pino, W.J.A., Bodlaender, H.L., van Rooij, J.M.M.: Cut and count and representative sets on branch decompositions. In: Proceedings of IPEC (2016, to appear)
16. Reed, B.: Tree width and tangles: a new connectivity measure and some applications. Surv. Comb. **241**, 87–162 (1997)

Tight Inefficiency Bounds
for Perception-Parameterized Affine
Congestion Games

Pieter Kleer[1] and Guido Schäfer[1,2(✉)]

[1] Centrum Wiskunde & Informatica (CWI), Amsterdam, The Netherlands
{kleer,schaefer}@cwi.nl
[2] Vrije Universiteit Amsterdam, Amsterdam, The Netherlands

Abstract. We introduce a new model of congestion games that captures several extensions of the classical congestion game introduced by Rosenthal in 1973. The idea here is to parameterize both the perceived cost of each player and the social cost function of the system designer. Intuitively, each player perceives the load induced by the other players by an extent of $\rho \geq 0$, while the system designer estimates that each player perceives the load of all others by an extent of $\sigma \geq 0$. For specific choices of ρ and σ, we obtain extensions such as altruistic player behavior, risk sensitive players and the imposition of taxes on the resources. We derive tight bounds on the price of anarchy and the price of stability for a large range of parameters. Our bounds provide a complete picture of the inefficiency of equilibria for these games. As a result, we obtain tight bounds on the price of anarchy and the price of stability for the above mentioned extensions. Our results also reveal how one should "design" the cost functions of the players in order to reduce the price of anarchy. Somewhat counterintuitively, if each player cares about all other players to the extent of $\rho = 0.625$ (instead of 1 in the standard setting) the price of anarchy reduces from 2.5 to 2.155 and this is best possible.

1 Introduction

Congestion games constitute an important class of non-cooperative games which was introduced by Rosenthal in 1973 [13]. In a congestion game, we are given a set of resource from which a set of players can choose. Each resource is associated with a cost function which specifies the cost of this resource depending on the total number of players using it. Every player chooses a subset of resources (from a set of resource subsets available to her) and experiences a cost equal to the sum of the costs of the chosen resources. Congestion games are both theoretically appealing and practically relevant. For example, they have applications in network routing, resource allocation and scheduling problems.

Rosenthal [13] proved that every congestion game has a pure Nash equilibrium, i.e., a strategy profile such that no player can decrease her cost by unilaterally deviating to another feasible set of resources. This result was established through the use of an exact potential function (known as *Rosenthal potential*)

© Springer International Publishing AG 2017
D. Fotakis et al. (Eds.): CIAC 2017, LNCS 10236, pp. 381–392, 2017.
DOI: 10.1007/978-3-319-57586-5_32

satisfying that the cost difference induced by a unilateral player deviation is equal to the potential difference of the respective strategy profiles. In fact, Monderer and Shapley [11] showed that the class of games admitting an exact potential function is isomorphic to the class of congestion games.

One of the main research directions in algorithmic game theory focusses on quantifying the inefficiency caused by selfish behavior. The idea is to assess the quality of a Nash equilibrium relative to an optimal outcome. Here the quality of an outcome is measured in terms of a given *social cost* objective (e.g., the sum of the costs of all players). Koutsoupias and Papadimitriou [10] introduced the *price of anarchy* as the ratio between the worst social cost of a Nash equilibrium and the social cost of an optimum. Anshelevich et al. [1] defined the *price of stability* as the ratio between the best social cost of a Nash equilibrium and the social cost of an optimum.

In recent years, several extensions of Rosenthal's congestion games were proposed to incorporate aspects which are not captured by the standard model. For example, these extensions include risk sensitivity of players in uncertain settings [12], altruistic player behavior [4,5] and congestion games with taxes [3]. We elaborate in more detail on these extensions in Sect. 2. These games were studied intensively with the goal to obtain a precise understanding of the price of anarchy.

In this paper, we introduce a new model of congestion games, which we term *perception-parameterized congestion games*, that captures all these extensions (and more) in a unifying way. The key idea here is to parameterize both the perceived cost of each player and the social cost function. Intuitively, each player perceives the load induced by the other players by an extent of $\rho \geq 0$, while the system designer estimates that each player perceives the load of all others by an extent of $\sigma \geq 0$. The above mentioned extensions reduce to special cases of our model by choosing the parameters ρ and σ accordingly.

Despite the fact that we deal with a more general class of congestion games, we manage to derive tight bounds on the price of anarchy and the price of stability for a large range of parameters. Our bounds provide a complete picture of the inefficiency of equilibria for these perception-parameterized congestion games. As a consequence, we obtain tight bounds on the price of anarchy and the price of stability for the above mentioned extensions. While the price of anarchy bounds are (mostly) known from previous results, the price of stability results are new. As in [3–5,12], we focus on congestion games with affine cost functions.

We illustrate our model by means of a simple example; formal definitions of our perception-parameterized congestion games are given in Sect. 2. Suppose we are given a set of m resources and that every player has to choose precisely one of these resources. The cost of a resource $e \in [m]^1$ is given by a cost function c_e that maps the *load* on e to a real value. In the classical setting, the load of a resource e is defined as the total number of players x_e using e. That is, the cost that player i experiences when choosing resource e is $c_e(x_e)$. In contrast, in

---

[1] Given a positive integer m, we use $[m]$ to refer to the set $\{1, \ldots, m\}$.

Table 1. An overview of (tight) price of anarchy and price of stability results for certain values of ρ and σ. Here $h(1) \approx 0.625$ (see Theorem 1 for a formal definition). The respective references where these bounds were established first are given in the column "Ref."; an asterisk indicates that this result is new.

Model	Parameters	PoA	Ref.	PoS	Ref.
Classical	$\rho = \sigma = 1$	$5/2$	[6]	1.577	[3]
Altruism (1)	$\sigma = 1,\ 1 \le \rho \le 2$	$\frac{4\rho+1}{1+\rho}$	[4,5]	$\frac{\sqrt{3}+1}{\sqrt{3}+\rho-1}$	[*]
Altruism (2)	$\sigma = 1,\ 2 \le \rho \le \infty$	$\rho + 1$	[5]	–	–
Risk-neutral players	$\sigma = \rho = 1/2$	$5/3$	[12]	1.447	[*]
Wald's minimax	$\sigma = 1/2,\ \rho = 1$	2	[2,12]	1	[*]
Constant universal taxes	$\sigma = 1,\ \rho = h(1)$	2.155	[3]	2.013	[*]
Generalized affine CG	–	∞	[*]	2	[*]

our setting players have different perceptions of the load induced by the other players. More precisely, the *perceived load* of player i choosing resource e is $1 + \rho(x_e - 1)$, where $\rho \ge 0$ is a parameter. Consequently, the perceived cost of player i for choosing e is $c_e(1 + \rho(x_e - 1))$. Note that as ρ increases players care more about the presence of other players.[2] In addition, we introduce a similar parameter $\sigma \ge 0$ for the social cost objective. Intuitively, this can be seen as the system designer's estimate of how each player perceives the load of the other players. In our example, the social cost is defined as $\sum_{e \in [m]} c_e(1 + \sigma(x_e - 1))x_e$.

Our Results. We prove the following bounds on the price of anarchy (PoA) and the price of stability (PoS) of affine congestion games for a large range of parameters (ρ, σ) (specified below):

$$\text{PoA} \le \max\left\{\rho + 1, \frac{2\rho(1 + \sigma) + 1}{\rho + 1}\right\} \quad \text{and} \quad \text{PoS} \le \frac{\sqrt{\sigma(\sigma + 2)} + \sigma}{\sqrt{\sigma(\sigma + 2)} + \rho - \sigma}. \quad (1)$$

We prove that these bounds are tight for general affine congestion games. Further, for the special case of symmetric network congestion games we show that the bound of $(2\rho(1 + \sigma) + 1)/(\rho + 1)$ on the price of anarchy is asymptotically tight. In contrast, for this case we derive a better (tight) bound on the price of stability for $\sigma = 1$ and $\rho \ge 0$. An overview of the price of anarchy and the price of stability results that we obtain from (1) for several applications known in the literature is given in Table 1; see Fig. 2 for an illustration of our PoA bound. The connection between these applications and our model is discussed at the end of Sect. 2.

In light of the above bounds, we obtain an (almost) complete picture of the inefficiency of equilibria (parameterized by ρ and σ); for example, see Fig. 1 for

---

[2] In this work, we concentrate on the homogeneous player case.

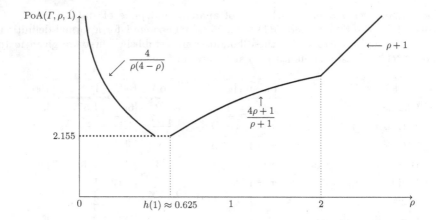

Fig. 1. Lower bounds on the price of anarchy for $\sigma = 1$. The bounds $(4\rho + 1)/(\rho + 1)$ and $\rho + 1$ are also tight upper bounds. The dotted horizontal line indicates the lower bound following from [4, Theorem 3.7]. The bound $4/(\rho(4 - \rho))$ is a lower bound for symmetric singleton congestion games given in the proof of Theorem 5. A tight bound for $0 < \rho \leq h(1)$ remains an open problem.

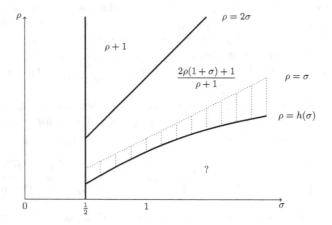

Fig. 2. The bound $\rho + 1$ holds for $\rho \geq 2\sigma \geq 1$. The bound $(2\rho(1+\sigma)+1)/(1+\rho)$ holds for $\sigma \leq \rho \leq 2\sigma$. Roughly speaking, this bound also holds for $h(\sigma) \leq \rho \leq \sigma$, but our proof of Theorem 1 only works for a discretized range of σ (hence the vertical dotted lines in this area). The function h is given in Theorem 1.

an illustration of the price of anarchy if $\sigma = 1$. Note that the price of anarchy *decreases* from $\frac{5}{2}$ for $\rho = 1$ to 2.155 for $\rho = h(1) \approx 0.625$.[3]

---

[3] The price of anarchy for $\rho = h(1)$ was first established by Caragiannis et al. [3]. However, our bounds reveal that the price of anarchy is in fact minimized at $\rho = h(1)$ (see also Fig. 1).

2 Our Model, Applications and Related Work

We first formally introduce our model of congestion games with parameterized perceptions. We then show that our model subsumes several other models that were studied in the literature as special cases.

A *congestion game* Γ is given by a tuple $(N, E, (\mathcal{S}_i)_{i \in N}, (c_e)_{e \in E})$ where $N = [n]$ is the set of players, E the set of resources (or facilities), $\mathcal{S}_i \subseteq 2^E$ the set of strategies of player i, and $c_e : \mathbb{R}_{\geq 0} \to \mathbb{R}_{\geq 0}$ the cost function of facility e. Given a strategy profile $s = (s_1, \ldots, s_n) \in \times_i \mathcal{S}_i$, we define x_e as the number of players using resource e, i.e., $x_e = x_e(s) = |\{i \in N : e \in s_i\}|$. If $\mathcal{S}_i = \mathcal{S}_j$ for all $i, j \in N$, the game is called *symmetric*. For a given graph $G = (V, E)$, we call Γ a *(directed) network congestion game* if for every player i there exist $s_i, t_i \in V$ such that $\mathcal{S}_i$ is the set of all (directed) (s_i, t_i)-paths in G. An *affine congestion game* has cost functions of the form $c_e(x) = a_e x + b_e$ with $a_e, b_e \geq 0$. If $b_e = 0$ for all $e \in E$, the game is called *linear*.

We introduce our unifying model of *perception-parameterized congestion games* with affine latency functions. For a fixed parameter $\rho \geq 0$, we define the cost of player $i \in N$ by

$$C_i^\rho(s) = \sum_{e \in s_i} c_e(1 + \rho(x_e - 1)) = a_e[1 + \rho(x_e - 1)] + b_e \tag{2}$$

for a given strategy profile $s = (s_1, \ldots, s_n)$. For a fixed parameter $\sigma \geq 0$, the social cost of a strategy profile s is given by

$$C^\sigma(s) = \sum_{i \in N} C_i^\sigma(s) = \sum_{e \in E} x_e(a_e[1 + \sigma(x_e - 1)] + b_e). \tag{3}$$

We refer to the case $\rho = \sigma = 1$ as the *classical congestion game* with cost functions $c_e(x) = a_e x + b_e$ for all $e \in E$.

A strategy profile s is a *Nash equilibrium* if for all players $i \in N$ it holds that $C_i^\rho(s) \leq C_i^\rho(s_i', s_{-i})$ for all $s_i' \in \mathcal{S}_i$, where (s_i', s_{-i}) denotes the strategy profile in which player i plays s_i' and all the other players their strategy in s. The price of anarchy (PoA) and price of stability (PoS) of a game Γ are defined as

$$\text{PoA}(\Gamma, \rho, \sigma) = \frac{\max_{s \in \text{NE}} C^\sigma(s)}{\min_{s^* \in \times_i \mathcal{S}_i} C^\sigma(s^*)} \quad \text{and PoS } (\Gamma, \rho, \sigma) = \frac{\min_{s \in \text{NE}} C^\sigma(s)}{\min_{s^* \in \times_i \mathcal{S}_i} C^\sigma(s^*)},$$

where $\text{NE} = \text{NE}(\rho)$ denotes the set of Nash equilibria with respect to the player costs as defined in (2). For a collection of games $\mathcal{H}$, $\text{PoA}(\mathcal{H}, \rho, \sigma) = \sup_{\Gamma \in \mathcal{H}} \text{PoA}(\Gamma, \rho, \sigma)$ and $\text{PoS}(\mathcal{H}, \rho, \sigma) = \sup_{\Gamma \in \mathcal{H}} \text{PoS}(\Gamma, \rho, \sigma)$. Unless stated otherwise, our results refer to the class of perception-parameterized congestion games with affine latency functions; we therefore drop the parameter $\mathcal{H}$ below. Rosenthal [13] shows that classical congestion games (i.e., $\rho = \sigma = 1$) have an exact potential function: $\Phi : \times_i \mathcal{S}_i \to \mathbb{R}$ is an *exact potential function* for a congestion game Γ if for every strategy profile s, for every $i \in N$ and every $s_i' \in \mathcal{S}_i$: $\Phi(s) - \Phi(s_{-i}, s_i') = C_i(s) - C_i(s_{-i}, s_i')$. The Rosenthal potential

$\Phi(s) = \sum_{e \in E} \sum_{k=1}^{x_e} c_e(k)$ is an exact potential function for classical congestion games.

We review various models that fall within, or are related to, the framework proposed above (for certain values of ρ and σ). These models sometimes interpret the parameters differently than explained above.

Altruism [4,5]. We can rewrite the cost of player i as $C_i^\rho(s) = \sum_{e \in s_i}(a_e x_e + b_e) + (\rho - 1)a_e(x_e - 1)$. The term $(\rho - 1)a_e(x_e - 1)$ can be interpreted as a "dynamic" (meaning load-dependent) tax that players using resource e have to pay. For $1 \leq \rho \leq \infty$ and $\sigma = 1$, this model is equivalent to the altruistic player setting proposed by Caragiannis et al. [4]. Chen et al. [5] also study this model of altruism for $1 \leq \rho \leq 2$ and $\sigma = 1$.

Constant Taxes [3]. We can rewrite the cost of player i as $C_i^\rho(s) = \sum_{e \in s_i} \rho a_e x_e + (1 - \rho)a_e + b_e$. Dividing by ρ gives that s is a Nash equilibrium with respect to C_i^ρ if and only if s is a Nash equilibrium with respect to $T_i^\rho(s) = C_i^\rho/\rho = \sum_{e \in s_i}(a_e x_e + b_e/\rho) + \sum_{e \in s_i}(1 - \rho)/\rho a_e$. That is, s is a Nash equilibrium in a classical congestion game in which players take into account constant resource taxes of the form $(1 - \rho)/\rho \cdot a_e$. Caragiannis, Kaklamanis and Kanellopoulos [3] study this type of taxes, which they call *universal tax functions*, for ρ satisfying $(1 - \rho)/\rho = 3/2\sqrt{3} - 2$. They consider these taxes to be *refundable*, i.e., they are not taken into account in the social cost, which is equivalent to the case $\sigma = 1$. Note that the function $\tau : (0, 1] \to [0, \infty)$ defined by $\tau(\rho) = (1-\rho)/\rho$ is bijective.[4] Caragiannis et al. [3] showed that the price of anarchy can be decreased to 2.155 by the usage of universal tax functions, which improves significantly the classical bound of 2.5. Furthermore, from [3, Theorem 3.7] it follows that the price of anarchy can never be better than 2.155 for $0 \leq \rho \leq h(1)$. However, in this work we show that the price of stability increases from 1.577 (for classical games) to 2.013, for this specific set of tax functions.

Risk Sensitivity Under Uncertainty [12]. Nikolova, Piliouras and Shamma [12] consider congestion games in which there is a (non-deterministic) order of the players on every resource. A player is only affected by players in front of her. That is, the load on resource e for player i in a strict ordering r, where $r_e(i)$ denotes the position of player i, is given by $x_e(i) = |\{j \in N : r_e(j) \leq r_e(i)\}|$. The cost of player i is then $C_i(s) = \sum_{e \in s_i} c_e(x_e(i))$. Note that $x_e(i)$ is a random variable if the ordering is non-deterministic. The social cost of the model is defined by the sum of all player costs $C^{\frac{1}{2}}(s) = \sum_{e \in E} \frac{1}{2}a_e x_e(x_e + 1) + b_e$ which is independent of the ordering r.[5] Note that the social cost corresponds to the case $\sigma = \frac{1}{2}$ in our framework. Nikolova et al. [12] study various risk attitudes towards the ordering r that is assumed to have a uniform distribution over all possible orderings. The two relevant attitudes are that of *risk-neutral* players and players applying *Wald's minimax principle*. Risk-neutral players define their cost as the expected cost under the ordering r, which correspond to the case $\rho = \frac{1}{2}$ in (2). This

---

[4] This relation between altruism (or spite) and constant taxes is also mentioned by Caragiannis et al. [4].

[5] In every ordering there is always one player first, one player second, and so on.

can roughly be interpreted as that players expect to be scheduled in the middle on average. Wald's minimax principle implies that players assume a worst-case scenario, i.e., being scheduled last on all the resources. This corresponds to the case $\rho = 1$.

Approximate Nash Equilibria [7]. Suppose that s is a Nash equilibrium under the cost functions defined in (2). Then, in particular, we have $C_i^1(s) \leq C_i^\rho(s) \leq C_i^\rho(s_i', s_{-i}) \leq \rho C_i^1(s_i', s_{-i})$ for any player i and $s_i' \in \mathcal{S}_i$ and $\rho \geq 1$. That is, we have $C_i^1(s) \leq \rho C_i^1(s_i', s_{-i})$ which means that the profile s is a ρ-approximate equilibrium, as studied by Christodoulou, Koutsoupias and Spirakis [7]. In particular, this implies that any upper bound on the price of anarchy, or price of stability, in our framework yields an upper bound on the price of stability for ρ-approximate equilibria for the same class of games. For $\sigma = 1$ and $1 \leq \rho \leq 2$, we obtain a bound of $(\sqrt{3} + 1)/(\sqrt{3} + \rho - 1)$ on the price of stability. In particular, this also yields the same bound on the price of stability for ρ-approximate equilibria. This bound was previously obtained by Christodoulou et al. [7]. Conceptually our approach is different: We prove our bound by observing that every Nash equilibrium in our framework yields an approximate equilibrium. In particular, this gives rise to a potential function that can be used to carry out the technical details (namely the potential function that is exact for our congestion game).[6]

Generalized Affine Congestion Games. Let $\mathcal{A}'$ denote the class of all congestion games Γ for which all resources have the same cost function $c(x) = ax + b$, where $a = a(\Gamma)$ and $b = b(\Gamma)$ satisfy $a \geq 0$ and $a + b > 0$. The class of affine congestion games with non-negative coefficients is contained in $\mathcal{A}'$ since every such game can always be transformed[7] into a game Γ' with $a_e = 1$ and $b_e = 0$ for all resources $e \in E'$, where E' is the resource set of Γ'. Without loss of generality we can assume that $a + b = 1$, since the cost functions can be scaled by $1/(a + b)$. The cost functions of $\Gamma \in \mathcal{A}'$ can then equivalently be written as $c(x) = \rho x + (1 - \rho)$ for $\rho \geq 0$. This is precisely the definition of $C_i^\rho(s)$ (with $a_e = 1$ and $b_e = 0$ taken there). In particular, if we take $\sigma = \rho$, meaning that $C^\rho(s) = \sum_{i \in N} C_i^\rho(s)$, we have $\mathrm{PoA}(\mathcal{A}') = \sup_{\rho \geq 0} \mathrm{PoA}(\mathcal{A}, \rho, \rho)$ and $\mathrm{PoS}(\mathcal{A}') = \sup_{\rho \geq 0} \mathrm{PoS}(\mathcal{A}, \rho, \rho)$, where $\mathcal{A}$ denotes the class of affine congestion games with non-negative coefficients.

Due to page limitations some material is omitted below. All missing details can be found in the full version of this paper [9].

3 Price of Anarchy

We derive the upper bound on the price of anarchy given in (1). We start with the bound of $(2\rho(1 + \sigma) + 1)/(\rho + 1)$.

---

[6] Nevertheless, the framework of Christodoulou et al. [7] is somewhat more general and might be used to obtain a tight bound for the price of stability of approximate equilibria (which is not known to the best of our knowledge).

[7] This transformation can be done in such a way that both PoA and PoS of the game do not change. For a proof the reader is referred to, e.g., [5, Lemma 4.3].

We need the following technical lemma for the proof of Theorem 1:

Lemma 1. *Let s be a Nash equilibrium under the cost functions $C_i^\rho(s)$ and let s^* be a minimizer of $C^\sigma(\cdot)$. For $\rho, \sigma \geq 0$ fixed, if there exist $\alpha(\rho, \sigma), \beta(\rho, \sigma) \geq 0$ such that*

$$(1 + \rho x)y - \rho(x - 1)x - x \leq -\beta(\rho, \sigma)(1 + \sigma(x - 1))x + \alpha(\rho, \sigma)(1 + \sigma(y - 1))y$$

for all non-negative integers x and y, then $\beta(\rho, \sigma)C^\sigma(s) \leq \alpha(\rho, \sigma)C^\sigma(s^*)$.

Theorem 1. *We have $PoA(\rho, \sigma) \leq (2\rho(1 + \sigma) + 1)/(\rho + 1)$ if*

(i) $\frac{1}{2} \leq \sigma \leq \rho \leq 2\sigma$, or
(ii) $\sigma = 1$ and $h(\sigma) \leq \rho \leq 2\sigma$, where $h(\sigma) = g(1 + \sigma + \sqrt{\sigma(\sigma + 2)}, \sigma)$ is the optimum of the function

$$g(a, \sigma) = \frac{\sigma(a^2 - 1)}{(1 + \sigma)a^2 - (2\sigma + 1)a + 2\sigma(\sigma + 1)}.$$

Further, there exists a function $\Delta = \Delta(\sigma)$ satisfying for every fixed $\sigma_0 \geq 1/2$: if $\Delta(\sigma_0) \geq 0$, then the stated bound is true for all $h(\sigma_0) \leq \rho \leq 2\sigma_0$.

Proof (Sketch). For the functions $\alpha(\rho, \sigma) = (2\rho(1+\sigma)+1)/(1+2\sigma)$ and $\beta(\rho, \sigma) = (1 + \rho)/(1 + 2\sigma)$, we prove the inequality in Lemma 1. We show that for certain functions f_1 and f_2, the smallest ρ satisfying the inequality of Lemma 1 is given by the quantity

$$h(\sigma) = \sup_{x,y \in \mathbb{N}: f_1(x,y,\sigma) > 0} -\frac{f_2(x, y, \sigma)}{f_1(x, y, \sigma)}.$$

We divide the set $(x, y) \in \mathbb{N} \times \mathbb{N}$ in lines of the form $x = ay$ and determine the supremum over every line. After that we take the supremum over all lines, which then gives the desired result. We first show that the case $x \leq y$ is trivial. We then focus on $y < x$. In this case, we show that $h(\sigma) = \max\{\gamma_1(\sigma), \gamma_2(\sigma)\}$ for certain functions γ_1 and γ_2. Numerical experiments suggest that $\Delta(\sigma) := \gamma_1(\sigma) - \gamma_2(\sigma) \geq 0$, that is, the maximum is always attained for γ_1 (which is the definition of h given in the statement). In particular, this means that if for a fixed σ the non-negativity of $\Delta(\sigma)$ is satisfied, then this yields an exact proof of the inequality of Lemma 1 for $h(\sigma) \leq \rho \leq 2\sigma$. The function Δ is specified in the full version of this paper [9]. □

Numerical experiments suggest that $\Delta(\sigma)$ is non-negative for all $\sigma \geq 1/2$. We emphasize that for a fixed σ, with $\Delta(\sigma) \geq 0$, the proof that the inequality holds for all $h(\sigma) \leq \rho \leq 2\sigma$ is exact in the parameter ρ. The first two cases of Theorem 1 capture all the price of anarchy results from the literature.

We next show that the bound of Theorem 1 is also an (asymptotic) lower bound for linear symmetric network congestion games.[8] This improves a result

---

[8] In the the full version [9] we show tightness for general congestion games.

in the risk-uncertainty model of Piliouras et al. [12], who only prove asymptotic tightness for symmetric linear congestion games (for their respective values of ρ and σ). It also improves a result in the altruism model by Chen et al. [5], who show tightness only for general congestion games.

Christodoulou and Koutsoupias [6] showed that for symmetric congestion games ($\rho = \sigma = 1$) the bound of $\frac{5}{2}$ on the price of anarchy is asymptotically tight. More recently, Correa et al. [8] proved that the bound of $\frac{5}{2}$ is tight for symmetric network congestion games. Our lower bound proof is a generalization of their construction.

Theorem 2. *For $\rho, \sigma > 0$ fixed, there exists a symmetric network linear congestion game such that for every $\epsilon > 0$, $PoA(\rho, \sigma) \geq (2\rho(1+\sigma)+1)/(\rho+1) - \epsilon$.*

For $\rho \geq 2\sigma$, we can obtain a tight bound of $\rho + 1$ on the price of anarchy. Remarkably, the bound itself does not depend on σ, only the range of ρ and σ for which it holds. For the parameters $\sigma = 1$ and $\rho \geq 2$ in the altruism model of Caragiannis et al. [4], this bound is known to be tight for non-symmetric singleton congestion games (where all strategies consist of a single resource). We only provide tightness for general congestion games, but the construction is significantly simpler.

Theorem 3. *We have $PoA(\rho, \sigma) \leq \rho+1$ for $1 \leq 2\sigma \leq \rho$ and this bound is tight.*

4   Price of Stability

We show the bound given in (1) to be an (asymptotically) tight bound for the price of stability for a large range of pairs (ρ, σ). We need the following technical lemma.

Lemma 2. *For all non-negative integers x and y, and $\sigma \geq 0$ arbitrary, we have*

$$\left(x - y + \frac{1}{2}\right)^2 - \frac{1}{4} + 2\sigma x(x - 1) + (\sqrt{\sigma(\sigma+2)} + \sigma)[y(y-1) - x(x-1)] \geq 0.$$

Theorem 4. *We have*

$$PoS(\rho, \sigma) \leq \frac{\sqrt{\sigma(\sigma+2)} + \sigma}{\sqrt{\sigma(\sigma+2)} + \rho - \sigma} \quad for\ \sigma > 0\ and\ \frac{2\sigma}{1 + \sigma + \sqrt{\sigma(\sigma+2)}} \leq \rho \leq 2\sigma$$

and this bound is asymptotically tight.

Our proof is similar to a technique of Christodoulou, Koutsoupias and Spirakis [7] for upper bounding the price of stability of ρ-approximate equilibria. However, for a general σ the analysis is more involved. The main technical contribution comes from establishing the inequality in Lemma 2. The proof of the asymptotic tightness is also based on a construction due to Christodoulou et al. [7] used for obtaining a (non-tight) lower bound on the price of stability of approximate equilibria. The lower bound proof is omitted here.

Proof. Note that we can write $C_i^\rho(s) = a_e x_e + b_e + (\rho - 1)a_e x_e$. By Rosenthal [13],

$$\Phi^\rho(s) := \sum_{e \in E} a_e \frac{x_e(x_e + 1)}{2} + b_e x_e + (\rho - 1) \sum_{e \in E} a_e \frac{(x_e - 1)x_e}{2}$$

is an exact potential for $C_i^\rho(s)$. The idea of the proof is to combine the Nash inequalities and the fact that the global minimum of $\Phi^\rho(\cdot)$ is a Nash equilibrium.

Let s denote the global minimum of Φ^ρ and s^* a socially optimal solution. We can without loss of generality assume that $a_e = 1$ and $b_e = 0$. The Nash inequalities (as in the price of anarchy analysis) yield

$$\sum_{e \in E} x_e(1 + \rho(x_e - 1)) \le \sum_{e \in E}(1 + \rho x_e)x_e^*.$$

The fact that s is a global optimum of $\Phi^\rho(\cdot)$ yields $\Phi^\rho(s) \le \Phi^\rho(s^*)$, which reduces to

$$\sum_{e \in E} \rho x_e^2 + (2 - \rho)x_e \le \sum_{e \in E} \rho(x_e^*)^2 + (2 - \rho)x_e^*.$$

If we can find $\gamma, \delta \ge 0$, and some $K \ge 1$, for which

$$(0 \le)\ \gamma \left[\rho(x_e^*)^2 + (2 - \rho)x_e^* - \rho x_e^2 - (2 - \rho)x_e \right] + \delta \left[(1 + \rho x_e)x_e^* - x_e(1 + \rho(x_e - 1)) \right]$$

$$\le K \cdot x_e^*[1 + \sigma(x_e^* - 1)] - x_e[1 + \sigma(x_e - 1)], \tag{4}$$

then this implies that $C^\sigma(s)/C^\sigma(s^*) \le K$. We take $\delta = (K - 1)/\rho$ and $\gamma = ((\rho - 1)K + 1)/(2\rho)$. It is not hard to see that $\delta \ge 0$ always holds, however, for γ we have to be more careful. We will later verify for which combinations of ρ and σ the parameter γ is indeed non-negative. Rewriting the expression in (4) yields that we have to find K satisfying $K \ge f_2(x_e, x_e^*, \sigma)/f_1(x_e, x_e^*, \rho, \sigma)$, where

$$f_2(x_e, x_e^*, \sigma) := (x_e^*)^2 - 2x_e x_e^* + (1 + 2\sigma)x_e^2 - x_e^* + (1 - 2\sigma)x_e$$

$$f_1(x_e, x_e^*, \rho, \sigma) := (1 - \rho + 2\sigma)(x_e^*)^2 - 2x_e x_e^* + (1 + \rho)x_e^2 + (\rho - 1 - 2\sigma)x_e^* - (\rho - 1)x_e.$$

Note that this reasoning is correct only if $f_1(x_e, x_e^*, \rho, \sigma) \ge 0$. This is true because

$$f_1(x_e, x_e^*, \rho, \sigma) = \left(x_e - x_e^* + \frac{1}{2} \right)^2 - \frac{1}{4} + (2\sigma - \rho)x_e^*(x_e^* - 1) + \rho x_e(x_e - 1)$$

is non-negative for all $x_e, x_e^* \in \mathbb{N}$, $\sigma \ge 0$ and $0 \le \rho \le 2\sigma$. Furthermore, the expression is zero if and only if $(x_e, x_e^*) \in \{(0, 1), (1, 1)\}$. But for these pairs the nominator is also zero, and hence the expression in (4) is therefore satisfied for those pairs. We can write

$$f_2(x_e, x_e^*, \sigma) = \left(x_e - x_e^* + \frac{1}{2} \right)^2 - \frac{1}{4} + 2\sigma x_e(x_e - 1)$$

and therefore $f_2/f_1 = \frac{A}{A+(2\sigma-\rho)B}$, where

$$A = \left(x_e - x_e^* + \frac{1}{2}\right)^2 - \frac{1}{4} + 2\sigma x_e(x_e - 1) \quad \text{and} \quad B = x_e^*(x_e^* - 1) - x_e(x_e - 1).$$

Note that if $\rho = 2\sigma$, we have $f_2/f_1 = 1$, and hence we can take $K = 1$. Otherwise,

$$\frac{A}{A + (2\sigma - \rho)B} \le \frac{\sqrt{\sigma(\sigma + 2)} + \sigma}{\sqrt{\sigma(\sigma + 2)} + \rho - \sigma} =: K \Leftrightarrow A + (\sqrt{\sigma(\sigma + 2)} + \sigma)B \ge 0.$$

The inequality on the right is true by Lemma 2.

To finish the proof, we determine the pairs (ρ, σ) for which the parameter γ is non-negative. This holds if and only if

$$(\rho - 1)K + 1 = (\rho - 1)\frac{\sqrt{\sigma(\sigma + 2)} + \sigma}{\sqrt{\sigma(\sigma + 2)} + \rho - \sigma} + 1 \ge 0.$$

Rewriting this yields the bound on ρ in the statement of the theorem.   □

The price of anarchy bound of $(1 + 2\rho(1 + \sigma))/(1 + \rho)$ is tight even for symmetric network congestion games with linear cost functions (see Theorem 2). In contrast, this is not true for the price of stability bound (for $\sigma = 1$):

Theorem 5. *Let Γ be a linear symmetric network congestion game, then*

$$PoS(\Gamma, \rho, 1) \le \begin{cases} 4/(\rho(4 - \rho)) & \text{if } 0 \le \rho \le 1 \\ 4/(2 + \rho) & \text{if } 1 \le \rho \le 2 \\ (2 + \rho)/4 & \text{if } 2 \le \rho < \infty. \end{cases}$$

In particular, if Γ is a symmetric congestion game on an extenstion-parallel[9] graph G, then the upper bounds even hold for the price of anarchy. All bounds are tight.

For $\rho \ge 1$, the bounds were previously shown by Caragiannis et al. [4] for the price of anarchy of singleton symmetric congestion games (which can be modeled on an extension-parallel graph).

Since any Nash equilibrium under the player cost $C_i^\rho(\cdot)$ is in particular a ρ-approximate Nash equilibrium, we also obtain the following result.

Corollary 1. *The price of stability for ρ-approximate equilibria, with $1 \le \rho \le 2$, is upper bounded by $4/(2 + \rho)$ for linear symmetric network congestion games.*

Acknowledgements. We thank the anonymous referees for their very useful comments.

---

[9] A graph G is *extension-parallel* if it consists of (i) a single edge, (ii) a single edge and an extension-parallel graph composed in series, or (iii) two extension-parallel graphs composed in parallel.

References

1. Anshelevich, E., Dasgupta, A., Kleinberg, J., Tardos, E., Wexler, T., Roughgarden, T.: The price of stability for network design with fair cost allocation. In: Proceedings of the 45th Annual IEEE Symposium on Foundations of Computer Science, FOCS 2004, pp. 295–304 (2004)
2. Caragiannis, I., Fanelli, A., Gravin, N., Skopalik, A.: Computing approximate pure Nash equilibria in congestion games. SIGecom Exch. **11**(1), 26–29 (2012)
3. Caragiannis, I., Kaklamanis, C., Kanellopoulos, P.: Taxes for linear atomic congestion games. ACM Trans. Algorithms **7**(1), 13:1–13:31 (2010)
4. Caragiannis, I., Kaklamanis, C., Kanellopoulos, P., Kyropoulou, M., Papaioannou, E.: The impact of altruism on the efficiency of atomic congestion games. In: Wirsing, M., Hofmann, M., Rauschmayer, A. (eds.) TGC 2010. LNCS, vol. 6084, pp. 172–188. Springer, Heidelberg (2010). doi:10.1007/978-3-642-15640-3_12
5. Chen, P.A., de Keijzer, B., Kempe, D., Schäfer, G.: Altruism and its impact on the price of anarchy. ACM Trans. Econ. Comput. **2**(4), 17:1–17:45 (2014)
6. Christodoulou, G., Koutsoupias, E.: The price of anarchy of finite congestion games. In: Proceedings of the Thirty-seventh Annual ACM Symposium on Theory of Computing, STOC 2005, pp. 67–73. ACM, New York (2005)
7. Christodoulou, G., Koutsoupias, E., Spirakis, P.G.: On the performance of approximate equilibria in congestion games. Algorithmica **61**(1), 116–140 (2011)
8. Correa, J., de Jong, J., de Keijzer, B., Uetz, M.: The curse of sequentiality in routing games. In: Markakis, E., Schäfer, G. (eds.) WINE 2015. LNCS, vol. 9470, pp. 258–271. Springer, Heidelberg (2015). doi:10.1007/978-3-662-48995-6_19
9. Kleer, P., Schäfer, G.: Tight inefficiency bounds for perception-parameterized affine congestion games. CoRR abs/1701.07614 (2017)
10. Koutsoupias, E., Papadimitriou, C.: Worst-case equilibria. In: Meinel, C., Tison, S. (eds.) STACS 1999. LNCS, vol. 1563, pp. 404–413. Springer, Heidelberg (1999). doi:10.1007/3-540-49116-3_38
11. Monderer, D., Shapley, L.S.: Potential games. Games Econ. Behav. **14**(1), 124–143 (1996)
12. Piliouras, G., Nikolova, E., Shamma, J.S.: Risk sensitivity of price of anarchy under uncertainty. In: Proceedings of the Fourteenth ACM Conference on Electronic Commerce, EC 2013, pp. 715–732. ACM, New York (2013)
13. Rosenthal, R.W.: A class of games possessing pure-strategy Nash equilibria. Int. J. Game Theory **2**, 65–67 (1973)

Perpetually Dominating Large Grids

Ioannis Lamprou$^{(\boxtimes)}$, Russell Martin, and Sven Schewe

Department of Computer Science, University of Liverpool, Liverpool, UK
{Ioannis.Lamprou,Russell.Martin,Sven.Schewe}@liverpool.ac.uk

Abstract. In the *Eternal Domination* game, a team of guard tokens initially occupies a dominating set on a graph G. A rioter then picks a node without a guard on it and attacks it. The guards defend against the attack: one of them has to move to the attacked node, while each remaining one can choose to move to one of his neighboring nodes. The new guards' placement must again be dominating. This attack-defend procedure continues perpetually. The guards win if they can eternally maintain a dominating set against any sequence of attacks, otherwise the rioter wins.

We study rectangular grids and provide the first known general upper bound for these graphs. Our novel strategy implements a square rotation principle and eternally dominates $m \times n$ grids by using approximately $\frac{mn}{5}$ guards, which is asymptotically optimal even for ordinary domination.

Keywords: Eternal domination · Combinatorial game · Two players · Graph protection · Grid

1 Introduction

Protection and security needs have always remained topical throughout human history. Nowadays, patrolling a network of premises, forcefully defending against attacks and ensuring a continuum of safety are top-level affairs in any military strategy or homeland security agenda.

Going back in time, the *Roman Domination* problem was introduced in [22]: where should Emperor Constantine the Great have located his legions in order to optimally defend against attacks in unsecured locations without leaving another location unsecured? In computer science terms, the interest is in producing a placement of guards on a graph such that any node without a guard has at least one neighbor with two guards on it. In other words, we are looking for a *dominating set* of the graph (i.e. each node must have a guard on it or on at least one of its neighbors), but with some extra qualities. Some seminal work on this topic includes [15,21].

The above modeling caters only for a single attack on an unsecured node. A natural question is to consider special domination strategies against a sequence of attacks on the same graph [5]. In this setting, (some of) the guards are allowed

---

Work partially supported by EPSRC grant EP/M027287/1 (Energy Efficient Control).

D. Fotakis et al. (Eds.): CIAC 2017, LNCS 10236, pp. 393–404, 2017.
DOI: 10.1007/978-3-319-57586-5_33

to move after each attack to defend against it and modify their overall placement. The difficulty here lies in establishing a robust guards' placement in order to retain domination after coping with each attack. Such a sequence of attacks can be of finite (i.e. a set of k consecutive attacks) or even *infinite* length.

In this paper, we focus on the latter. We wish to protect a graph against attacks happening indefinitely on its nodes. Initially, the guards are placed on some nodes of the graph such that they form a dominating set (a simple one; not a Roman one). Then, an attack occurs on an unguarded node. All the guards (may) now move in order to counter the attack: one of them moves to the attacked node, while each of the others moves to one of his neighboring nodes such that the new guards' placement forms again a dominating set. This scenario takes place ad infinitum. The attacker's objective is to devise a sequence of attacks which leads the guards to a non-dominating placement. On the other hand, the guards wish to maintain a sequence of dominating sets without any interruption. The *Eternal Domination* problem, studied in this paper, deals with determining the minimum number of guards such that they perpetually protect the graph in the above fashion. The focus is on rectangular grids, where we provide a first, up to our knowledge, upper bound.

Related Work. Infinite order domination was originally considered by Burger et al. [4] as an extension to finite order domination. Later on, Goddard et al. [12] proved some first bounds with respect to some other graph-theoretic notions (like independence and clique cover) for the one-guard-moves and all-guards-move cases. The relationship between eternal domination and clique cover is examined more carefully in [1]. There exists a series of other papers with several combinatorial bounds, e.g. see [13,16,17,20].

Regarding the special case of grid graphs, Chang [6] gave many strong upper and lower bounds for the domination number. Indeeed, Gonçalves et al. [14] proved Chang's construction optimal for rectangular grids where both dimensions are greater or equal to 16. Moving onward to eternal domination, bounds for $3 \times n$ [8], $4 \times n$ [2] and $5 \times n$ [23] grids have been examined, where for $3 \times n$ the bounds are almost tight and for $4 \times n$ exactly tight.

Due to the mobility of the guards in eternal domination and the breakdown into alternate turns (guards vs attacker), one can view this problem as a pursuit-evasion combinatorial game in the same context as *Cops & Robber* [3] and the *Surveillance Game* [10,11]. In all three of them, there are two players who take turns alternately with one of them pursuing the other possibly indefinitely.

Besides, an analogous *Eternal Vertex Cover* problem has been considered [9,18], where attacks occur on the edges of the graph. In that setting, the guards defend against an attack by traversing the attacked edge, while they move in order to preserve a vertex cover after each turn.

For an overall picture and further references on the topic, the reader is suggested to tend to a recent survey on graph protection [19].

Our Result. We make a first step towards answering an open question in [19] and show that, in order to ensure eternal domination in rectangular grids, only a linear number of extra guards is needed compared to domination.

To obtain this result, we devise an elegantly unraveling strategy of successive (counter) clockwise rotations for the guards to perpetually dominate an infinite grid. This strategy is referred to as the *Rotate-Square* strategy. Then, we apply the same strategy to finite grids with some extra guards to ensure the boundary remains always guarded.

Overall, we show $\lceil \frac{mn}{5} \rceil + \mathcal{O}(m+n)$ guards suffice to perpetually dominate a big enough $m \times n$ grid.

Outline. In Sect. 2, we define some basic graph-theoretic notions and *Eternal Domination* as a two-player combinatorial pursuit-evasion game. Forward, in Sect. 3, we describe the basic components of the Rotate-Square strategy and prove it can be used to dominate an infinite grid forever. Later, in Sect. 4 we show how the strategy can be adjusted to perpetually dominate finite grids by efficiently handling moving near the boundary and the corners. Finally, in Sect. 5, we shortly mention some concluding remarks and open questions. Due to space requirements, some proofs are omitted from this version.

2   Preliminaries

Let $G = (V(G), E(G))$ be a simple connected graph. We denote an edge between two connected vertices, namely v and u, as $(u, v) \in E(G)$ (or equivalently (v, u)). The *open-neighborhood* of a subset of vertices $S \subseteq V(G)$ is defined as $N(S) = \{v \in V(G) \setminus S : \exists u \in S$ such that $(u, v) \in E(G)\}$ and the *closed-neighborhood* as $N[S] = S \cup N(S)$. A *path* of length $n - 1 \in \mathbb{N}$, namely P_n, is a graph where $V(P_n) = \{v_0, v_1, \ldots, v_{n-1}\}$ and $E(P_n) = \{(v_0, v_1), (v_1, v_2), \ldots (v_{n-2}, v_{n-1})\}$. The *Cartesian product* of two graphs G and H is another graph denoted $G \square H$ where $V(G \square H) = V(G) \times V(H)$ and two vertices (v, v') and (u, u') are adjacent if either $v = u$ and $(v', u') \in E(H)$ or $v' = u'$ and $(v, u) \in E(G)$. A *grid*, namely $P_m \square P_n$, is the Cartesian product of two paths of lengths $m, n \in \mathbb{N}$.

A set of vertices $S \subseteq V(G)$ is called a *dominating set* of G if $N[S] = V(G)$. That is, for each $v \in V(G)$ either $v \in S$ or there exists a node $u \in S$ ($u \neq v$) such that $(u, v) \in E(G)$. A minimum-size such set, say S^*, is called a *minimum dominating set* of G and $\gamma(G) = |S^*|$ is defined as the *domination number* of G. For grids, we simplify the notation $\gamma(P_m \square P_n)$ to $\gamma_{m,n}$.

Eternal Domination can be regarded as a combinatorial pursuit-evasion game played on a graph G. There exist two players: one of them controls the *guards*, while the other controls the *rioter* (or *attacker*). The game takes place in *rounds*. Each round consists of two *turns*: one for the guards and one for the rioter.

Initially (round 0), the guard tokens are placed such that they form a dominating set on G. Then, without loss of generality, the rioter attacks a node without a guard on it. A guard, dominating the attacked node, must now move on it to counter the attack. Notice that at least one such guard exists because their initial placement is dominating. Moreover, the rest of the guards may move; a guard on node v can move to any node in $N[\{v\}]$. The guards wish to ensure that their modified placement is still a dominating set for G. The game proceeds

in a similar fashion in any subsequent rounds. Guards win if they can counter
any attack of the rioter and perpetually maintain a dominating set; that is, for
an infinite number of attacks. Otherwise, the rioter wins if she manages to force
the guards to reach a placement that is no longer dominating; then, an attack
on an undominated node suffices to win.

Definition 1. $\gamma^\infty(G)$ *stands for the* eternal domination number *of a graph* G,
i.e. the minimum size of a guards' team that can eternally dominate G *(when
all guards can move at each turn).*

As above, we simplify $\gamma^\infty(P_m \square P_n)$ to $\gamma_{m,n}^\infty$. Since the initial guards' placement is dominating, we get $\gamma^\infty(G) \geq \gamma(G)$ for any graph G. By a simple rotation,
we get $\gamma_{m,n} = \gamma_{n,m}$ and $\gamma_{m,n}^\infty = \gamma_{n,m}^\infty$. Finally, multiple guards are *not* allowed
to lie on a single node, since this could provide an advantage for the guards [7].

3   Eternally Dominating an Infinite Grid

In this section, we describe a strategy to eternally dominate an *infinite grid*. We
denote an infinite grid as G_∞ and define it as a pair $(V(G_\infty), E(G_\infty))$, where
$V(G_\infty) = \{(x,y) : x, y \in \mathbb{Z}\}$ and any node $(x,y) \in V(G_\infty)$ is connected to
$(x, y-1)$, $(x, y+1)$, $(x-1, y)$ and $(x+1, y)$. In Fig. 1 (and those to follow),
we depict the grid as a square mesh where each *cell* corresponds to a node of
$V(G_\infty)$ and neighbors only four other cells: the one above, below, left and right
of it. We assume row x is *above* row $x+1$ and column y is *left* of column $y+1$.

Initially, let us consider a family of dominating sets for
G_∞. In the following, let $\mathbb{Z}^2 := \mathbb{Z} \times \mathbb{Z}$ and $\mathbb{Z}_5 := \{0, 1, 2, 3, 4\}$
stand for the group of integers modulo 5. We then define the
function $f : \mathbb{Z}^2 \to \mathbb{Z}_5$ as $f(x,y) = x + 2y \pmod 5$ for any
$(x,y) \in \mathbb{Z}^2$. This function appears in [6] and is central to
providing an optimal dominating set for sufficiently large
finite grids. Now, let $D_t = \{(x,y) \in V(G_\infty) : f(x,y) = t\}$ for
$t \in \mathbb{Z}_5$ and $\mathcal{D}(G_\infty) = \{D_t : t \in \mathbb{Z}_5\}$. For purposes of symmetry, let us define $f'(x,y) = f(y,x)$ and then $D'_t = \{(x,y) \in$
$V(G_\infty) : f'(x,y) = t\}$ and $\mathcal{D}'(G_\infty) = \{D'_t : t \in \mathbb{Z}_5\}$.

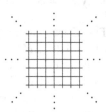

Fig. 1. The infinite
grid G_∞

Proposition 1. *Any* $D \in \mathcal{D}(G_\infty) \cup \mathcal{D}'(G_\infty)$ *is a dominating set for* G_∞.

Notice that the above constructions form *perfect* dominating sets, i.e. dominating sets where each node is dominated by *exactly* one other node, since for
each node $v \in V(G_\infty)$ exactly one node from $N[\{v\}]$ lies in D_t (respectively D'_t)
by the definition of D_t (respectively D'_t).

3.1   A First Eternal Domination Strategy

Let us now consider a *shifting-style* strategy as the simplest and most straightforward strategy to eternally dominate G_∞. The guards initially pick a placement

D_t for some $t \in \mathbb{Z}_5$. Next, an attack occurs on some unguarded node. Since the D_t placement perfectly dominates G_∞, there exists exactly one guard adjacent to the attacked node. Thence, it is mandatory for him to move onto the attacked node. His move defines a direction in the grid: left, right, up or down. The rest of the strategy reduces to each guard moving according to the defined direction.

Altogether, the guards are all shifting toward the same direction. Therefore, they abandon their original D_t placement, but end up in a $D_{t'}$ placement where t' depends on t and the direction coerced by the attack. The last holds since moving toward the same direction has the same effect to the outcome of the $f(\cdot)$ function found in the definition of D_t. It is easy to see that the above strategy can be repeated after any attack of the rioter. Thus, the guards always occupy a placement in $\mathcal{D}(G_\infty)$ and, by Proposition 1, they dominate G_∞ perpetually.

The aforementioned strategy works fine for the infinite grid, as demonstrated above. Nonetheless, applying it (directly or modified) to a finite grid encounters many obstacles. Shifting the guards toward one course leaves some nodes in the very end of the opposite course (near the boundary) undominated, since there is no longer an unlimited supply of guards to ensure protection. To overcome this problem, we propose a different strategy whose main aim is to redistribute the guards without creating any bias to a specific direction.

3.2   Empty Squares

The key idea toward another eternal domination strategy is to *rotate* the guards' placement around some *squares* (i.e. subgrids of size 2×2) such that, intuitively, the overall movement is zero and the guards always occupy a placement in $\mathcal{D}(G_\infty) \cup \mathcal{D}'(G_\infty)$ after an attack is defended.

Consider a node $(x, y) \in V(G_\infty)$, where $(x, y) \in D_t$ for some value t. Now, assume that the guards lie on the nodes dictated in D_t and thence form a dominating set. By looking around (x, y), we identify the existence of 4 *empty squares* (i.e. sets of 4 cells with no guard on them):

- $SQ_0 = \{(x-1, y+1), (x-1, y+2), (x, y+1), (x, y+2)\}$
- $SQ_1 = \{(x+1, y), (x+1, y+1), (x+2, y), (x+2, y+1)\}$
- $SQ_2 = \{(x, y-2), (x, y-1), (x+1, y-2), (x+1, y-1)\}$
- $SQ_3 = \{(x-2, y-1), (x-2, y), (x-1, y-1), (x-1, y)\}$

One can verify that, for every $(w, z) \in \bigcup_{i=0}^{3} SQ_i$, we get $f(w, z) \neq f(x, y)$ and thus $(w, z) \notin D_t$. Figure 2a demonstrates the above observation. Notice that (x, y) has exactly one neighbor in each of these squares and is the only guard who dominates these 4 neighbors, since the domination is perfect. Furthermore, an attack on the neighbor lying in SQ_i would mean the guard moves there and *slides* along an edge of $SQ_{(i+1) \bmod 4}$, i.e. both its current and previous position is neighboring to a node in $SQ_{(i+1) \bmod 4}$. For example, in Fig. 2a, an attack on the bottom-right cell of SQ_3 would mean the guard slides along SQ_0. Finally, each square is protected by exactly 4 guards around it (one for each of its vertices) in a formation as seen in Fig. 2a.

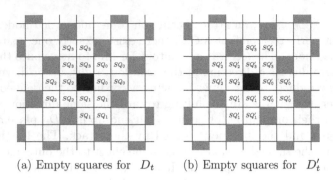

(a) Empty squares for D_t (b) Empty squares for D_t'

Fig. 2. Empty squares

The aforementioned observations also extend to a node (x, y) lying on a dominating set D_t'. We now define the 4 empty squares as follows (see Fig. 2b):

- $SQ_0' = \{(x, y + 1), (x, y + 2), (x + 1, y + 1), (x + 1, y + 2)\}$
- $SQ_1' = \{(x + 1, y - 1), (x + 1, y), (x + 2, y - 1), (x + 2, y)\}$
- $SQ_2' = \{(x - 1, y - 2), (x, y - 1), (x, y - 2), (x, y - 1)\}$
- $SQ_3' = \{(x - 2, y), (x - 2, y + 1), (x - 1, y), (x - 1, y + 1)\}$

Similarly to before, the squares are empty, since for every $(w, z) \in \bigcup_{i=0}^{3} SQ_i'$ we get $f'(w, z) \neq f'(x, y)$ and thus $(w, z) \notin D_t'$. The (x, y)-guard has exactly one neighbor in each of these squares and protecting an attack on SQ_i' now means sliding along the edge of $SQ_{(i-1) \bmod 4}'$. Finally, each square is protected by exactly 4 guards in a formation that looks like a clockwise step of the formation seen before for D_t.

3.3   The Rotate-Square Strategy

We hereby describe the *Rotate-Square* strategy and prove that it perpetually dominates G_∞. The strategy makes use of the empty squares idea and, once an attack occurs, the square along which the defence-responsible guard slides is identified as the *pattern square*. Then, the other 3 guards corresponding to the pattern square perform a (counter) clockwise step depending on the move of the defence-responsible guard. Let us break the guards' turn down into some distinct components to facilitate a formal explanation. Of course, the guards are always assumed to move concurrently during their turn. That is, they *centrally* compute the whole strategy move and then each one moves to the position dictated by the strategy at the same time.

Initially, the guards are assumed to occupy a dominating set D in $\mathcal{D}(G_\infty) \cup \mathcal{D}'(G_\infty)$. Then, an attack occurs on a node in $V(G_\infty) \setminus D$. To defend against it, the guards apply *Rotate-Square*:

(1) Identify the defence-responsible guard; there is exactly one since the domination is perfect.

(2) Identify the pattern square SQ_j from the 4 empty squares around this guard.
(3) Rotate around SQ_j according to the defence-responsible guard's move.
(4) Repeat the rotation pattern in horizontal and vertical lanes in hops of distance 5.

Let us examine each of these strategy components more carefully. Step (1) requires looking at the grid and spotting the guard who lies on a neighboring node of the attack. In step (2), the pattern square is identified as described in the previous subsection following the $(i \pm 1) \mod 4$ rule depending on the current dominating set (Figs. 2a and b). In step (3), the 4 guards around the pattern square (including the defence-responsible guard) take a (counter) clockwise step based on the node to be defended. For an example, see Fig. 3a: the defence-responsible guard (in black) defends against an attack on the bottom-right cell of SQ_3 by sliding along SQ_0 in clockwise fashion. Then, the other 3 guards around SQ_0 (in gray) take a clockwise step sliding along an SQ_0-edge as well. The latter happens in order to preserve that SQ_0 remains empty. Eventually, in step (4), the pattern square (SQ_0) is used as a guide for the move of the rest of the guards. Consider an SQ_0-guard initially lying on node (w, z). By construction of D_t, guards lie on all nodes $(w \pm 5\alpha, z \pm 5\beta)$ for $\alpha, \beta \in \mathbb{N}$, since adding multiples of 5 in both dimensions does not affect the outcome of $f(\cdot)$. In the end, all these corresponding guards mimic the move of (w, z), i.e. they move toward the same direction. This procedure is executed for all the guards of SQ_0. The rest of the guards, i.e. guards that do not correspond to any SQ_0-guard, remain still during this turn. We vizualise such an example in Fig. 3b. The circles enclose the repetitions of the pattern square, where the original pattern square is given in black. The dotted nodes remain still during this turn.

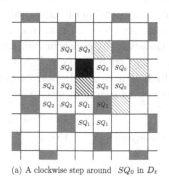

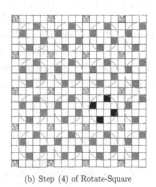

(a) A clockwise step around SQ_0 in D_t

(b) Step (4) of Rotate-Square

Fig. 3. Step (3) and (4) of rotate-square

Lemma 1. *Assume the guards occupy a dominating placement $D \subseteq V(G_\infty)$ in $\mathcal{D}(G_\infty) \cup \mathcal{D}'(G_\infty)$ and an attack occurs on a node in $V(G_\infty) \setminus D$. After applying the Rotate-Square strategy, the guards successfully defend against the attack and again form a dominating set in $\mathcal{D}(G_\infty) \cup \mathcal{D}'(G_\infty)$.*

Proof. In this proof, we are going to demonstrate that any of the 4 possible attacks (one per empty square) around a node in a D_t (or D'_t) placement can be defended by Rotate-Square and, most importantly, the guards still occupy a placement in $\mathcal{D}(G_\infty) \cup \mathcal{D}'(G_\infty)$ after their turn. Below, in Fig. 4, we provide pictorial details for 1 out of 8 cases (4 for D_t and 4 for D'_t); we need not care about the value of t, since all D_t (respectively D'_t) placements are mere shifts to each other. The defence-responsible guard is given in black, while the rest in gray. Their previous positions are observable by a slight shade. The guards with no shade around them are exactly the ones who do not move during their turn. Also, notice that the guards who are mimicking the strategy of the pattern square occupy positions $(w \pm 5\alpha, z \pm 5\beta)$ for $\alpha, \beta \in \mathbb{N}$, where (w, z) is the new position of a pattern square guard. Then, $f(w, z) = f(w \pm 5\alpha, z \pm 5\beta)$ and $f'(w, z) = f'(w \pm 5\beta, z \pm 5\alpha)$ since the modulo 5 operation cancels out the addition (subtraction) of 5α and 5β. A similar observation holds for the set of guards that stand still during their turn. We identify a model guard, say on position (a, b), and then the rest of such guards are given by $(a \pm 5\alpha, b \pm 5\beta)$. Again, the $f(\cdot)$ (respectively $f'(\cdot)$) values of all these nodes remain equal. For this reason, we focus below only on the pattern square and the model guards and demonstrate that they share the same value of $f(\cdot)$ (respectively $f'(\cdot)$).

We hereby consider a potential attack around a node $(x, y) \in D_t$.

Attack on $(x - 1, y)$ (i.e. on SQ_3). We apply Rotate-Square around SQ_0. The four guards around SQ_0 and the model guard standing still move as follows (Fig. 4): Let P stand for the set of new positions given in Table 1. The guards now occupy positions $(w, z) \in P$ where $f'(w, z) = 2x + y - 2 \pmod 5 = 2x + y + 3 \pmod 5 = t'$. By this fact, we get $P \subseteq D'_{t'}$. Now, assume there exists a node $(w, z) \notin P$, but $(w, z) \in D'_{t'}$. Without loss of generality, we assume $w \in [x-3, x+1]$ and $z \in [y-1, y+3]$, since the configuration of the guards in this window is copied all over the grid by the symmetry of D_t or D'_t placements. Since $(w, z) \notin P$, this is a node with no guard on it. However, by construction, any such node is dominated by a *neighboring* node (w_1, z_1) with $f'(w_1, z_1) = t'$. Then, by assumption, $f'(w, z) = f'(w_1, z_1) = t'$, which is a contradiction because, by definition of $f'(\cdot)$, two neighboring nodes never have equal values.

Table 1. Attack on $(x - 1, y)$ (rotate around SQ_0); Fig. 4

Old position (w, z)	New position (w', z')	$f'(w', z')$ (mod 5)
(x, y)	$(x - 1, y)$	$2x + y - 2$
$(x - 2, y + 1)$	$(x - 2, y + 2)$	$2x + y - 2$
$(x - 1, y + 3)$	$(x, y + 3)$	$2x + y + 3$
$(x + 1, y + 2)$	$(x + 1, y + 1)$	$2x + y + 3$
$(x - 3, y - 1)$	$(x - 3, y - 1)$	$2x + y - 2$

All other cases can be proved in a similar fashion. Notice that an attack against a D_t placement leads to a $D'_{t'}$ placement for some t' and vice versa.   □

Theorem 1. *The guards perpetually dominate G_∞ by following the Rotate-Square strategy starting from an initial dominating set in $\mathcal{D}(G_\infty) \cup \mathcal{D}'(G_\infty)$.*

4   Eternally Dominating Finite Grids

We now apply the Rotate-Square strategy to finite grids, i.e. graphs of the form $P_m \square P_n$. The idea is to follow the rules of the strategy, but to never leave any boundary or corner node without a guard on it. A finite $m \times n$ grid consists of nodes (i, j) where $i \in \{0, 1, 2, \ldots, m-1\}$ and $j \in \{0, 1, 2, \ldots, n-1\}$. Nodes $(0, x), (m-1, x), (y, 0), (y, n-1)$ for $x \in \{1, 2, \ldots, n-2\}$ and $y \in \{1, 2, \ldots, m-2\}$ are called *boundary nodes*, while nodes $(0, 0), (0, n-1), (m-1, 0), (m-1, n-1)$ are called *corner nodes*. Connectivity is similar to the infinite grid. However, boundary nodes only have three neighbors, while corner nodes only have two.

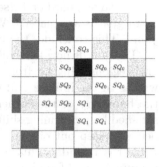

Fig. 4. Attack on SQ_3

Let us consider $V(t) = D_t \cap (P_m \square P_n)$ and $V'(t) = D'_t \cap (P_m \square P_n)$, respectively. We cite the following counting lemma from [6].

Lemma 2 (Lemma 2.2 [6]). $\lfloor \frac{mn}{5} \rfloor \leq |V(t)| \leq \lceil \frac{mn}{5} \rceil$ *holds for all t, and there exist t_0, t_1, such that $|V(t_0)| = \lfloor \frac{mn}{5} \rfloor$ and $|V(t_1)| = \lceil \frac{mn}{5} \rceil$ hold.*

The main observation in the proof of the above lemma is that there exist either $\lfloor \frac{m}{5} \rfloor$ or $\lfloor \frac{m}{5} \rfloor + 1$ D_t-nodes in one column of a $P_m \square P_n$ grid. Then, a case-analysis counting provides the above bounds. The same observation holds for D'_t, since f' is defined based on the same function $f : \mathbb{Z}^2 \to \mathbb{Z}_5$. Thence, we can extend the above lemma for D'_t cases with the proof being identical.

Lemma 3. $\lfloor \frac{mn}{5} \rfloor \leq |V'(t)| \leq \lceil \frac{mn}{5} \rceil$ *holds for all t, and there exist t_0, t_1, such that $|V'(t_0)| = \lfloor \frac{mn}{5} \rfloor$ and $|V'(t_1)| = \lceil \frac{mn}{5} \rceil$ hold.*

In order to study the domination of $P_m \square P_n$, the analysis is based on examining $V(t)$, but for an extended $P_{m+2} \square P_{n+2}$ mesh. Indeed, Chang [6] showed:

Lemma 4 (Theorem 2.2 [6]). *For any $m, n \geq 8$, $\gamma_{m,n} \leq \lfloor \frac{(m+2)(n+2)}{5} \rfloor - 4$.*

The result follows by picking an appropriate D_t placement and forcing into the boundary of $P_m \square P_n$ the guards on the boundary of $P_{m+2} \square P_{n+2}$. Moreover, Chang showed how to eliminate another 4 guards; one near each corner.

Below, to facilitate the readability of our analysis, we focus on a specific subcase of finite grids. We demonstrate an eternal dominating strategy for $m \times n$ finite grids where $m \bmod 5 = n \bmod 5 = 2$. Later, we extend to the general case.

The Strategy. Initially, we place our guards on nodes belonging to $V(t) = D_t \cap (P_m \square P_n)$ for some value of t. Unlike the approach in [6], we do not force inside any guards lying outside the boundary of $P_m \square P_n$. Since a sequence of attacks may force the guards to any $V(t)$ or $V'(t)$ placement (i.e. for any value of t), we pick an initial placement (say $V(t_1)$) for which $|V(t_1)| = \lceil \frac{mn}{5} \rceil$ to make sure there are enough guards to maintain domination while transitioning from one placement to the other. By Lemma 2, there exists such a placement. Moreover, we cover the whole boundary by placing a guard on each boundary or corner node with no guard on it (see Fig. 5a; the gray nodes denote the places where the extra guards are placed). We refer to any of these added guards as a *boundary guard*. This concludes the initial placement of the guards.

The guards now follow Rotate-Square limited within the grid boundaries. For grid regions lying far from the boundary, Rotate-Square is applied in the same way as in the infinite grid case. For pattern square repetitions happening near the boundary or the corners, Rotate-Square's new placement demands can be satisfied by performing shifts of boundary guards. In other words, when a guard needs to step out of the boundary, another guard steps inside to replace him, while the boundary guards between them shift one step on the boundary. An example can be found in Fig. 5b depicting a step of our strategy (from the black to the dark gray placement). Let us examine the designated window at the top of the boundary. Non-boundary guards move from the black to the dark gray positions, while boundary guards (in light gray) take a step rightward to make room for the dark gray guard moving in at the left and cover the black guard leaving the boundary at the right. Finally, black to dark gray transitions, where both nodes are on the boundary, mean the corresponding guards there simply do not move; there is no need to swap them. Overall, we refer to this slightly modified version of Rotate-Square as *Finite Rotate-Square*.

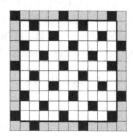
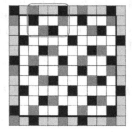

(a) An initial placement for the guards   (b) Boundary guards' shifting

Fig. 5. Finite rotate-square

Lemma 5. *Assume $m \bmod 5 = n \bmod 5 = 2$ and that the guards follow Finite Rotate-Square, for an Eternal Domination game in $P_m \square P_n$. Then, after every turn, their new placement P is dominating, all boundary and corner nodes have a guard on them and, for some t, there exists a set $V(t)$ (or $V'(t)$) such that $V(t) \subseteq P$ (or $V'(t) \subseteq P$).*

Proof. Consider the $(m-2) \times (n-2)$ subgrid remaining if we remove the boundary. Since $m \bmod 5 = n \bmod 5 = 2$, $(m-2)$ and $(n-2)$ perfectly divide 5. The latter means that each row (respectively column) of the subgrid has exactly $\frac{n-2}{5}$ (respectively $\frac{m-2}{5}$) guards on it. Now, without loss of generality, consider row one neighboring the upper boundary row, which is row zero. Let us assume that a pattern square propagation obligates a row-one guard to move to the boundary. Then, by symmetry of the pattern square, there exists another guard on the boundary who needs to move downward to row one. Notice that the same holds for each of the $\frac{n-2}{5}$ guards lying on row one, since the pattern square move propagates in hops of distance 5. Movements in and out of the boundary alternate due to the shape of the pattern square. Moreover, we need not care about where the pattern square is "cut" by the left/right boundary since, due to $n-2$ perfectly dividing 5, there are exactly $\frac{n-2}{5}$ full pattern squares occuring subject to shifting. Thence, we can apply the shifting procedure demonstrated in Fig. 5b to apply the moves and maintain a full boundary, while preserving the number of guards on row one. Notice that it suffices to look at 12×12 grids since for larger $m \times n$ grids with this property the patterns evolve similarly and so we can omit grid regions in the middle.

The new placement P is dominating, since the $(m-2) \times (n-2)$ subgrid is dominated by any $V(t)$ or $V'(t)$ placement and the boundary is always full of guards. Moreover, since we follow a modified Rotate-Square, P contains as a subset a node set $V(t)$ or $V'(t)$ after each guards' turn. $\qquad \square$

Lemma 6. *For $m, n \geq 7$ such that $m \bmod 5 = n \bmod 5 = 2$, $\gamma_{m,n}^{\infty} \leq \frac{mn}{5} + \frac{8}{5}(m+n) - \frac{24}{5}$ holds.*

So far, we focused on the special case where $m \bmod 5 = n \bmod 5 = 2$ and provided an upper bound for the eternal domination number. It is easy to generalize this bound for arbitrary values of m and n.

Lemma 7. *For $m, n \geq 7$, $\gamma_{m,n}^{\infty} \leq \frac{mn}{5} + \mathcal{O}(m+n)$ holds.*

Gonçalves et al. [14] showed $\gamma_{m,n} \geq \lfloor \frac{(m+2)(n+2)}{5} \rfloor - 4$ for any $m, n \geq 16$. By combining this with Lemma 4, we get the exact domination number $\gamma_{m,n} = \lfloor \frac{(m+2)(n+2)}{5} \rfloor - 4$ for $m, n \geq 16$. Then, by using Lemma 7, our main result follows.

Theorem 2. *For any $m, n \geq 16$, $\gamma_{m,n}^{\infty} \leq \gamma_{m,n} + \mathcal{O}(m+n)$ holds.*

5 Conclusions

We demonstrated a first strategy to eternally dominate general rectangular grids based on the repetition of a rotation pattern. Regarding further work, a more careful case-analysis of the boundary may lead to improvements regarding the coefficient of the linear term. On the bigger picture, it remains open whether this strategy can be used to obtain a constant additive gap between domination and eternal domination in large grids. Furthermore, the existence of a stronger lower bound than the trivial $\gamma_{m,n}^{\infty} \geq \gamma_{m,n}$ one also remains open.

References

1. Anderson, M., et al.: Maximum demand graphs for eternal security. J. Comb. Math. Comb. Comput. **61**, 111–128 (2007)
2. Beaton, I., Finbow, S., MacDonald, J.A.: Eternal domination numbers of $4 \times n$ grid graphs. J. Combin. Math. Combin. Comput. **85**, 33–48 (2013)
3. Bonato, A., Nowakowski, R.J.: The Game of Cops and Robbers on Graphs. American Mathematical Society, Providence (2011)
4. Burger, A.P., et al.: Infinite order domination in graphs. J. Comb. Math. Comb. Comput. **50**, 179–194 (2004)
5. Burger, A.P., Cockayne, E.J., Gründlingh, W.R., Mynhardt, C.M., van Vuuren, J.H., Winterbach, W.: Finite order domination in graphs. J. Combin. Math. Combin. Comput. **49**, 159–175 (2004)
6. Chang, T.Y.: Domination numbers of grid graphs. Ph.D. thesis, Department of Mathematics, University of South Florida (1992)
7. Finbow, S., Gaspers, S., Messinger, M.E., Ottaway, P.: A note on the eternal dominating set problem (2016, submitted)
8. Finbow, S., Messinger, M.-E., Bommel, M.: Eternal domination in $3 \times n$ grids. Australas. J. Combin. **61**, 156–174 (2015)
9. Fomin, F., et al.: Parameterized algorithm for eternal vertex cover. Inf. Process. Lett. **110**, 702–706 (2010)
10. Fomin, F.V., Giroire, F., Jean-Marie, A., Mazauric, D., Nisse, N.: To satisfy impatient web surfers is hard. Theoret. Comput. Sci. **526**, 1–17 (2014)
11. Giroire, F., Lamprou, I., Mazauric, D., Nisse, N., Pérennes, S., Soares, R.: Connected surveillance game. Theoret. Comput. Sci. **584**, 131–143 (2015)
12. Goddard, W., Hedetniemi, S.M., Hedetniemi, S.T.: Eternal security in graphs. J. Comb. Math. Comb. Comput. **52**, 169–180 (2005)
13. Goldwasser, J., Klostermeyer, W.F.: Tight bounds for eternal dominating sets in graphs. Discret. Math. **308**, 2589–2593 (2008)
14. Gonçalves, D., et al.: The domination number of grids. SIAM J. Discret. Math. **25**(3), 1443–1453 (2011)
15. Henning, M.A., Hedetniemi, S.T.: Defending the Roman Empire: a new strategy. Discret. Math. **266**, 239–251 (2003)
16. Henning, M.A., Klostermeyer, W.F.: Trees with large m-eternal domination number. Discret. Appl. Math. **211**, 79–85 (2016)
17. Henning, M.A., Klostermeyer, W.F., MacGillivray, G.: Bounds for the m-eternal domination number of a graph, Manuscript (2015)
18. Klostermeyer, W.F., Mynhardt, C.M.: Edge protection in graphs. Australas. J. Comb. **45**, 235–250 (2009)
19. Klostermeyer, W.F., Mynhardt, C.M.: Protecting a graph with mobile guards. Appl. Anal. Discret. Math. **10**, 1–29 (2016)
20. Klostermeyer, W.F., Mynhardt, C.M.: Vertex covers and eternal dominating sets. Discret. Appl. Math. **160**, 1183–1190 (2012)
21. ReVelle, C.S., Rosing, K.E.: Defendens imperium Romanum: a classical problem in military strategy. Am. Math. Mon. **107**, 585–594 (2000)
22. Stewart, I.: Defend the Roman Empire. Scientific American, New York City (1999). pp. 136–138
23. van Bommel, C.M., van Bommel, M.F.: Eternal domination numbers of $5 \times n$ grid graphs. J. Combin. Math. Combin. Comput., to appear

Rooted Uniform Monotone Minimum Spanning Trees

Konstantinos Mastakas[(✉)] and Antonios Symvonis

School of Applied Mathematical and Physical Sciences,
National Technical University of Athens, Athens, Greece
{kmast,symvonis}@math.ntua.gr

Abstract. We study the construction of the minimum cost spanning geometric graph of a given rooted point set P where each point of P is connected to the root by a path that satisfies a given property. We focus on two properties, namely the monotonicity w.r.t. a single direction (y-monotonicity) and the monotonicity w.r.t. a single pair of orthogonal directions (xy-monotonicity). We propose algorithms that compute the rooted y-monotone (xy-monotone) minimum spanning tree of P in $O(|P| \log^2 |P|)$ (resp. $O(|P| \log^3 |P|)$) time when the direction (resp. pair of orthogonal directions) of monotonicity is given, and in $O(|P|^2 \log |P|)$ time when the optimum direction (resp. pair of orthogonal directions) has to be determined. We also give simple algorithms which, given a rooted connected geometric graph, decide if the root is connected to every other vertex by paths that are all monotone w.r.t. the same direction (pair of orthogonal directions).

1 Introduction

A geometric path $W = (w_0, w_1, \ldots, w_l)$ is *monotone in the direction of* y, also called *y-monotone*, if it is *y-decreasing*, i.e. $y(w_0) \geq y(w_1) \geq \ldots \geq y(w_l)$ or if it is *y-increasing*, i.e. $y(w_0) \leq y(w_1) \leq \ldots \leq y(w_l)$, where $y(p)$ denotes the y coordinate of a point p. W is *monotone* if there exists an axis y' s.t. W is y'-monotone. Arkin et al. [5] proposed a polynomial time algorithm which connects two given points by a geometric path that is monotone in a given (an arbitrary) direction and does not cross a set of obstacles, if such a path exists. Furthermore, the problem of drawing a directed graph as an *upward graph*, i.e. a directed geometric graph such that each directed path is y–increasing, has been studied in the field of graph drawing, e.g. see [9,11].

A geometric graph $G = (P, E)$ is *monotone* if every pair of points of P is connected by a monotone geometric path, where the direction of monotonicity does not need to be the same for each pair. If there exists a single direction of monotonicity, we denote the graph as *uniform monotone*. Uniform monotone graphs were also denoted as 1–*monotone graphs* by Angelini [2]. When the direction of monotonicity is known, say y, the graph is called *y–monotone*. Monotone graphs were introduced by Angelini et al. [3]. The problem of drawing a graph as a monotone graph has been studied in the field of graph drawing; e.g. see

© Springer International Publishing AG 2017
D. Fotakis et al. (Eds.): CIAC 2017, LNCS 10236, pp. 405–417, 2017.
DOI: 10.1007/978-3-319-57586-5_34

[2–4,12]. The reverse problem, namely, given a point set P we are asked to construct a monotone spanning geometric graph on the points of P, has trivial solutions, i.e. the complete graph $K_{|P|}$ on the points of P as well as the path graph $W_{|P|}$ which visits all points of P in increasing order of their y coordinates are both y–monotone spanning geometric graphs of P.

The *Euclidean minimum spanning tree problem*, i.e. the problem of constructing the minimum cost spanning geometric tree of a plane point set P (where the cost of the tree is taken to be the sum of the Euclidean lengths of its edges), has also received attention [22]. Shamos and Hoey [26] showed that it can be solved in $\Theta(|P| \log |P|)$ time.

Combining the Euclidean minimum spanning tree problem with the notion of monotonicity leads to a large number of problems that, to the best of our knowledge, have not been previously investigated. The most general problem can be stated as follows: *"Given a point set P find the minimum cost monotone spanning geometric graph of P, i.e. the geometric graph such that every pair of points of P is connected by a monotone path"*. Since in a monotone graph the direction of monotonicity need not be the same for all pairs of vertices, it is not clear whether the minimum cost monotone spanning graph is a tree. We call this problem the *Monotone Minimum Spanning Graph problem*. We note that there exist point sets for which the Euclidean minimum spanning tree is not monotone and hence does not coincide with the monotone minimum spanning graph. Consider for example a point set with four points for which the Euclidean minimum spanning tree is a geometric path that is not monotone.

We focus on a simple variant of the general monotone minimum spanning graph problem. Let P be a rooted point set, i.e. a point set having a designated point, say r, as its root. We do not insist on having monotone paths between every pair of points of P but rather only between the root r with all other points of P. Moreover, we insist that all paths are *uniform* in the sense that they are all monotone with respect to the same direction, i.e. we build *rooted uniform monotone graphs*. Actually, as it turns out (Corollary 2), in this problem the sought graphs are trees and, thus, we refer to it as the *rooted Uniform Monotone Minimum Spanning Tree* (for short, *rooted UMMST*) *problem*. In the rooted UMMST problem we have the freedom to select the direction of monotonicity. When we are restricted to have monotone paths in a specific direction, say y, we have the *rooted y-Monotone Minimum Spanning Tree* (for short, *rooted y-MMST*) *problem*. Figure 1(a) illustrates a rooted y–monotone spanning graph of a rooted point set P, while the rooted y–MMST of P is given in Fig. 1(b).

Rooted point sets have been previously studied in the context of minimum spanning trees. The *capacitated minimum spanning tree* is a tree that has a designated vertex r (its root) and each of the subtrees attached to r contains no more than c vertices. c is called the *tree capacity*. Solving the capacitated minimum spanning tree problem optimally has been shown by Jothi and Raghavachari to be NP-hard [14]. In the same paper, they have also presented approximation algorithms for the case where the vertices correspond to points on the Euclidean plane.

Fig. 1. Illustration of rooted y–monotone spanning graphs

If the geometric path W is both x–monotone and y–monotone then it is denoted as xy–*monotone*. Furthermore, if there exists a Cartesian System $x'y'$ s.t. W is $x'y'$–monotone then W is *2D-monotone*. Based on xy–monotone geometric paths and in analogy to the (rooted) monotone, uniform monotone and y–monotone graphs, we define the (*rooted*) *2D-monotone, uniform 2D-monotone* and xy–*monotone graphs*. 2D-monotone paths/graphs were also recently denoted by Bonichon et al. [7] as *angle-monotone paths/graphs*. Bonichon et al. [7] gave a $O(|P| \cdot |E|^2)$ time algorithm that decides if a geometric graph $G = (P, E)$ is 2D-monotone. In order to do so, Bonichon et al. [7] gave a $O(|E|^2)$ time algorithm which is used as a subroutine and decides if the graph is rooted 2D-monotone, where the root is a specified vertex. Bonichon et al. [7] also noted that it is not always feasible to construct a planar 2D-monotone spanning geometric graph of a given point set. Similarly to the rooted UMMST and rooted y-MMST problems we define the corresponding *rooted Uniform 2D-Monotone Minimum Spanning Tree* (for short, *rooted 2D-UMMST*) and *rooted xy-Monotone Minimum Spanning Tree* (for short, *rooted xy-MMST*) problems, which ask for the minimum cost rooted Uniform 2D-Monotone spanning tree and rooted xy-monotone spanning tree of a given rooted point set, respectively.

A path/curve W is *increasing-chord* (see [17,23]) if for every four points p_1, p_2, p_3, p_4 traversed in this order along it, it holds that $d(p_2, p_3) \le d(p_1, p_4)$ where $d(p, q)$ denotes the Euclidean distance between the points p and q. A geometric graph $G = (P, E)$ is *increasing-chord* if each two points of P are connected by an increasing-chord path. Increasing-chord graphs were introduced by Alamdari et al. [1]. Alamdari et al. [1] noted that any 2D-monotone path/graph is also increasing-chord. Drawing a graph as an increasing-chord graph is studied in [1, 21]. On the other hand, constructing increasing-chord graphs that span a given point set is studied in [1, 8, 19]. In all the papers that construct increasing-chord spanning graphs of a point set, i.e. in [1, 8, 19], the constructed increasing-chord paths connecting the vertices are additionally 2D-monotone.

Our Contribution

Let P be a rooted point set. We give algorithms that produce the rooted y-MMST of P and the rooted xy–MMST of P in $O(|P| \log^2 |P|)$ time and $O(|P| \log^3 |P|)$ time, respectively. We also propose algorithms that build the rooted UMMST of P and the rooted 2D-UMMST of P in $O(|P|^2 \log |P|)$ time when the optimum direction and the optimum pair of directions has to be determined, respectively. For all these four problems, we provide a $\Omega(|P|\log|P|)$ time lower bound which is easily derived.

We also propose simple algorithms that decide whether a given connected geometric graph on a rooted point set is (i) rooted y–monotone, (ii) rooted uniform monotone, (iii) rooted xy-monotone and (iv) rooted uniform 2D-monotone.

Due to space constraints, we omit a lot of proofs, pseudocodes and details. They can be viewed in the full version of this article [20].

2   Definitions and Preliminaries

In this article we deal with the Euclidean plane, i.e. every point set is a subset of $\mathbb{R}^2$, and we consider only rooted point sets. Let x, y be the axes of a Cartesian System. The x and y coordinates of a point p are denoted by $x(p)$ and $y(p)$, respectively. W.l.o.g., we assume that the root r of the point sets coincides with the origin of the Cartesian System, i.e., $x(r) = y(r) = 0$. We also assume that the point sets are *in general position*, i.e. no three points are collinear.

Let P be a point set. P is called *positive* (*negative*) w.r.t. the direction of y or y–positive (y–negative) if for each $p \in P, y(p) \geq 0$ (resp. $y(p) \leq 0$). Let a be a real number then by $P_{y\leq a}$ we denote the set of points of P that have y coordinate less than or equal to a. Subsets $P_{y\geq a}, P_{x\leq a}, P_{x\geq a}, P_{|y|\leq a}, P_{|y|\geq a}, P_{|x|\leq a}$ and $P_{|x|\geq a}$ are similarly defined.

Let P be a point set and p be a point of the plane, then $d(p, P)$ denotes the Euclidean distance from p to the point set P, i.e. $d(p, P) = \min_{q \in P} d(p, q)$.

The line segment with endpoints p and q is denoted as $\overline{pq}$. The *slope* of a line L is the angle that we need to rotate the x axis counterclockwise s.t. the x axis becomes parallel to L. Each slope belongs to the range $[0, \pi)$.

A *geometric graph* $G = (P, E)$ consists of a set P of points which are denoted as its vertices and a set E of line segments with endpoints in P which are denoted as its edges. If P is rooted then G is a *rooted geometric graph*. The cost of a geometric graph $G = (P, E)$, denoted as cost(G), is the sum of the Euclidean lengths of its edges. Let $G_1 = (P_1, E_1), G_2 = (P_2, E_2), \ldots, G_n = (P_n, E_n)$ be n geometric graphs then the union $G_1 \cup G_2 \cup \ldots \cup G_n$ is the geometric graph $G = (P, E)$ s.t. $P = P_1 \cup P_2 \cup \ldots \cup P_n$ and $E = E_1 \cup E_2 \cup \ldots \cup E_n$.

The *closest point* (or *nearest neighbor*) *problem* is an important problem in computational geometry. It was initially termed as the *post-office problem* by Knuth [16]. In this problem there exists a set S of points that is static (it cannot be changed by inserting points to it or deleting points from it) and the goal is to find the closest point (or nearest neighbor) from S to a given query point. This problem is usually reduced to the problem of locating in which region of a planar subdivision the query point is located [10,24]. Efficient static data structures have been constructed to answer these queries in logarithmic time by performing fast preprocessing algorithms, e.g. see [15,18]. Concerning the semi-dynamic version of the closest point problem, in which insertions of points to S are allowed, Bentley [6] gave a very useful semi-dynamic data structure.

Fact 1 (Bentley [6]). *There exists a semi-dynamic data structure that allows only two operations, the insertion of a point and a closest point query. Where, a closest point query takes $O(\log^2 n)$ time (with n denoting the size of the structure) and inserting n elements in the structure takes $O(n \log^2 n)$ total time.*

3 The Rooted y-Monotone Minimum Spanning Tree (rooted y–MMST) Problem

In this section we study the construction of the rooted y–MMST of a rooted point set P. We initially show that we can deal with $P_{y \leq 0}$ and $P_{y \geq 0}$ separately. Then, we provide a characterization of the rooted y–MMST of rooted y–positive (or y–negative) point sets. Using the previous two, we develop an algorithm that constructs the rooted y–MMST of P.

Observation 1. Let P be a rooted point set and $G = (P, E)$ be a rooted y–monotone spanning graph of P and let $\overline{p_d p_u} \in E$ with $y(p_d) < 0 < y(p_u)$. Then, every path from the root r to a point $p \in P \setminus \{r\}$ that contains $\overline{p_d p_u}$ is not y–monotone since it moves "south" to p_d and then "north" to p_u, or vice versa.

Corollary 1. *Let P be a rooted point set, $G^{opt} = (P, E)$ be the rooted y–monotone minimum spanning graph of P and $p_d, p_u \in P$ such that $y(p_d) < 0 < y(p_u)$. Then, $\overline{p_d p_u} \notin E$.*

Lemma 1. *Let P be a rooted point set and G^{opt} be the rooted y–monotone minimum spanning graph of P. Furthermore, let $G^{opt}_{y \leq 0}$ and $G^{opt}_{y \geq 0}$ be the rooted y–monotone minimum spanning graphs of $P_{y \leq 0}$ and $P_{y \geq 0}$, respectively. Then, G^{opt} is the union of $G^{opt}_{y \leq 0}$ and $G^{opt}_{y \geq 0}$.*

We now study the construction of the rooted y–monotone minimum spanning graph of a rooted y–positive (or y–negative) point set P with root r. We define $S[P, y]$ to be the sequence of points of P ordered by the following rule: *"The points of $S[P, y]$ are ordered w.r.t. their absolute y coordinates and, if two points have the same y coordinate, then they are ordered w.r.t. their distance from the preceding points in $S[P, y]$."*. More formally, $S[P, y] = (r = p_0, p_1, p_2, \ldots, p_n)$ s.t. $|y(p_0)| \leq |y(p_1)| \leq |y(p_2)| \leq \ldots \leq |y(p_n)|$ and $|y(p_i)| = |y(p_{i+1})|$ implies that $d(p_i, \{p_0, p_1, \ldots, p_{i-1}\}) \leq d(p_{i+1}, \{p_0, p_1, \ldots, p_{i-1}\})$ and $P = \{p_0, p_1, p_2, \ldots, p_n\}$. We now give a characterization of the rooted y–monotone minimum spanning graph of P.

Lemma 2. *Let $G = (P, E)$ be a rooted geometric graph where P is a rooted y–positive (or y–negative) point set with $S[P, y] = (r = p_0, p_1, p_2, \ldots, p_n)$. Then, G is the rooted y–monotone minimum spanning graph of P if and only if (i) p_n is connected in G only with its closest point (or nearest neighbor) from $\{p_0, p_1, \ldots, p_{n-1}\}$, i.e. the point p_j such that $d(p_n, p_j) = d(p_n, \{p_0, p_1, \ldots, p_{n-1}\})$, and (ii) $G \setminus \{p_n\}$ is the rooted y–monotone minimum spanning graph of $P \setminus \{p_n\}$.*

Lemmas 1 and 2 lead to the next Corollary.

Corollary 2. *The rooted y–monotone minimum spanning graph of a rooted point set P is a geometric tree.*

Let P be a rooted y–positive (or y–negative) point set and $S[P, y] = (r = p_0, p_1, \ldots, p_n)$. We call the closest point to p_i from $\{p_0, p_1, \ldots, p_{i-1}\}$ the *parent* of p_i and we denote it as $par(p_i)$. More formally, $par(p_i) = p_j$ if and only if $p_j \in \{p_0, p_1, \ldots, p_{i-1}\}$ and $d(p_i, p_j) = d(p_i, \{p_0, p_1, \ldots, p_{i-1}\})$. Then, Lemma 2 implies the following Corollary.

Corollary 3. *The edges of the rooted y–MMST of P are exactly the line segments $\overline{par(p_i)p_i}$, for $i = 1, 2, \ldots, n$.*

Corollary 3 implies a $O(|P|^2)$ time algorithm for producing the rooted y–MMST of P. However, using the semi-dynamic data structure for closest point queries given by Bentley [6], the time complexity of our rooted y–MMST algorithm becomes $O(|P| \log^2 |P|)$.

Theorem 1. *The rooted y–MMST of a rooted point set P can be computed in $O(|P| \log^2 |P|)$ time.*

Proof. We first construct $P_{y \leq 0}$ and $P_{y \geq 0}$. Then, we apply our rooted y–MMST algorithm on $P_{y \leq 0}$ and $P_{y \geq 0}$ constructing $T_{y \leq 0}$ and $T_{y \geq 0}$, respectively. By Corollary 3, $T_{y \leq 0}$ and $T_{y \geq 0}$ are the rooted y–MMSTs of $P_{y \leq 0}$ and $P_{y \geq 0}$, respectively. Using Fact 1 and since $O(|P|)$ insertions and $O(|P|)$ closest point queries are performed, computing $T_{y \leq 0}$ and $T_{y \geq 0}$ takes $O(|P| \log^2 |P|)$ time. By Lemma 1, $T_{y \leq 0} \cup T_{y \geq 0}$ is the rooted y–MMST of P. $\qquad \square$

In the next Theorem, we give a lower bound for the time complexity of any algorithm which given a rooted point set P produces the rooted y–MMST of P.

Theorem 2. *Any algorithm which given a rooted point set P, produces the rooted y–MMST of P requires $\Omega(|P| \log |P|)$ time.*

Proof. We use the reduction from sorting that was given by Shamos [25]. Let $(a_1, a_2, \ldots, a_n)$ be a sequence of nonnegative integers. We reduce this sequence to the rooted point set $P = \{r = (0, 0), (a_1, a_1^2), (a_2, a_2^2), \ldots, (a_n, a_n^2)\}$. Then, the rooted y–MMST of P contains exactly the edges $\overline{rp_1}, \overline{p_1 p_2}, \ldots, \overline{p_{n-1} p_n}$ s.t. $a_i' = x(p_i), i = 1, 2, \ldots, n$, where $(a_1', a_2', \ldots, a_n')$ is the sorted permutation of $(a_1, a_2, \ldots, a_n)$. The lower bound follows since sorting n numbers requires $\Omega(n \log n)$ time. $\qquad \square$

We note that using the same reduction, i.e. the reduction from sorting that was given by Shamos [25], the same lower bound can be easily obtained for the rooted UMMST and (rooted) Monotone Minimum Spanning Graph Problem.

We conclude this section, by showing that rooted y–monotone graphs can be efficiently recognized. Our approach is similar to the approach employed in the third section of the article of Arkin et al. [5].

Theorem 3. *Let $G = (P, E)$ be a rooted connected geometric graph. Then, we can decide in $O(|E|)$ time if G is rooted y–monotone.*

Proof. We first transform G into a directed geometric graph $\overrightarrow{G}$ in $O(|E|)$ time, by assigning direction to the edges and removing some of them. Let $\overline{pq}$ be an edge of G. If p and q belong to opposite half planes w.r.t. the x axis then $\overline{pq}$ cannot be used in a y–monotone path from the root r to a point of $P \setminus \{r\}$ (see Observation 1). Hence, we remove the edge $\overline{pq}$ from the graph. If $y(p) = y(q)$ then we insert both $\overrightarrow{pq}$ and $\overrightarrow{qp}$ in $\overrightarrow{G}$. Otherwise, assuming w.l.o.g., that $|y(p)| < |y(q)|$, we insert $\overrightarrow{pq}$ in $\overrightarrow{G}$. G is rooted y–monotone if and only if r is connected with all other points of P in $\overrightarrow{G}$. The latter can be easily decided in $O(|E|)$ time by a breadth first search or a depth first search traversal. $\qquad\square$

4 Rooted Uniform Monotone Graphs: Minimum Spanning Tree Construction, and Recognition

4.1 Building the Rooted UMMST

In this subsection, we focus on the rooted UMMST problem. We tackle the problem by giving a rotational sweep algorithm. Rotational sweep is a well known technique in computational geometry in which a (directed) line is rotated counterclockwise (or clockwise) and during this rotation important information about the solution of the problem is updated.

Observation 2. Let y' be an axis. If we rotate the y' axis counterclockwise then the sequence $S[P_{y' \geq 0}, y']$ or the sequence $S[P_{y' \leq 0}, y']$ changes only when the y' axis reaches (moves away from) a line perpendicular to a line passing through two points of P. Then, by Lemma 2 and Corollary 3, the rooted y'–MMST of P may only change at the same time.

Observation 3. Let y' and y'' be axes of opposite directions. Then, the rooted y'–MMST of P is the same as the rooted y''–MMST of P. Hence, when computing the rooted UMMST of P we only need to take into account the y' axes such that the angle that we need to rotate the x axis counterclockwise to become codirected with the y' is less than π.

Based on Observations 2 and 3, we define the set $\Theta = \{\theta \in [0, \pi) : \theta$ is the slope of a line perpendicular to a line passing through two points of $P\}$. We also define $S[\Theta]$ to be the sorted sequence that contains the slopes of Θ in increasing order, i.e. $S[\Theta] = (\theta_0, \theta_1, \ldots, \theta_{m-1}), \theta_i < \theta_{i+1}, i = 0, 1, \ldots, m-2$ and $m \leq \binom{|P|}{2}$. We further define the set $\Theta_{\text{critical}} = \{\theta_0, \theta_1, \ldots, \theta_{m-1}\} \cup \{\frac{\theta_0 + \theta_1}{2}, \frac{\theta_1 + \theta_2}{2}, \ldots,$ $\frac{\theta_{m-2} + \theta_{m-1}}{2}, \frac{\theta_{m-1} + \pi}{2}\}$ which we call the *critical set of slopes* since examining the axes with slope in Θ_{critical} is sufficient for computing the rooted UMMST of P (see Lemma 3). $|\Theta_{\text{critical}}| = O(|P|^2)$. We now assign "names" to the axes with slopes in Θ_{critical}. Let y_{2i} be the axis with slope θ_i, $i = 0, 1, \ldots, m-1$ and let y_{2i+1} be the axis with slope $\frac{\theta_i + \theta_{i+1}}{2}$, $i = 0, 1, \ldots, m-2$, and y_{2m-1} be the axis with slope $\frac{\theta_{m-1} + \pi}{2}$.

Lemma 3. *The rooted UMMST of P is one of the rooted y'-MMST of P over all axes y' with slope in $\Theta_{critical}$ and, more specifically, the one of minimum cost.*

We now describe our algorithm which produces the rooted UMMST of a rooted point set P. Our rooted UMMST algorithm is a rotational sweep algorithm. It considers an axis y', which initially coincides with y_0, and then it rotates it counterclockwise until y' becomes opposite to the x axis. Throughout this procedure, it updates the rooted y'-MMST of P. By Lemma 3, it only needs to obtain each rooted y_i-MMST of P, where y_i is an axis with slope in $\Theta_{critical}$, $0 \leq i \leq 2m - 1$. Let T_i^{opt} be the rooted y_i-MMST of P, $0 \leq i \leq 2m - 1$. Our rooted UMMST algorithm can now be stated as follows: It initially constructs T_0^{opt} using Theorem 1. Then, it iterates for $i = 1, 2, \ldots, 2m - 1$ obtaining at the end of each iteration T_i^{opt} by modifying T_{i-1}^{opt}. In order to do this efficiently, it maintains a tree T which is initially equal to T_0^{opt} and throughout its operation it evolves to $T_1^{opt}, T_2^{opt}, \ldots, T_{2m-1}^{opt}$. Similarly, it maintains the sequences S^- and S^+ which are initially equal to $S[P_{y_0 \leq 0}, y_0]$ and $S[P_{y_0 \geq 0}, y_0]$, respectively, and evolve to $S[P_{y_i \leq 0}, y_i]$ and $S[P_{y_i \geq 0}, y_i]$, respectively, $i = 1, 2, \ldots, 2m - 1$. Our algorithm stores the axis which corresponds to the produced rooted UMMST of P, so far, in the variable "minAxis". In its final step, it recomputes the rooted "minAxis"-MMST of P using Theorem 1 and returns this tree.

Theorem 4. *The rooted UMMST of a rooted point set P can be computed in $O(|P|^2 \log |P|)$ time.*

Proof. By Lemma 3, our rooted UMMST algorithm produces the rooted UMMST of P. We now show that its time complexity is $O(|P|^2 \log |P|)$. The axes $y_0, y_1, \ldots, y_{2m-1}$ with slopes in $\Theta_{critical}$ can be computed in $O(|P|^2 \log |P|)$ time. Let k_i be the number of pairs of points of P that have the same projection onto the y_{2i} axis, $0 \leq i \leq m-1$. Then $\sum_{i=0}^{m-1} k_i = \binom{|P|}{2}$. For each $i = 0, 1, \ldots, m-1$, we compute a list L_i which contains these k_i pairs. All $L_i, 0 \leq i \leq m - 1$, can be computed in $O(|P|^2 \log |P|)$ total time.

For each point p in $S^- \setminus \{r\}$ (resp. in $S^+ \setminus \{r\}$) we maintain a data structure $PD(p)$ which is a self-balancing binary search tree that contains all the points that precede p in S^- (resp. S^+) accompanied with their distance from p. More formally, let S^- be equal to $(r = p_0, p_1, \ldots, p_s)$. Then, for each $p_j, j = 1, 2, \ldots, s$, $PD(p_j)$ contains the pairs $(p_0, d(p_0, p_j)), (p_1, d(p_1, p_j)), \ldots, (p_{j-1}, d(p_{j-1}, p_j))$. The key of each $(p_l, d(p_l, p_j)), l = 0, 1, \ldots, j - 1$, is the distance $d(p_l, p_j)$. Similarly, we define $PD(p)$ for each $p \in S^+ \setminus \{r\}$. We employ these $PD(p), p \in P \setminus \{r\}$, data structures since using the information stored in them we can obtain the parent of each p efficiently. In more detail, for each $p \in P \setminus \{r\}$ the par(p) in T can be obtained or updated in $O(\log |P|)$ time by taking into account the $PD(p)$, since the pair $(par(p), d(par(p), p))$ is the element with the minimum key in $PD(p)$.

Computing the initial values of T, S^-, S^+ and $PD(p), p \in P \setminus \{r\}$ can be done in $O(|P|^2 \log |P|)$ time. This is true since $T_0^{opt}, S[P_{y_0 \leq 0}, y_0]$ and $S[P_{y_0 \geq 0}, y_0]$ are computed in $O(|P| \cdot \log^2 |P|)$ time (see Theorem 1). Furthermore, computing $PD(p_j)$ for some p_j in $S^- \setminus \{r\}$ (resp. in $S^+ \setminus \{r\}$), when S^- (resp. S^+) equals to

$S[P_{y_0 \le 0}, y_0]$ (resp. $S[P_{y_0 \ge 0}, y_0]$) takes $O(|P| \log |P|)$ time since we have to insert each $(p_i, d(p_i, p_j))$, with $i < j$, to $\text{PD}(p_j)$ and each such insertion takes $O(\log |P|)$ time. Hence, the total running time for initially computing all $\text{PD}(p), p \in P \setminus \{r\}$, is $O(|P|^2 \log |P|)$.

Let T be equal to T_{i-1}^{opt} and let S^- (resp. S^+) be equal to $S[P_{y_{i-1} \le 0}, y_{i-1}]$ (resp. $S[P_{y_{i-1} \ge 0}, y_{i-1}]$) then T and S^- (resp. S^+) can be updated such that T becomes equal to T_i^{opt} and S^- (resp. S^+) becomes equal to $S[P_{y_i \le 0}, y_i]$ (resp. $S[P_{y_i \ge 0}, y_i]$) in:

1. $O(k_{\lfloor \frac{i}{2} \rfloor} \log |P|)$ time if i is even and y_i is not perpendicular to a line passing through the root r and another point in P.
2. $O(k_{\lfloor \frac{i}{2} \rfloor} \log |P|)$ time if i is odd and y_{i-1} is not perpendicular to a line passing through the root r and another point in P.
3. $O(|P| \log |P|)$ time if i is even and y_i is perpendicular to a line passing through the root r and another point $q \in P \setminus \{r\}$.
4. $O(|P| \log |P|)$ time if i is odd and y_{i-1} is perpendicular to a line passing through r and another point $q \in P \setminus \{r\}$.

Since $\sum_{i=0}^{m-1} k_i = O(|P|^2)$, the total running time of our algorithm is $O(|P|^2 \log |P|)$. $\qquad\square$

4.2 Recognizing Rooted Uniform Monotone Graphs

We now proceed to the problem of deciding if a given rooted connected geometric graph is rooted uniform monotone. Like we did for the rooted UMMST problem, we tackle this decision problem with a rotational sweep algorithm.

We first define some auxiliary sets. Let $G = (P, E)$ be a rooted connected geometric graph with root r and p be a point of $P \setminus \{r\}$. Let $A(p, y)$ be the set that contains all the adjacent points to p that are on the same side with p w.r.t. the x axis and are strictly closer to the x axis than p. More formally, $A(p, y) = \{q : q \in \text{Adj}(p), q$ lies on the same half plane with p w.r.t. the x axis and $|y(q)| < |y(p)|\}$. Let $B(y)$ denote the set $\{p : p \in P \setminus \{r\}$ and $A(p, y) \ne \emptyset\}$. Let $C(y)$ be the set that consists of the points $p \in P \setminus \{r\}$ that (i) do not belong to $B(y)$ and (ii) are connected with some other point q with the same y coordinate such that $A(q, y) \ne \emptyset$. More formally, $C(y) = \{p : p \in P \setminus (B(y) \cup \{r\})$ such that there exists $q \in \text{Adj}(p)$ with $y(q) = y(p)$ and $A(q, y) \ne \emptyset\}$. An example of a rooted geometric graph and the corresponding sets is given in Fig. 2.

Lemma 4. *Let $G = (P, E)$ be a rooted connected geometric graph such that for each $p \in P \setminus \{r\}$, $y(p) \ne 0$. Then, G is rooted y–monotone if and only if $|B(y)| + |C(y)| = |P| - 1$.*

Remark 1. If there exists a point $p \in P \setminus \{r\}$ with $y(p) = 0$ then G is rooted y–monotone if and only if (i) p is connected with r and (ii) $|B(y)| + |C(y)|$ equals to $|P| - 2$.

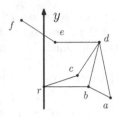

Fig. 2. $A(a, y) = \{b\}$, $A(b, y) = \emptyset$, $A(c, y) = \{r\}$, $A(d, y) = \{b, c\}$, $A(e, y) = \emptyset$ and $A(f, y) = \{e\}$. $B(y) = \{a, c, d, f\}$ and $C(y) = \{e\}$.

Remark 2. If we know $B(y)$, $C(y)$ and whether there exists a point $p \in P \setminus \{r\}$ with $y(p) = 0$ connected to r, we can decide if G is rooted y–monotone. This implies a $O(|E|)$ time algorithm, different from the one given in Theorem 3, which recognizes rooted y–monotone geometric graphs.

Observation 4. Let $G = (P, E)$ be a rooted connected geometric graph. If we rotate an axis y' counterclockwise, then $B(y')$, $C(y')$ and the points p in $P \setminus \{r\}$ with $y'(p) = 0$, change only when the y' axis reaches (or moves away from) a line that is perpendicular to an edge of G or that is perpendicular to a line passing through the root r and a point of $P \setminus \{r\}$.

Using similar arguments to the ones employed for solving the rooted UMMST problem, we define a set of critical slopes and appropriate axes which we have to test in order to decide if the rooted connected geometric graph $G = (P, E)$ is rooted uniform monotone. Let $\Theta = \{\theta \in [0, \pi) : \theta$ is the slope of a line perpendicular to either an edge of G or to a line passing through the root r and another point of $P\}$. $S[\Theta]$ is the sorted sequence that contains the slopes of Θ in increasing order, i.e. $S[\Theta] = (\theta_0, \theta_1, \ldots, \theta_{m-1})$, $\theta_i < \theta_{i+1}$, $i = 0, 1, \ldots, m - 2$ and $m < |E| + |P|$. We define the critical set of slopes, $\Theta_{\text{critical}} = \{\theta_0, \theta_1, \ldots, \theta_{m-1}\} \cup \{\frac{\theta_0 + \theta_1}{2}, \frac{\theta_1 + \theta_2}{2}, \ldots, \frac{\theta_{m-2} + \theta_{m-1}}{2}, \frac{\theta_{m-1} + \pi}{2}\}$. We now assign "names" to the axes with slope in Θ_{critical}. Let y_{2i} be the axis with slope θ_i, $0 \leq i \leq m - 1$. Moreover, let y_{2i+1} be the axis of slope $\frac{\theta_i + \theta_{i+1}}{2}$, $0 \leq i \leq m - 2$ and y_{2m-1} be the axis of slope $\frac{\theta_{2m-1} + \pi}{2}$. In analogy to Lemma 3 we obtain the next Lemma.

Lemma 5. *G is rooted uniform monotone if and only if it is rooted y'–monotone for some y' axis of slope in Θ_{critical}.*

We now give a rotational sweep algorithm that tests whether a given rooted connected geometric graph $G = (P, E)$ is rooted uniform monotone. Our rooted uniform monotone recognition algorithm rotates an axis y' which initially coincides with y_0 until it becomes opposite to the x axis. Throughout this rotation, it checks if G is rooted y'–monotone. Taking into account Lemma 5, our algorithm only needs to test if G is rooted y_i–monotone for some $i = 0, 1, \ldots, 2m - 1$.

Theorem 5. *Let $G = (P, E)$ be a rooted connected geometric graph. Then, we can decide in $O(|E| \log |P|)$ time if G is rooted uniform monotone.*

Proof. By Lemma 5, it is immediate that our rooted uniform monotone recognition algorithm decides if G is rooted uniform monotone. We now show that its time complexity is $O(|E| \log |P|)$. Computing the axes $y_0, y_1, \ldots, y_{2m-1}$, with slope in Θ_{critical} can be done in $O(|E| \log |P|)$ time. Let k_i be the number of pairs of points of P connected by an edge perpendicular to $y_{2i}, 0 \leq i \leq m - 1$. Then, $\sum_{i=0}^{m-1} k_i = |E|$. For each $i = 0, 1, \ldots, m - 1$, we construct a list L_i containing the k_i pairs of points of P that are connected by an edge perpendicular to y_{2i}. All L_i, $0 \leq i \leq m - 1$, can be computed in $O(|E| \log |P|)$ total time.

Let y_i be the last axis taken into account. Our algorithm maintains for each $p \in P \setminus \{r\}$ a data structure $A(p)$ which represents the set $A(p, y_i)$ (which is a subset of the $\text{Adj}(p)$). $A(p)$ contains the indices of the points of P that belong to $A(p, y_i)$. $A(p)$ can be implemented by any data structure which supports insert, delete and retrieve operations in $O(\log |P|)$ time (e.g. a 2–3 tree). Our algorithm also maintains the data structure B that represents the $B(y_i)$. In order to performing the insert and delete operations in $O(1)$ time, B is implemented as an array of boolean with size $O(|P|)$.

Computing all $A(p)$ s.t. $A(p)$ equals to $A(p, y_0)$, $p \in P \setminus \{r\}$, takes $O(|E| \log |P|)$ total time. After that, computing B s.t. B equals to $B(y_0)$ takes $O(|P|)$ time.

Similarly to Theorem 4, all the necessary updates we need to make in each $A(p), p \in P \setminus \{r\}$, and B during the algorithm take $O(|E| \log |P|)$ total time.

Given $B(y_{2i})$ and each $A(p, y_{2i}), p \in P \setminus \{r\}$, computing $C(y_{2i}), 0 \leq i \leq m-1$, takes $O(k_i)$ time using the list L_i. Then, having B and C and knowing if y_i is perpendicular to some line passing through r and another point p of P with $\overline{pr} \in E$, we can test if G is y_i–rooted-monotone in $O(1)$ time (see Lemma 4 and Remark 2).

From all the previous, the time complexity of the algorithm is $O(|E| \log |P|)$.
$\square$

We note that the approach we took for deciding if a given rooted connected geometric graph is rooted uniform monotone has some similarities with the approach employed in the third section of the article of Arkin et al. [5].

5 Rooted Uniform 2D-Monotone Graphs: Minimum Spanning Tree Production, and Recognition

In this section we study monotonicity w.r.t. two perpendicular axes. Due to space constraints but also due to the fact that our methods are analogous to the ones developed for the one direction case, we only state our results. For the details see the full version of this article [20].

We use (in our rooted xy–MMST algorithm) the semi-dynamic data structure for closest point with attribute value in specified range queries that was implicitly produced by Bentley [6], i.e it was implicitly produced from his corresponding static data structure (Sect. 4 of [6]) and his results about the decomposable problems (Sect. 3 of [6]).

Theorem 6. *The rooted xy–MMST of a rooted point set P can be computed in $O(|P| \cdot \log^3 |P|)$ time.*

Theorem 7. *Producing the rooted xy–MMST (or the rooted 2D-UMMST or the rooted 2D-MMST) of a rooted point set P requires $\Omega(|P|\log|P|)$ time.*

Theorem 8. *Let $G = (P, E)$ be a rooted connected geometric graph. Then, we can decide in $O(|E|)$ time if G is rooted xy–monotone.*

Theorem 9. *We can produce the rooted 2D-UMMST of a rooted point set P in $O(|P|^2 \log|P|)$ time.*

Theorem 10. *Given a rooted connected geometric graph $G = (P, E)$, we can decide in $O(|E|\log|P|)$ time if G is rooted uniform 2D-monotone.*

6 Conclusions and Future Work

In this article we studied the problem of constructing the minimum cost spanning geometric graph of a given rooted point set in which the root is connected to all other vertices by paths that are monotone w.r.t. a single direction, i.e. they are y-monotone (or w.r.t. a pair of orthogonal directions, i.e. they are xy-monotone). We showed that the minimum cost spanning geometric graph is actually a tree and we proposed polynomial time algorithms that construct it for the case where the direction (the pair of orthogonal directions) of monotonicity is given or remains to be determined.

Several directions for further research are open including: (1) computing the minimum cost spanning geometric graph of a given k-rooted point set, i.e. a point set with k designated points as its roots, containing monotone paths w.r.t. a single direction from each root to every other point in the point set, (2) determining whether the (rooted) monotone (or 2D-monotone) minimum spanning graph problem is solvable in polynomial time or if it is NP-hard and (3) replacing the property of monotonicity with another property, e.g. the increasing-chord property or the self-approaching (see [1,13]) property.

References

1. Alamdari, S., Chan, T.M., Grant, E., Lubiw, A., Pathak, V.: Self-approaching graphs. In: Didimo, W., Patrignani, M. (eds.) GD 2012. LNCS, vol. 7704, pp. 260–271. Springer, Heidelberg (2013). doi:10.1007/978-3-642-36763-2_23
2. Angelini, P.: Monotone drawings of graphs with few directions. In: 6th International Conference on Information, Intelligence, Systems and Applications, IISA , Corfu, Greece, 6–8 July, pp. 1–6. IEEE (2015)
3. Angelini, P., Colasante, E., Di Battista, G., Frati, F., Patrignani, M.: Monotone drawings of graphs. J. Graph Algorithms Appl. **16**(1), 5–35 (2012)
4. Angelini, P., Didimo, W., Kobourov, S., Mchedlidze, T., Roselli, V., Symvonis, A., Wismath, S.: Monotone drawings of graphs with fixed embedding. Algorithmica **71**(2), 233–257 (2015)

5. Arkin, E.M., Connelly, R., Mitchell, J.S.B.: On monotone paths among obstacles with applications to planning assemblies. In: Mehlhorn, K. (ed.) Proceedings of the Fifth Annual Symposium on Computational Geometry, SCG 1989, pp. 334–343. ACM (1989)

6. Bentley, J.L.: Decomposable searching problems. Inf. Process. Lett. **8**(5), 244–251 (1979)

7. Bonichon, N., Bose, P., Carmi, P., Kostitsyna, I., Lubiw, A., Verdonschot, S.: Gabriel triangulations and angle-monotone graphs: local routing and recognition. In: Hu, Y., Nöllenburg, M. (eds.) GD 2016. LNCS, vol. 9801, pp. 519–531. Springer, Cham (2016). doi:10.1007/978-3-319-50106-2_40

8. Dehkordi, H.R., Frati, F., Gudmundsson, J.: Increasing-chord graphs on point sets. J. Graph Algorithms Appl. **19**(2), 761–778 (2015)

9. Di Battista, G., Tamassia, R.: Algorithms for plane representations of acyclic digraphs. Theoret. Comput. Sci. **61**, 175–198 (1988)

10. Dobkin, D.P., Lipton, R.J.: Multidimensional searching problems. SIAM J. Comput. **5**(2), 181–186 (1976)

11. Garg, A., Tamassia, R.: On the computational complexity of upward and rectilinear planarity testing. SIAM J. Comput. **31**(2), 601–625 (2001)

12. He, D., He, X.: Nearly optimal monotone drawing of trees. Theoret. Comput. Sci. **654**, 26–32 (2016)

13. Icking, C., Klein, R., Langetepe, E.: Self-approaching curves. Math. Proc. Cambridge Philos. Soc. **125**, 441–453 (1999)

14. Jothi, R., Raghavachari, B.: Approximation algorithms for the capacitated minimum spanning tree problem and its variants in network design. ACM Trans. Algorithms **1**(2), 265–282 (2005)

15. Kirkpatrick, D.G.: Optimal search in planar subdivisions. SIAM J. Comput. **12**(1), 28–35 (1983)

16. Knuth, D.E.: The Art of Computer Programming, Volume III: Sorting and Searching. Addison-Wesley, Boston (1973)

17. Larman, D.G., McMullen, P.: Arcs with increasing chords. Math. Proc. Cambridge Philos. Soc. **72**, 205–207 (1972)

18. Lipton, R.J., Tarjan, R.E.: Applications of a planar separator theorem. SIAM J. Comput. **9**(3), 615–627 (1980)

19. Mastakas, K., Symvonis, A.: On the construction of increasing-chord graphs on convex point sets. In: 6th International Conference on Information, Intelligence, Systems and Applications, IISA, Corfu, Greece, 6–8 July, pp. 1–6. IEEE (2015)

20. Mastakas, K., Symvonis, A.: Rooted uniform monotone minimum spanning trees. CoRR, abs/1607.03338v2 (2017)

21. Nöllenburg, M., Prutkin, R., Rutter, I.: On self-approaching and increasing-chord drawings of 3-connected planar graphs. J. Comput. Geom. **7**(1), 47–69 (2016)

22. Preparata, F.P., Shamos, M.I.: Computational Geometry: An Introduction. Texts and Monographs in Computer Science. Springer, Heidelberg (1988)

23. Rote, G.: Curves with increasing chords. Math. Proc. Cambridge Philos. Soc. **115**, 1–12 (1994)

24. Shamos, M.I.: Geometric complexity. In: Rounds, W.C., Martin, N., Carlyle, J.W., Harrison, M.A. (eds.) Proceedings of the 7th Annual ACM Symposium on Theory of Computing, STOC 1975, pp. 224–233. ACM (1975)

25. Shamos, M.I.: Computational geometry. Ph.D. thesis, Yale University, USA (1978)

26. Shamos M.I., Hoey, D.: Closest-point problems. In: 16th Annual Symposium on Foundations of Computer Science, pp. 151–162. IEEE Computer Society (1975)

Existence of Evolutionarily Stable Strategies Remains Hard to Decide for a Wide Range of Payoff Values

Themistoklis Melissourgos[1(✉)] and Paul Spirakis[1,2]

[1] Department of Computer Science, University of Liverpool, Liverpool, UK
{T.Melissourgos,P.Spirakis}@liverpool.ac.uk
[2] Computer Technology Institute and Press "Diophantus" (CTI), Patras, Greece

Abstract. The concept of an *evolutionarily stable strategy* (ESS), introduced by Smith and Price [4], is a refinement of Nash equilibrium in 2-player symmetric games in order to explain counter-intuitive natural phenomena, whose existence is not guaranteed in every game. The problem of deciding whether a game possesses an ESS has been shown to be Σ_2^P-complete by Conitzer [1] using the preceding important work by Etessami and Lochbihler [2]. The latter, among other results, proved that deciding the existence of ESS is both **NP**-hard and **coNP**-hard. In this paper we introduce a *reduction robustness* notion and we show that deciding the existence of an ESS remains **coNP**-hard for a wide range of games even if we arbitrarily perturb within some intervals the payoff values of the game under consideration. In contrast, ESS exist almost surely for large games with random and independent payoffs chosen from the same distribution [11].

Keywords: Game theory · Computational complexity · Evolutionarily stable strategies · Robust reduction

1 Introduction

1.1 Concepts of Evolutionary Games and Stable Strategies

Evolutionary game theory has proven itself to be invaluable when it comes to analysing complex natural phenomena. A first attempt to apply game theoretic tools to evolution was made by Lewontin [3] who saw the evolution of genetic mechanisms as a game played between a species and nature. He argued that a species would adopt the "maximin" strategy, i.e. the strategy which gives it the best chance of survival if nature does its worst. Subsequently, his ideas were improved by the seminal work of Smith and Price in [4] and Smith in [12] where the study of natural selection's processes through game theory was triggered.

The work of the second author was partially supported by the ERC Project ALGAME.

For a full version with detailed examples and proofs see https://arxiv.org/abs/1701.08108 [5].

© Springer International Publishing AG 2017
D. Fotakis et al. (Eds.): CIAC 2017, LNCS 10236, pp. 418–429, 2017.
DOI: 10.1007/978-3-319-57586-5_35

They proposed a model in order to decide the outcome of groups consisting of living individuals, conflicting in a specific environment.

The key insight of evolutionary game theory is that a set of behaviours depends on the interaction among multiple individuals in a population, and the prosperity of any one of these individuals depends on that interaction of its own behaviour with that of the others. An **evolutionarily stable strategy (ESS)** is defined as follows: An infinite population consists of two types of infinite groups with the same set of pure strategies; the *incumbents*, that play the (mixed) strategy s and the *mutants*, that play the (mixed) strategy $t \neq s$. The ratio of mutants over the total population is ϵ. A pair of members of the total population is picked uniformly at random to play a finite symmetric bimatrix game Γ with payoff matrix A_Γ. Strategy s is an ESS if for every $t \neq s$ there exists a constant ratio ϵ_t of mutants over the total population, such that, if $\epsilon < \epsilon_t$ the expected payoff of an incumbent versus a mutant is strictly greater than the expected payoff of a mutant versus a mutant. For convenience, we say that "s is an ESS of the game Γ".

The concept of ESS tries to capture resistance of a population against invaders. This concept has been studied in two main categories: infinite population groups and finite population groups. The former was the one where this Nash equilibrium refinement was first defined and presented by [4]. The latter was studied by Schaffer [10] who shows that the finite population case is a generalization of the infinite population one. The current paper deals with the infinite population case which can be mathematically modelled in an easier way and in addition, its results may provide useful insight for the finite population case. (For an example of ESS analysis in an infinite population game see the full version [5].)

1.2   Previous Work

Searching for the exact complexity of deciding if a bimatrix game possesses an ESS, Etessami and Lochbihler [2] invent a nice reduction from the complement of the CLIQUE problem to a specific game with an appointed ESS, showing that the ESS problem is **coNP**-hard. They also accomplish a reduction from the SAT problem to ESS, thus proving that ESS is **NP**-hard too. This makes impossible for the ESS to be **NP**-complete, unless $\mathbf{NP} = \mathbf{coNP}$. Furthermore, they provide a proof for the general ESS being contained in Σ_2^P, the second level of the polynomial-time hierarchy, leaving open the question of what is the complexity class in which the problem is complete.

A further improvement of those results was made by Nisan [8], showing that, given a payoff matrix, the existence of a mixed ESS is **coDP**-hard. (See Papadimitriou and Yannakakis [9] for background on this class.) A notable consequence of both [2] and [8] is that the problem of *recognizing* a mixed ESS, once given along with the payoff matrix, is **coNP**-complete. However, the question of the exact complexity of ESS existence, given the payoff matrix, remained open. A few years later, Conitzer finally settles this question in [1], showing that ESS is actually Σ_2^P-complete.

On the contrary, Hart and Rinott [11] showed that if the symmetric bimatrix game is defined by a $n \times n$ payoff matrix with elements independently randomly chosen according to a distribution F with exponential and faster decreasing tail, such as exponential, normal or uniform, then the probability of having an ESS with just 2 pure strategies in the support tends to 1 as n tends to infinity. In view of this result, and since the basic reduction of [2] used only 3 payoff values, it is interesting to consider whether ESS existence remains hard for arbitrary payoffs in some intervals.

1.3   Our Results

In the reduction of Etessami and Lochbihler that proves **coNP**-hardness of ESS the values of the payoffs used, are 0, $\frac{k-1}{k}$ and 1, for $k \in \mathbb{N}$. A natural question is if the hardness results hold when we **arbitrarily** perturb the payoff values within respective intervals (in the spirit of smoothed analysis [13]). In our work we extend the aforementioned reduction and show that the specific reduction remains valid even after significant changes of the payoff values.

We can easily prove that the evolutionarily stable strategies of a symmetric bimatrix game remain the exact same if we add, subtract or multiply (or do all of them) with a positive value its payoff matrix. However, that kind of value modification forces the entries of the payoff matrix to change in an entirely correlated manner, hence it does not provide an answer to our question. In this work, we prove that if we have partitions of entries of the payoff matrix with the same value for each partition, independent arbitrary perturbations of those values within certain intervals do not affect the validity of our reduction. In other words, we prove that determining ESS existence remains hard even if we perturb the payoff values associated with the reduction. En route we give a definition of "reduction robustness under arbitrary perturbations" and show how the reduction under examination adheres to this definition.

In contrast, [11] show that if the payoffs of a symmetric game are random and independently chosen from the same distribution F with "exponential or faster decreasing tail" (e.g. exponential, normal or uniform), then an ESS (with support of size 2) exists with probability that tends to 1 when n tends to infinity.

One could superficially get a non-tight version of our result by saying that (under supposed continuity assumptions in the ESS definition) any small perturbation of the payoff values will not destroy the reduction. However, in such a case (a) the continuity assumptions have to be precisely stated and (b) this does not explain why the ESS problem becomes easy when the payoffs are random [11].

In fact, the value of our technique is, firstly, to get as tight as possible ranges of the perturbation that preserve the reduction (and the ESS hardness) without any continuity assumptions, secondly, to indicate the basic difference from random payoff values (which is exactly the notion of partition of payoffs into groups in our definition of robustness, and the allowance of arbitrary perturbation within some interval in each group), and finally, the ranges of the allowed perturbations that we determine are quite tight. For the reduction to be preserved when we independently perturb the values (in each of our partitions arbitrarily), one must show that a system of inequalities has always a feasible solution, and we manage to show this in our final theorem. Our result seems to indicate that existence of an ESS remains hard despite a smoothed analysis [13].

An outline of the paper is as follows: In Sect. 2 we define the robust reduction notion and we provide a reduction, based on the one from [2], that is essentially modified in order to be robust. In Sect. 3 we give our main result and Sect. 4 refers to further work and conclusions.

1.4 Definitions and Notation

Background from Game Theory. A *finite two-player strategic form game* $\Gamma = (S_1, S_2, u_1, u_2)$ is given by finite sets of pure strategies S_1 and S_2 and utility, or *payoff*, functions $u_1 : S_1 \times S_2 \mapsto \mathbb{R}$ and $u_2 : S_1 \times S_2 \mapsto \mathbb{R}$ for the row-player and the column-player, respectively. Such a game is called *symmetric* if $S_1 = S_2 =: S$ and $u_1(i,j) = u_2(j,i)$ for all $i,j \in S$.

In what follows, we are only concerned with finite symmetric two-player strategic form games, so we write (S, u_1) as shorthand for (S, S, u_1, u_2), with $u_2(j,i) = u_1(i,j)$ for all $i,j \in S$. For simplicity assume $S = 1, \ldots, n$, i.e., pure strategies are identified with integers $i, 1 \leq i \leq n$. The row-player's *payoff matrix* $A_\Gamma = (a_{i,j})$ of $\Gamma = (S, u_1)$ is given by $a_{i,j} = u_1(i,j)$ for $i,j \in S$, so $B_\Gamma = A_\Gamma^T$ is the payoff matrix of the column-player. Note that A_Γ is not necessarily symmetric, even if Γ is a symmetric game.

A *mixed strategy* $s = (s(1), \ldots, s(n))^T$ for $\Gamma = (S, u_1)$ is a vector that defines a probability distribution on s and, in the sequel, we will denote by $s(i)$ the probability assigned by strategy s on the pure strategy $i \in S$. Thus, $s \in X$, where $X = \left\{ s \in \mathbb{R}^n_{\geq 0} : \sum_{i=1}^{n} s(i) = 1 \right\}$ denotes the set of mixed strategies in Γ, with $\mathbb{R}^n_{\geq 0}$ denoting the set of non-negative real number vectors $(x_1, x_2, \ldots, x_n)$. s is called *pure* iff $s(i) = 1$ for some $i \in S$. In that case we identify s with i. For brevity, we generally use "strategy" to refer to a mixed strategy s, and indicate otherwise when the strategy is pure. In our notation, we alternatively view a mixed strategy s as either a vector $(s_1, \ldots, s_n)^T$, or as a function $s : S \mapsto \mathbb{R}$, depending on which is more convenient in the context.

The *expected payoff* function, $U_k : X \times X \mapsto \mathbb{R}$ for player $k \in 1, 2$ is given by $U_k(s, t) = \sum_{i,j \in S} s(i)t(j)u_k(i,j)$, for all $s,t \in X$. Note that $U_1(s,t) = s^T A_\Gamma t$ and $U_2(s,t) = s^T A_\Gamma^T t$. Let s be a strategy for $\Gamma = (S, u_1)$. A strategy $t \in X$ is a *best response* to s if $U_1(t,s) = \max_{t' \in X} U_1(t',s)$. The *support* supp$(s)$ of s is the set $\{i \in S : s(i) > 0\}$ of pure strategies which are played with non-zero probability. The *extended support* ext-supp(s) of s is the set $\{i \in S : U_1(i,s) = \max_{x \in X} U_1(x,s)\}$ of all pure best responses to s.

A pair of strategies (s,t) is a *Nash equilibrium* (NE) for Γ if s is a best response to t and t is a best response to s. Note that (s,t) is a NE if and only if supp$(s) \subseteq$ ext-supp(t) and supp$(t) \subseteq$ ext-supp(s). A NE (s,t) is *symmetric* if $s = t$.

Definition 1 (Symmetric Nash equilibrium). *A strategy profile (s,s) is a symmetric NE for the symmetric bimatrix game $\Gamma = (S, u_1)$ if $s^T A_\Gamma s \geq t^T A_\Gamma s$ for every $t \in X$.*

A definition of ESS equivalent to that presented in Subsect. 1.1 is:

Definition 2 (Evolutionarily stable strategy). *A (mixed) strategy $s \in X$ is an evolutionarily stable strategy (ESS) of a two-player symmetric game Γ if:*

1. *(s, s) is a symmetric NE of Γ, and*
2. *if $t \in X$ is any best response to s and $t \neq s$, then $U_1(s, t) > U_1(t, t)$.*

Due to [7], we know that every symmetric game has a symmetric Nash equilibrium. The same does not hold for evolutionarily stable strategies (for example "rock-paper-scissors" does not have any pure or mixed ESS).

Definition 3 (ESS problem). *Given a symmetric two-player normal-form game Γ, we are asked whether there exists an evolutionarily stable strategy of Γ.*

Background from Graph Theory. An *undirected graph* G is an ordered pair (V, E) consisting of a set V of *vertices* and a set E, disjoint from V, of *edges*, together with an *incidence function* ψ_G that associates with each edge of G an unordered pair of distinct vertices of G. If e is an edge and u and v are vertices such that $\psi_G(e) = \{u, v\}$, then e is said to *join* u and v, and the vertices u and v are called the *ends* of e. We denote the numbers of vertices and edges in G by $v(G)$ and $e(G)$; these two basic parameters are called the *order* and *size* of G, respectively.

Definition 4 (Adjacency matrix). *The adjacency matrix of the above undirected graph G is the $n \times n$ matrix $A_G := (a_{uv})$, where a_{uv} is the number of edges joining vertices u and v and $n = v(G)$.*

Definition 5 (Clique). *A clique of an undirected graph G is a complete subgraph of G, i.e. one whose vertices are joined with each other by edges.*

Definition 6 (CLIQUE problem). *Given an undirected graph G and a number k, we are asked whether there is a clique of size k.*

As mentioned earlier, in what follows, $\mathbb{R}_{\geq 0}^n$ denotes the set of non-negative real number vectors $(x_1, x_2, \ldots, x_n)$ and $n = |V|$.

Theorem 1 (Motzkin and Straus [6]). *Let $G = (V, E)$ be an undirected graph with maximum clique size d. Let $\Delta_1 = \left\{ x \in \mathbb{R}_{\geq 0}^n : \sum_{i=1}^n x_i = 1 \right\}$. Then $\max_{x \in \Delta_1} x^T A_G x = \frac{d-1}{d}$.*

Corollary 1. *Let $G = (V, E)$ be an undirected graph with maximum clique size d. Let $A_G^{\tau, \rho}$ be a modified adjacency matrix of graph G where its entries with value 0 are replaced by $\tau \in \mathbb{R}$ and its entries with value 1 are replaced by $\rho \in \mathbb{R}$. Let $\Delta_1 = \left\{ x \in \mathbb{R}_{\geq 0}^n : \sum_{i=1}^n x_i = 1 \right\}$. Then $\max_{x \in \Delta_1} x^T A_G^{\tau, \rho} x = \tau + (\rho - \tau) \frac{d-1}{d}$.*

Proof. $x^T A_G^{\tau, \rho} x = x^T \left[\tau \cdot \mathbf{1} + (\rho - \tau) \cdot A_G \right] x = \tau + (\rho - \tau) \cdot x^T A_G x$, where $\mathbf{1}$ is the $n \times n$ matrix with value 1 in every entry. By Theorem 1 the result follows. $\square$

Corollary 2 (Etessami and Lochbihler [2]). *Let $G = (V, E)$ be an undirected graph with maximum clique size d and let $l \in \mathbb{R}_{\geq 0}$. Let $\Delta_l = \{x \in \mathbb{R}_{\geq 0}^n : \sum_{i=1}^n x_i = l\}$. Then $\max_{x \in \Delta_l} x^T A_G x = \frac{d-1}{d} l^2$.*

2 Robust Reductions

Definition 7 (Neighbourhood). *Let $v \in \mathbb{R}$. An (open) interval $I(v) = [a, b]$ ($I(v) = (a, b)$) with $a < b$ where $a \leq v \leq b$, is called a neighbourhood of v of range $|b - a|$.*

Definition 8 (Robust reduction under arbitrary perturbations of values). *We are given a valid reduction of a problem to a strategic game that involves a real matrix A of payoffs as entries a_{ij}. A consists of m partitions, with each partition's entries having the same value $v(t)$, for $t \in \{1, 2, \ldots, m\}$. Let $I(v(t)) \neq \emptyset$ be a neighbourhood of $v(t)$ and $w(t) \in I(v(t))$ be an arbitrary value in that neighbourhood. The reduction is called robust under arbitrary perturbations of values if it is valid for all the possible matrices W with entries $w(t)$.*

For a first extension based on the reduction of [2], see the full version [5].

2.1 A Robust Reduction from the Complement of CLIQUE to ESS

In the sequel we extend the idea of Etessami and Lochbihler [2] by replacing the constant payoff values they use with variables, and finding the intervals they belong to in order for the reduction to hold. We replace the zeros and ones of their reduction with $\tau \in \mathbb{R}$ and $\rho \in \mathbb{R}$ respectively. We also replace their function $\lambda'(k) = 1 - \frac{1}{k}$ with $\lambda(k) = 1 - \frac{1}{k^x}$, where $k \in \mathbb{N}$ and $x \geq 3$. Note that we can normalize the game's payoff values in $[0, 1]$ and retain the exact same ESSs.

Given an undirected graph $G = (V, E)$ we construct the following game $\Gamma_{k,\tau,\rho}^x(G) := (S, u_1)$ for suitable $\tau < \rho$ to be determined later. Note that from now on we will only consider rational τ and ρ so that every payoff value of the game is rational.

$S = V \cup \{a, b, c\}$ are the strategies for the players where $a, b, c \notin V$.
$n = |V|$ is the number of nodes.

- $u_1(i, j) = \rho$ for all $i, j \in V$ with $(i, j) \in E$.
- $u_1(i, j) = \tau$ for all $i, j \in V$ with $(i, j) \notin E$.
- $u_1(z, a) = \rho$ for all $z \in S - \{b, c\}$.
- $u_1(a, i) = \lambda(k) = 1 - \frac{1}{k^x}$ for all $i \in V$.
- $u_1(y, i) = \rho$ for all $y \in \{b, c\}$ and $i \in V$.
- $u_1(y, a) = \tau$ for all $y \in \{b, c\}$.
- $u_1(z, y) = \tau$ for all $z \in S$ and $y \in \{b, c\}$.

Theorem 2. *Let $G = (V, E)$ be an undirected graph. The game $\Gamma_{k,\tau,\rho}^x(G)$ with*

$$- \rho \in \left(1 + \frac{n^{x-1} - 2^x}{2^x n^{x-1}(n-1)}, \quad 1 + \frac{(n+1)^x - n2^x}{2^x(n+1)^x(n-1)}\right] \quad and$$
$$\tau \in \left[(1 - \rho)(n - 1) + 1 - \frac{1}{n^{x-1}}, \quad 1 - \frac{1}{2^x}\right)$$

or

$$- \rho \in \left(1 + \frac{(n+1)^x - n2^x}{2^x(n+1)^x(n-1)}, \quad +\infty\right) \quad and$$
$$\tau \in \left[(1 - \rho)(n - 1) + 1 - \frac{1}{n^{x-1}}, \quad (1 - \rho)(n - 1) + 1 - \frac{n}{(n+1)^x}\right)$$

has an ESS if and only if G has no clique of size k.

Proof. Let $G = (V, E)$ be an undirected graph with maximum clique size d. We consider the game $\Gamma_{k,\tau,\rho}^x(G)$ defined above. Suppose s is an ESS of $\Gamma_{k,\tau,\rho}^x(G)$.

For the reduction we will prove three claims by using contradiction, that taken together show that the only possible ESS s of $\Gamma_{k,\tau,\rho}(G)$ is the pure strategy a. Here we should note that these three claims hold not only for the aforementioned intervals of τ and ρ, but for any $\tau, \rho \in \mathbb{R}$ for which $\tau < \rho$.   □

Claim 1. The support of any possible ESS of $\Gamma_{k,\tau,\rho}^x(G)$ does not contain b or c ($supp(s) \cap \{b, c\} = \emptyset$).

Suppose $supp(s) \cap \{b, c\} \neq \emptyset$.

Let $t \neq s$ be a strategy with $t(i) = s(i)$ for $i \in V, t(y) = s(b) + s(c)$ and $t(y') = 0$ where $y, y' \in \{b, c\}$ such that $y \neq y'$ and $s(y) = min\{s(b), s(c)\}$. Since $u_1(b, z) = u_1(c, z)$ for all $z \in S$,

$$U_1(t, s) = \sum_{i \in V} t(i)U_1(i, s) + (t(b) + t(c))U_1(b, s) + t(a)U_1(a, s),$$

$$U_1(s, s) = \sum_{i \in V} s(i)U_1(i, s) + (s(b) + s(c))U_1(b, s) + s(a)U_1(a, s),$$

which yields $U_1(t, s) = U_1(s, s)$ and so t is a best response to s. Also,

$$U_1(s, t) = \sum_{i \in V} s(i)U_1(i, t) + (s(b) + s(c))U_1(b, t) + s(a)U_1(a, t),$$

$$U_1(t, t) = \sum_{i \in V} t(i)U_1(i, t) + (t(b) + t(c))U_1(b, t) + t(a)U_1(a, t),$$

which yields $U_1(s, t) = U_1(t, t)$. But this is a contradiction since it should be $U_1(s, t) > U_1(t, t)$ as s is an ESS.

Claim 2. The support of any possible ESS of $\Gamma_{k,\tau,\rho}^x(G)$ contains a ($supp(s) \not\subseteq V$).

Suppose $supp(s) \subseteq V$.

Then, we denote by A_G the adjacency matrix of the graph G.

$$U_1(s, s) = \sum_{i,j \in V} s(i)s(j)u_1(i, j) = x^T A_G^{\tau,\rho} x$$

$$\leq \tau + (\rho - \tau)\frac{d - 1}{d} \quad \text{(by Corollary 1)}$$
$$< \rho = U_1(b, s) \quad \text{for every } \rho > \tau.$$

But this is a contradiction since s is an ESS and therefore a NE. From Claim 1 and Claim 2, it follows that $a \in supp(s)$, i.e. $s(a) > 0$.

Claim 3. $s(a) = 1$.

Suppose $s(a) < 1$.

Since (s, s) is a NE, a is a best response to s and $a \neq s$. Then $U_1(s, a) = \sum_{z \in supp(s)} s(z) u_1(s, a) = \rho = U_1(a, a)$. But this is also a contradiction since it should be $U_1(s, a) > U_1(a, a)$ as s is an ESS.

Therefore, the only possible ESS of $\Gamma^x_{k,\tau,\rho}(G)$ is the pure strategy a. Now we show the following lemma, which concludes also the proof of Theorem 2.

Lemma 1. *The game $\Gamma^x_{k,\tau,\rho}(G)$ with the requirements of Theorem 2 has an ESS (strategy a) if and only if there is no clique of size k in graph G.*

Proof. We consider two cases for k:

Case 1: $d < k$. Let $t \neq a$ be a best response to a. Then $supp(t) \subseteq V \cup \{a\}$.

Let $r = \sum_{i \in V} t(i)$. So $r > 0, (t \neq a)$ and $t(a) = 1 - r$. Combining Corollaries 1 and 2 we get,

$$
\begin{aligned}
U_1(t, t) - U_1(a, t) &= \sum_{i,j \in V} t(i) t(j) u_1(i, j) + r \cdot t(a) \cdot \rho \\
&\quad + t(a) \cdot r \cdot \frac{k^x - 1}{k^x} + t(a)^2 \cdot \rho - \left[r \cdot \frac{k^x - 1}{k^x} + t(a) \cdot \rho \right] \\
&\leq \left[\tau + (\rho - \tau) \frac{d - 1}{d} \right] r^2 + r(1 - r) \cdot \rho \\
&\quad + (1 - r) r \frac{k^x - 1}{k^x} + (1 - r)^2 \cdot \rho - r \frac{k^x - 1}{k^x} - (1 - r) \cdot \rho \\
&= \frac{r^2}{d} E, \qquad \text{where } E = \tau - (1 - \rho)(d - 1) - \left(1 - \frac{d}{k^x}\right)
\end{aligned}
$$

If we can show that $E < 0$ then strategy a is an ESS. We show why $E < 0$:

Let's define the following function: $f(k, d, \rho) = (1 - \rho)(d - 1) + 1 - \frac{d}{k^x}$ with the restrictions: $k \geq d + 1, 1 \leq d \leq n, x \geq 3$.

By minimizing $f(k, d, \rho)$ with respect to k and d, we end up to 2 cases determined by the interval to which ρ belongs. So,

$$
\tau^* = \min_{k,d} f(k, d, \rho) = \begin{cases} 1 - \frac{1}{2^x}, & \text{if } \rho \leq 1 + \frac{(n+1)^x - n2^x}{2^x (n+1)^x (n-1)} \\[2mm] (1 - \rho)(n - 1) + 1 - \frac{n}{(n+1)^x}, & \text{if } \rho > 1 + \frac{(n+1)^x - n2^x}{2^x (n+1)^x (n-1)} \end{cases}
$$

Therefore, we can demand τ to be strictly less than τ^*, making $U_1(t, t) - U_1(a, t)$ negative. We conclude that when $d < k$ then strategy a is an ESS.

Case 2: $d \geq k$. Let $C \subseteq V$ be a clique of G of size k. Then t with $t(i) = \frac{1}{k}$ for $i \in C$ and $t(j) = 0$ for $j \in S \setminus C$ is a best response to a and $t \neq a$, and

$$U_1(t,t) = \sum_{i,j \in C} t(i)t(j)u_1(i,j) = \frac{(k-1)\rho + \tau}{k},$$

$$U_1(a,t) = \frac{k^x - 1}{k^x}. \quad \text{Then,}$$

$$U_1(t,t) - U_1(a,t) = \frac{1}{k}\left[\tau - (1-\rho)(k-1) - (1 - \frac{1}{k^{x-1}})\right]$$

$$= \frac{1}{k}E', \text{ where } E' = \tau - (1-\rho)(k-1) - (1 - \frac{1}{k^{x-1}})$$

If $E' \geq 0$ then a cannot be an ESS. We explain why $E' \geq 0$:
Let's define the following function:

$$y(k,\rho) = (1-\rho)(k-1) + 1 - \frac{1}{k^{x-1}}, \quad \text{with the restrictions: } k \leq d.$$

Then we define the function $z(d,\rho)$:

$$z(d,\rho) = \max_k y(k,\rho) = (1-\rho)(d-1) + 1 - \frac{1}{d^{x-1}}$$

$$\text{so,} \quad \tau^{**} = \max_d z(d,\rho) = (1-\rho)(n-1) + 1 - \frac{1}{n^{x-1}},$$

Now, given that τ needs to be at least τ^{**} but strictly less than τ^* the following should hold:

$$(1-\rho)(n-1) + 1 - \frac{1}{n^{x-1}} < 1 - \frac{1}{2^x}, \text{ or equivalently, } \rho > 1 + \frac{n^{x-1} - 2^x}{2^x n^{x-1}(n-1)}$$

So we conclude that when $d \geq k$ then strategy a is not an ESS. This completes the proof of Lemma 1 and Theorem 2. $\qquad \square$

Corollary 3. *The* ESS *problem with payoff values in the domains given in Theorem 2 is **coNP-hard**.*

3   Our Main Result

Now we can prove our main theorem:

Theorem 3. *Any reduction as in Theorem 2 for $x = x_0 \geq 3$ from the complement of the* CLIQUE *problem to the* ESS *problem is robust under arbitrary perturbations of values in the intervals:*

$$\tau \in \left[1 - \frac{1}{2^{x_0}} - D, 1 - \frac{1}{2^{x_0}} - D + B\right),$$

$$\rho \in \left(1 + \frac{(n+1)^{x_0} - n2^{x_0}}{2^{x_0}(n+1)^{x_0}(n-1)}, 1 + \frac{(n+1)^{x_0} - n2^{x_0}}{2^{x_0}(n+1)^{x_0}(n-1)} + A\right),$$

$$\lambda \in \left[1 - \frac{1}{k^{x_0}}, 1 - \frac{1}{k^{x_1}}\right],$$

where $x_1 \in (x_0, x_0 \log_n(n+1))$, $C = \frac{(n+1)^{x_0} - n^{x_1}}{n^{x_1-1}(n+1)^{x_0}(n-1)}$, $D = C(n-1)$, any $A \in (0, C)$ and $B = (C - A)(n-1)$.

Proof. We denote three partitions of the game's payoff matrix $U : U_\tau, U_\rho, U_\lambda$ disjoint sets, with $U_\tau \cup U_\rho \cup U_\lambda = U$ and values τ, ρ, λ of their entries respectively. Each set's entries have the same value. For every $\lambda \in \left[1 - \frac{1}{k^{x_0}}, 1 - \frac{1}{k^{x_1}}\right]$ there is a $x = -\log_k(1 - \lambda)$ in the interval $[x_0, x_1]$ such that $\lambda = 1 - \frac{1}{k^x}$, where $x_0 \geq 3$ and $x_1 \in (x_0, x_0 \log_n(n+1))$. We will show that, for this x, any reduction with the values of τ, ρ in the respective intervals stated in Theorem 2, is valid.

In Fig. 1, we show the validity area of τ depending on ρ with parameter x, due to Theorem 2. The thin and thick plots bound the validity area (shaded) for $x = x_0$ and $x = x_1$ respectively.

While x increases, the parallel lines of the lower and upper bound of τ move to the right, the horizontal line of the upper bound of τ moves up, and the left acute angle as well as the top obtuse angle of the plot move to the left (by examination of the monotonicity of those bounds with respect to x).

The lower bound of τ for an $x = x' > x_0$ equals the upper bound of τ for $x = x_0$, when $x' = x_0 \log_n(n+1)$. **Thus, for all $\mathbf{x} \in (\mathbf{x_0}, \mathbf{x_0} \log_\mathbf{n}(\mathbf{n+1}))$ there is a non-empty intersection between the validity areas.** We have picked an $x = x_1 \in (x_0, x_0 \log_n(n+1))$.

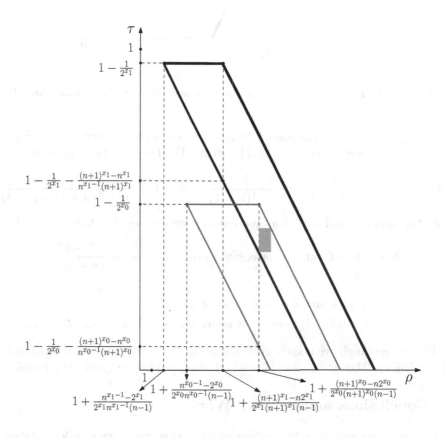

Fig. 1. The validity area of τ and ρ with parameter x.

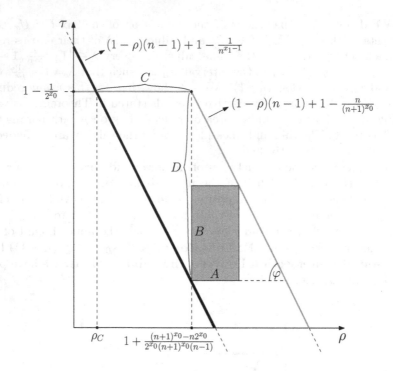

Fig. 2. Detail of the validity areas' intersection and the ρ, τ robust area (shaded).

In Fig. 2, we show a zoom-in of the intersection of the validity areas of Fig. 1. Let the intersection of lines: $1 - \frac{1}{2^{x_0}}, (1-\rho)(n-1)+1-\frac{1}{n^{x_1-1}}$ be at point $\rho = \rho_C$.

Then, $\rho_C = 1 - \dfrac{1}{2^{x_0}(n-1)} - \dfrac{1}{n^{x_1-1}(n-1)}$. So, $C = \dfrac{(n+1)^{x_0} - n^{x_1}}{n^{x_1-1}(n+1)^{x_0}(n-1)}$.

From the upper bound of τ as a function of ρ we can see that $\tan\varphi = n - 1$.

Thus, $D = C\tan\varphi$, or equivalently, $D = \dfrac{(n+1)^{x_0} - n^{x_1}}{n^{x_1-1}(n+1)^{x_0}}$.

Now we can pick any $A \in (0, C)$. So, it must be

$$B = (C - A)\tan\varphi, \quad \text{or equivalently,} \quad B = (n-1)(C - A).$$

For the rectangle with sides A, B shown in Fig. 2, the reduction is valid for all $x \in [x_0, x_1]$, thus for all $\lambda \in \left[1 - \frac{1}{k^{x_0}}, 1 - \frac{1}{k^{x_1}}\right]$. This completes the proof. □

4   Conclusions and Further Work

In this work we introduce the notion of reduction robustness under arbitrary perturbations within an interval and we provide a generalized reduction based on the one in [2] that proves **coNP**-hardness of ESS. We demonstrate that our generalised reduction is robust, thus showing that the hardness of the problem is preserved even after certain arbitrary perturbations of the payoff values of the derived game. As a future work we would like to examine the robustness of reductions for other hard problems, especially game-theoretic ones.

References

1. Conitzer, V.: The exact computational complexity of evolutionarily stable strategies. In: Chen, Y., Immorlica, N. (eds.) WINE 2013. LNCS, vol. 8289, pp. 96–108. Springer, Heidelberg (2013). doi:10.1007/978-3-642-45046-4_9
2. Etessami, K., Lochbihler, A.: The computational complexity of evolutionarily stable strategies. Int. J. Game Theory **37**(1), 93–113 (2008). doi:10.1007/s00182-007-0095-0
3. Lewontin, R.: Evolution and the theory of games. J. Theor. Biol. **1**(3), 382–403 (1961)
4. Smith, J.M., Price, G.R.: The logic of animal conflict. Nature **246**(5427), 15–18 (1973)
5. Melissourgos, T., Spirakis, P.: Existence of evolutionarily stable strategies remains hard to decide for a wide range of payoff values. ArXiv e-prints, January 2017
6. Motzkin, T.S., Straus, E.G.: Maxima for graphs and a new proof of a theorem of Turán. Can. J. Math. **17**, 533–540 (1965)
7. Nash, J.: Non-cooperative games. Ann. Math. **54**(2), 286–295 (1951)
8. Nisan, N.: A note on the computational hardness of evolutionary stable strategies. Electron. Colloq. Comput. Complex. (ECCC) **13**(076) (2006)
9. Papadimitriou, C., Yannakakis, M.: The complexity of facets (and some facets of complexity). J. Comput. Syst. Sci. **28**(2), 244–259 (1984)
10. Schaffer, M.E.: Evolutionarily stable strategies for a finite population and a variable contest size. J. Theor. Biol. **132**(4), 469–478 (1988)
11. Sergiu Hart, B.W., Rinott, Y.: Evolutionarily stable strategies of random games, and the vertices of random polygons. Ann. Appl. Probab. **18**(1), 259–287 (2008)
12. Smith, J.M.: The theory of games and the evolution of animal conflicts. J. Theor. Biol. **47**(1), 209–221 (1974)
13. Spielman, D.A., Teng, S.: Smoothed analysis: an attempt to explain the behavior of algorithms in practice. Commun. ACM **52**(10), 76–84 (2009)

Linear Search with Terrain-Dependent Speeds

Jurek Czyzowicz[1], Evangelos Kranakis[2]([✉]), Danny Krizanc[3], Lata Narayanan[4],
Jaroslav Opatrny[4], and Sunil Shende[5]

[1] Dép. d'informatique, Université du Québec en Outaouais, Gatineau, Canada
Jurek.Czyzowicz@uqo.ca
[2] School of Computer Science, Carleton University, Ottawa, ON, Canada
kranakis@scs.carleton.ca
[3] Department of Mathematics and Computer Science, Wesleyan University,
Middletown, CT, USA
krizanc@wesleyan.edu
[4] Department of Computer Science and Software Engineering, Concordia University,
Montreal, QC, Canada
lata@encs.concordia.ca, opatrny@cs.concordia.ca
[5] Department of Computer Science, Rutgers University, Camden, USA
shende@camden.rutgers.edu

Abstract. We revisit the *linear search* problem where a robot, initially placed at the origin on an infinite line, tries to locate a stationary target placed at an unknown position on the line. Unlike previous studies, in which the robot travels along the line at a constant speed, we consider settings where the robot's speed can depend on the direction of travel along the line, or on the profile of the terrain, e.g. when the line is inclined, and the robot can accelerate. Our objective is to design search algorithms that achieve good *competitive ratios* for the *time* spent by the robot to complete its search versus the time spent by an omniscient robot that knows the location of the target.

We consider several new robot mobility models in which the *speed* of the robot depends on the terrain. These include (1) different constant speeds for different directions, (2) speed with constant acceleration and/or variability depending on whether a certain segment has already been searched, (3) speed dependent on the incline of the terrain. We provide both upper and lower bounds on the competitive ratios of search algorithms for these models, and in many cases, we derive *optimal* algorithms for the search time.

Keywords: Search algorithm · Zig-zag algorithm · Competitive ratio · Linear terrain · Robot · Speed of movement

1 Introduction

Searching and exploration are fundamental problems in the areas of robotics and autonomous mobile agents. The objective for searching is to find a target

J. Czyzowicz, E. Kranakis and L. Narayanan—Research supported in part by NSERC Discovery grant.

D. Fotakis et al. (Eds.): CIAC 2017, LNCS 10236, pp. 430–441, 2017.
DOI: 10.1007/978-3-319-57586-5_36

placed at an unknown location in the domain in a provably optimal manner. In the *linear search* problem, the target is placed at a location on the infinite line unknown to the robot. The robot moves with uniform speed, and the goal is to find the target in minimum time. This problem was first proposed by Bellman [6] and independently by Beck [4].

Previous studies on the linear search problem generally assume that the robot moves with constant speed that is independent of the terrain. In this paper, we study a generalization of the problem where the horizontal line may be replaced by a more complicated (and hence, more realistic) continuous *linear terrain*. Moreover, the speed of the robot may depend in various ways on the nature or profile of the terrain. The robot initiates the search for the unknown target on the terrain from a reference starting point (without loss of generality, the origin). In our models, the robot can move with *different speeds* depending on its position on the terrain, its direction of movement, its exploration history etc. We also assume that the robot starts moving initially in the positive x-direction, or more informally, moving to the *right* (the *leftward* movement is in the negative x-direction).

Consider the linear search problem with a single robot. Since the position of the target is unknown to the robot, the robot cannot proceed indefinitely in just one direction and is forced to turn around and explore the terrain in the opposite direction as well; this *zig-zag* movement is inevitable and must be repeated periodically. The **canonical zig-zag search algorithm** is described below: note that the algorithm is parametrized by an infinite sequence of positive distances $X = \{x_k\}_{k \geq 1}$ from the origin that specifies the turning points. We refer to the sequence X as the *strategy*. To ensure progress in searching along a given direction, each trip away from the origin must cover *more* distance along the line than the previous trip in the same direction: this is formalized in the requirement that $x_k < x_{k+2}$ for all $k \geq 1$.

Input: Infinite sequence of distances $X = \{x_1, x_2, \ldots\}$ with $0 < x_k < x_{k+2}$ for all $k \geq 1$

for $k \leftarrow$ 1, 2, ... **do**
  if k *is odd (resp. even)* **then**
    move right (*resp.* left) a distance of x_k unless the target is found enroute;
    if *target found* **then**
      quit search
    end
    turn; then move left (*resp.* right), return to origin;
  end
end

Algorithm 1. Zig-Zag Search

A natural measure of the efficacy of the zig-zag search algorithm with strategy X, is how well it performs in competition with an omniscient adversary that knows the exact location of the target. Let $\sigma_X(d)$ be the ratio between the *time* taken by the robot using the zig-zag strategy X to reach an unknown target at distance d from the origin versus the *time* taken by the adversary to proceed directly to the target. Then, $\sigma_X \triangleq \sup_{d>1} \sigma_X(d)$ denotes the **competitive ratio** of the algorithm. We denote the optimal competitive ratio by σ^*.

For strategies where $x_k = \alpha r^{k-1}$ for some constant $\alpha > 0$, we call r the *expansion factor* of the strategy. Let D denote the *doubling strategy* that is a strategy with expansion factor 2 and $\alpha = 1$. Thus, $D = \{1, 2, 2^2, \ldots\}$. When the robot moves with unit speed in both directions, it is well-known that the doubling strategy is optimal: $\sigma^* = \sigma_D = 9$, see for example [2].

1.1   Our Results

A natural point of departure from traditional unit-speed models is to considering *linear terrains* in which the speed of the robot *depends* on the nature of the terrain or the environment. Two kinds of models are considered:

(1) **Two-speed models of linear search:** The robot can operate at two distinct constant speeds 1 and $s > 1$ in the following models.

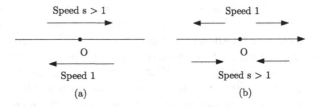

Fig. 1. Two-speed models based on (a) absolute direction and (b) direction relative to origin

- The absolute direction or *tailwind* model, viz. unit speed going left and tailwind speed $s > 1$ going right (see Fig. 1(a))
- The direction relative to the origin or the *beacon* model, viz. unit speed moving away from the origin and speed s moving towards it (see Fig. 1(b))
- The *exploration history* model, where the robot explores unknown regions slowly and deliberately with unit speed, but is able to search faster (with speed s) when it encounters a region already seen earlier in its search.

For the tailwind model, we analyze a *time-based* zig-zag search strategy in Subsect. 2.1, which is provably better than the doubling strategy. It turns out that the doubling zig-zag strategy is *optimal* for the beacon model; we prove this in Subsect. 2.2. We also show in Subsect. 2.3 that the exploration history model admits an asymptotically optimal strategy, whose expansion factor depends on the speed s.

(2) **Constant acceleration models for linear terrain search:** We first consider a linear search model with the property that whenever the robot starts

from rest (i.e. either initially from the origin, or when it turns around in the zig-zag search), its speed increases at a constant rate c until the next turn, i.e. at time t after starting from rest, the robot's speed is given by $s(t) = ct$, see Fig. 2(a). In Subsect. 3.1, we show that for this model, $6.36 < \sigma^* < \sigma_D \approx 11.1$.

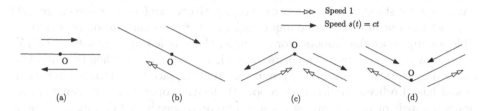

Fig. 2. Constant acceleration models: (a) Line (b) Inclined line (c) Hill (d) Valley

We then study search on *inclined linear terrains*. The robot can operate in two modes where it is moving with unit speed *going uphill* and with constant acceleration going *downhill.* The different terrains include an inclined line, a symmetric hill with the hill-top at the origin, or a symmetric valley with the valley-bottom at the origin as shown in Fig. 2(b), (c) and (d) respectively. Again, at time t from rest, the robot's speed *going downhill* is given by $s(t) = ct$. The increase in speed due to constant acceleration going downhill on a slope is a very natural manifestation of Newtonian physics: for example, we could interpret the constant c as the gravitational acceleration along the incline.

We analyze the doubling zig-zag strategy and lower bounds on the optimal strategy for the inclined line, hill, and valley models in Subsects. 3.2, 3.3, and 3.4 respectively. There are surprising differences in the nature of the results: while the competitive ratio of the doubling strategy is unbounded in the inclined line and hill models, we show that in the valley model, the competitive ratio is constant.

Due to space limitations all omitted proofs appear in the full version of the paper [16].

1.2 Related Work

Searching an environment or terrain with one or more searchers, possibly moving at different speeds, has the objective of localizing a hidden target in the minimum amount of time. Numerous variants of the search problem have been considered, e.g. with static or moving targets, multiple searchers with or without communication capabilities, and in environments that may not be fully known in advance.

The search problem has been extensively studied, e.g. see the survey by Benkoski *et al.* [7]; deterministic algorithms for optimal linear search [2]; incorporating a *turn cost* when a robot changes direction during the search [18]; when bounds on the distance to the target are known in advance [9]; and for moving targets or more general linear cost functions [8]. Other approaches include optimal randomized algorithms for the related *cow-path problem* [20], and stochastic and game theoretic investigations [1,5].

The search problem has also been studied in environments where search occurs in graphs (see, e.g. [19]) or along dynamically evolving links of networks [10,21]. More recently, variants of search using collections of *collaborating* robots have been investigated. The robots can employ either *wireless* communication (at any distance) or *face-to-face* communication, where communication is only possible among co-located robots. For example, the problem of *evacuation* [13,15] is essentially a search problem where search is completed only when the target is reached by the last robot. Linear group search in the face-to-face communication model has also been studied with robots that either operate at the same speed or with a pair of robots having distinct maximal speeds [3,11]. Finally, a new direction of research seeks to analyze linear search with multiple robots where some fraction of the robots may exhibit either *crash faults* [14] or *Byzantine faults* [12].

2   Two-Speed Models of Linear Search

In this section we consider linear search problems where the robot can switch between two different constant speeds depending on its absolute direction of movement (the *tailwind* model) or its direction of movement relative to the origin (the *beacon* model).

2.1   The Tailwind Model

In this model, the robot moves at speed $s > 1$ in the positive (right) direction and at unit speed in the negative (left) direction as depicted in Fig. 1(a). Observe that if we use the doubling strategy, the size of the explored segment expands by a factor of 2 in each iteration (i.e. between turns). However, the strategy favours the negative direction of the line in the sense that *it spends less time exploring the positive direction of the line* because the speed is higher when moving right.

To account for this, we propose a different strategy, *viz.* one that *balances* the search time on *both sides* of the origin. In other words, we expand the *time* spent on each side of the origin, rather than the distance travelled as follows. Fix two parameters $r > 1$ and $\alpha > 0$. Then, our strategy is defined as sequence $X = \{x_1, x_2, \ldots\} = \{s, \alpha r, r^2 s, \alpha r^3, \ldots\}$, i.e. with $x_{2k-1} = r^{2k-2}s$ and $x_{2k} = \alpha r^{2k-1}$ for $k \geq 1$.

Thus, strategy X spends *even powers* of r time moving to the right from the origin, and α times *odd powers* of r time moving the left. In the next theorem we show how to select the parameters α, r so as to optimize the search time. In particular, we prove the following result.

Theorem 1. *Assume the robot has speed $s \geq 1$ when moving left to right and speed 1 otherwise. For α, r such that $\alpha = (1 - s + \sqrt{(s-1)^2 + 4r^2 s})/(2r)$, $r = \sqrt{2 + (s+1)/\sqrt{s}}$, and $X = \{s, \alpha r, r^2 s, \alpha r^3, \ldots\}$ we have that $2 + 1/s \leq \sigma^* \leq \sigma_X$,*

$$\sigma_X \leq 1 + \frac{s + 2\sqrt{s} + 1}{s + \sqrt{s} + 1} \cdot \frac{s+1}{2s} \cdot \left(s + 1 + \sqrt{(s-1)^2 + 8s + 4\sqrt{s}(s+1)}\right) \quad (1)$$

Remark 1. Note that as $s \to \infty$, the righthand side of Inequality (1) approaches $\frac{3}{2} + s + o(s)$.

2.2 The Beacon Model

In this model the robot moves with speed 1 away from the origin and constant speed $s > 1$ towards the origin of the line.

Theorem 2. *The doubling strategy is **optimal** for the beacon model, i.e.*

$$\sigma^* = \sigma_D = 5 + \frac{4}{s}. \quad (2)$$

Proof. First we prove the upper bound. Assume the robot executes Algorithm 1 with the doubling strategy, and let the target be at distance d from the origin. Let k be such that $2^k < d \leq 2^{k+1}$. Since $2^k < d$, starting from the origin, by the k-th iteration of the algorithm the robot spends search time $2^0 + 2^0/s + 2^1 + 2^1/s + \cdots + 2^k + 2^k/s = (2^{k+1} - 1)(1 + 1/s)$ and returns to the origin without having found the target. In the next turn, the robot again starts from the origin, spends time $2^{k+1} + 2^{k+1}/s$ and returns back to the origin, since the adversary could place the target to the other side of the origin. Hence the total time spent so far is $(2^{k+2} - 1)(1 + 1/s)$. Finally, since $d \leq 2^{k+1}$, in the last turn the robot finds the target in time d. It follows that

$$\sigma_D = \sup_{d > 0} \frac{(2^{k+2} - 1)(1 + 1/s) + d}{d} < \frac{4d(1 + 1/s) + d}{d} = 5 + 4/s.$$

For the lower bound, we use a lower bounding technique [17] (itself based on [2]) used to obtain a lower bound of 9 for the unit speed model. Consider a deterministic strategy $X = (x_1, x_2, \ldots)$ with $x_i > 0$, for all $1 \leq i < \infty$. We consider several cases depending on the position of the target.

Assume the target is between x_k and x_{k+2}. Then the time it takes to find the target is equal to

$$x_1 + x_1/s + x_2 + x_2/s + \cdots + x_{k+1} + x_{k+1}/s + d = (1 + 1/s)\sum_{i=1}^{k+1} x_i + d.$$

It follows that the competitive ratio is

$$\sigma_X = \sup_k \sup_{d > x_k} \left\{1 + (1 + 1/s)\frac{\sum_{i=1}^{k+1} x_i}{d}\right\} = \sup_k \left\{1 + (1 + 1/s)\frac{\sum_{i=1}^{k+1} x_i}{x_k}\right\}$$

As a consequence it is easily seen that

$$\sigma_X \geq 2 + 1/s + (1 + 1/s)\frac{\sum_{i=1}^{k-1} x_i}{x_k} + (1 + 1/s)\frac{x_{k+1}}{x_k}$$

and hence, $\sigma_X x_k \geq (2 + 1/s)x_k + (1 + 1/s)\sum_{i=1}^{k-1} x_i + (1 + 1/s)x_{k+1}$. Separating x_{k+1} from the last inequality we derive

$$x_{k+1} \leq \frac{\sigma_X - (2 + 1/s)}{1 + 1/s} x_k - \sum_{i=1}^{k-1} x_i.$$

If we now put $\mu_0 = (\sigma_X - (2 + 1/s))/(1 + 1/s)$ and $\nu_0 = 1$, the last inequality can be rewritten as

$$x_{k+1} \leq \mu_0 x_k - \nu_0 \sum_{i=1}^{k-1} x_i, \tag{3}$$

At this point, we can use the technique suggested in [2,17] as follows. Induction on the Inequality (3) can be used to construct two infinite sequences of **positive** integers $\{\mu_i : i \geq 0\}$ and $\{\nu_i : i \geq 0\}$ defined via a system of recurrences of the form:

$$\mu_{m+1} = \mu_0 \mu_m - \nu_m \tag{4}$$
$$\nu_{m+1} = \nu_0 \mu_m + \nu_m. \tag{5}$$

for all $m \geq 0$. The recurrences (4), (5) can be solved using difference equations that yield the characteristic polynomial $z^3 - \mu_0 z^2 + (\mu_0 + 1) = 0$. This polynomial has $z = -1$ as one of its roots. Dividing by $z + 1$ we obtain the polynomial

$$z^2 - (\mu_0 + 1)z + (\mu_0 + 1) = 0 \tag{6}$$

whose two roots are $\rho_1, \rho_2 := \frac{\mu_0 + 1 \pm \sqrt{D}}{2}$, where $D := (\mu_0 - 1)^2 - 4$ is the discriminant of the quadratic Eq. (6). Note that $D < 0$ if and only if $\mu_0 < 3$. In turn, $\mu_0 < 3$ is equivalent to σ_X being less than a certain constant c, since σ_X and μ_0 are related (in the original proof of the lower bound, for instance, this yields $\sigma_X < c = 9$).

On the other hand, it can be shown that D being less than 0 implies that for some $k \geq 1$, the value μ_k is negative. This is a contradiction since μ_k must be positive by the construction for all k. Hence, D must in fact be greater than or equal to zero, and it follows that the CR σ_X is greater than or equal to c.

By applying the above lower bound technique to the present case, we conclude that the roots of the resulting quadratic equation are conjugate complex numbers with non-zero imaginary parts iff $\mu_0 < 3$, which is equivalent to $\sigma_X < 5 + 4/s$. This completes the proof of the lower bound of Theorem 2.   □

Remark 2. Assume that the robot moves with speed 1 towards the origin, and speed s away from the origin of the line. A proof similar to that of Theorem 2 shows that Algorithm 1 with the doubling strategy is optimal and its competitive ratio is also $5 + \frac{4}{s}$. Details are left to the reader.

2.3   The Exploration History Model

We consider a robot that moves at speed 1 when searching for the target, but the robot can move at speed $s > 1$ when moving over a part of the line already explored. For example, the robot's attention to identifying the target limits the speed at which the robot can move.

Theorem 3. *Let* $r = 1 + \sqrt{2/(s+1)}$, *and* $X = (r^0, r^1, r^2, \ldots)$ *be an expansion strategy. Then, with this strategy, the zig-zag algorithm's competitive ratio satisfies*

$$2 + 1/s \leq \sigma^* \leq \sigma_X = 2 + \frac{1}{s}\left(3 + 2\sqrt{2s+2}\right). \tag{7}$$

Proof. Consider first the lower bound. The robot must visit both points $+d$ and $-d$. Without loss of generality assume that $-d$ is the first point to be visited by the robot. Then the adversary will place the target at $+d$. Therefore the robot will traverse the segment $[-d, 0]$ once with speed at least 1 to reach $-d$ from 0 and a second time with speed s on its way to $+d$ from $-d$. The resulting competitive ratio is at least $(2d + d/s)/d = 2 + 1/s$. This proves the lower bound.

Next we look at the upper bound. Consider a robot following zig-zag strategy $X = (r^1, r^2, r^3, \ldots)$ and that the first move of the robot is to the right with the target located at distance d with $r^k < d \leq r^{k+2}$. The time needed by the robot to find the target is equal to

$$r^1 + r^1/s + r^2 + (r^2 + r^1)/s + r^3 - r^1 + \cdots + (r^k + r^{k-2})/s + r^{k+1} - r^{k-1}$$

$$+ (r^{k+1} + r^k)/s + d - r^k = \left(1 + \frac{1}{s}\right)r^{k+1} + \frac{2}{s}\sum_{i=1}^{k}\frac{r^i}{s} + d$$

It follows that the competitive ratio of this strategy σ_X is

$$\sigma_X = \sup_{k \geq 1}\left(\left(\left(1 + \frac{1}{s}\right)r^{k+1} + \frac{2}{s}\sum_{i=1}^{k}\frac{r^i}{s} + r^k\right)/r^k\right) = \left(1 + \frac{1}{s}\right)r + \frac{2r}{s(r-1)} + 1.$$

To find the optimal value of r we put the derivative $d\sigma_X/dr = 1 + 1/s - 2/(s(r-1)^2)$ equal to 0, which gives us that the competitive ratio is optimized for $r = 1 + \sqrt{2/(s+1)}$ and for this r, we obtain $\sigma_X = 2 + \frac{1}{s}(3 + 2\sqrt{2s+2})$. $\square$

Remark 3. For example, if $s = 1$ then $\sigma = 9$, if $s = 2$ then $\sigma \approx 5.95$, if $s = 3$ then $\sigma \approx 4.88$, and if $s = 4$ then $\sigma \approx 4.33$. Thus as $s \to \infty$ the value of r approaches 1 and the competitive ratio σ as given in Theorem 3 approaches 2. Therefore, the strategy is asymptotically optimal in s.

3 Searching with Constant Acceleration

In this framework, the robot exhibits constant acceleration $c > 0$ in some part of the linear terrain when starting from rest. As is well known from Newtonian physics, at time t after the robot accelerates from rest, it will be moving with speed $s = ct$ and would have covered a distance of $x(t) = ct^2/2$. Thus, to cover distance x we need time $\sqrt{2x/c}$.

3.1   Constant Acceleration in both Directions

Here we assume that the constant acceleration applies in both directions through-out the entire terrain (see Fig. 2(a)).

Theorem 4. *Assume the robot is searching with constant acceleration c in either direction, starting from rest initially, as well as at turning points. Then:*

$$3(\sqrt{2} + 1/\sqrt{2}) \leq \sigma^* \leq \sigma_D \leq \frac{2\sqrt{3}}{\sqrt{2} - 1} + \sqrt{3} + 1 \tag{8}$$

3.2   Moving on an Inclined Line

In this section we consider the situation where the robot has unit speed in one direction, but in the other direction, due to the inclination of the line, the robot is subjected to a constant acceleration c.

Consider a target at distance $d > 1$ from the origin. In the theorem below we show that the doubling strategy has unbounded competitive ratio.

Theorem 5. *Assume the robot moves with acceleration c in the positive direction, and constant speed 1 in the negative direction using the doubling strategy. Then for any $d \geq 1$,*

$$\sqrt{2c}\sqrt{d} < \sigma_D(d) \leq \sqrt{8c} \cdot \sqrt{d} + O(1).$$

Furthermore, $\sigma^* \geq \sup_{d>1} \min\{2 + \sqrt{2/(cd)}, \sqrt{2} + \sqrt{cd/2}\}$.

Proof. Consider the lower bound first. The robot must visit both points $+d$ and $-d$. Assume that $-d$ is the first point to be visited by the robot. Then the adversary will place the target at $+d$. Therefore the robot will traverse the segment $[-d, 0]$ in time at least d and then move downhill a distance $2d$ to the target. Thus $\sigma^* \geq (d + \sqrt{4d/c})/\sqrt{2d/c} = \sqrt{cd/2} + \sqrt{2}$. Now assume that $+d$ is the first point to be visited by the robot. Then the adversary will place the target at $-d$. Therefore the robot will traverse the segment $[0, +d]$ in time at least $\sqrt{2d/c}$ and then move uphill a distance $2d$ to the target. Thus $\sigma^* \geq (\sqrt{2d/c} + 2d)/d = 2 + \sqrt{2/(cd)}$, which proves the lower bound.

Next we look at the upper bound. Assume the robot executes the doubling strategy and let the target be at distance d from the origin. Let k be such that $2^k < d \leq 2^{k+1}$ and $b = \sqrt{2/c}$. There are two cases to consider depending on the parity of k.

The Target is Uphill and k is Even. Since $2^k < d$, starting from the origin, the robot spends search time

$$b + (2^0 + 2^1) + b\sqrt{2^1 + 2^2} + \cdots + (2^{k-1} + 2^k) + b\sqrt{2^k + 2^{k+1}} + 2^{k+1} + d$$

$$= 2^{k+2} - 1 + b + b\sqrt{2^1 + 2^2} + \cdots + b\sqrt{2^k + 2^{k+1}} + d$$

$$< 5d + b + b\sqrt{3}(\sqrt{2^0} + \sqrt{2^2} + \cdots + \sqrt{2^k}) = 5d + b\sqrt{3}(2^{k/2+1} - 1)$$

The target is uphill at distance d; therefore if we divide the above expression by d, we conclude that in this case $\sigma_D(d) \leq 5 + O(d^{-1/2})$.

The Target is Downhill and k is Odd. Starting from the origin, the robot spends search time

$$T = 1 + b\sqrt{2^0 + 2^1} + 2^1 + 2^2 + \cdots + b\sqrt{2^{k-1} + 2^k} + 2^k + 2^{k+1} + b\sqrt{2^{k+1} + d}$$
$$= 2^{k+2} - 1 + b\sqrt{2^0 + 2^1} + \cdots + b\sqrt{2^{k-1} + 2^k} + \sqrt{2^{k+1} + d}$$
$$= 2^{k+2} - 1 + b\sqrt{3}(2^{(k+1)/2} - 1) + \sqrt{2^{k+1} + d}$$

Since $2^k < d \leq 2^{k+1}$, we have that

$$2d - 1 + b\sqrt{3}(\sqrt{d} - 1) < T < 4d + b\sqrt{3}\sqrt{2d} + b\sqrt{3d}.$$

The target is downhill at distance d; if we divide the above expression by $b\sqrt{d}$, we conclude that in this case

$$\frac{2d}{b\sqrt{d}} = \sqrt{2c}\sqrt{d} < \sigma_D(d) \leq \frac{4}{b}\sqrt{d} + O(1) = \sqrt{8c} \cdot \sqrt{d} + O(1).$$

$$\square$$

3.3   Starting at the Top of a Hill

This model differs from the previous one by having the origin of the line located on the top of a hill. Thus, the speed of a robot increases when going downhill from the origin. Namely it travels with constant speed 1 uphill but has a constant acceleration when going downhill. The main result here is that the competitive ratio of the optimal search algorithm is unbounded. Notice that if a robot has initial speed 1 at the top of the hill then when going downhill with constant acceleration c it has at time t speed $1 + ct$, and to covers distance x it needs time $(\sqrt{1 + 2cx} - 1)/c$.

Theorem 6. *Assume that the robot travels with constant acceleration c away from the origin, and with unit speed towards the origin. Then $\sigma_D(d) = \Theta(\sqrt{d})$ and this is optimal.*

Proof. The upper bound proof uses the main idea of the upper bound in Theorem 5. However, unlike in Theorem 5, the analysis of the algorithm is now symmetric. As before, let k be such that $2^k < d \leq 2^{k+1}$ and $b = \sqrt{2/c}$. Since $2^k < d$, starting from the origin, the robot spends search time

$$b + 1 + (\sqrt{1 + c2^2} - 1)/c + 2^1 + (\sqrt{1 + c2^3} - 1)/c + 2^2$$
$$+ \cdots + (\sqrt{1 + c2^{k+2}} - 1)/c + 2^{k+1} + (\sqrt{1 + cd} - 1)/c$$
$$= b + \frac{1}{c}(-(k+1) + \sqrt{1 + cd} + \sum_{i=2}^{k+2} \sqrt{1 + c2^i}) + \sum_{i=0}^{k+1} 2^i$$
$$< b + \sqrt{1 + cd}/c + b\sum_{i=3}^{k+3} \sqrt{2^i} + \sum_{i=0}^{k+1} 2^i < b + 2^{k+2} + b2^{(k+4)/2} + \sqrt{1 + cd}/c$$
$$< b + 4d + b(1 + 4\sqrt{d}) + \sqrt{1 + cd}/c.$$

The target is downhill at distance d; if we divide the above expression by $b\sqrt{d}$, we get that the competitive ratio in this case is at most $\frac{b+4d+b(1+4\sqrt{d})+\sqrt{1+cd}/c}{b\sqrt{d}} = \frac{4}{b}\sqrt{d} + O(1)$.

To see the lower bound, observe that the robot must visit both points $+d$ and $-d$. Assume that $-d$ is the first point to be visited by the robot. Then the adversary will place the target at $+d$. Therefore the robot will traverse the segment $[-d, 0]$ in time at least $b\sqrt{d}$. To get to $+d$, the robot needs time at least d to get to the origin and another $(\sqrt{1 + 2cd} - 1)/c = \sqrt{1/c^2 + b^2 d}$ to reach the target.

The omniscient optimal algorithm needs time $b\sqrt{d}$. Thus, for any strategy X,

$$\sigma_X(d) \geq \frac{b\sqrt{d} + d + \sqrt{1/c^2 + b^2 d}}{b\sqrt{d}} = 1 + \sqrt{d}/b + \sqrt{\frac{1}{c^2 b^2 d} + 1}.$$

$\square$

3.4   Starting at the Bottom of a Valley

An interesting situation occurs if we reverse the speeds, i.e., the origin is located at the bottom of a valley and thus we have constant acceleration when moving towards the origin, but the robot moves at unit speed away from the origin (see Fig. 2). In this case can prove the following theorem:

Theorem 7. *Assume that the robot travels with constant acceleration c towards the origin, and with unit speed away from the origin. Then for any $d \geq 1$:*

$$\sigma_D(d) \leq 5 + O(d^{-1/2})$$

Furthermore, $\sigma^* \geq 5$.

4   Discussion

In this paper we have considered and analyzed several zig-zag strategies for search on a linear terrain for cases when the speed of the robot is not constant. Our work provides an initial step for the study of a robot searching terrains of different profiles for a target placed at an unknown location. We study two kinds of models of speed: two-speed models, and constant acceleration models. An interesting observation is in our two-speed models, as in the traditional one-speed model, the performance of the doubling algorithm vis-a-vis an omniscient optimal algorithm gets worse as d (the distance of the target to the initial location) increases and converges to some maximum value as $d \to \infty$. However, in the constant acceleration models that we studied, either the competitive ratio is unbounded, or the performance of the doubling algorithm improves vis-a-vis an omniscient optimal algorithm as d increases.

References

1. Alpern, S., Gal, S.: The Theory of Search Games and Rendezvous, vol. 55. Kluwer Academic Publishers, Dordrecht (2002)

2. Baeza Yates, R., Culberson, J., Rawlins, G.: Searching in the plane. Inf. Comput. **106**(2), 234–252 (1993)
3. Bampas, E., Czyzowicz, J., Gasieniec, L., Ilcinkas, D., Klasing, R., Kociumaka, T., Pajak, D.: Linear search by a pair of distinct-speed robots. In: Suomela, J. (ed.) SIROCCO 2016. LNCS, vol. 9988, pp. 195–211. Springer, Cham (2016). doi:10.1007/978-3-319-48314-6_13
4. Beck, A.: On the linear search problem. Israel J. Math. **2**(4), 221–228 (1964)
5. Beck, A., Warren, P.: The return of the linear search problem. Israel J. Math. **14**(2), 169–183 (1973)
6. Bellman, R.: An optimal search. SIAM Review **5**(3), 274–274 (1963)
7. Benkoski, S., Monticino, M., Weisinger, J.: A survey of the search theory literature. Naval Res. Logist. (NRL) **38**(4), 469–494 (1991)
8. Bose, P., De Carufel, J.-L.: A general framework for searching on a line. In: Kaykobad, M., Petreschi, R. (eds.) WALCOM 2016. LNCS, vol. 9627, pp. 143–153. Springer, Cham (2016). doi:10.1007/978-3-319-30139-6_12
9. Bose, P., De Carufel, J.-L., Durocher, S.: Revisiting the problem of searching on a line. In: Bodlaender, H.L., Italiano, G.F. (eds.) ESA 2013. LNCS, vol. 8125, pp. 205–216. Springer, Heidelberg (2013). doi:10.1007/978-3-642-40450-4_18
10. Casteigts, A., Flocchini, P., Quattrociocchi, W., Santoro, N.: Time-varying graphs and dynamic networks. In: Frey, H., Li, X., Ruehrup, S. (eds.) ADHOC-NOW 2011. LNCS, vol. 6811, pp. 346–359. Springer, Heidelberg (2011). doi:10.1007/978-3-642-22450-8_27
11. Chrobak, M., Gasieniec, L., Gorry, T., Martin, R.: Group search on the line. In: Italiano, G.F., Margaria-Steffen, T., Pokorný, J., Quisquater, J.-J., Wattenhofer, R. (eds.) SOFSEM 2015. LNCS, vol. 8939, pp. 164–176. Springer, Heidelberg (2015). doi:10.1007/978-3-662-46078-8_14
12. Czyzowicz, J., Georgiou, K., Kranakis, E., Krizanc, D., Narayanan, L., Opatrny, J., Shende, S.: Search on a line with Byzantine robots. In: ISAAC, LIPCS (2016)
13. Czyzowicz, J., Georgiou, K., Kranakis, E., Narayanan, L., Opatrny, J., Vogtenhuber, B.: Evacuating robots from a disk using face-to-face communication (extended abstract). In: Paschos, V.T., Widmayer, P. (eds.) CIAC 2015. LNCS, vol. 9079, pp. 140–152. Springer, Cham (2015). doi:10.1007/978-3-319-18173-8_10
14. Czyzowicz, J., Kranakis, E., Krizanc, D., Narayanan, L., Opatrny. J.: Search on a line with faulty robots. In: PODC, pp. 405–414 (2016)
15. Czyzowicz, J., Kranakis, E., Krizanc, D., Narayanan, L., Opatrny, J., Shende, S.: Wireless autonomous robot evacuation from equilateral triangles and squares. In: Papavassiliou, S., Ruehrup, S. (eds.) ADHOC-NOW 2015. LNCS, vol. 9143, pp. 181–194. Springer, Cham (2015). doi:10.1007/978-3-319-19662-6_13
16. Czyzowicz, J., Kranakis, E., Narayanan, L., Opatrny, J., Shende, S.: Linear search with terrain-dependent speeds (2017). http://arxiv.org/abs/1701.03047
17. De Carufel, J.-L.: Personal communication
18. Demaine, E.D., Fekete, S.P., Gal, S.: Online searching with turn cost. Theoret. Comput. Sci. **361**(2), 342–355 (2006)
19. Fomin, F.V., Thilikos, D.M.: An annotated bibliography on guaranteed graph searching. Theoret. Comput. Sci. **399**(3), 236–245 (2008)
20. Kao, M.-Y., Reif, J.H., Tate, S.R.: Searching in an unknown environment: an optimal randomized algorithm for the cow-path problem. Inf. Comput. **131**(1), 63–79 (1996)
21. Kuhn, F., Lynch, N., Oshman, R.: Distributed computation in dynamic networks. In: Proceedings of the Forty-Second ACM Symposium on Theory of Computing, pp. 513–522. ACM (2010)

Linear-Time Generation of Random Chordal Graphs

Oylum Şeker[1]($\boxtimes$), Pinar Heggernes[2], Tınaz Ekim[1], and Z. Caner Taşkın[1]

[1] Department of Industrial Engineering, Boğaziçi University, Istanbul, Turkey
{oylum.seker,tinaz.ekim,caner.taskin}@boun.edu.tr
[2] Department of Informatics, University of Bergen, Bergen, Norway
pinar.heggernes@uib.no

Abstract. Chordal graphs form one of the most well studied graph classes. Several graph problems that are NP-hard in general become solvable in polynomial time on chordal graphs, whereas many others remain NP-hard. For a large group of problems among the latter, approximation algorithms, parameterized algorithms, and algorithms with moderately exponential or sub-exponential running time have been designed. Chordal graphs have also gained increasing interest during the recent years in the area of enumeration algorithms. Being able to test these algorithms on instances of chordal graphs is crucial for understanding the concepts of tractability of hard problems on graph classes. Unfortunately, only few published papers give algorithms for generating chordal graphs. Even in these papers, only very few methods aim for generating a large variety of chordal graphs. Surprisingly, none of these methods is based on the "intersection of subtrees of a tree" characterization of chordal graphs. In this paper, we give an algorithm for generating chordal graphs, based on the characterization that a graph is chordal if and only if it is the intersection graph of subtrees of a tree. The complexity of our algorithm is linear in the size of the produced graph. We give test results to show the variety of chordal graphs that are produced, and we compare these results to existing results.

1 Introduction

Algorithms particularly tailored to exploit properties of various graph classes have formed an increasingly important area of graph algorithms during the last five decades. With the introduction of relatively new theories for coping with NP-hard problems, like parameterized algorithms, algorithmic research on graph classes has become even more popular recently, and the number of results in this area appearing at international conferences and journals is now higher than ever. One of the most studied graph classes in this context is the class of chordal graphs, i.e., graphs that contain no induced cycle of length 4 or more. Chordal graphs arise in practical applications from a wide variety of unrelated fields,

This work is supported by the Research Council of Norway, Bogazici University Research Fund (grant 11765), and Turkish Academy of Sciences GEBIP award.

© Springer International Publishing AG 2017
D. Fotakis et al. (Eds.): CIAC 2017, LNCS 10236, pp. 442–453, 2017.
DOI: 10.1007/978-3-319-57586-5_37

like sparse matrix computations, database management, perfect phylogeny, VLSI, computer vision, knowledge based systems, and Bayesian networks [6,13,21,24,26]. This graph class that first appeared in the literature as early as 1958 [14], has steadily increased its popularity, and there are now more than 20 thousand references on chordal graphs according to Google Scholar.

With a large number of existing algorithms specially tailored for chordal graphs, it is interesting to note that not much has been done to test these algorithms in practice. Very few such tests are available as published articles [2,18,22]. In particular, there seems to be no efficient and all-purpose chordal graph generator available. Most of the work in this direction involves generating chordal graphs tailored to test a particular algorithm or result [2,22]. This is a clear shortcoming for the field, and it was even mentioned as an important open task at a Dagstuhl Seminar [16]. Until some years ago, most of the algorithms tailored for chordal graphs had polynomial running time, and testing was perhaps not crucial. Now, however, many parameterized and exponential-time algorithms exist for chordal graphs, for problems that remain hard on this graph class, see e.g., [4,12,19,20]. The proven running times of such algorithms might often be too high compared to the practical running time. Just to give some examples from the field of enumeration, there are now several algorithms and upper bounds on the maximum number of various objects in chordal graphs [1,11,12]. However, the lower bound examples at hand usually do not match these upper bounds. Tests on random chordal graphs is a good way of getting better insight about whether the known upper bounds are too high or tight.

In this paper we present an algorithm for generating random chordal graphs. The algorithm is based on the characterization that a graph is chordal if and only if it is the intersection graph of subtrees of a tree. Surprisingly, this characterization does not seem to have been exploited for chordal graph generation earlier. The running time of our algorithm is linear in the size of the generated graph, and it generates a large variety of chordal graphs, where the variety is measured using the characteristics of maximal cliques as already used in [22]. After proving the correctness and the time complexity, we give extensive tests to demonstrate the kind of chordal graphs that our algorithm generates. We compare our tests with existing test results, and we implement one of the earlier proposed methods and include this in our tests. According to these tests, our algorithm outperforms previous algorithms both with respect to complexity and with respect to the richness of the family of the generated chordal graphs. Observe that graph isomorphism is as hard on chordal graphs as on general graphs [17], which adds to the difficulty of producing chordal graphs uniformly random. Still our algorithm is able to generate every chordal graph with positive probability.

2   Background, Terminology and Existing Algorithms

In this section we give the necessary background on chordal graphs, as well as a short review of the existing algorithms for chordal graph generation. We work with simple and undirected graphs, and we use standard graph terminology.

We let n denote the number of vertices and m denote the number of edges of a graph. A *maximal clique* is an inclusion-wise maximal set of vertices that are pairwise adjacent. An ordering $(v_1, v_2, \ldots, v_n)$ of the vertices of a graph is a *perfect elimination order (peo)* if the set of higher numbered neighbors of each vertex forms a clique.

Let $F = \{S_1, S_2, \ldots, S_n\}$ be a family of sets from the same universe. A graph G is called an *intersection graph of F* if there is a bijection between the set of vertices $\{v_1, v_2, \ldots, v_n\}$ of G and the sets in F such that v_i and v_j are adjacent if and only if $S_i \cap S_j \neq \emptyset$, for $1 \leq i, j \leq n$. In the special case where there is a tree T such that each set in F corresponds to the vertex set of a subtree of T, then G is called the *intersection graph of subtrees of a tree*. In this case, we call T a *host tree* for G.

A tree T with a bijection between its vertex set and the set of maximal cliques of a graph G, is called a *clique tree* of G if, for every vertex v of G, the set of vertices of T that correspond to the cliques containing v induce a connected subtree of T.

A graph is *chordal* if it contains no induced cycle of length 4 or more. A chordal graph on n vertices has at most n maximal cliques [7]. Chordal graphs have many different characterizations. For our purposes, the following will be sufficient.

Theorem 1 [5,8–10]. *Let G be a graph. The following are equivalent.*

- *G is chordal.*
- *G has a perfect elimination order.*
- *G is the intersection graph of subtrees of a tree.*
- *G has a clique tree.*

Especially the last two points of Theorem 1 are crucial for our algorithm and its implementation. To make sure that there is no confusion between the vertices of G and the vertices of a host tree or a clique tree, we will from now on refer to vertices of a tree as *nodes*.

Rose, Tarjan, and Lueker [25] gave an algorithm called Maximal Cardinality Search (MSC) that creates a perfect elimination order of a chordal graph in time $O(n + m)$. Blair and Peyton [3] gave a modification of MCS to list all the maximal cliques of a chordal graph in time $O(n + m)$. Implicit in their proofs is the following well-known fact that is not often highlighted on its own.

Lemma 1 [3,25]. *The sum of the sizes of the maximal cliques of a chordal graph is $O(n + m)$.*

Next, we briefly mention the algorithms for generating chordal graphs from the works of Andreou, Papadopoulou, Spirakis, Theodorides, and Xeros [2]; Pemmaraju, Penumatcha, and Raman [22]; and Markenzon, Vernet, and Araujo [18]. Some of these algorithms create very limited chordal graphs, which is either mentioned by the authors or clear from the algorithm. Thus, in the following we only mention the algorithms that are general enough to be interesting in our context.

It should also be noted that the purpose of Andreou et al. [2] is not to obtain general chordal graphs, but rather chordal graphs with a known bound on some parameter. One of the algorithms that they propose starts from an arbitrary graph and adds edges to obtain a chordal graph. How the edges are added is not given in detail, but note that there are many algorithms for generating a chordal graph from a given graph by adding a minimal set of edges and their running time is usually $O(nm)$, far from linear [15]. Andreou et al. [2] do not report on the quality of chordal graphs obtained by this method.

We highlight below the algorithms that are the most promising with respect to generating random chordal graphs. In addition to these, there is an $O(n^2)$-time algorithm by Markenzon et al. [18] that generates a random tree and adds edges to this tree until a chordal graph with desired edge density is obtained. However, no test results about the quality of the generated graphs is given.

Alg 1 [2]. The algorithm constructs a chordal graph by using a peo. At every iteration, a new vertex is added and made adjacent to a random selection of already existing vertices. Then necessary edges are added to turn the neighborhood of the new vertex into a clique. No test results are given in the paper about the quality of the chordal graphs this algorithm produces. As we found the algorithm interesting, we have implemented it, and we compare the resulting graphs to those generated by our algorithm in Sect. 4.

Alg 2 [18, 22]. The algorithm starts from a single vertex. At each subsequent step, a clique C in the existing graph is chosen at random, and a new vertex is added adjacent to exactly the vertices of C. The inverse of the order in which the vertices are added is a peo of the final graph. It is observed by the authors of both papers that this procedure results in chordal graphs with approximately $2n$ edges experimentally. They propose the following changes:

Alg 2a [18] modifies the above generated graph by randomly choosing maximal cliques that are adjacent according to the clique tree and merging these until desired edge density is obtained. Some test results about the graphs generated by Alg 2a are provided in [18]. Although these tests are not as comprehensive as the tests we give on our algorithm in Sect. 4, we compare our results to those of [18] as best we can. The running time of Alg 2a is $O(m + n\alpha(2n, n))$.

Alg 2b [22] is a modification of Alg 2 in a different way: instead of randomly choosing a clique, a maximum clique is chosen and a random subset of it is made adjacent to the new vertex. Although test results for Alg 2b are provided in [22], the authors acknowledge that the produced graphs are still very particular with very few large maximal cliques and many very small maximal cliques. For this reason, we do not include Alg 2b in our comparisons.

3   Generating Chordal Graphs Using Subtrees of a Tree

We find it surprising that the intersection graph of subtrees of a tree characterization of chordal graphs has not been used for generation. One reason could be that this characterization does not give a direct way to decide the number of edges. However, as we will see, edge density can be regulated by adjusting the sizes of the generated subtrees. We are now ready to present our main algorithm for generating chordal graphs on n vertices:

Algorithm ChordalGen

Input: Two integers n and k

Output: A chordal graph G on n vertices and m edges

Generate a tree T on n nodes uniformly at random
Create n random subtrees of $T : \{T_1, \ldots, T_n\}$ of average size k
Output as G the intersection graph of the trees $\{T_1, \ldots, T_n\}$.

By Theorem 1 the graph generated by Algorithm ChordalGen is chordal. We want to show that this high level definition of the algorithm is general and can create any chordal graph. The proof of the following lemma is already implicit in the proofs of the relevant parts of Theorem 1. We give it here, as it will also be of help in the explanation of the running time of our algorithm.

Lemma 2. *Let G be a chordal graph on n vertices and m edges. There is an execution of Algorithm ChordalGen that generates G.*

Proof. First of all we want to show that there is a host tree T on exactly n nodes, and a set of n subtrees of T, such that G is the intersection graph of these subtrees. Let T' be a clique tree of G. Let us call the vertices of G: $v_1, v_2, \ldots, v_n$. Define subtree T_i' to be the subtree of T' that corresponds to the nodes (maximal cliques) that contain vertex v_i, for $1 \leq i \leq n$. By the definition of a clique tree, T' has at most n nodes and each T_i' is a connected subgraph of T'. If T' has less than n nodes, we can add new nodes adjacent to arbitrary nodes of T' until we get a new tree T with exactly n nodes. The subtrees stay the same. As two vertices are adjacent in G if and only if they appear together in a clique, G is the intersection graph of subtrees $T_1', \ldots, T_n'$ of T. Finally, we simply let k be the average size of the subtrees T_i.   □

The most interesting part of the algorithm is the generation of the subtrees of T. For this, we propose an algorithm called SubtreeGen as follows.

Algorithm SubtreeGen

Input: A tree T on n nodes and an integer k

Output: A set of n subtrees of T of average size k

for $i = 1$ **to** n **do**
  Select a random node x of T and set $T_i = \{x\}$
  Select a random integer $k_i \leq n$ between 1 and $2k - 1$
  for $j = 1$ **to** $k_i - 1$ **do**
    Select a random node y of T_i that has neighbors in T outside of T_i
    Select a random neighbor z of y outside of T_i and add z to T_i
  Output $\{T_1, T_2, \ldots, T_n\}$

Lemma 3. *The running time of Algorithm SubtreeGen is* $O(n + \sum_{1=i}^{n} |T_i|)$.

Proof. Observe first that each subtree T_i is simply a list of nodes of T. We show that after an initial $O(n)$ preprocessing time, each subtree T_i can be generated in time $O(|T_i|)$. For this, we need to be able to add a new node to T_i in constant time, at each of the $k_i - 1$ steps.

As selecting random elements in constant time is easier when accessing the elements of an array directly by indices, we start with copying the nodes of T into an array A of size n, and copying the adjacency list of each node x into an array A_x of size $deg(x)$. This can clearly be done in total time $O(n)$ since T is a tree.

In general, selecting an unselected element of a set at random can be done easily in constant time if the set is represented with an array. Let us say we have an array S of t elements. We keep a separation index s that separates the selected elements from the not selected ones. At the beginning s is 1. At each step, we generate a random integer r between s and t. $S[r]$ is our randomly selected element. Then we swap the elements $S[s]$ and $S[r]$ and increase s by 1.

We can use this method both for selecting a node y of T_i that still has neighbors outside and for selecting a neighbor z of y that has not yet been selected. For the latter, whenever we select a neighbor z of y, we move z to the first part of the array A_x using swap. When the separation index reaches the degree of y then we know that y should not be selected to grow the subtree T_i at later steps. Representing T_i with an array of size k_i, we can use the same trick to move y to a part of the array that we will not select from. Also, when z is added, we can check whether it is a leaf in T in constant time, and immediately move it to the irrelevant part of the array for T_i if so, since z can then not be used for growing T_i at later steps. It is sufficient to check that z is a leaf of T, because otherwise it must have neighbors outside of T_i, since T is a tree and we cannot have cycles. When the generation of T_i is finished, the separation indices of each of its nodes should be reset before we start generating T_{i+1}. The adjacency arrays need not be reorganized, as we will anyway be selecting neighbors at random.

Note that we do not need this trick to select an initial node x of each subtree T_i, because we should indeed be able to select the same node several times (and grow another subtree from it perhaps in a different way).

With the described method, each step of Algorithm SubtreeGen takes $O(1)$ time, in addition to initial $O(n)$ time to copy the information into appropriate arrays. Thus the total running time is $O(n + \sum_{1=i}^{n} |T_i|)$. □

We can now prove the total running time for chordal graph generation.

Theorem 2. *Algorithm ChordalGen generates a chordal graph with n vertices and m edges in time* $O(n + m)$.

Proof. Rodionov and Choo [23] prove that the following procedure which runs in $O(n)$ time generates a tree T on n nodes uniformly at random: start with a tree T that contains only one node. Then repeat $n - 1$ times the following: pick a random node x of T and add a new node adjacent to it.

We use Algorithm SubtreeGen to generate n subtrees of T. By Lemma 3, this adds to our running time an order of the sum of the sizes of the generated subtrees.

To each subtree T_i, we associate a vertex v_i of G. In addition to storing the node lists T_i, we also store in the nodes of T information about which subtrees contain that node. More precisely, at node x of T, we store the following list: $\{v_j \mid T_j$ contains $x\}$. Observe that this is equivalent to each node of T representing a clique by storing the list of graph vertices that are contained in this clique. By Lemma 2, T then contains the information that corresponds to a clique-tree of G. By Lemma 1 the sum of the sizes of the lists contained at the nodes is $O(n+m)$. As we only used the lists T_i to generate this information, the sum of the sizes of the subtrees is also $O(n + m)$. By methods described by Blair and Peyton [3] it is possible to turn T into a proper clique tree for G in time $O(n + m)$. Thus, in total $O(n + m)$ time we both have a representation of our output graph G and a list of maximal cliques of it. It could, however, be desirable to output an adjacency list representation for G. Markenzon et al. [18], using the methods of Blair and Peyton [3], explain how this can be done in $O(n + m)$ time.   □

As argued in the proof of Theorem 2, the sum of the sizes of the generated subtrees is $O(n+m)$. In our test results, we give both k and the resulting number of edges, m, to give an indication of how k affects the density of the generated graph. It is also possible to supply Algorithm SubtreeGen with a vector of n subtree sizes $\{k_1, k_2, \ldots, k_n\}$ to generate subtrees of exactly desired size. This does not change the running time of the algorithm. Within the same running time, even more user control is possible, like limiting the maximum degree of each subtree, if so desired, for instance to generate intersection graphs of paths in a tree. In fact, a completely different method for subtree generation can be plugged in instead of SubtreeGen in Algorithm ChordalGen. This gives the possibility of fine-tuning the generation towards designated purposes. In the concluding section, we mention a few other ideas for subtree generation.

4   Experimental Results

In this section, we give extensive test results to show what kind of chordal graphs are generated by Algorithm ChordalGen. In Table 1 we show how the selection of parameter k affects the number of resulting edges and connected components. We also present the number of maximal cliques, and the minimum, maximum, and mean size for the maximal cliques, along with their average standard deviation. For each parameter pair n and average subtree size k, we performed ten independent runs and report the average values across those then runs. For each n, we tuned the average subtree sizes in order to approximately achieve some selected average edge density values of 0.01, 0.1, 0.5, and 0.8, where edge density is defined as $\frac{m}{n(n-1)/2}$.

We want to compare our results to the results showing the kind of chordal graphs that are generated by Alg 2a [18]. Note, however that, the results given by [18] only contain graphs on 10000 vertices, with varying number of edges. Most metrics presented in [18] are about the number of edges. When it comes to the maximal cliques, they present only the average maximum clique size over the generated graphs for each edge density. Comparing these to our numbers we

Table 1. Experimental results of Algorithm ChordalGen

n	Avg subtree size (k)	Density	# edges	# conn. comp.s	# maximal cliques	Min clique size	Max clique size	Mean clique size	Sd of clique sizes
1000	4.0	0.011	5646.6	16.5	355.7	1.0	21.4	6.16	3.41
1000	17.0	0.101	50374.8	1.0	169.6	5.2	134.1	30.62	20.26
1000	70.0	0.505	252237.8	1.0	77.6	29.6	474.0	140.24	92.23
1000	162.5	0.801	399906.4	1.0	49.1	74.8	726.2	313.27	163.59
2500	7.0	0.011	35289.6	3.0	680.5	1.1	54.0	11.58	6.96
2500	32.0	0.103	322433.4	1.0	299.0	10.5	344.8	62.07	44.47
2500	135.0	0.503	1572067.0	1.0	134.5	50.0	1196.3	291.23	206.63
2500	318.0	0.803	2509818.4	1.0	88.4	115.0	1866.1	639.35	385.94
5000	10.5	0.010	130255.1	1.2	1092.9	1.8	97.1	18.15	11.47
5000	50.5	0.098	1229487.3	1.0	476.4	15.3	650.9	100.99	77.59
5000	225.0	0.509	6361645.4	1.0	199.5	76.9	2531.3	504.05	381.98
5000	549.0	0.809	10114806.0	1.0	122.0	163.6	3695.8	1217.44	756.58
10000	16.0	0.010	506598.1	1.0	1745.7	3.4	203.5	29.05	19.98
10000	85.0	0.107	5338077.0	1.0	706.5	25.0	1366.8	181.86	148.12
10000	377.0	0.497	24832462.0	1.0	312.6	103.0	4871.6	861.59	681.79
10000	926.0	0.802	40101492.0	1.0	191.7	236.6	7294.4	2109.65	1394.89

see that graphs corresponding to edge densities 0.01, 0.1, 0.5, and 0.8 of Alg 2a have average maximum clique sizes 727, 2847, 6875, and 8760, respectively. As can be seen from Table 1, these numbers are quite higher than the corresponding numbers for the graphs generated by Algorithm ChordalGen. In fact, studying the numbers more carefully, we can conclude that the maximum clique of a graph generated by Alg 2a contains almost all the edges of the graph. In the case of density 0.01, such a clique contains more than half of the edges, whereas in the case of higher densities, the largest clique contains more than 80, 94, and 95 percent of the edges, respectively. Thus there does not seem to be an even distribution of the sizes of maximal cliques of graphs generated by Alg 2a.

As we mentioned in Sect. 2, we also implemented Alg 1 [2]. In Table 2 we give results analogous to Table 1 for 1000, 2500, and 5000 vertices. In order to obtain results for Table 2 comparable to those given in Table 1, we wanted to have approximately the same edge density values. For this purpose, when determining in Alg 2 the number of neighbors of a vertex at each step, we multiplied the total number of candidate vertices with a coefficient between 0 and 1, which we call *upper bound coefficient*. A running time analysis for this algorithm has not been given [2]. With our implementation, this algorithm turned out to be too slow to allow testing graphs on 10000 vertices in reasonable time. However, already from the obtained numbers, we can reach a conclusion for Alg 1 similar to that on Alg 2a. Observe that the maximum clique sizes obtained for 5000 vertices by Alg 1, are comparable to the maximum clique sizes obtained for 10000 vertices by Algorithm ChordalGen. Hence, like Alg 2a, also Alg 1 seems to generate graphs with few big maximal cliques. As can be seen in Table 2, Alg 1 outputs connected

Table 2. Experimental results of our implementation of Alg 1 [2]

n	Upper bound coef.	Density	# edges	#conn. comp.s	# maximal cliques	Min clique size	Max clique size	Mean clique size	Sd of clique sizes
1000	0.00130	0.011	5368.6	1.0	935.1	2.0	56.0	4.7	8.74
1000	0.00300	0.103	51519.5	1.0	755.3	2.0	219.2	31.3	63.85
1000	0.01100	0.499	249288.5	1.0	405.2	2.0	561.8	184.3	231.81
1000	0.03500	0.808	403436.2	1.0	185.8	2.0	793.8	394.5	346.26
2500	0.00053	0.010	31978.1	1.0	2322.9	2.0	154.2	8.5	25.39
2500	0.00120	0.101	316107.4	1.0	1882.4	2.0	549.3	71.3	160.68
2500	0.00440	0.501	1565224.3	1.0	1005.7	2.0	1401.1	458.8	592.22
2500	0.01400	0.807	2519340.1	1.0	470.0	2.0	1980.7	988.2	887.66
5000	0.00027	0.011	134535.7	1.0	4628.4	2.0	320.7	16.0	56.66
5000	0.00062	0.107	1331285.2	1.0	3717.0	2.0	1144.5	144.1	339.34
5000	0.00220	0.503	6289143.9	1.0	2001.7	2.0	2804.3	919.5	1195.48
5000	0.00700	0.804	10049827	1.0	938.5	2.0	3945.8	1945.0	1787.89

chordal graphs for the selected set of average edge density values and number of vertices. The minimum size of the maximal cliques did not show any variation throughout our experiments and always turned out to be two. The consistency in this measure may be an additional indication of the lack of potential to produce a diverse range of chordal graphs.

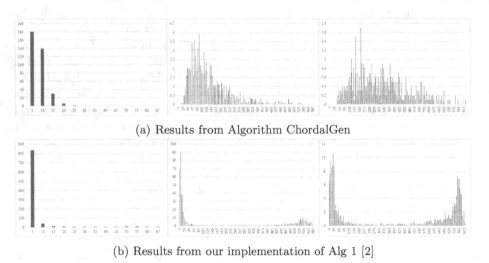

(a) Results from Algorithm ChordalGen

(b) Results from our implementation of Alg 1 [2]

Fig. 1. Histograms of maximal clique sizes for $n = 1000$ and average edge densities 0.01, 0.5, and 0.8 (from left to right)

We wanted to evidence the above conclusions by investigating how the sizes of the maximal cliques are distributed. Figures 1, 2 and 3 show the average number of maximal cliques across ten independent runs in intervals of width five, for 1000, 2500, and 5000 vertices and varying edge densities.

These figures consist of two sub-figures, and each subfigure is comprised of three histograms for three different average edge density values. The top sub-figures show the results from Algorithm ChordalGen and the second those from our implementation of Alg 1 [2]. For a given n and average density value, the ranges of x-axes are kept the same in order to render the histograms comparable. The y-axes, however, have different ranges because maximum frequencies in histograms corresponding to Alg 1 and Algorithm ChordalGen vary drastically. The ratios of the maximum frequencies of Alg 1 to those of Algorithm ChordalGen range roughly from 4 to 160.

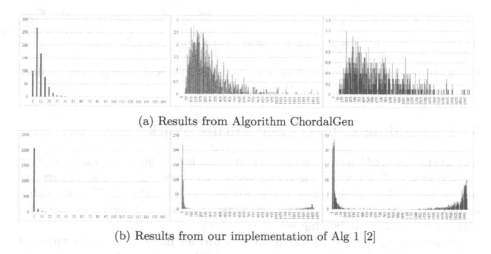

(a) Results from Algorithm ChordalGen

(b) Results from our implementation of Alg 1 [2]

Fig. 2. Histograms of maximal clique sizes for $n = 2500$ and average edge densities 0.01, 0.5, and 0.8 (from left to right)

As Figs. 1, 2 and 3 reveal, the vast majority of maximal cliques of graphs output by Alg 1 have sizes of 2 to 15. With the increase in edge densities, frequencies of large-size maximal cliques become visible relative to the dominant small clique frequencies; however, all but the extremes of the range is barely used regardless of selection of n and edge density. The restricted shape of the distribution of clique sizes indicates that Alg 1 likely produces chordal graphs of limited structure in general. Algorithm ChordalGen, however, does not demonstrate such bias toward the extremes over the range of its maximal clique sizes; output graphs contain maximal cliques of many different sizes. The fair dispersion of clique sizes of Algorithm ChordalGen suggests diversity of its output graphs, which is a desired characteristic of a random chordal graph generator.

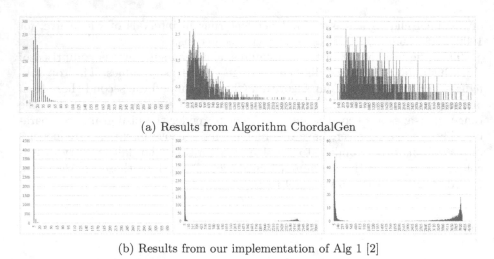

(a) Results from Algorithm ChordalGen

(b) Results from our implementation of Alg 1 [2]

Fig. 3. Histograms of maximal clique sizes for $n = 5000$ and average edge densities 0.01, 0.5, and 0.8 (from left to right)

5   Concluding Remarks and Future Work

Algorithm ChordalGen is the first linear-time method in literature for generating chordal graphs. Furthermore, as it can be seen from the test results, it generates the most varied chordal graphs, compared to existing methods. The algorithm is also very general and flexible in the sense that many different methods for subtree generation can be plugged in.

As already mentioned in Sect. 3, we can further fine-tune the generation of subtrees for special purposes, and we can use other methods for subtree generation in Algorithm ChordalGen instead of SubtreeGen. Two such possible ideas are generating the subtrees by selecting a set of random nodes and connecting them via the paths in the host tree, and selecting a random set of edges to remove from the host tree and selecting one of the resulting connected components as a subtree. We will revisit these methods in the long version of this work.

References

1. Abu-Khzam, F., Heggernes, P.: Enumerating minimal dominating sets in chordal graphs. Inf. Process. Lett. **116**(12), 739–743 (2016)
2. Andreou, M.I., Papadopoulou, V.G., Spirakis, P.G., Theodorides, B., Xeros, A.: Generating and radiocoloring families of perfect graphs. In: Nikoletseas, S.E. (ed.) WEA 2005. LNCS, vol. 3503, pp. 302–314. Springer, Heidelberg (2005). doi:10. 1007/11427186_27
3. Blair, J.R.S., Peyton, B.W.: An introduction to chordal graphs, clique trees. In: George, A., Gilbert, J.R., Liu, J.W.H. (eds.) Graph Theory, Sparse Matrix Computations. The IMA Volumes in Mathematics and Its Applications, vol. 56, pp. 1–29. Springer, New York (1993)

4. Bougeret, M., Bousquet, N., Giroudeau, R., Watrigant, R.: Parameterized complexity of the sparsest k-subgraph problem in chordal graphs. In: Geffert, V., Preneel, B., Rovan, B., Štuller, J., Tjoa, A.M. (eds.) SOFSEM 2014. LNCS, vol. 8327, pp. 150–161. Springer, Cham (2014). doi:10.1007/978-3-319-04298-5_14

5. Buneman, P.: A characterisation of rigid circuit graphs. Disc. Math. **9**(3), 205–212 (1974)

6. Brandstädt, A., Le, V.B., Spinrad, J.: Graph Classes: A Survey. SIAM Monographs on Discrete Mathematics and Applications. SIAM, Philadelphia (1999)

7. Dirac, G.A.: On rigid circuit graphs. Ann. Math. Semin. Univ. Hamburg **25**, 71–76 (1961)

8. Fulkerson, D., Gross, O.: Incidence matrices and interval graphs. Pac. J. Math. **15**(3), 835–855 (1965)

9. Gavril, F.: Algorithms for minimum coloring, maximum clique, minimum covering by cliques, and maximum independent set of a chordal graph. SIAM J. Comput. **1**(2), 180–187 (1972)

10. Gavril, F.: The intersection graphs of subtrees in trees are exactly the chordal graphs. J. Comb. Theory Ser. B **16**(1), 47–56 (1974)

11. Golovach, P., Heggernes, P., Kratsch, D.: Enumerating minimal connected dominating sets in graphs of bounded chordality. Theor. Comput. Sci. **630**, 63–75 (2016)

12. Golovach, P., Heggernes, P., Kratsch, D., Saei, R.: An exact algorithm for subset feedback vertex set on chordal graphs. J. Discret. Algorithms **26**, 7–15 (2014)

13. Golumbic, M.C.: Algorithmic Graph Theory and Perfect Graphs. Annals of Discrete Mathematics. Elsevier, Amsterdam (2004)

14. Hajnal, A., Surányi, J.: Über die Ausflösung von Graphen in vollständige Teilgraphen. Ann. Univ. Sci. Bp. **1**, 113–121 (1958)

15. Heggernes, P.: Minimal triangulations of graphs: a survey. Discret. Math. **306**(3), 297–317 (2006)

16. Loksthanov, D.: Dagstuhl Seminar 14071 "Graph Modification Problems" (2014)

17. Lueker, G.S., Booth, K.S.: A linear time algorithm for deciding interval graph isomorphism. JACM **26**(2), 183–195 (1979)

18. Markenzon, L., Vernet, O., Araujo, L.H.: Two methods for the generation of chordal graphs. Ann. Oper. Res. **157**(1), 47–60 (2008)

19. Marx, D.: Parameterized coloring problems on chordal graphs. Theor. Comput. Sci. **351**(3), 407–424 (2006)

20. Misra, N., Panolan, F., Rai, A., Raman, V., Saurabh, S.: Parameterized algorithms for MAX COLORABLE INDUCED SUBGRAPH problem on perfect graphs. In: Brandstädt, A., Jansen, K., Reischuk, R. (eds.) WG 2013. LNCS, vol. 8165, pp. 370–381. Springer, Heidelberg (2013). doi:10.1007/978-3-642-45043-3_32

21. Pearl, J.: Probabilistic Reasoning in Intelligent Systems: Networks of Plausible Inference. Morgan Kaufmann, Burlington (2014)

22. Pemmaraju, S.V., Penumatcha, S., Raman, R.: Approximating interval coloring and max-coloring in chordal graphs. J. Exp. Algorithmics **10**, 2–8 (2005)

23. Rodionov, A.S., Choo, H.: On generating random network structures: trees. In: Sloot, P.M.A., Abramson, D., Bogdanov, A.V., Gorbachev, Y.E., Dongarra, J.J., Zomaya, A.Y. (eds.) ICCS 2003. LNCS, vol. 2658, pp. 879–887. Springer, Heidelberg (2003). doi:10.1007/3-540-44862-4_95

24. Rose, D.J.: A graph-theoretic study of the numerical solution of sparse positive definite systems of linear equations. Graph Theory Comput. **183**, 217 (1972)

25. Rose, D.J., Tarjan, R.E., Lueker, G.S.: Algorithmic aspects of vertex elimination on graphs. SIAM J. Comput. **5**(2), 266–283 (1976)

26. Spinrad, J.P.: Efficient Graph Representations. Fields Institute Monograph Series, vol. 19. AMS, Providence (2003)

Population Protocols with Faulty Interactions: The Impact of a Leader

Giuseppe Antonio Di Luna[1]([⊠]), Paola Flocchini[1], Taisuke Izumi[2],
Tomoko Izumi[3], Nicola Santoro[4], and Giovanni Viglietta[1]

[1] University of Ottawa, Ottawa, Canada
{gdiluna,paola.flocchini,gvigliet}@uottawa.ca
[2] Nagoya Institute of Technology, Nagoya, Japan
t-izumi@nitech.ac.jp
[3] Ritsumeikan University, Kyoto, Japan
izumi-t@fc.ritsumei.ac.jp
[4] Carleton University, Ottawa, Canada
santoro@scs.carleton.ca

Abstract. We consider the problem of simulating traditional population protocols under weaker models of communication, which include
one-way interactions (as opposed to two-way interactions) and omission
faults (i.e., failure by an agent to read its partner's state during an interaction), which in turn may be detectable or undetectable. We focus on
the impact of a leader, and we give a complete characterization of the
models in which the presence of a unique leader in the system allows the
construction of simulators: when simulations are possible, we give explicit
protocols; when they are not, we give proofs of impossibility. Specifically,
if each agent has only a finite amount of memory, the simulation is possible only if there are no omission faults. If agents have an unbounded
amount of memory, the simulation is possible as long as omissions are
detectable. If an upper bound on the number of omissions involving the
leader is known, the simulation is always possible, except in the one-way
model in which one side is unable to detect the interaction.

1 Introduction

1.1 Framework

Consider a system of *interacting computational entities*, called agents, whose
interaction is however not under their control but decided by an external scheduler. Such are for example systems of wireless mobile entities where two entities can interact (i.e., exchange information) when their movement brings them
into communication range of each other. However, their movements, and thus
their interactions, are unpredictable. Systems satisfying this condition, sometimes called *opportunistic mobility* or *passive mobility*, have been extensively

This work was supported in part by NSERC Discovery Grants, and by KAKENHI
No. 15H00852 and 25289227.

© Springer International Publishing AG 2017
D. Fotakis et al. (Eds.): CIAC 2017, LNCS 10236, pp. 454–466, 2017.
DOI: 10.1007/978-3-319-57586-5_38

examined under a variety of assumptions, especially within the context of distributed computing in highly dynamic networks and time-varying graphs (for recent surveys see [5,9]).

In particular, in the *population protocol* model (PP), introduced in the seminal paper [1], the entities are assumed to be finite-state and anonymous (i.e., identical), execute the same protocol, and interactions are always between pairs of agents. The roles of the two agents involved in an interaction are asymmetric: one agent is considered the *starter* and the other is the *reactor*. Still, the communication is *two-way*: each agent receives the state of the other and executes the protocol to update its own state based on the received information and its own state. Furthermore, in the selection of the occurrences of the interactions, the scheduler is constrained to satisfy some fairness assumption.

The restricted computational universe defined by the basic assumptions of PP has been subsequently expanded in an attempt to overcome the inherent computability limitations and to examine the computational impact of factors such as non-constant memory (e.g., [6]), presence of a leader (e.g., [4]), storage of information on edges (e.g., [7,8]), etc.

In all these models, including the original one, the interaction is assumed to be fault-free. An immediate important question is what happens if interactions are subject to *failures*.

Very little is know in this regard. An insight comes from the study of the so-called *one-way interaction models* [2], where the starter of an interaction is not able to see the state of the reactor (*immediate transmission*), or it is not even able to detect that the interaction has taken place (*immediate observation*). This study showed that, under one-way interactions, the computational power of the agents is strictly weaker than with the usual bidirectional interactions. In particular, if the interactions are not detectable by the starter (i.e., immediate observation), the agents can compute only simple threshold predicates [2].

The complete range of *omission failures* has been classified in [11], where the following general question was posed: under what additional system capabilities is it possible to correctly execute every traditional two-way population protocol in spite of dynamic omission failures? More specifically, under what conditions (if any) it is possible to *simulate* the execution of every two-way population protocol for a given class of omission failures? The simulator should be a population protocol that, in each execution in the model defined by the considered class of omissions, produces a correct execution of any traditional two-way population protocol $\mathcal{P}$ given as input, regardless of the nature of $\mathcal{P}$ and the constraints its execution might have; and it does so unobtrusively at each agent, interacting with $\mathcal{P}$ only by observing its internal state, and providing to it the internal state of another agent. In other words, a simulator provides an interface between the simulated protocol $\mathcal{P}$ and the physical communication layer, giving the system the illusion of being in a fault-free two-way environment.

The existence of simulators is important in scenarios in which we do not only concern ourselves with the final output of a population protocol, but also with the *execution* that leads to the result. We may want, for instance, to guarantee that our simulating agents enter some critical states exactly as many times as they would if they were actually executing the protocol that is being simulated.

The existence of fault-tolerant one-way simulators of two-way protocols has been investigated in [11] in terms of the amount *memory* required by the agents to perform such simulations, and a variety of models and results were established. It is shown that, with no *a-priori* knowledge, the simulation of two-way protocols in the presence of omissions is impossible even if the agents have infinite memory. In the weakest models investigated, this impossibility holds even if the number of omission failures in each execution is limited to one. On the other hand, it is also shown that simulation is possible if agents have unique IDs or the total number of agents is known. Moreover, in some restricted models, simulation is possible when an upper bound on the number of omission faults is known.

In this paper we continue this general line of research and investigate how the presence in the system of a distinguished agent, a *leader*, can impact the capability of the system to tolerate dynamic omission failures. More precisely, we study the possibility and impossibility of simulation of two-way protocols with the aid of a leader, with respect to the different classes of omission failures and one-way interactions.

1.2   Main Contributions

As in [11], we consider all the computationally distinct models that arise from the introduction of omission faults and/or one-way interactions in two-way protocols: TW, IT, IO, T_i ($i = 1, 2, 3$), and I_j ($j = 1, 2, 3, 4$); see Fig. 2, where the transition function δ, detailed in Sect. 2, uniquely identifies each model. In particular, TW refers to two-way protocols without omissions; IT and IO refer to the one-way models *immediate transmission* and *immediate observation*, introduced in [2]; the T_i's and I_i's refer to the distinct two-way and one-way models with omissions, respectively.

We consider two types of omission adversaries: informally, a "malignant" one (**UO**), which is able to arbitrarily insert omission faults into "globally fair" sequences of interactions, and a "benign" one ($\diamond$**NO**), which inserts some omission faults, but eventually stops. To make our results stronger, we always assume the benign adversary in the impossibility proofs and the malignant one in the possibility proofs.

We study the negative impact that omissions have on computability, and we show that the simulation of two-way protocols is impossible even with the aid of a leader (Theorem 1), assuming that the amount of memory is bounded.

On the other hand, we show that the presence of both a leader and infinite memory on each agent makes the simulation possible in the weak intermediate one-way models I_1 and I_2 (Theorem 4), and thus in all the upper models of Fig. 2. The fact that this possibility does not apply to IO and T_1 is not accidental: indeed we prove that, for these two models, the simulation is impossible even with both a leader and infinite memory, even against the benign omission adversary (Theorem 2).

Finally, we study what happens when a bound on the omission failures involving the leader is known, and essentially we show that simulators exists for models I_1 and I_2 (Theorem 5) and model T_1 (Theorem 6), and these imply the possibility of simulations in all other omissive models.

For non-omissive models, we show that two-way simulation is possible in the IT model (Theorem 7). In light of the fact that with constant memory, in absence of additional capabilities, IT protocols are strictly less powerful than TW (see [11]), our results show that this computational gap can be overcome by using a leader.

Our main results are summarized in Fig. 1, where white blobs represent possibilities, and gray blobs impossibilities. As a consequence of these results, we have a complete characterization of the feasibility of simulations with respect to the presence of a leader. Due to lack of space, some proofs are omitted, they can be found in [12].

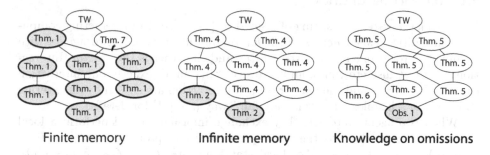

Finite memory   Infinite memory   Knowledge on omissions

Fig. 1. Map of results (cf. Fig. 2). White blobs denote the existence of simulators; gray blobs indicate that simulations are not possible.

1.3   Related Work

Starting with the seminal paper [1], there have been extensive investigations on population protocols (e.g., see [9]). In order to overcome the inherent computability restrictions of the model, several extensions have been proposed. For example, endowing each agent with non-constant memory [6], assuming the presence of a leader [4], allowing a certain amount of information to be stored on the edges of the "communication graph" [7,8], etc.

The possibility of reliable computations in PP has been studied only with respect to processors' faults, and the basic model has necessarily been expanded. In [10] it has been shown how to compute functions tolerating $\mathcal{O}(1)$ crash stops and transient failures, assuming that the number of failures is bounded and known. In [3] the majority problem under $\mathcal{O}(\sqrt{n})$ Byzantine failures, assuming a fair probabilistic scheduler, has been studied. In [13] unique IDs are assumed, and it is shown how to compute functions tolerating a bounded number of Byzantine faults, under the assumption that Byzantine agents cannot forge IDs.

Finally, to the best of our knowledge, the one-way model, without omissions, has been studied only in [2], where it is shown that IT and IO, when equipped with constant memory, can compute a set of functions that is strictly included in that of TW. The omission models that we consider have been introduced for the first time in [11], where a characterisation of what can be simulated without a leader is given. Our paper complements and enriches the results of [11], showing what additional power is obtained assuming the presence of a leader.

2   Model and Terminology

In this section we briefly define the computation model, the notion of omission, and the notion of simulator. Due to space constraints, we do not include all the formal definitions, which can be found in [11].

2.1   Interacting Entities

We consider a system consisting of a set $A = \{a_1, \ldots, a_n\}$ of interacting computational entities, called *agents*. Each *interaction* involves only two agents with asymmetric roles: one agent is the *starter* and the other is the *reactor*. Interactions occur at discrete times, and at every "time unit" exactly one interaction occurs. The starter and the reactor of each interaction are chosen by an external "adversarial scheduler" in a "globally fair" way (see [11] for details).

When two agents interact, they exchange information and perform a local computation according to the same protocol $\mathcal{P}$. A protocol is a pair $\mathcal{P} = (Q_\mathcal{P}, \delta_\mathcal{P})$, where $Q_\mathcal{P}$ is a set of local states and $\delta_\mathcal{P} : Q_\mathcal{P} \times Q_\mathcal{P} \to Q_\mathcal{P} \times Q_\mathcal{P}$ is the transition function defining the states of the two interacting agents at the end of their local computation. Some elements of $Q_\mathcal{P}$ are labeled as "initial states"; when the execution of the protocol begins, all agents have (any combination of) initial states. With a small abuse of notation, and when no ambiguity arises, we will use the same literal (e.g., a_i) to indicate both an agent and its local state. A *configuration* of $\mathcal{P}$ is a multiset of local states of $\mathcal{P}$.

We can model the presence of a *leader* in the system by stipulating that, in every initial configuration, there is exactly one agent in a distinguished state (or set of states).

Depending on the conditions imposed on the transition function, three main models of interactions have been identified: the standard *two-way* model and the one-way models, *immediate transmission* and *immediate observation*, presented in [2].

Two-Way Interaction Model (TW). The state transition function consists of two functions $f_s : Q_\mathcal{P} \times Q_\mathcal{P} \to Q_\mathcal{P}$ and $f_r : Q_\mathcal{P} \times Q_\mathcal{P} \to Q_\mathcal{P}$, one for the starter and the other for the receiver respectively, with $\delta_\mathcal{P}(a_s, a_r) = (f_s(a_s, a_r), f_r(a_s, a_r))$.

Immediate Transmission Model (IT). The state transition function consists of two functions $g : Q_\mathcal{P} \to Q_\mathcal{P}$ and $f : Q_\mathcal{P} \times Q_\mathcal{P} \to Q_\mathcal{P}$, with $\delta_\mathcal{P}(a_s, a_r) = (g(a_s), f(a_s, a_r))$. Note that, in the IT interaction model, the starter does not read the state of the reactor but it explicitly detects the interaction, as it applies function g to its own state.

Immediate Observation Model (IO). The state transition function has the form $\delta_\mathcal{P}(a_s, a_r) = (a_s, f(a_s, a_r))$. Note that, in the IO model, there is no detection of the interaction (or proximity) by the starter.

2.2 Omissions

An *omission* is a fault involving a single interaction. In an omissive interaction, an agent does not receive any information about the state of the other. If omissions can occur in the system, then transition functions become more general relations.

Two-Way Omissive Models. In the most general omissive model, T_3, the transition relation has the form

$$\delta(a_s, a_r) = \{(f_s(a_s, a_r), f_r(a_s, a_r)), \ (o(a_s), f_r(a_s, a_r)),$$
$$(f_s(a_s, a_r), h(a_r)), \ (o(a_s), h(a_r))\}.$$

The first pair is the outcome of an interaction when no omission is present; the other three pairs represent all possible outcomes when there is an omission: respectively, an omission on the starter's side, on the reactor's side, and on both sides. The functions o and h represent the detection capabilities of each agent: if one of these is the identity, then omissions are *undetectable* on the respective side. This gives rise to the weaker models T_2 and T_1 depicted in Fig. 2 (see [11] for more details).

One-Way Omissive Models. These models are defined by the transition relation

$$\delta(a_s, a_r) = \{(g(a_s), f(a_s, a_r)), \ (o(a_s), h(a_r))\}.$$

The first pair is the outcome of an interaction when no omission is present, and the second pair when there is an omission. Note that the IO model corresponds to the case in which g is the identity function and there are no omissions. Once again, omissions are undetectable starter-side if o is the identity function or if $o = g$. Moreover, if $h = g$, the reactor has detected the *proximity* of another agent, but is unable to read its state or even determine who is the starter and who is the reactor. Collectively, these variations give rise to models I_1 to I_4 in Fig. 2. Other combinations of omissions and detections are possible, but they are provably equivalent to some of the aforementioned ones (see [11] for more details).

Omissions are introduced by an adversarial entity. We consider two types of adversaries:

(1) the *Unfair Omissive Adversary* (**UO**), which arbitrarily inserts omissive interactions in any execution, and
(2) the *Eventually Non-Omissive Adversary* ($\diamond$**NO**), which can only insert finitely many omissions in an execution.

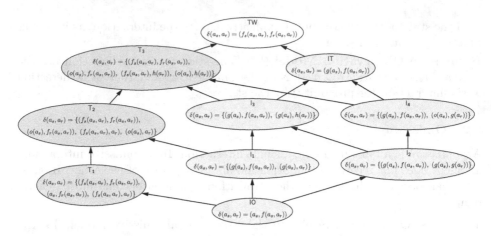

Fig. 2. Interaction models (up to equivalence) and their computational relationships. An arrow between two blobs indicates that the class of solvable problems in the source blob is included in that of the destination blob. The models on the left, T_1, T_2, T_3, are the two-way models with omissions. The models on the right, I_1, I_2, I_3, I_4, are the one-way models with omissions.

2.3 Simulation of Two-Way Protocols

Let $\mathcal{P}$ be a two-way protocol, and let $\mathcal{S}(\mathcal{P})$ be any protocol (which could be one-way, omissive, or both). Next we are going to informally define what it means for $\mathcal{S}(\mathcal{P})$ to *simulate* $\mathcal{P}$ (for a formal definition, refer to [11]).

We want the set of local states of $\mathcal{S}(\mathcal{P})$ to be of the form $Q_{\mathcal{P}} \times Q_{\mathcal{S}}$, where $Q_{\mathcal{P}}$ is the set of local states of $\mathcal{P}$ (the "simulated states"), and $Q_{\mathcal{S}}$ is some additional memory space used in the simulation. Suppose now to start an execution of $\mathcal{S}(\mathcal{P})$ on a system of $n > 2$ agents from a given initial configuration. Agents are allowed to freely change the $Q_{\mathcal{S}}$ component of their local states; but when they change their $Q_{\mathcal{P}}$ component, we want the change to reflect the transition function of $\mathcal{P}$. That is, if $\delta_{\mathcal{P}}(a_s, a_r) = (f_s(a_s, a_r), f_r(a_s, a_r))$, then for every agent whose simulated state changes from a_s to $f_s(a_s, a_r)$, there must be some other agent (at some point in time) whose simulated state changes from a_r to $f_r(a_s, a_r)$. Moreover, there must be a perfect matching between such transitions, in such a way that each starter of a simulated transition can be implicitly mapped to an appropriate reactor. Also, such a perfect matching must be "temporally consistent", i.e., there must be an ordering of the simulated two-way interactions that respects the order of the local state changes of each agent.

We additionally require that, if the execution of $\mathcal{S}(\mathcal{P})$ is globally fair (in the sense defined in [11]), then also the resulting simulated execution of $\mathcal{P}$ is globally fair.

3   Simulation with a Leader in Omissive Models: Impossibility

In this section we prove that the presence of a leader, alone, might not be sufficient to overcome dynamic omissions. Indeed, we prove that there are two-way protocols that cannot be simulated with omissive interactions even if a leader is present.

3.1   Impossibility with Finite Memory

We investigate what happens when we restrict the memory of agents to be bounded by some function of $|A|$. We show our impossibility results directly for the T_3 omissive model. The results clearly carry over to all the less powerful omissive models.

Theorem 1. *A system of agents, each of which has a finite amount of memory, cannot simulate every two-way protocol in the T_3 model (hence in all the omissive models), even with the presence of a leader and under the $\Diamond\mathbf{NO}$ adversary.*

3.2   Impossibility with Infinite Memory

For this case we can show that simulation is impossible in the omissive two-way model without detection, and thus in IO.

Theorem 2. *A system of agents, each of which has an infinite amount of memory, cannot simulate every two-way protocol in the T_1 model (hence in IO), even with the presence of a leader and under the $\Diamond\mathbf{NO}$ adversary.*

Observation 1. *Since in IO there are no omissions, the statement of Theorem 2 for the IO model trivially extends to the scenario in which the number of omissions in the sequence of interactions is known in advance by the agents.*

4   Simulation in Omissive Models

In this section we are going to make use of a result that appears in [11] as Theorem 4.5. This theorem assumes each agent to have a unique ID, which is a non-negative integer, as part of its local state.

Theorem 3. *Assuming IO, unique IDs, and $\mathcal{O}(\log(\max \mathrm{ID}))$ bits of memory on each agent (where $\max \mathrm{ID}$ is the maximum ID in the system), there exists a simulator for every two-way protocol, even under the $\mathbf{UO}$ adversary.*   □

What this theorem says is that, if the agents initially have unique IDs, they can perform a simulation of any two-way protocol, even if the simulation runs in the weakest model, IO, and against the strongest adversary, **UO**.

In this section we show that, in certain models, we can implement a *naming algorithm*, i.e., an algorithm that assign unique IDs to all agents. Once an ID has been assigned to an agent, it cannot change. Therefore, the naming algorithm and the simulator of Theorem 3 can be combined into a single protocol and can even run in parallel: if an agent has no ID yet, the simulator simply ignores every interaction involving this agent. By global fairness, eventually all agents will have unique IDs, and the simulation will finally involve the entire system, producing a globally fair simulated execution.

The protocols will be presented using an algorithmic style: for each interaction of the form (a_s, a_r), the starter agent a_s executes function Upon Event Starter sends() and the reactor agent a_r executes Upon Event Reactor delivers (var^s), where var^s is the variable var in the local state of agent a_s.

4.1   Naming Algorithm with Infinite Memory

If the leader has infinite memory, it can implement a simple naming algorithm under certain models. Since Theorem 2 already states the impossibility of simulation under models T_1 and IO, we will assume model I_1 or model I_2. Constructing a simulator for these models will imply the existence of a simulator for all other models except T_1 and IO (refer to Fig. 1).

Theorem 4. *Assuming* I_1 *or* I_2, *the presence of a leader, and an infinite amount of memory on each agent, there exists a simulator for every two-way protocol, even under the* **UO** *adversary.*

4.2   Naming Algorithms with Knowledge on Omissions

Now we assume that agents have only a finite amount of memory, but they know in advance a finite upper bound L on the number of omission faults that the adversary is going to insert *in interactions that involve the leader*. Note that the adversary can still be **UO** even if only finitely many omissive interactions involve the leader.

Naming Algorithm for I_1 **and** I_2**.** In this case the memory is bounded by a function of L and the size of the system, n. It is worth mentioning that the precise value of n is not known to the agents, and L is only an upper bound on the number of omissions involving the leader, not necessarily the exact number.

Theorem 5. *Assuming* I_1 *or* I_2, *the presence of a leader, knowledge of an upper bound* L *on the number of omission failures in interactions that involve the leader, and* $\Theta(L \log nL)$ *bits of memory on each agent (where n is the number of agents), there exists a simulator for every two-way protocol, even under the* **UO** *adversary.*

Proof. We implement the naming algorithm presented in Fig. 3. The leader has an array of $L+1\,next\_ID$ variables. This array is initialized to $[1, 2, \ldots, L+1]$ and, when an ID is assigned, the corresponding entry of the array will be incremented by $L + 1$, so that no two equal IDs are ever be generated.

All entries of *next_ID* are initially *unlocked*: this information is stored in the leader's Boolean array *locked*. The *active ID* is defined as the unlocked entry of *next_ID* having minimum index, if there is any (line 10). This is the ID that will tentatively be assigned next. Whenever the leader detects the proximity of another agent (i.e., it executes function g on its own state, or function Upon Event Starter sends in the algorithm of Fig. 3), it locks the active entry of *next_ID* (line 12). The purpose of locking an entry of *next_ID* is that the leader cannot allow its value to grow indefinitely, because now memory is limited. Instead, the leader will make the entry temporarily inactive, and will keep it on hold until it gathers more information in the following interactions.

On the other hand, if an agent a sees the leader (i.e., it executes function f or function Upon Event Reactor delivers in Fig. 3), and a does not have an ID yet, then it assigns itself the active ID from the leader's *next_ID* variable (line 30). So, the next time the leader sees a, it will read its new ID and it will know that the corresponding entry of *next_ID* can be unlocked (line 23) and its value can be incremented by $L + 1$ (line 24). It may happen that the leader is involved in an omissive interaction, and therefore the entry of *next_ID* that it locks will never be unlocked again. However, this can happen at most L times, while the array has $L + 1$ entries.

This is not sufficient yet, because the same agent a may see the leader multiple times in a row and cause all entries of *next_ID* to become locked. If a only stores one ID, it will have no way to tell the leader that more than one entry of *next_ID* has to be unlocked. This is why a also has a variable called *redundant*, which is a Boolean array that will store information on all the active entries of *next_ID* that a sees after receiving an ID. So, if the agent a already has an ID and it sees the leader again, it sets to *true* the entry of *reduntant* corresponding to the active ID of the leader (line 32).

Now, suppose that the leader sees that a has an entry of *redundant* set to *true*. This implies that the corresponding entry of *next_ID* is currently locked and should be unlocked. However, this cannot be done right away: the leader wants to give a an "acknowledgment", so that a will set the entry of *redundant* to *false* first. This is to prevent the scenario in which the entry of *next_ID* gets unlocked, becomes active, another agent b sees it, and takes it as its own ID. If then the leader sees a again (still with *redundant* on *true*), it will unlock the entry of *next_ID*. Then perhaps yet another agent c will see the leader, getting the same ID as b.

To prevent such an incorrect behavior, the leader has another variable array called *waiting*, in which it stores the IDs of the agents that should reset their *redundant* variables. So, when the leader sees that a has some entry of *redundant* set to *true*, it stores the (unique) ID of a in the corresponding entry of *waiting* (line 18). Then, when a sees the leader again and reads its own ID in the *waiting* array, it knows that it has to set to *false* the corresponding entries of *redundant* (line 34). Finally, when the leader sees a again and notices that the entry of *redundant* has been set to *false*, it can reset the corresponding entry of *waiting* (line 20) and unlock the entry of *new_ID* (line 21).

```
 1: Variables
 2: my_ID                              ▷ the leader has this variable initialized to 0, non-leaders to ⊥
 3: next_ID[] := [1, 2, . . . , L + 1]                                          ▷ leader variable
 4: locked[] := [false, false, . . . , false]                                   ▷ leader variable
 5: waiting[] := [⊥, ⊥, . . . , ⊥]                                              ▷ leader variable
 6: redundant[] := [false, false, . . . , false]                           ▷ non-leader variable
 7:
 8: Upon Event Starter sends()
 9: if my_ID = 0 then                                                          ▷ I am the leader
10:     j := min{j | locked[j] = false, L + 2}
11:     if j < L + 2 then
12:         locked[j] := true
13:
14: Upon Event Reactor delivers (my_ID^s, next_ID^s[], locked^s[], waiting^s[], redundant^s[])
15: if my_ID = 0 then                                                          ▷ I am the leader
16:     for all j ∈ {1, 2, . . . , L + 1} do
17:         if redundant^s[j] = true then
18:             waiting[j] := my_ID^s
19:         else if waiting[j] = my_ID^s then
20:             waiting[j] := ⊥
21:             locked[j] := false
22:         if ∃j, next_ID[j] = my_ID^s then
23:             locked[j] := false
24:             next_ID[j] := next_ID[j] + L + 1
25: else                                                                     ▷ I am not the leader
26:     if my_ID^s = 0 then                                           ▷ my partner is the leader
27:         j = min{j | locked^s[j] = false, L + 2}
28:         if j < L + 2 then
29:             if my_ID = ⊥ then
30:                 my_ID := next_ID^s[j]
31:             else
32:                 redundant[j] := true
33:         if my_ID ≠ ⊥ ∧ ∃j, waiting^s[j] = my_ID then
34:             redundant[j] := false
```

Fig. 3. Naming algorithm for l_1 and l_2 with knowledge on omissions, used in Theorem 5

The fact that the algorithm does not give the same ID to two different agents follows from the observation that at most one agent can keep an entry of new_ID locked at any given time, which in turn follows from the way the two variables $redundant$ and $waiting$ function together. If no omission occurs and the leader is observed by some agent a, then a will store information about the currently active ID. If a takes this ID for itself, that entry of $next\_ID$ will be incremented before any other agent can get the same ID. If a has already an ID, the entry of $next\_ID$ will remain locked until a has reset its own $redundant$ variable. Moreover, the fact that the algorithm will eventually assign every agent an ID immediately follows from the global fairness of the adversarial scheduler.

Since the IDs in the $next\_ID$ array increase by $L + 1$ every time one is assigned, and since there are n agents in total, the value of every ID is $\mathcal{O}(nL)$. Hence, $\mathcal{O}(L \log nL)$ bits of memory are required to store each agent's arrays, and $\mathcal{O}(\log nL)$ more bits are required to run the simulator of Theorem 3.

Naming Algorithm for T_1. Observe that the previous naming algorithm does not work for model T_1, and Theorem 2 does not hold when some kind of upper bound on omissions is known.

Theorem 6. *Assuming T_1, the presence of a leader, knowledge of an upper bound L on the number of omission failures in interactions that involve the leader, and $\Theta(L \log nL)$ bits of memory on each agent (where n is the number of agents), there exists a simulator for every two-way protocol, even under the* **UO** *adversary.*

5   Simulation for IT

Notice that IT is the only finite-memory model for which the impossibility result of Theorem 1 does not hold (see Fig. 1). It turns out that in this model we can implement a simulator that sequentializes the simulated two-way interactions via a token-passing technique.

Theorem 7. *Assuming IT, the presence of a leader, and a constant amount of memory on each agent, there exists a simulator for every two-way protocol, even under the* **UO** *adversary.*

References

1. Angluin, D., Aspnes, J., Diamadi, Z., Fischer, M.J., Peralta, R.: Computation in networks of passively mobile finite-state sensors. Distrib. Comput. **18**(4), 235–253 (2006)
2. Angluin, D., Aspnes, J., Eisenstat, D., Ruppert, E.: On the power of anonymous one-way communication. In: Anderson, J.H., Prencipe, G., Wattenhofer, R. (eds.) OPODIS 2005. LNCS, vol. 3974, pp. 396–411. Springer, Heidelberg (2006). doi:10.1007/11795490_30
3. Angluin, D., Aspnes, J., Eisenstat, D.: A simple population protocol for fast robust approximate majority. Distrib. Comput. **21**(2), 87–102 (2008)
4. Angluin, D., Aspnes, J., Eisenstat, D.: Fast computation by population protocols with a leader. Distrib. Comput. **21**(3), 61–75 (2008)
5. Casteigts, A., Flocchini, P., Quattrociocchi, W., Santoro, N.: Time-varying graphs and dynamic networks. Int. J. Parallel Emergent Distrib. Syst. **27**(5), 387–408 (2012)
6. Chatzigiannakis, I., Michail, O., Nikolaou, S., Pavlogiannis, A., Spirakis, P.G.: Passively mobile communicating machines that use restricted space. Theoret. Comput. Sci. **412**(46), 6469–6483 (2011)
7. Chatzigiannakis, I., Michail, O., Nikolaou, S., Pavlogiannis, A., Spirakis, P.G.: All symmetric predicates in $NSPACE(n^2)$ are stably computable by the mediated population protocol model. In: Hliněný, P., Kučera, A. (eds.) MFCS 2010. LNCS, vol. 6281, pp. 270–281. Springer, Heidelberg (2010). doi:10.1007/978-3-642-15155-2_25
8. Michail, O., Chatzigiannakis, I., Spirakis, P.G.: Mediated population protocols. Theoret. Comput. Sci. **412**(22), 2434–2450 (2011)

9. Michail, O., Chatzigiannakis, I., Spirakis, P.G.: New Models for Population Protocols. Synthesis Lectures on Distributed Computing Theory. Morgan & Claypool,
San Rafael (2011)
10. Delporte-Gallet, C., Fauconnier, H., Guerraoui, R., Ruppert, E.: When birds die:
making population protocols fault-tolerant. In: Gibbons, P.B., Abdelzaher, T.,
Aspnes, J., Rao, R. (eds.) DCOSS 2006. LNCS, vol. 4026, pp. 51–66. Springer,
Heidelberg (2006). doi:10.1007/11776178_4
11. Di Luna, G.A., Flocchini, P., Izumi, T., Izumi, T., Santoro, N., Viglietta, G.: On
the power of weaker pairwise interaction: fault-tolerant simulation of population
protocols. arXiv:1610.09435 [cs.DC] (2016)
12. Di Luna, G.A., Flocchini, P., Izumi, T., Izumi, T., Santoro, N., Viglietta, G.: Population protocols with faulty interactions: the impact of a leader. arXiv:1611.06864
[cs.DC] (2016)
13. Guerraoui, R., Ruppert, E.: Names trump malice: tiny mobile agents can tolerate
Byzantine failures. In: Albers, S., Marchetti-Spaccamela, A., Matias, Y., Nikoletseas, S., Thomas, W. (eds.) ICALP 2009. LNCS, vol. 5556, pp. 484–495. Springer,
Heidelberg (2009). doi:10.1007/978-3-642-02930-1_40

Paper Dedicated to Stathis Zachos on the
Occasion of his 70th Birthday

Stathis Zachos at 70!

Eleni Bakali[1], Panagiotis Cheilaris[1], Dimitris Fotakis[1] (iD), Martin Fürer[2],
Costas D. Koutras[3], Euripides Markou[4], Christos Nomikos[5],
Aris Pagourtzis[1(✉)] (iD), Christos H. Papadimitriou[6], Nikolaos S. Papaspyrou[1],
and Katerina Potika[7]

[1] School of Electrical and Computer Engineering,
National Technical University of Athens, 15780 Athens, Greece
mpakali@corelab.ntua.gr, {philaris,fotakis,pagour}@cs.ntua.gr,
nickie@softlab.ntua.gr
[2] Department of Computer Science and Engineering,
Pennsylvania State University, University Park, PA 16802, USA
furer@cse.psu.edu
[3] Department of Informatics and Telecommunications,
University of Peloponnese, 22100 Tripoli, Hellas
ckoutras@uop.gr
[4] Department of Computer Science and Biomedical Informatics,
University of Thessaly, 35100 Lamia, Greece
emarkou@ucg.gr
[5] Department of Computer Science and Engineering, University of Ioannina,
45110 Ioannina, Greece
cnomikos@cs.uoi.gr
[6] EECS Deparment, University of California at Berkeley, Berkeley, CA 94720, USA
christos@cs.berkeley.edu
[7] Department of Computer Science, San Jose State University,
San Jose, CA 95192, USA
katerina.potika@sjsu.edu

Abstract. This year we are celebrating the 70th birthday of Stathis!
We take this chance to recall some of his remarkable contributions to
Computer Science.

1 Introduction

Stathis Zachos was born in Athens in 1947. His father Kyriakoulis Zachos was
a distinguished professor of Mining Engineering and Geophysics in the National
Technical University of Athens and his mother Evangelie Spanidis-Zachos was a
physicist, PhD advisee of Hans Geiger. He is the brother of theoretical physicist
Cosmas Zachos. His first son Kyriakos tragically died in a car accident in 2001.
He is now married to Sofia Chatzilambrou and they have a son, Konstantinos,
and two daughters, Christina and Katerina.

Stathis studied in Eidgenössische Technische Hochschule Zürich (ETHZ),
from where he obtained a Diploma in Mathematics (1972) and a PhD in the area
of Mathematical Logic and Foundations of Computer Science (1978), advised by

© Springer International Publishing AG 2017
D. Fotakis et al. (Eds.): CIAC 2017, LNCS 10236, pp. 469–484, 2017.
DOI: 10.1007/978-3-319-57586-5_39

Erwin Engeler and Ernst Specker. He has held the posts of professor in Computer Science at the University of California in Santa Barbara, the City University of New York, and the National Technical University of Athens. He has also served as an adjunct professor at ETHZ and as a visiting scholar at MIT.

Stathis Zachos has conducted pioneering research in randomized complexity classes, probabilistic quantifiers, Arthur-Merlin games, and interactive proof systems. His uniform description of structural complexity has influenced many researchers, and has provided key insights in proving important theorems in computational complexity. His work has been often cited in main textbooks and papers. Among his most cited contributions are: (a) Proving that the GRAPH ISOMORPHISM (GI) problem is unlikely to be NP-complete (with Ravi Boppana and Johan Håstad [8]); notably, a remarkable recent result states that GI is of quasi-polynomial complexity [1], and (b) Introducing and proving properties of the class $\oplus$P (with Papadimitriou [29]), later employed in Toda's celebrated result that the Polynomial Time Hierarchy is contained in $P^{\#P}$. Stathis has also contributed to several other fields such as graph coloring, randomized and approximation algorithms, logics, and computational geometry, to name a few. He has Erdős number 2.

Stathis has been a restless teacher in all departments where he has served. Especially in the National Technical University of Athens (NTUA) he has set a remarkable almost 30-year long teaching record. During his first ten years at NTUA he had struggled almost alone to set up a descent computer science curriculum in the Department of Electrical and Computer Engineering. He completely revised the content and teaching of introductory programming courses, and of the 'Algorithms and Complexity' course, and introduced many undergraduate and graduate courses in computer science, computability, complexity, and cryptography. A little later he organized a similar curriculum for the Department of Applied Mathematics and Physical Sciences, where he also taught for many years. On top of those, he has been a founding member and teacher of the renowned "Graduate Programme in Logic, Algorithms and Computation (MPLA)" for about 20 years. Recently, he actively participated in the establishment of the "Graduate Programme in Algorithms, Logic and Discrete Mathematics (ALMA)" which was launched in 2016. During his presence at NTUA he has taught well over ten thousand students, introducing them to the concepts of programming, computer science and theoretical computer science.

Stathis has also been an inspiring mentor and advisor for hundreds of students that have been involved in research projects, diploma theses and doctoral studies under his supervision. Stathis has always been there for them, enlightening their path to knowledge, encouraging their quests for further studies, and advising them not only on academic but also on personal matters. Through the years he has supervised more than 20 doctoral and more than 80 diploma theses. Notably, many of his advisees pursue academic careers in renowned institutions all over the world.

Stathis has always been actively promoting activities that strengthen the interaction among the community, especially between students and senior or junior researchers. He co-organized international conferences such as STOC, ICALP, CiE (Computability in Europe), and ASL (Association for Symbolic Logic) European Summer Meeting. In addition, he has been a founding member and regular organizer for various local conferences and workshops such as ACAC (Athens Colloquium on Algorithms and Complexity), NYCAC (New York Colloquium on Algorithms and Complexity), PLS (Panhellenic Logic Symposium), AtheCrypt (Athens Cryptography day) and AGaThA (Algorithmic Game Theory Athens).

In the rest of this article we present in more detail some of Stathis's impressive contributions – and our own memories from our collaboration with him.

2 Playing Games with Generalized Quantifiers (by Martin Fürer)

I have been lucky to start my studies at ETH in the same year as Stathis Zachos. We both have been fascinated by Ernst Specker, and therefore we took all courses offered in mathematical logic and regularly participated in the logic seminar, where Paul Bernays was an attentive participant. One of the first results studied was the solution of Hilbert's 10th problem. In the early years at ETH, we got involved in the theory and practice of games. Politically, we have been shaped by the protests against the war in Vietnam and the resistance to the military dictatorship in Greece.

Very soon after the P versus NP question was formulated, we started as graduate students to participate in a second weekly two hour seminar with Volker Strassen and Ernst Specker. We have been impressed with all the exciting open problems in this area, and we slowly switched from mathematics to theoretical computer science.

When I decided to choose a postdoc position with Les Valiant instead of a regular position at UC Santa Barbara, I "sent" Stathis there instead. Our common interests in complexity had much to do with randomized computations and quantifier alternations. We embraced the Sipser-Gács result on the containment of BPP in the polynomial hierarchy in its simplified version of Lautemann. At least I thought so. But Stathis told me once that the version with moving combs with irregular teeth might be partially my invention, as he could not find it in the paper. I am not sure, because I never read these papers, but maybe Clemens Lautemann has used this illustration in an oral presentation at Oberwolfach.

Certainly, our earlier interest in games provided us a good foundation to investigate the power of interchanging quantifiers [35]. We showed how complexity classes like BPP and RP can be characterized by pairs of generalized quantifiers. Arthur-Merlin classes involved pairs of alternating sequences of such quantifiers. We also studied more involved complexity classes obtained by also using such classes as oracles, focusing on results produced by investigating the effect of changing the quantifier order in a definition. Due to the interest of such results in interactive proof systems, we also had a joint paper with Goldreich, Mansour and Sipser [12].

It is always a pleasure to see and interact with Stathis, be it mathematical or related to any other subject. I almost share a birthday with Stathis. But "almost" means that I will always be ahead, and he will never quite catch up. I keep telling him that he is still so young, even at 70.

3   On the Power of Counting – and Friendship (by Christos Papadimitriou)

The summer of 1982 I returned to MIT from my stormy first year at the National Technical University of Athens. Meeting with Stathis Zachos was one of the human and cultural delights of that year; we had agreed to meet again at MIT in the summer to do research together (I believe he was on his way from ETH to Santa Barbara). The result of that collaboration was a little paper that was pivotal for my thinking, and presaged major advances in Complexity.

In 1982, the P vs NP problem was already a decade-old classic, the Karp-Lipton Theorem in 1980 had brought to the forefront complexity hierarchies and the eventuality of their collapse, while Valiant had established counting as a novel computational modality, inviting comparison with the polynomial hierarchy. Our paper with Stathis identified two low-hanging fruits in this research direction: First, counting encompasses a modest layer of the polynomial hierarchy (namely, polynomial computation with a logarithmic number of NP oracle calls). Second, hierarchies of certain computational modalities - such as the parity of solutions to NP problems - do collapse. Both proofs were among the first instances of a methodology that would become a pillar of Complexity, namely encoding computation in the number of branches of a nondeterministic machine. (We note at the end of our paper that Les Valiant told us that he had independently observed the second result.)

Stathis went on to write much more on the field of "structural complexity", while all my previous and subsequent studies of Complexity were of the kind that is directly motivated by problems. I am grateful that, through this collaboration with Stathis Zachos, I was brought closer to understanding the other side, a fusion that eventually resulted in my Complexity book a dozen years later. A bit before that, in 1991 Seinosuke Toda famously came up with towering generalizations of both of our remarks, concluding that the polynomial hierarchy would collapse before it could encompass counting. A quarter century later, this remains one of the most incisive and overarching results in all of Complexity.

Our paper [29] was presented at the 6th GI (a conference I believe no longer exists), which, if memory serves, took place in Bremen where Stathis and I met again that winter, just before my move to Stanford. Today the paper looks quaint, in typescript and with the # sign handwritten, and yet I had no trouble retrieving it in the Internet (where I learned that it has been cited a few hundred times). Things seem to be changing at a remarkable pace – in fact, Stathis Zachos is turning seventy! And yet, great intellectual quests, and good old friendships, prevail.

4   An Advisor that Counts (by Aris Pagourtzis)

I first met Stathis back in 1989–90 during a graduate seminar organized by Foto Afrati at the National Technical University of Athens. Stathis, with his friendly and passionate character quickly earned my attention and admiration. Around summer of 1990 I was lucky enough to be offered a PhD position under his supervision, which I accepted, not without any hesitation. At that time I was not entirely sure whether it would be a good idea to pursue doctoral studies; however, Stathis with his energy and enthusiasm quickly dissolved my doubts and dragged me to this strange and fascinating world of academic research and to the even more strange and fascinating topic of structural complexity. The way Stathis was approaching the subject was unique and very influential to me: he always had a very pictorial, intuitive approach, mostly playing with trees neatly depicting both nondeterministic computations and randomized ones in a uniform manner. Stathis was constantly trying to pursue his unifying, path counting approach towards both nondeterminism and randomization; his view led to great results and insights on classes BPP [34, 36], ⊕P [29] and the GRAPH ISOMORPHISM problem [8] which inspired lots of further research by numerous scientists, including myself and several other students of him.

In winter of 1995 Stathis invited me to New York where I had a chance to closely work with him and his doctoral student Kiron Sharma. Our collaboration was quite fruitful and led to various results on counting using computation trees some of which appeared in Kiron's doctoral thesis and were presented in workshops and conferences. One of the most interesting outcomes was the invariance of the probabilistic/counting class PP under threshold variations [27], which simplified and strengthened earlier results of Beigel and Gill contained in [7]. A couple of years later, Aggelos Kiayias joined the team and worked with us towards a better understanding of the difficulty of counting. This collaboration led to definitions of new, somewhat esoteric' complexity classes capturing aspects of this difficulty. We proved by structural arguments that all these classes are interreducible with #P under Cook[1] reductions, that is, one oracle call is enough in order to reduce a problem of any these classes to some problem of any other in polynomial time. The most interesting among these classes so far turned to be the class TotP having a simple and natural definition: it is the class of functions that count the total number of computation paths of a nondeterministic polynomial time Turing machine, regardless of being accepting or rejecting. These results appeared in various fora as well as in my doctoral dissertation (see [14] and references therein).

Along the way we realized (thanks to a comment of Phokion Kolaitis during a early presentation of some of these results [15]) that the new classes, and especially TotP, contain counting functions for which it is easy to check whether the function value is nonzero. But this is a property shared by some of the most interesting counting problems: those that admit an efficient approximation scheme! Indeed, we were soon able to show that problems such as #PERFECT MATCHINGS (aka 0–1 PERMANENT) and #DNF-SAT belong to this new class. This sparkled a fresh interest in TotP and later on, Stathis and myself managed to show that in fact, TotP is the class of counting functions with easy decision that possess a natural self-reducibility property [28]. This has been a major outcome of our collaboration with Stathis, which continued to trigger new research efforts of our group [3] comparing TotP with interval size function classes of Hemaspaandra *et al.* [13]. Our most recent results, which Stathis significantly helped to derive and formulate correctly, concern TotP-complete problems under parsimonious reductions and appear in this proceedings ([2], see also Sect. 12).

Apart from counting, we collaborated with Stathis in several papers in the field of path coloring algorithms (together with Christos Nomikos and Katerina Potika – see Sects. 6 and 9 for more details). This collaboration inspired further research within our group [4,5] leading, among others, to one more doctorate advised by Stathis, namely that of Evangelos Bampas. We also had extensive collaboration in teaching, student mentoring and organizing scientific and social events. In all these activities he has been a 'driving force' always setting the bar higher – and doing his best to make things happen!

Stathis has influenced my research, teaching, and my life in general, in so many ways that it would be impossible to describe in few lines. I will only say that I completely owe him my decision to pursue an academic career. This is no surprise: he has set such an example of scientific quality, passion and dignity that makes people around him believe that the best one can do in life is to follow in his footsteps. Indeed, most of his doctoral students and many of his diploma students have become researchers and/or teachers all around the globe. In my opinion, this is his greatest achievement and contribution.

5   Logic, History, Politics and Basketball (by Costas Koutras)

I met Stathis in Fall 1990. He had just joined the faculty at the Dept. of Electrical Engineering of NTUA. I was a beginner in my PhD studies, trying to find my way into research. I felt strongly inclined towards logic and its applications, holding a strong fascination with the problems of *logic-based Knowledge representation*, which was very trendy at the time. In our first meetings, I was a bit surprised to find that Stathis kept asking on my interests and my motivation, much like as if he was 'interviewing' me. Suddenly, one morning I was awaken by his phone call. He called to say that he was planning a '*Logic in Computer Science*' seminar, aiming to educate young researchers in logic and its intimate relation to computing. This was it. In the months that followed, Stathis organized and held a series of seminars on *Mathematical Logic* and *Computability Theory, Automated Theorem Proving, λ-calculus and combinatory logic*. These seminars revealed to me a new, beautiful world; we went through the fundamentals of (classical and non-classical) logic and recursion theory. Stathis even took me by the hand and introduced me to *Modal Logic*, a topic on which there was practically no knowledge and/or research activity in Greece till that time.

Stathis taught me so many things, in so many areas, that I would need a lot of space to describe. He kept mentoring me and gave me the way to the *Graduate Programme on Logic, Algorithms and the Theory of Computation* (MPLA), definitely a tremendous environment for people working in *logic and theoretical computer science* and one of the happiest moments in the greek academic space. I have to apologize for the late return in his investment, as it took me some time before I started producing my own results. Some time later, he called me and handed me a bunch of papers written by M. Fitting on a family of Heyting-valued modal logics, encouraging me to take a look. Studying this family of logics I was introduced to the machinery of lattice theory and universal algebra and I gained useful insight to the metamathematics of Modal Logic. We wrote two papers together, both in this area [16,17].

Yes, I totally agree with something Martin Fürer wrote (Sect. 2): '*he is still so young, even at 70*'. That's good news. We may find the time to do some joint work on Complexity (to which he also introduced me) and I may somehow convince him that it is *my* view on Greek history which is correct and not *his*. There is absolutely no chance that we can agree on politics; he keeps making me angry. I know that I can't compete with him in many areas. It is only in basketball that he will never beat me. But only in this.

6   A Colorful Time (by Christos Nomikos)

I met Stathis in 1990, when I attended his undergraduate course on "Models of Computation, Formal Languages and Automata Theory". Very soon I realized that Stathis was an excellent teacher, not only because of his deep knowledge in Theoretical Computer Science and Mathematical Logic, but also due to his exceptional ability to make complicated notions intuitively clear for the average student. The impact of this course on me was the main reason for my decision to continue my studies at postgraduate level in NTUA.

My interest in algorithms for path coloring – a problem with applications to bandwidth allocation in all-optical networks – started while Aris (Sect. 4) and myself were PhD-students in NTUA, under the supervision of Stathis. We were actually influenced by a series of talks given by Milena Mihail in 1994, while she had been visiting NTUA invited by Stathis. We continued the research in this area in collaboration with Stathis for several years after we had received our PhD.

The optimal use of the available bandwidth is a crucial issue in optical networks. Optimization problems relevant to bandwidth allocation can be modeled as path coloring problems in graphs. In the Path Coloring problem the goal is to color a set of given paths using a minimum number of colors, so that paths passing through the same edge have different colors. This corresponds to satisfying a given set pre-routed requests, using a minimum number of wavelengths. Path Coloring is NP-hard, even when it is restricted to simple graph topologies, including stars and rings. The main direction of our research was the development of approximation algorithms for some variations of Path Coloring for special classes of graphs.

Since the number of available wavelengths is actually limited by technology, we considered variations of PATH COLORING (PC) in which the number of wavelengths is a part of the input rather than a parameter that has to be optimized.

In the MAX-PC problem the goal is to satisfy a maximum number of pre-routed requests, using the available wavelengths. In graph theoretic terms, the goal is to color a maximum number of paths, using a given number of colors. We proposed a $\frac{2}{3}$-factor approximation algorithm for MAX-PC in rings, which was presented in the Workshop on Algorithmic Aspects of Communication (AlAsCo 1997), and was later published in Computer Networks [24]. We also studied a variation of this problem, in which the routing is not given and the goal is to assign a communication path and a wavelength to each request, so that a maximum number of requests are satisfied using the available wavelengths. For ring graphs we proposed algorithms for symmetric and one-way communication, with approximation factors $\frac{2}{3}$ and $\frac{7}{11}$, respectively. This work was presented in INFOCOM 2003 [23].

In the case that the number of available wavelengths is limited and we still want to satisfy all the requests, we can use parallel fiber links between each pair of nodes and allow multiple requests to be transmitted between a pair of nodes in the same wavelength, by using different links. Based on the above observation we introduced and studied the PATH MULTICOLORING problem, in which, given a set of requests and the number of available wavelengths per link, the goal is to satisfy all the requests by activating a minimum total number of parallel links. For this problem, we developed an exact polynomial time algorithm for chain graphs, and 2-factor approximation algorithms for rings and stars. A preliminary version of this paper was presented in AlAsCo 1997, and the full paper was published in Information Processing Letters [22].

A very interesting problem that emerged during our research in the area of optimization problems for optical networks is BLUE-RED MATCHING, which is a variation of EXACT-MATCHING introduced by Papadimitiou and Yannakakis in 1982. In our paper presented in MFCS 2007 [25], we proved that BLUE-RED MATCHING cannot be solved in polynomial time unless EXACT-MATCHING is in P, and we proposed a polynomial time randomized algorithm and an approximation algorithm with factor $\frac{3}{4}$ for this problem.

7   On Art and Increased Visibility (by Euripides Markou)

I met Stathis Zachos during the spring of 1994. It was just a few months after I had graduated with a bachelor in Physics and I was searching my way towards Computer Science. Along this quest I had registered and attended a 3-month postgraduate seminar on Computational Mathematics and Graph Algorithms at the National Technical University of Athens. It was there when I first met Stathis who was one of the instructors of the seminar. I was impressed and further motivated by his teaching. It was just after that seminar that I decided to take the exams for entering NTUA as a PhD student. I succeeded and early in 1995 I am a PhD student and Stathis is my advisor.

Our relation is highly influential for me. Stathis is an unlimited new world of knowledge for me. I am always looking forward to discussing, working, and just hearing him. His teaching is not only very interesting and motivating but also joyful. Even when you are very familiar with the topic of his talk, you always get a new view and build a new interest. You also learn about the history of everything related and even more, you really enjoy it. It is not an exaggeration to say that his talks resemble influential and joyful theatrical performances: they have everything, you really enjoy and you leave the room full of new thoughts, ideas and questions. And he also enjoys so much to teach! No matter how tired he is or what is the time! I attended some courses he gave ending after midnight and some times even starting around midnight!

My relation with Stathis during my PhD studies, changed me not only scientifically but also changed and shaped my personality. Stathis opened for me a window through which I could see a different view of the world and through this procedure I rediscovered my country, my friends and even myself. I remember him at his home telling us stories about important people of science, but also about world and greek history while we were eating or while he was taking us small trips with his boat.

Stathis projects the mathematical way of thinking, in seeking the truth in aspects of life far beyond mathematics and he can communicate this in such a joyful and deeply motivating way. I learned from him (and loved) to question, argue and prove.

In the scientific part of our relation during my PhD studies, I remember him thinking together with me and other researchers in front of the whiteboard, proposing and testing new ideas that would give us the nice and elegant proofs we were seeking. It was not rare that we worked this way all night until dawn at his office at NTUA or sometimes even at his home.

In our joint work we mainly studied variations of the ART GALLERY problem. Until then it was known that finding a minimum number of guards that can cover a polygon is an NP-hard (even APX-hard) problem, while a polynomial time approximation algorithm with a logarithmic ratio was known. We studied a somewhat opposite version: how to place a fixed number of guards on the boundary of a polygon so that to guard a maximum length (or value) of the boundary. We proved that those problems are also APX-hard [11,19] and we designed polynomial time algorithms that achieve a constant approximation ratio [18].

We also studied problems on visibility graphs like for example how to select a fixed number of cliques so that to maximize the total weight of covered vertices. We proved that this problem does not admit a FPTAS, and we gave a polynomial time algorithm achieving a constant approximation ratio [18]. Notice that finding a maximum weighted clique was known to be in P, while partitioning the graph into a minimum number of cliques was known to be APX-hard and logn-approximable. It is a nice coincidence that the preliminary version of our main result (a gap-preserving reduction which was a fairly new method at that time) had been presented in the 5th CIAC 2003 in Rome [19]. The full version of the paper had been published in Computational Geometry:Theory and Applications in 2007 [11].

We published one journal paper and five conference papers. Especially two of the conference papers and the journal paper cover most part of my PhD thesis. However the number of our joint publications does not agree with the number of papers in which I consider Stathis as my co-author, since he is mentally my co-author in every paper I have done and every paper I will do in the future. I can hear him helping me to find an elegant proof!

Stathis is turning seventy and he is full of energy as ever (or even more). He is teaching, researching and advising more and more people on how to argue, question and prove the truth!

8   Teaching Programming Through Problem Solving (by Nikos Papaspyrou)

Quite often our lives are shaped by coincidences, smaller or larger. Meeting and working with influential people is a prominent life-changing factor and it does not come as a surprise that most of the colleagues who have written this article's sections describe how their relationship with Stathis has changed their lives, to a smaller or larger extent.

So, all this said, it was definitely my bad luck that I entered the NTUA as a freshman in 1988, one year before Stathis started teaching there. My first formal introduction to programming occurred as part of a FORTRAN 77 course, rather classic in engineering departments at that time (in a few stubborn ones even today) and taught in a rather unimaginative fashion. Fortunately, I already knew how to program, in fact I had learned a few programming languages on my own in my high-school years, and this experience did not manage to put me off the fun of programming. Neither was I lucky enough to have Stathis teach me algorithms and complexity; the one course where I was lucky to have him before I graduated was on formal languages and computability.

My academic relationship with Stathis really started around 1995, when I came back from Cornell, after hastily finishing a MSc there, and started a PhD in Athens. Stathis was a member of my PhD defense committee. He was also the primary responsible for the opening of a faculty position in programming languages at the NTUA (a field that was rather underrepresented back then), the same position in which I was elected in 2001. Since then, Stathis and I have collaborated in the teaching of several courses related to programming for more than 15 years, together with Aris (Sect. 4) and, more recently, Dimitris (Sect. 11).

Although the ways Stathis and I teach are very different (mainly because I cannot remember all the Swiss jokes that Stathis is fond of using), our teaching approach towards "Introduction to Programming" all these years shares a main characteristic, in which all four involved instructors fundamentally agree. We put a huge emphasis on *problem solving* and efficient algorithm design, both in the lectures and in the lab, instead of merely teaching the syntax and semantics of a programming language. Stathis and I have written a paper on this topic (the only paper in which we are co-authors), discussing the role of an "educational" programming language and its desired characteristics for this purpose [30].

So, coming back to coincidence, I am convinced that my life would have been poorer without Stathis as part of it. On the occasion of his 70th birthday, I want to thank him for the fantastic journey we've shared so far and wish him many more healthy and productive years, full of energy like his previous ones.

9 Saving Colors and Memories (by Katerina Potika)

While being a PhD candidate at the National Technical University a group of people has helped me to reach my first goal, which was to successfully finish my PhD, and various important secondary goals, like gain confidence on doing research, learn effective teaching methods and build successful relationships in a small social group, like our Corelab.

The central person of this group of people was my advisor, Professor Stathis Zachos. He showed me a new world of academic presence, where you have to work hard (sometimes to the extreme), do research by taking small steps every day, be a good teacher and at the same time have a genuine interest for the well being of your students. Getting my PhD was a hard process and I didn't anticipate the numerous challenges while juggling multiple roles. My advisor Professor Zachos was there for me and taught me how to work hard and to circumvent difficulties in order to finish tasks. In his own words he is the "spiritual" father of his students.

I have many fond and challenging memories from that period, but one that stands above all was in Spring 2001. It was a stressful period for me, as I was getting some preliminary results on my research, but still had not a clear view of were I was heading. A small note on that period is that Professor Zachos taught (and teaches) gladly a lot of courses and there was always an undeclared competition between his, mostly, graduate students and his courses over who will get most of his time. However, the Spring semester supposed to be the "lighter" teaching loaded one, which meant more time to work with him on research.

During that time we had frequent meetings in preparation of a paper to be submitted. Unfortunately, we had to overcome additionally difficulties, like were to meet, since the campus was closed for a few weeks. He suggested we meet at his house in order to also be close to his young son, Kyriakos. When I arrived at his house, I realized that the house was full with students, undergraduate and graduate, working with the Professor. He just moved his work office to his living room. His home was always open to his students, we all remember fondly Thursday night "taratsa" (roof top) parties at his primary home in Athens, usually late May, and "Kinetta" parties in July at his vacation home. Coming back to my story, during the meeting I needed to print a paper and Kyriakos volunteered to help me find the printer. While the paper was printing, we chatted a little bit about school and I had the opportunity to know him a little bit better before he left, a few months later, sadly forever.

The preliminary results on PATH MULTICOLORING for special tree graphs we got during that time were accepted to the Networking Conference of 2004 [20]. Later, these results were enhanced and appeared in two journal publications [21, 26].

10 Combinatory Logic and Graph Coloring (by Panagiotis Cheilaris)

I started collaborating with Stathis Zachos in 2001, still as an undergraduate student. During the years that he was my advisor, I became aware of his breadth and depth of knowledge in Computer Science, but also in many other fields. His dedication to teaching and disseminating this knowledge is exemplary.

Our first collaboration in 2001 was on a problem that he had studied in his thesis [33], concerning a subsystem of Haskell Curry's combinatory logic, where only the combinator S with reduction rule $Sxyz \rightarrow xz(yz)$ is allowed.[1] In his thesis, Stathis, among other things, had proposed a general method for proving that some terms in this subsystem have infinite reductions. The termination problem, i.e., whether a given term has an infinite reduction, was only solved twenty years after, by Waldmann [32]. In fact, Waldmann proved that the above problem can be solved in linear time on the size of the input S-term, but his proof relied on checking some subcases with the help of a computer program. Stathis, always striving for elegance, believed that the case analysis could be simplified and in [9], together with Ramirez, we gave a proof without the use of computers.

With Stathis, we also worked on colorings of the vertices of a graph that have to satisfy some property for every simple path of the graph. In the survey paper [10], together with Ernst Specker, we gave a landscape of such colorings. Such a coloring is the following: Given a graph $G = (V, E)$, an *ordered coloring* is an assignment $C \colon V \rightarrow \mathbb{Z}^+$ such that for every simple path p in the graph, the maximum color that occurs in the vertices of p occurs in exactly one vertex. The problem of computing ordered colorings is a well-known and widely studied problem with many applications in VLSI design, parallel Cholesky factorization of matrices, and planning efficient assembly of products in manufacturing systems. In [6], together with Bar-Noy, Lampis, and Mitsou, we gave the best known upper bound on the number of colors needed to color the $m \times m$ grid graph.

11 An Amazing Teacher (by Dimitris Fotakis)

I still have vivid memories from the spring of 2009, when I joined NTUA and started collaborating with Stathis. Stathis was very supportive, open and enthusiastic, as always. He took over the role of my mentor at NTUA and was trying to help me with my new teaching obligations and to guide me through my new working environment. Among his many wise advices and funny quotes, I still remember his warning: "Most of our students are more clever than us!". I smiled back in disbelief. It took me just a couple of months to realize how amazingly right Stathis was (as always) and to share his great enthusiasm about introducing all our extremely talented students to the beautiful and exciting world of Theoretical Computer Science.

---

[1] In Smullyan's book *To mock a mockingbird* [31], combinator S is called the *starling*.

Since then, Stathis has been always around with his friendship, his wise word, his unconditional support, his funny jokes and his juicy gossip about almost everything and everybody. In these eight years, Stathis and I have been teaching together "Algorithms and Complexity", "Introduction to Computer Programming" (Nikos is also actively involved in this one) and several graduate courses (Aris was also part of most of them). Collaborating with Stathis, I learnt a lot from his example, experience and deep knowledge, and became part of his legacy. He deeply affected me with his great enthusiasm about teaching and exposed me to a virtually uncountable ways to motivate our students. I still remember how much proud I felt when Stathis first expressed his satisfaction about our Algorithms course. And I am still trying to grasp the basic principles behind his extremely efficient interviewing technique.

Stathis, thank you for everything and happy birthday!

12 Completeness, at Last! (by Eleni Bakali)

I first met Prof. Stathis Zachos when I was an undergraduate student at NTUA. He was my professor on multiple courses on computer science and he was my favorite among the others. What did I become fascinated with? Stathis Zachos is able not only to impart knowledge, but to kindle enthusiasm; during the courses, except for the particular material for each subject, we were given a general picture of his field, like a huge wonderful garden that we were invited to explore. Additionally, in every theorem and proof he used to emphasize a couple of crucial points without penetrating technical details that we could conclude on our own. Prof. Zachos has been an inspiring teacher to me. I recollect his remarks outside the strict course framework, e.g. "Gödel formulated a statement saying 'I am not provable'" or "PCP theorem was proven by some magic technique smearing a small error all around"; These ideas made me search for hours trying to fully understand what was going on and thus sparkled my interest in theoretical computer science.

As a researcher and an advisor as well, he leaves me astonished every time that we discuss about a scientific matter, not only by the breadth of his knowledge but also by the clarity and the purity of his thought process. Nevertheless, there is one thing that I didn't realize from the very beginning. His way of thinking is unique. Instead of making a definition and trying to prove a theorem in one go, he makes dozens of definitions and shows how they are related. This way, the theorem reveals itself, almost "by definition". For example, using his characterizations of some probabilistic classes via computation trees [36], he was able to prove new theorems and at the same time to simplify some older proofs significantly.

As a part of my PhD, we (Zachos and our team) co-authored a paper titled "Completeness Results for Counting Problems with Easy Decision" [2]. Zachos along with some of his collaborators had previously defined the class TotP [14,28] in some papers regarding subclasses of #P, where these classes were defined in terms of computation trees of nondeterministic polynomial-time Turing machines

(NPTMs). In particular, TotP is the class of functions counting the total number of computation paths of an NPTM. TotP seems particularly interesting, since it contains every self-reducible problem in #P which has an easy decision version. Thus, it contains the only problems that may admit a good approximation (unless $NP = P$ or $NP = RP$). As my advisor, Prof. Zachos suggested as a research topic to search for TotP-complete problems, a question that had been open for a while. We discussed various properties of the class, and we managed to encode them into properties which a TotP-complete problem should have. Interestingly, our results provide new insights and understanding of the hardness of SAT as well as the reasons of failure of some algorithms which are based only on the hypercube structure. Moreover, our work suggests new research directions for studying not only TotP but also #P. This shows once again how powerful both structural complexity and Zachos's way of thinking are in tackling problems, using adequate definitions which reveal facts you wouldn't even imagine otherwise.

Stathis Zachos besides being my professor, collaborator and mentor, he is one of the people that I appreciate and trust the most. He stood by my side and he helped me overcome difficult situations in both academic and non academic life. For all these reasons, I cordially thank him.

13   Epilogue

One of Stathis's greatest talents is to bring people together. He has spent endless hours communicating, interrogating colleagues and students, gossiping, joking, organizing social gatherings, and showing uncountable photos of friends and colleagues to each other, each one fully commented in his unique style – piquant details included. This way we got to know well people that we had never met before, gaining a pleasant feeling of extended community. In a way, Stathis had invented social networks well before the advent of Facebook and cellphones. In this social network he has integrated research and teaching, well balanced with social and family life. Indeed, his family and home have always been open to us and our families, making the 'network' even stronger and academic life more meaningful and attractive. We feel happy, honoured and proud to be part of Stathis's network and we only wish Stathis to continue his great job for many years to come.

Acknowledgments. We would like to apologize to numerous friends and colleagues of Stathis that would deserve to appear in this article if time and space permitted. We are grateful to them for being part of this great community, influencing Stathis and all of us in so many positive ways.

We are also thankful to all Stathis's family members we have met—including those, fondly remembered and greatly missed, who are not with us anymore—for their tolerance, support and friendship throughout so many, long-lasting research and social meetings in their house in Lycabettus and, of course, in Kinetta.

References

1. Babai, L.: Graph isomorphism in quasipolynomial time (extended abstract). In: Wichs, D., Mansour, Y. (eds.) Proceedings of the 48th Annual ACM SIGACT Symposium on Theory of Computing, STOC 2016, Cambridge, MA, USA, 18–21 June 2016, pp. 684–697. ACM (2016)
2. Bakali, E., Chalki, A., Pagourtzis, A., Pantavos, P., Zachos, S.: Completeness results for counting problems with easy decision. In: Fotakis, D., Pagourtzis, A., Paschos, V.T. (eds.) CIAC 2017. LNCS, vol. 10236, pp. 55–66. Springer, Cham (2017)
3. Bampas, E., Göbel, A.-N., Pagourtzis, A., Tentes, A.: On the connection between interval size functions and path counting. Comput. Complex. (2016). doi:10.1007/s00037-016-0137-8
4. Bampas, E., Pagourtzis, A., Pierrakos, G., Potika, K.: On a noncooperative model for wavelength assignment in multifiber optical networks. IEEE/ACM Trans. Netw. **20**(4), 1125–1137 (2012)
5. Bampas, E., Pagourtzis, A., Pierrakos, G., Syrgkanis, V.: Selfish resource allocation in optical networks. In: Spirakis, P.G., Serna, M. (eds.) CIAC 2013. LNCS, vol. 7878, pp. 25–36. Springer, Heidelberg (2013). doi:10.1007/978-3-642-38233-8_3
6. Bar-Noy, A., Cheilaris, P., Lampis, M., Mitsou, V., Zachos, S.: Ordered coloring of grids and related graphs. Theoret. Comput. Sci. **444**, 40–51 (2012)
7. Beigel, R., Gill, J.: Counting classes: thresholds, parity, mods, and fewness. Theor. Comput. Sci. **103**(1), 3–23 (1992)
8. Boppana, R.B., Håstad, J., Zachos, S.: Does co-NP have short interactive proofs? Inf. Process. Lett. **25**(2), 127–132 (1987)
9. Cheilaris, P., Ramirez, J., Zachos, S.: Checking in linear time if an S-term normalizes. In: Proceedings of the 8th Panhellenic Logic Symposium (2011)
10. Cheilaris, P., Specker, E., Zachos, S.: Neochromatica. Commentationes Math. Univ. Carol. **51**(3), 469–480 (2010)
11. Fragoudakis, C., Markou, E., Zachos, S.: Maximizing the guarded boundary of an Art Gallery is APX-complete. Comput. Geom.: Theory Appl. **38**(3), 170–180 (2007)
12. Fürer, M., Goldreich, O., Mansour, Y., Sipser, M., Zachos, S.: On completeness and soundness in interactive proof systems. Adv. Comput. Res. **5**, 249–442 (1989)
13. Hemaspaandra, L.A., Homan, C.M., Kosub, S., Wagner, K.W.: The complexity of computing the size of an interval. SIAM J. Comput. **36**(5), 1264–1300 (2007)
14. Kiayias, A., Pagourtzis, A., Sharma, K., Zachos, S.: Acceptor-definable counting classes. In: Manolopoulos, Y., Evripidou, S., Kakas, A.C. (eds.) PCI 2001. LNCS, vol. 2563, pp. 453–463. Springer, Heidelberg (2003). doi:10.1007/3-540-38076-0_29
15. Kiayias, A., Pagourtzis, A., Zachos, S.: Cook reductions blur structural differences between functional complexity classes. In: Proceedings of the 2nd Panhellenic Logic Symposium, Delphi, 13–17 July 1999, pp. 132–137 (1999)
16. Koutras, C.D., Koletsos, G., Zachos, S.: Many-valued modal non-monotonic reasoning: sequential stable sets and logics with linear truth spaces. Fundam. Inform. **38**(3), 281–324 (1999)
17. Koutras, C.D., Zachos, S.: Many-valued reflexive autoepistemic logic. Logic J. IGPL **8**(1), 33–54 (2000)
18. Markou, E., Fragoudakis, C., Zachos, S.: Approximating visibility problems within a constant. In: Proceedings of the 3rd Workshop on Approximation and Randomization Algorithms in Communication Networks, pp. 91–103 (2002)

19. Markou, E., Zachos, S., Fragoudakis, C.: Maximizing the guarded boundary of an Art Gallery is APX-complete. In: Petreschi, R., Persiano, G., Silvestri, R. (eds.) CIAC 2003. LNCS, vol. 2653, pp. 24–35. Springer, Heidelberg (2003). doi:10.1007/3-540-44849-7_10

20. Nomikos, C., Pagourtzis, A., Potika, K., Zachos, S.: Fiber cost reduction and wavelength minimization in multifiber WDM networks. In: Mitrou, N., Kontovasilis, K., Rouskas, G.N., Iliadis, I., Merakos, L. (eds.) NETWORKING 2004. LNCS, vol. 3042, pp. 150–161. Springer, Heidelberg (2004). doi:10.1007/978-3-540-24693-0_13

21. Nomikos, C., Pagourtzis, A., Potika, K., Zachos, S.: Routing and wavelength assignment in multifiber WDM networks with non-uniform fiber cost. Comput. Netw. **50**(1), 1–14 (2006)

22. Nomikos, C., Pagourtzis, A., Zachos, S.: Routing and path multicoloring. Inf. Process. Lett. **80**(5), 249–256 (2001)

23. Nomikos, C., Pagourtzis, A., Zachos, S.: Minimizing request blocking in all-optical rings. In: Proceedings of the 22nd Annual Joint Conference of the IEEE Computer and Communications Societies (IEEE INFOCOM 2003), San Franciso, CA, USA, 30 March–3 April 2003. IEEE (2003)

24. Nomikos, C., Pagourtzis, A., Zachos, S.: Satisfying a maximum number of prerouted requests in all-optical rings. Comput. Netw. **42**(1), 55–63 (2003)

25. Nomikos, C., Pagourtzis, A., Zachos, S.: Randomized and approximation algorithms for blue-red matching. In: Kučera, L., Kučera, A. (eds.) MFCS 2007. LNCS, vol. 4708, pp. 715–725. Springer, Heidelberg (2007). doi:10.1007/978-3-540-74456-6_63

26. Pagourtzis, A., Potika, K., Zachos, S.: Path multicoloring with fewer colors in spiders and caterpillars. Computing **80**(3), 255–274 (2007)

27. Pagourtzis, A., Sharma, K., Zachos, S.: Tree models, probabilistic polynomial time computations. In: Lin, X. (ed.) Proceedings of Computing: The Fourth Australasian Theory Symposium (CATS 1998). Australian Computer Science Communications, vol. 20, pp. 291–304. Springer-Verlag Singapore Pte. Ltd., Singapore (1998)

28. Pagourtzis, A., Zachos, S.: The complexity of counting functions with easy decision version. In: Královič, R., Urzyczyn, P. (eds.) MFCS 2006. LNCS, vol. 4162, pp. 741–752. Springer, Heidelberg (2006). doi:10.1007/11821069_64

29. Papadimitriou, C.H., Zachos, S.K.: Two remarks on the power of counting. In: Cremers, A.B., Kriegel, H.-P. (eds.) GI-TCS 1983. LNCS, vol. 145, pp. 269–275. Springer, Heidelberg (1982). doi:10.1007/BFb0036487

30. Papaspyrou, N.S., Zachos, S.: Teaching programming through problem solving: the role of the programming language. In: Proceedings of the 4th Workshop on Advances in Programming Languages (WAPL 2013), Federated Conference on Computer Science and Information Systems (FedCSIS 2013), pp. 1533–1536, Kraków, Poland, September 2013

31. Smullyan, R.: To Mock a Mockingbird and Other Logic Puzzles Including an Amazing Adventure in Combinatory Logic. Knopf, New York (1985)

32. Waldmann, J.: The combinator S. Inf. Comput. **159**, 2–21 (2000)

33. Zachos, S.: Kombinatorische Logik und S-Terme. Ph.D. thesis, ETH Zürich (1978)

34. Zachos, S.: Robustness of probabilistic computational complexity classes under definitional perturbations. Inf. Control **54**(3), 143–154 (1982)

35. Zachos, S., Furer, M.: Probabilistic quantifiers vs. distrustful adversaries. In: Nori, K.V. (ed.) FSTTCS 1987. LNCS, vol. 287, pp. 443–455. Springer, Heidelberg (1987). doi:10.1007/3-540-18625-5_67

36. Zachos, S., Heller, H.: A decisive characterization of BPP. Inf. Control **69**(1–3), 125–135 (1986)

Author Index

Printed in the United States
By Bookmasters